VICTORIAN SCIENCE and RELIGION

A Bibliography with Emphasis on Evolution, Belief, and Unbelief, Comprised of Works Published from c. 1900-1975

by
Sydney Eisen
Professor of History and Humanities
York University

and

Bernard V. Lightman
Assistant Professor of History
Honors college
University of Oregon

ARCHON BOOKS
1984

Library of Congress Cataloging in Publication Data

Eisen, Sydney.
    Victorian science and religion.

    1. Biology—England—History—Bibliography.
2. Evolution—History—Bibliography. 3. Evolution and
religion—History—Bibliography. 4. Religion and
science—History—Bibliography. 5. Geology—England—
History—Evolution. I. Lightman, Bernard,
1950-   . II. Title.
Z5320.E57 1983 [QH305.2.E5] 016.575   82-24497
ISBN 0-208-02010-1

For
DORIS
and
MERLE

CONTENTS

PART C   Religion—Ideas and Institutions

NOTE ON SOURCES

Bibliographical sources for this period may be found in two very useful works: David J. DeLaura (Ed.), Victorian Prose: A Guide to Research (New York: The Modern Language Association of America, 1973) and Lionel Madden, How to Find Out about the Victorian Period: A Guide to Sources of Information (Oxford: Pergamon Press, 1970). Titles listed below should be used in conjunction with the guides and indexes cited in these volumes.

The following will provide bibliographical information on a wide range of subjects: Josef L. Altholz, Victorian England, 1837-1901 (Cambridge: For the Conference on British Studies at the University Press, 1970); Lucy M. Brown and Ian R. Christie (Eds.), Bibliography of British History, 1789-1851 (Oxford: Clarendon Press, 1977); H.J. Hanham, British History, 1851-1914 (Oxford: At the Clarendon Press, 1976); George Watson (Ed.), The New Cambridge Bibliography of English Literature, Vol. 3 (Cambridge: The University Press, 1969); Writings on British History, 1934 (London: Jonathan Cape; later, London: University of London, Institute of Historical Research, 1937-); "Victorian Bibliography," in Victorian Studies (annually from 1957-58; previously in Modern Philology; combined periodically under different editors as Bibliographies in Victorian Literature); Annual Bibliography of Victorian Studies (1980-), edited by Brahma Chaudhuri, in which retrospective volumes are also being published and which offers database services; Victorian Periodicals Review (1979-; formerly Victorian Periodicals Newsletter); Victorian Newsletter; the MLA International Bibliography; and (in the same series as DeLaura's Victorian Prose) Frederic E. Faverty (Ed.), The Victorian Poets: A Guide to Research (Second ed., Cambridge, Mass.: Harvard University Press, 1968), George H. Ford (Ed.), Victorian Fiction: A Second Guide to Research (New York: The Modern Language Association of America, 1978), and J. Don Vann and Rosemary T. VanArsdel (Eds.), Victorian Periodicals: A Guide to Research (New York: The Modern Language Association of America, 1978).

The relationship between science and religion is dealt with in a brief general bibliography by Arthur Maltby, Religion and Science (London: The Library Association, 1965), and in a very comprehensive one with a particular emphasis in James R. Moore, The Post-Darwinian Controversies: A Study of the Protestant Struggle to Come to Terms with Darwin in Great Britain and America 1870-1900 (Cambridge: Cambridge University Press, 1979).

For references on science see: the Dictionary of Scientific Biography; the annual critical bibliography in Isis as well as the Cumulative Bibliography; and "Relations of Literature and Science: A Bibliography of Scholarship," in Clio: An International Journal of Literature, History and Philosophy of History (from 1974-75; previously distributed at meetings of the Modern Language Association, with some reprinted in Symposium: A Journal Devoted to Modern Foreign Languages and Literatures; see also Fred A. Dudley, The Relations of Literature and Science: A Selected Bibliography, 1930-1967, Ann Arbor, Mich.: University Microfilms, 1968).

For religion and theology see: Owen Chadwick, The History of the Church: A Select Bibliography (Second ed., London: Historical Association, 1966) and the bibliography in Chadwick, The Victorian Church (Second ed., Part 1, London: Adam & Charles Black, 1970; Third ed., Black, 1971); John Kent, "The Study of Modern Ecclesiastical History since 1930," in J. Daniélou, A.H. Courtain, and John Kent, Historical Theology, Vol. 2 of The Pelican Guide to Modern Theology (Harmondsworth, Middlesex: Penguin Books, Ltd., 1969); The Catholic Periodical and Literature Index; The Historical Magazine of the Protestant Episcopal Church: Episcopal and Anglican History (annotated bibliographies from 1966; also William Wilson Manross, "Catalog of Articles in the 'Historical Magazine' . . .," HMPEC 23, 1954, 367-420, and 33, 1964, 355-384); Index to Religious Periodicals; and the Revue d'Histoire Ecclésiastique.

INTRODUCTION

The relationship between science and religion, and particularly between Darwinian evolution and theology, was one of the critical issues of Victorian intellectual life. In the past, historians dealt with this theme in terms of "conflict" or "warfare," though many Victorians would have recoiled from the possibility of such disharmony in their mental universe. In our own day, the complexities inherent in that relationship are more fully appreciated, as is the role of "non-scientific" factors (such as clashing professional ambitions) in engendering hostility between scientists and theologians. Moreover, scholars now tend to place increasing emphasis on the continuity between the pre-Darwinian world of Natural Theology and the world shaped by Darwinian evolution.

If science and religion were in many respects inseparable, they nevertheless could boast their own intellectual and institutional traditions, and as the century wore on, their professional independence and mutual isolation became increasingly pronounced. While we need to understand the source and nature of both the harmony and the bitterness, we must also be able to see science and religion as multifaceted individual categories touching almost everything around them. Our bibliography reflects this general perspective: we treat science and religion both as separate entities and as they relate to one another, and we are also concerned with their wider intellectual importance.

This bibliography consists of studies dealing with ideas and institutions in Victorian science and religion, published between about 1900 and 1975. While making some allowance for various scientific and religious developments in the late 18th and early 19th centuries, it concentrates on the period from about 1830 to 1900. It is divided into three "parts": A. MAIN CURRENTS; B. NATURAL

THEOLOGY, GEOLOGY, AND EVOLUTION; and C.   RELIGION - IDEAS AND INSTITUTIONS. There are altogether twenty-two "categories" (or chapters), some of which are, in turn, subdivided into "sections." While our emphasis has been on evolution, belief, and unbelief, we have developed a number of more or less related areas (e.g., education). We have tried to be as thorough as possible in the core of the bibliography; it will be obvious, however, that usefulness rather than completeness has been our guiding principle and that some categories, such as Evolution and History, were intended more as introductions than as exhaustive surveys of the field. Separate indexes of authors and subjects complete the work.

We should note the limits we have imposed on our undertaking. "Science," here, refers mainly to geology and biology; our emphasis, moreover, is on ideas and institutions, with little attention given to technical aspects of the subject. In the area of "religion" we have only occasionally ventured beyond the borders of England, and we have barely touched upon non-Christian faiths.

While some works are cited in more than one place, we have also made internal economies. Where a dissertation has been turned into a monograph, we have listed only the latter, unless the published work differs considerably from the dissertation. Our use of biographies has been sparing and not always consistent; we have generally listed them in those areas in which our findings were meager, or where they made up a significant portion of the available literature (e.g., in theology). In the case of major literary and philosophical figures, where we have had to be very selective in choosing our material, only the occasional biography appears (both in the interests of space and because biographies and "life and letters" are so easily located). When work on this project began some years ago, 1975 was chosen as a cutoff date; as the scope of the research became apparent, it seemed wise to retain

that date so that we could complete our task. We are,
however, contemplating doing another volume to include the
large and growing body of literature that has since
appeared.

A number of technical matters, some of greater import
than others, should be mentioned. To achieve a measure of
uniformity in titles, we have generally used a colon to
separate title and subtitle. We have also used Arabic
numerals for all volume numbers; thus, in the case of a
journal followed by "37 (1964), 77-80," the first number
refers to the volume. Broad works on the period were not
broken down into their component chapters, but a relevant
chapter in a mixed collection would be listed separately.
In adding parenthetical names or comments at the end of
entries, we have usually confined ourselves to expanding
upon non-specific or ambiguous titles (e.g., Biographies
of Scientific Men), though we have not attempted to
enlarge upon the great number of general titles that refer
to categories of individuals, ideas, or events in our
period. Many portions of the bibliography overlap with
others; we trust that the headnotes as well as the subject
index will prove useful in directing the reader to complementary areas. For purposes of economy and convenience, a
section on an individual thinker generally includes all
works about him, though the individual might well have
appeared under more than one heading; for example,
writings on Herbert Spencer are all assembled under
Evolution and Social and Political Thought, rather than
being divided between this category and Varieties of
Belief, Doubt, and Unbelief, not to mention other possi-
bilities. Some exceptions have been made, as in the case
of John Stuart Mill, whose philosophic ideas are dealt
with under Science - Method and Philosophy, while his
views on religion appear under Varieties of Belief, Doubt,
and Unbelief. In all these areas, we have been guided as

much by what seemed to be reasonable in each instance, as by any formal criteria.

This bibliography grew out of an undergraduate seminar on "Science and Religion in Victorian England." While it now ranges well beyond the boundaries of that seminar, its eccentricities may still reflect its origin. During the years in which it was put together, I was engaged in university administration, from which the gathering of titles and the never-ending redefinition of categories provided a pleasant diversion.

I have received assistance from many individuals and I wish to thank them, while in no way making them responsible for any deficiencies. I am grateful to my students, Robert McMillan, who helped me get the project under way, William Perry, who checked the manuscript at an early stage, and Michele Green, who tracked down numerous entries as it was being completed; to Sheldon Levy, my assistant and colleague in the Office of the Dean of the Faculty of Arts, who spent countless hours making it a more serviceable instrument; to the librarians of the Scott Library, particularly Joan Carruthers, Joanne Chumakov, and Grace Heggie, for advice and assistance, and to Mary Hudecki (Supervisor), Gary MacDonald, and Susan Partridge, all of the Interlibrary Loan Office, for countless, cheerfully-rendered services; to the reference librarians in many libraries, who were good enough to respond to our queries; to the secretaries in the History Department, particularly Enid Grant and Michelle Srebrolow, who graciously met numerous unreasonable deadlines; to Sarah Eisen for checking numbers and citations; to Lynn Herbert for reviewing the final copy; and to Heather Wardle who, with the assistance of Midge Rouse, made a major contribution in the formidable task of preparing the author and subject indexes. I should like to express my appreciation to Rayann Bell, Eileen Comba,

Pat Humenyk, Susan Masters, Diane Pantele, Grace Baxter, and Rhoda Rilling of the Secretarial Services Department (Faculty of Arts), who, in producing several drafts and the final copy, brought to the task the highest professional standards as well as endless patience; and I want to acknowledge my great debt to Doris Rippington (Director) for her careful supervision of the project over several years and for the benefit of her judgement and experience in determining its format.

For their encouragement and for giving me the benefit of their knowledge, I am indebted to friends and colleagues, particularly Professor Murray Sachs of Brandeis University, Professors Trevor Levere, Richard Helmstadter, and John Robson of the University of Toronto, and Professors Elio Costa, Gwenda Echard, William Echard, Michiel Horn, Roxanne Marcus, and Mark Webber of York. I am very grateful to Dr. Robert M. Young, of London, England, from whom I have learned a great deal about this subject, for reviewing the whole manuscript at an earlier stage and for his suggestions with reference to its publication. It would be difficult to conclude any acknowledgement of debts without paying due homage to those authors and bibliographers whose works we have mined in the course of preparing this one.

Bernie Lightman, who shares the authorship, was a student in my seminar over a decade ago. He worked with me as a research assistant while he was pursuing his M.A. (in the York University-University of Toronto Victorian Studies Option) and, intermittently, while he was advancing towards the completion of his Ph.D. at Brandeis University. At some point we became collaborators, and we have completed the work together. It has been an enjoyable partnership.

York University   Sydney Eisen
Downsview, Ontario
January, 1984

PART A. MAIN CURRENTS

I. Science

1. General

    This category includes general histories of science
and scientific ideas, histories of botany, zoology, biology, and geology, and histories of scientific institutions and journals. It also lists works on the status
of science and on the careers of individual scientists
e.g., J. D. Hooker, P. H. Gosse, and Hugh Miller. See I.
Science, 2. Method and Philosophy; VI. Geology . . .;
VII. Evolution.

1. ADAMS, ALEXANDER B. Eternal Quest: The Story of the
    Great Naturalists. New York: Putnam, 1968.
2. ADAMS, FRANK DAWSON. The Birth and Development of
    the Geological Sciences. Baltimore: Williams &
    Wilkins Company, 1938.
3. ALDINGTON, RICHARD. The Strange Life of Charles
    Waterton. London: Evan Bros., 1949.
4. ALEXANDER, EDWARD. "Ruskin and Science." Modern
    Language Review 64 (July 1969), 508-521.
5. ALLAN, MEA. The Hookers of Kew, 1785-1911. London:
    Joseph, 1967.
6. ANDRADE, E. N. A Brief History of the Royal Society.
    London: The Royal Society, 1960.
7. ANDREWS, JOHN S. "Philip Henry Gosse, F.R.S." Li-
    brary Association Record 28 (June 1961), 197-201.
8. ANNAN, NOEL GILROY. "Books in General." New Statesman
    and Nation 32 (Dec. 7, 1946), 423. (Lives of William
    Buckland and his son, Frank)
9. ANON. "The Expositors of Science: Growth of a New
    Public (1837-1937)." Times Literary Supplement No.
    1839 (May 1, 1937), 336.

10.   ANON.   Obituary Notices of Fellows of the Royal Soci-
      ety.   9 vols.   London: Royal Society, 1932-54.
      (Continued as: Biographical Memoirs of Fellows of
      the Royal Society.   London: Royal Society, 1955-   )

11.   ANTICH, BARBARA.   "The Royal Society - 300 Years of
      Science."   Journal of Chemical Education 39 (1962),
      588-589.

12.   ARMSTRONG, E. F.   "The Influence of the Prince Con-
      sort on Science."   Journal of the Royal Society
      of Arts 94, No. 4705 (Nov. 23, 1945), 4-15.

13.   BADASH, LAWRENCE.   "The Completeness of 19th-Century
      Science."   Isis 63 (1972), 48-58.

14.   BAILEY, SIR EDWARD BATTERSBY.   "Hugh Miller: 1852-
      1856.   Commemoration at Cromarty."   Nature 170 (1952),
      790-791.

15.   BARBER, BERNARD.   "Resistance by Scientists to Sci-
      entific Discovery."   Science 134 (1961), 596-602.

16.   BARR, E. SCOTT.   "Nature's 'Scientific Worthies'."
      Isis 56 (1965), 354-356.   (List of portraits pub-
      lished between 1873 and 1934 in Nature)

17.   BASALLA, GEORGE.   "Science and Government in England:
      1800-1870."   Ph.D. Dissertation, Harvard University,
      1964.

18.   BASALLA, GEORGE, WILLIAM R. COLEMAN AND ROBERT H.
      KARGON.   (Eds.)   Victorian Science: A Self-portrait
      from the Presidential Addresses of the British Asso-
      ciation for the Advancement of Science.   Garden City,
      N.Y.:   Doubleday, 1970.

19.   BASTIN, JOHN.   "The First Prospectus of the Zoologi-
      cal Society of London: New Light on the Society's
      Origins."   Journal of the Society for the Biblio-
      graphy of Natural History 5 (1970), 369-388.

20.   BASTIN, JOHN.   "A Further Note on the Origins of the
      Zoological Society of London."   Journal of the Soci-
      ety for the Bibliography of Natural History 6 (1973),
      236-241.

21. BAYFORD, E. G. "Elihu Berry: A Little Known York-
    shire Botanist." Naturalist (July-Sept. 1946),
    113-114.

22. BAYFORD, E. G. "The Predecessors of the 'Natural-
    ist': A Critical Survey." Naturalist No. 782 (Sept.
    1940), 228-232.

23. BAYFORD, E. G. "Some Notes on the History of 'The
    Naturalist,' 1875-1940." Naturalist No. 781 (Aug.
    1940), 203-204.

24. BEAN, WILLIAM JACKSON. The Royal Botanic Gardens,
    Kew: Historical and Descriptive. London, New York:
    Cassell and Company, Limited, 1908.

25. BERMAN, MORRIS. "The Early Years of the Royal In-
    stitution, 1799-1810: A Re-evaluation." Science
    Studies (London) 2 (1972), 205-240.

26. BERMAN, MORRIS. "'Hegemony' and the Amateur Tradi-
    tion in British Science." Journal of Social History
    8 (Winter 1975), 30-50.

27. BERMAN, MORRIS. Social Change and Scientific Organi-
    zation: The Royal Institution, 1799-1810. Ann
    Arbor: University Microfilms, 1971.

28. BERNAL, JOHN DESMOND. Science in History. London:
    Watts, 1954.

29. BINGHAM, MADELEINE. The Making of Kew. London:
    Joseph, 1975.

30. BOND, ELISABETH. "A Maker of Botanical History: The
    Reverend Miles Joseph Berkeley (1803-1889)." Country
    Life 156 (1974), 639-640.

31. BORGES, JORGE LUIS. "The Creation and P. H. Gosse."
    In Other Inquisitions 1937-1952. Trans. Ruth L. C.
    Simms. New York: Washington Square Press, 1965,
    22-25.

32. BOWER, FREDERICK ORPEN. Joseph Dalton Hooker.
    London: S.P.C.K.; New York: The Macmillan Company,
    1919.

33. BOWER, FREDERICK ORPEN. Sixty Years of Botany in
     Britain, 1875-1935: Impressions of an Eye-witness.
     London: Macmillan and Co., Limited, 1938.

34. BOYD, ROBERT. "The British Museum Forty Years Ago.
     Mr. Gosse's Reminiscences." Daily Graphic (Oct. 11,
     1906), 10.

35. BRADSHAW, ABEL P. "Jethro Tinker: A Stalybridge
     Naturalist of the 18th-19th Century." North Western
     Naturalist 20, No. 3-4 (1946), 229-232.

36. BRAGG, SIR WILLIAM LAWRENCE. "History in the Ar-
     chives of the Royal Society." Science 89 (1939),
     445-453. (Also in Nature 144, 1939, 21-28)

37. BRUSH, STEPHEN G. "Science and Culture in the Nine-
     teenth Century." The Graduate Journal 7 (1967),
     477-565.

38. BRUTON, F. A. "Philip Henry Gosse's Entomology of
     Newfoundland." Entomological News 41 (Feb. 1930),
     34-37.

39. BURGESS, G. H. O. The Curious World of Frank Buck-
     land. London: Baker, 1967.

40. BUSH, R. C. "The Development of Geological Mapping
     in Britain, 1790-1825." Ph.D. Dissertation, Uni-
     versity College, University of London, 1974.

41. BUTCHER, R. W. "Faraday as a Botanist." Nature 128
     (1931), 820-821.

42. CAMERON, HECTOR CHARLES. Sir Joseph Banks, K.B.,
     P.R.S.: The Autocrat of the Philosophers. London:
     Batchworth Press, 1952.

43. CANNON, WALTER F. "The Normative Role of Science in
     Early Victorian Thought." Journal of the History of
     Ideas 25 (1964), 487-502.

44. CANNON, WALTER F. "The Role of the Cambridge Move-
     ment in Early Nineteenth Century Science." Proceed-
     ings of the Tenth International Congress of the His-
     tory of Science (Ithaca, 1962), 1 (1964), 317-320.

45. CARDWELL, DONALD STEPHEN LOWELL. The Organization of
     Science in England: A Retrospect. London: William
     Heinemann, 1957.

46. CAWS, PETER. "Evidence and Testimony: Philip Henry
     Gosse and the 'Omphalos' Theory." In Harold Orel and
     George J. Worth (Eds.), Six Studies in Nineteenth-
     Century English Literature and Thought. Lawrence:
     University of Kansas, 1962, 69-90.

47. CHALLINOR, JOHN. "The Beginnings of Scientific
     Paleontology in Britain." Annals of Science 6 (1948),
     46-53. (James Parkinson and William Martin)

48. CHALLINOR, JOHN. "From Whitehurst's 'Inquiry' to
     Farey's Derbyshire." Transactions and Annual Report:
     North Staffordshire Field Club 81 (1947), 52-88.

49. CHALLINOR, JOHN. The History of British Geology: A
     Bibliographical Study. New York: Barnes and Noble,
     1971.

50. CHALLINOR, JOHN. "The Progress of British Geology
     during the Early Part of the 19th Century."
     Annals of Science 26 (1970), 177-234.

51. CHALLINOR, JOHN. "Some Correspondence of Thomas
     Webster, Geologist (1773-1844)." Annals of Science
     17 (1961), 175-195; 18 (1962), 147-175; 19 (1963),
     49-79, 285-297; 20 (1964), 59-79, 143-164.

52. CHALLINOR, JOHN. "Thomas Webster's Letters on the
     Geology of the Isle of Wight, 1811-1813." Proceed-
     ings of the Isle of Wight Natural History and Arch-
     aeological Society 4 (1949), 108-122.

53. CHESSON, WILFRID HUGH. (Ed.) Eliza Brightwen: The
     Life and Thoughts of a Naturalist. With Introduction
     and Epilogue by Sir Edmund William Gosse. London and
     Leipsic: T. F. Unwin, 1909.

54. CHORLEY, RICHARD J., ANTHONY J. DUNN AND ROBERT P.
     BECKINSALE. The History of the Study of Landforms:
     Or the Development of Geomorphology. London:
     Methuen; New York: Wiley, 1964.

55. CLEEVELY, R. J. "A Provisional Bibliography of Natural History Works by the Sowerby Family." Journal of the Society for the Bibliography of Natural History 6 (1974), 482-559.

56. CLEEVELY, R. J. "The Sowerbys, the 'Mineral Conchology,' and their Fossil Collection." Journal of the Society for the Bibliography of Natural History 6 (1974), 418-481.

57. CLEMENT, ARCHIBALD GEORGE AND ROBERT HUGH STANNUS ROBERTSON. Scotland's Scientific Heritage. Edinburgh: Oliver and Boyd, 1961.

58. COLEMAN, WILLIAM R. Biology in the Nineteenth Century: Problems of Form, Function, and Transformation. New York: Wiley, 1972.

59. CORNISH, CHARLES JOHN. Sir William Henry Flower. London, New York: Macmillan and Co., Limited, 1904.

60. CROWTHER, JAMES GERALD. Scientific Types. London, Barrie and Rockliff: Cresset Press, 1968. (Includes C. T. R. Wilson, J. W. Strutt, Lord Rayleigh, T. Young, T. H. Huxley, J. Tyndall, A. de Morgan, J. Dewar, O. Reynolds, C. Babbage, W. M. Fletcher, A. Schuster, G. B. Airy)

61. CROWTHER, JAMES GERALD. Statesmen of Science. London: Cresset Press, 1965. (Includes H. Brougham, W. R. Grove, L. Playfair, the Prince Consort, the Seventh Duke of Devonshire, A. Strange, R. B. Haldane, H. T. Tizard, F. A. Lindemann)

62. CURWEN, ELIOT CECIL. (Ed.) The Journal of Gideon Mantell, Surgeon and Geologist, Covering the Years 1818-1852. London, New York: Oxford University Press, 1940.

63. CUTTER, ERIC. "Sir Archibald Geikie: A Bibliography." Journal of the Society for the Bibliography of Natural History 7 (1974), 1-18.

64. DAMPIER, SIR WILLIAM CECIL AND MARGARET DAMPIER-WHETHAM. Readings in the Literature of Science. New York: Harper, 1959.

65. DAMPIER, SIR WILLIAM CECIL. A Shorter History of
     Science. Cambridge: The University Press, 1944.

66. DANCE, S. PETER. "Further Additions to Stageman's
     'Bibliography of the First Editions of Philip Henry
     Gosse (1955)'." Journal of the Society for the
     Bibliography of Natural History 7 (1974), 87.

67. DARLINGTON, C. D. "The Early Hybridizers and the
     Origins of Genetics." Herbertia 4 (1937), 63-69.

68. DAWES, BEN. A Hundred Years of Biology. London:
     Gerald Duckworth and Co., 1952.

69. DAWSON, WARREN ROYAL. "Sir Joseph Hooker and Dawson
     Turner." Journal of the Society for the Bibliography
     of Natural History 2 (1950), 218-222.

70. DEACON, MARGARET. Scientists and the Sea, 1650-1900:
     A Study of Marine Science. London, New York: Aca-
     demic Press, 1971.

71. DINGLE, HERBERT. (Ed.) A Century of Science, 1851-
     1951. New York: Roy Publishers, 1951.

72. DOBREE, BONAMY. "Science and Poetry in England."
     Sewanee Review 61 (1953), 658-664.

73. DONCASTER, ISLAY. In the Footsteps of the Natural-
     ists. London: Phoenix House, 1961. (Includes
     P. H. Gosse)

74. DONY, JOHN GEORGE. "Bedfordshire Naturalists: I,
     William Crouch, 1818-1846." Bedfordshire Natural
     History Society Journal 1 (1947), 50-52.

75. DONY, JOHN GEORGE. "Bedfordshire Naturalists: II,
     James Saunders, 1839-1925." Bedfordshire Naturalist
     2 (1948), 58-61.

76. DOUGLAS, J. A. AND J. M. EDMONDS. "John Phillips's
     Geological Maps of the British Isles." Annals of
     Science 6, No. 4 (1950), 361-375.

77. DUNN, LESLIE CLARENCE. "The Birth of Genetics." In
     A Short History of Genetics: The Development of Some
     of the Main Lines of Thought 1864-1939. New York:
     McGraw-Hill, 1965, 3-77.

78.   EDEES, E. S.   "Robert Garner, 1808-1890." Transac-
      tions of the North Staffordshire Field Club 74 (1950),
      13-45.

79.   EDWARDS, NICHOLAS.   "Some Correspondence of Thomas
      Webster (ca. 1772-1844) Concerning the Royal Insti-
      tution." Annals of Science 28 (1972), 43-60.

80.   EDWARDS, NICHOLAS.   "Thomas Webster (circa 1772-
      1844)." Journal of the Society for the Bibliography
      of Natural History 5 (1971), 468-473.

81.   ELLIS, E. S.   "The Gloucester Natural History Soci-
      ety." Cotteswold Natural Field Club Proceedings
      for 1941, 27, Pt. 3 (1942), 126-128.

82.   ELWES, E. V.   "P. H. Gosse as a Naturalist." Journal
      of the Torquay Natural History Society 1 (1914),
      259-266.

83.   ESSIG, EDWARD OLIVER.   History of Entomology. New
      York: The Macmillan Company, 1931.

84.   EVERITT, JUDITH.   "Philip Henry Gosse: A Victorian
      Case-History." Ph.D. Dissertation, University of
      Sussex, 1969.

85.   EYLES, VICTOR A.   "John Macculloch, F.R.S., and His
      Geological Map: An Account of the First Geological
      Survey of the British Isles." Annals of Science 2,
      No. 1 (Jan. 1937), 114-129.

86.   EYLES, VICTOR A.   "Macculloch's Geological Map of
      Scotland: An Additional Note." Annals of Science 4
      (1939), 107.

87.   EYLES, VICTOR A.   "Robert Jameson and the Royal
      Scottish Museum." Discovery 15, No. 4 (Apr. 1954),
      155-162.

88.   EYLES, VICTOR A.   "Sir James Hall, Bt. (1761-1832)."
      Endeavour 20 (1961), 210-216.

89.   FARLEY, JOHN.   "The Spontaneous Generation Contro-
      versy (1859-1880): British and German Reactions to
      the Problem of Abiogenesis." Journal of the History
      of Biology 5 (1972), 285-319.

90.   FLETCHER, HAROLD ROY. The Story of the Royal Horti-
      cultural Society, 1804-1968. London: Oxford Uni-
      versity Press for the Royal Horticultural Society,
      1969.

91.   FLETT, SIR JOHN SMITH. The First Hundred Years of
      the Geological Survey of Great Britain. London:
      H. M. Stationery Office, 1937.

92.   FOOTE, GEORGE A. "Mechanism, Materialism and Science
      in England, 1800-1850." Annals of Science 8 (1952),
      152-161.

93.   FOOTE, GEORGE A. "The Place of Science in the Brit-
      ish Reform Movement, 1830-50." Isis 42 (1951),
      192-208.

94.   FOOTE, GEORGE A. "Science and Its Function in Early
      Nineteenth Century England." Osiris 11 (1954),
      438-454.

95.   FOOTE, GEORGE A. "Sir Humphry Davy and His Audience
      at the Royal Institution." Isis 43 (1952), 6-12.

96.   FOOTE, GEORGE A. "A Study of Attitudes toward Sci-
      ence in Nineteenth Century England, 1800-1851."
      Ph.D. Dissertation, Cornell University, 1950.

97.   FORBES, THOMAS R. "William Yarrell, British Natural-
      ist." Proceedings of American Philosophical Society
      106 (1962), 505-515.

98.   FORD, TREVOR D. "The First Detailed Geological
      Sections Across England, by John Farey (1806-8)."
      Mercian Geologist 2 (1967), 41-49.

99.   FORD, TREVOR D. "White Watson (1760-1835) and His
      Geological Sections." Proceedings of the Geologists'
      Association 71 (1960), 346-363.

100. FRENCH, RICHARD D. Antivivisection and Medical
      Science in Victorian Society. Princeton: Princeton
      University Press, 1975.

101. FRENCH, RICHARD D. "Some Problems and Sources in
      the Foundations of Modern Physiology in Great Brit-
      ain." History of Science 10 (1971), 28-55.

102. FUSSELL, G. E. "Some Lady Botanists of the Nine-
     teenth Century: IV. Elizabeth and Sarah Mary
     Fitton; V. Jane Marcet (1769-1858)." Gardeners'
     Chronicle 130 (1951), 179-181, 238.
103. GAGE, ANDREW THOMAS. A History of the Linnean
     Society of London. London: The Linnean Society,
     1938.
104. GARDENER, W. "John Lindley." Gardeners' Chronicle
     158 (1965), 386, 406, 409, 430, 434, 451, 457, 481,
     502, 507, 526.
105. GELDART, A. M. "Sir James Edward Smith and Some of
     His Friends." Transactions of the Norfolk and
     Norwich Naturalists' Society 9 (1914), 645-692.
106. GIBBS, F. W. "John Gillies, M.D., Traveller and
     Botanist, 1792-1834." Notes and Records of the
     Royal Society of London 9, No. 1 (1951), 115-136.
107. GILLISPIE, CHARLES COULSTON. The Edge of Objec-
     tivity: An Essay in the History of Scientific
     Ideas. Princeton: Princeton University Press,
     1960.
108. GILLMOR, C. STEWART. "The Place of the Geophysical
     Sciences in 19th-Century Natural Philosophy." Eos
     56 (1975), 4-7.
109. GIZYCKI, RAINALD VON. "Centre and Periphery in the
     International Scientific Community: Germany, France
     and Great Britain in the Nineteenth Century."
     Minerva 11 (1973), 474-494.
110. GOING, WILLIAM T. "Philip Henry Gosse on the Old
     Southwest Frontier." Georgia Review 21 (1967),
     25-38.
111. GOLDFARB, STEPHEN JOEL. "Main Currents of British
     Natural Philosophy." Dissertation Abstracts Inter-
     national 34 (1974), 7149A (Case Western Reserve
     University, 1973).
112. GOSSE, PHILIP. The Squire of Walton Hall: The Life
     of Charles Waterton. London: Cassell and Company,
     Limited, 1940.

113. GREEN, JOSEPH REYNOLDS. History of Botany, 1860-
     1900. Oxford: Clarendon Press, 1909.

114. GREEN, JOSEPH REYNOLDS. A History of Botany in the
     United Kingdom from the Earliest Times to the End of
     the 19th Century. London and Toronto: J. M. Dent
     and Sons, Limited; New York: E. P. Dutton and Co.,
     1914.

115. GRIFFITHS, ARTHUR BOWER. Biographies of Scientific
     Men. London: R. Sutton, 1912. (Includes G. Cuvier,
     R. Owen, C. Lyell, J. Dalton, G. Buffon, H. Davy and
     Lord Kelvin)

116. GUIMOND, ALICE ALINE. "The Honorable and Very Rev-
     erend William Herbert, Amaryllis Hybridizer and
     Amateur Biologist." Dissertation Abstracts 28
     (1967), 1022A (University of Wisconsin, 1966).

117. GUNTHER, ALBERT E. A Century of Zoology at the
     British Museum through the Lives of Two Keepers
     1815-1914. London: Dawsons of Pall Mall, 1975.
     (J. E. Gray, A. Gunther)

118. GUNTHER, ALBERT E. "A Note on the Autobiographical
     Manuscripts of John Edward Gray (1800-1875)."
     Journal of the Society for the Bibliography of
     Natural History 7 (1974), 35-76.

119. GUNTHER, ROBERT WILLIAM THEODORE. "Oxford Colleges
     and Their Men of Science through the Centuries." In
     The Book of Oxford. Printed for the 104th Annual
     Meeting of the British Medical Association. Oxford:
     Oxford University Press, 1936, 25-129.

120. HARRE, ROMANO. (Ed.) Problems of Scientific Revolu-
     tion: Progress and Obstacles to Progress in the Sci-
     ences. Oxford: Clarendon Press, 1975.

121. HARRE, ROMANO. (Ed.) Some Nineteenth Century Scien-
     tists. Oxford, New York: Pergamon Press, 1969.
     (Includes C. W. Thomson, J. Murray, A. Cayley,
     F. Galton, Lord Kelvin, N. Lockyer, S. G. Thomas,
     Lord Ramsay)

122.   HARRIS, TOM M.   "The Minute Books of the Linnean
       Club, from 1811 to 1955."   Biological Journal of
       the Linnean Society 3 (1971), 343-368.   (The Linnean
       Society of London)

123.   HARVEY-GIBSON, ROBERT JOHN.   Outlines of the History
       of Botany.   London: A. and C. Black, Ltd., 1919.

124.   HAYS, J. N.   "Science in the City: The London
       Institution, 1819-40."   British Journal for the
       History of Science 7 (1974), 146-162.

125.   HAYS, J. N.   "Three London Popular Scientific Insti-
       tutions, 1799-1840."   Ph.D. Dissertation, University
       of Chicago, 1970.   (Royal Institution, London
       Institution, Mechanics' Institution)

126.   HEDGPETH, JOEL W.   "A Century at the Seashore."
       Scientific Monthly 61 (1945), 194-198.   (Particular-
       ly P. H. Gosse, E. Forbes)

127.   HIGHAM, NORMAN.   A Very Scientific Gentleman: The
       Major Achievements of Henry Clifton Sorby.   London:
       Pergamon Press, 1964.

128.   HILL, ARTHUR.   "Introduction."   Herbertia 4 (1937),
       3-4.   (William Herbert)

129.   HINDLE, EDWARD, F.R.S. "Biology and the Royal Soci-
       ety."   The New Scientist 8 (1960), 233-235.

130.   HOGBEN, LANCELOT THOMAS.   Science for the Citizen:
       A Self-Educator Based on the Social Background of
       Scientific Discovery.   London: G. Allen and Unwin
       Ltd., 1938.

131.   HOLMYARD, ERIC JOHN.   "British Contributions to
       Science, 1851-1951."   Endeavour 10 (1951), 117-118.

132.   HOOKER, SIR JOSEPH DALTON.   "A Sketch of the Life
       and Labours of Sir William Jackson Hooker."   Annals
       of Botany 16 (1902), ix-ccxx.

133.   HOOKER, SIR JOSEPH DALTON.   A Sketch of the Life and
       Labours of Sir William Jackson Hooker, K.H., D.C.L.,
       F.R.S., F.L.S..   Oxford: Clarendon Press, 1903.

134.   HOWARTH, OSBERT JOHN RADCLIFFE.   "The British Asso-
       ciation."   Endeavour 3, No. 10 (Apr. 1944), 57-61.

135. HOWARTH, OSBERT JOHN RADCLIFFE. The British Asso-
      ciation for the Advancement of Science: A Retro-
      spect, 1831-1921. London: The Association, 1922.
136. HOWARTH, OSBERT JOHN RADCLIFFE. The British Asso-
      ciation for the Advancement of Science: A Retro-
      spect, 1831-1931. Centenary (Second) Edition.
      London: The Association, 1931.
137. HOWE, BEA. "A Victorian Lady Naturalist." Country
      Life 145 (May 1969), 1194-95. (Eliza Brightwen)
138. HUGHES, ARTHUR. History of Cytology. London and
      New York: Abelard-Schuman, 1959.
139. HUGHES, ARTHUR. "Science in English Encyclopedias,
      1704-1875." Annals of Science 7 (Dec. 1951), 340-
      370; 8 (Dec. 1952), 323-367; 9 (Sept. 1953), 233-
      264; 11 (Mar. 1955), 74-92.
140. HUXLEY, LEONARD. "A Great Darwinian and His
      Friends." Cornhill Magazine 22 (1907), 690-703.
      (J. D. Hooker)
141. INKSTER, IAN. "The Development of a Scientific
      Community in Sheffield, 1790-1850: A Network of
      People and Interests." Hunter Archaeological Soci-
      ety Transactions 10 (1973), 99-131.
142. INNIS, MARY QUAYLE. "Philip Henry Gosse in Canada."
      Dalhousie Review 17 (1937), 55-60.
143. IRWIN, R. A. (Ed.) Letters of Charles Waterton of
      Walton Hall Near Wakefield: Naturalist, Taxider-
      mist, and Author of "Wanderings in South America"
      and "Essays on Natural History." London: Rockliff,
      1955.
144. JACKSON, BENJAMIN DAYDON. George Bentham. London:
      J. M. Dent and Co.; New York: E. P. Dutton and Co.,
      1906.
145. JAMES, M. J. The New Aurelians: A Centenary Histo-
      ry of the British Entomological and Natural History
      Society, with an Account of the Collection, by
      A. E. Gardner. London: British Entomological and
      Natural History Society, 1973.

146. JENSEN, JOHN VERNON. "Interrelationships within the
      Victorian X-Club." Dalhousie Review 51 (1971),
      539-552.

147. JENSEN, JOHN VERNON. "The X-Club: Fraternity of
      Victorian Scientists." British Journal for the
      History of Science 5 (June 1970), 63-72.

148. JESSOP, W. J. E. "Samuel Haughton (1821-1897): A
      Victorian Polymath." Hermathena 116 (1973), 5-26.

149. JONES, BESSIE ZABAN. (Ed.) The Golden Age of Sci-
      ence: Thirty Portraits of the Giants of 19th-Cen-
      tury Science by Their Scientific Contemporaries.
      New York: Simon and Schuster, 1967.

150. JONES, HOWARD MUMFORD AND I. BERNARD COHEN. (Eds.)
      Science before Darwin: A Nineteenth-Century Anthol-
      ogy. London: Andre Deutsch, 1963.

151. JONES, R. V. "Domesday Book of British Science."
      New Scientist 49 (1971), 481-483. (The Royal Com-
      mission on the Advancement of Science set up in 1871
      under the Seventh Duke of Devonshire)

152. JONES, R. V. "Lyon Playfair, 1818-98." Nature 200
      (1963), 105-111.

153. JUDD, JOHN WESLEY. "Henry Clifton Sorby, and the
      Birth of Microscopical Petrology." Geological Maga-
      zine, 5th Series, 5 (1908), 193-204.

154. KNIGHT, DAVID M. Natural Science Books in English
      1600-1900. New York: Praeger, 1972.

155. KNIGHT, DAVID M. Sources for the History of Sci-
      ence, 1660-1914. Ithaca: Cornell University Press,
      1975.

156. LANG, WILLIAM DICKSON. (Ed.) "James Harrison of
      Charmouth, Geologist, 1819-64." Dorset Natural
      History and Archaeological Society Proceedings 68
      (1947), 103-118.

157. LAWRIE, JAMES. "Hugh Miller: Geologist and Man of
      Letters." Proceedings of the Royal Institution of
      Great Britain 40 (1964), 92-103.

158. LODGE, SIR OLIVER JOSEPH. Advancing Science: Being
     Personal Reminiscences of the British Association in
     the Nineteenth Century. London: E. Benn Limited,
     1931.

159. LODGE, SIR OLIVER JOSEPH. "Science in the 'Sixties."
     In John Drinkwater (Ed.), The Eighteen-Sixties:
     Essays by Fellows of the Royal Society of Litera-
     ture. Cambridge: Cambridge University Press, 1932,
     245-269.

160. LYDEKKER, RICHARD. Sir William Flower. London: J.
     M. Dent and Co.; New York: E. P. Dutton and Co.,
     1906.

161. LYONS, SIR HENRY GEORGE. "The Officers of the Royal
     Society (1662-1860)." Notes and Records of the
     Royal Society of London 3 (1941), 116-140.

162. LYONS, SIR HENRY GEORGE. The Royal Society, 1660-
     1940: A History of Its Administration under Its
     Charters. Cambridge: University Press, 1944.

163. MACEACHEN, DOUGALD B. "Wilkie Collins' 'Heart and
     Science' and the Vivisection Controversy." Victor-
     ian Newsletter No. 29 (1966), 22-25.

164. MACKENZIE, WILLIAM MACKAY. Hugh Miller: A Critical
     Study. London: Hodder and Stoughton, 1905.

165. MCKIE, D. "1851-1951: A Century of British Sci-
     ence." Journal of the Royal Society of Arts 99
     (1951), 316-325.

166. MACLEOD, ROY M. "The Ayrton Incident: A Commentary
     on the Relations of Science and Government in
     England, 1870-73." In Arnold Thackray and Everett
     Mendelsohn (Eds.), Science and Values: Patterns of
     Tradition and Change. New York: Humanities Press,
     1974, 45-78.

167. MACLEOD, ROY M., JAMES R. FRIDAY AND C. GREGOR.
     The Corresponding Societies of the British Associa-
     tion for the Advancement of Science, 1833-1929: A
     Survey of Historical Records, Archives and Publica-
     tions. London: Mansell, 1975.

168. MACLEOD, ROY M. "Is it Safe to Look Back?" <u>Nature</u>
      224 (Nov. 1, 1969), 417-461. (Traces the history of
      <u>Nature</u> in honour of the centenary of its first
      issue)

169. MACLEOD, ROY M. "A Note on 'Nature' and the Social
      Significance of Scientific Publishing, 1850-1914."
      <u>Victorian Periodicals Newsletter</u> No. 3 (Nov. 1968),
      16-17.

170. MACLEOD, ROY M. "Of Medals and Men: A Reward
      System in Victorian Science." <u>Notes and Records of
      the Royal Society of London</u> 26 (1971), 81-105.

171. MACLEOD, ROY M. "Printing under the Golden Lamp:
      Taylor and Francis Ltd., and Work in Progress on
      Scientific Periodical Publishing." <u>Victorian Peri-
      odicals Newsletter</u> Nos. 5 and 6 (Sept. 1969), 11-12.

172. MACLEOD, ROY M. "The Royal Society and the Govern-
      ment Grant: Notes on the Administration of Scien-
      tific Research, 1849-1914." <u>Historical Journal</u> 14
      (1971), 323-358.

173. MACLEOD, ROY M. "Science and the Civil List, 1824-
      1914." <u>Technology and Science</u> 6 (1970), 47-55.

174. MACLEOD, ROY M. "The Support of Victorian Science:
      The Endowment of Research Movements in Great Britain,
      1868-1900." <u>Minerva</u> 9 (1971), 197-230. (Reprinted
      as "Resources of Science in Victorian England: The
      Endowment of Science Movement, 1868-1900." In Peter
      Mathias, Ed., <u>Science and Society, 1600-1900</u>.
      Cambridge University Press, 1972)

175. MARTIN, THOMAS. "Early Years at the Royal Institu-
      tion." <u>British Journal for the History of Science</u> 2
      (1964), 99-115.

176. MARTIN, THOMAS. <u>The Royal Institution</u>. London, New
      York: Published for the British Council by Longmans,
      Green and Company, 1942.

177. MARTIN, THOMAS. "The Royal Institution of Great
      Britain." <u>Endeavour</u> 3, No. 12 (Oct. 1944), 135-136.

178. MASON, S. F. Main Currents of Scientific Thought.
      New York: H. Schuman, 1953.

179. MATHIAS, PETER. (Ed.) Science and Society 1600-
      1900. Cambridge: University Press, 1972.

180. MENDELSOHN, EVERETT. "The Biological Sciences in
      the Nineteenth Century: Some Problems and Sources."
      History of Science 3 (1964), 39-59.

181. MENDELSOHN, EVERETT. "The Emergence of Science as a
      Profession in Nineteenth Century Europe." In K.
      Hill (Ed.), The Management of Scientists. Boston:
      Beacon Press, 1964, 3-48.

182. MENDELSOHN, EVERETT. "Physical Models and Physio-
      logical Concepts: Explanation in Nineteenth Century
      Biology." In Robert S. Cohen and Marx W. Wartofsky
      (Eds.), Boston Studies in the Philosophy of Science.
      New York: Humanities Press, 1965, Vol. 2, 127-155.
      (Also in British Journal for the History of Science
      2, 1964-65, 201-219)

183. MILLHAUSER, MILTON. "Dr. Newton and Mr. Hyde:
      Scientists in Fiction from Swift to Stevenson."
      Nineteenth-Century Fiction 28 (Dec. 1973), 287-304.

184. MILLS, ERIC L. "Amphipods and Equipoise: A Study
      of T. R. R. Stebbing." Transactions of the Connec-
      ticut Academy of Arts and Science 44 (Dec. 1972),
      239-256.

185. MITCHELL, PETER CHALMERS. Centenary History of the
      Zoological Society of London. London: Printed for
      the Society, 1929.

186. MORRELL, J. B. "Individualism and the Structure of
      British Science in 1830." Historical Studies in the
      Physical Sciences 3 (1971), 183-204.

187. MORRIS, A. D. "Gideon Algernon Mantell (1790-1852),
      Surgeon and Geologist: 'Wizard of the Weald'."
      Proceedings of the Royal Society of Medicine 65
      (1972), 215-221.

188.   MORSE, EDGAR WILLIAM. "Natural Philosophy, Hypothe-
        ses, and Impiety: Sir David Brewster Confronts the
        Undulatory Theory of Light." Dissertation Abstracts
        International 36 (1975), 2372A (University of Cali-
        fornia, Berkeley, 1972).

189.   MUNBY, ALAN NOEL LATIMER. The History and Biblio-
        graphy of Science in England: The First Phase, 1833-
        1845, to Which is Added a Reprint of a Catalogue of
        Scientific Manuscripts in the Possession of J. O.
        Halliwell. Berkeley: University of California
        School of Librarianship, 1968.

190.   MURRAY, REV. ROBERT HENRY. Science and Scientists
        in the Nineteenth Century. London: The Sheldon
        Press, 1925. (W. Jenner, J. Y. Simpson, C. Lyell,
        H. L. F. von Helmholtz, J. P. Joule, Darwin, L.
        Pasteur, J. Lister)

191.   NEAVE, SHEFFIELD AIREY AND FRANCIS JAMES GRIFFIN.
        The History of the Entomological Society of London,
        1833-1933. London, Bungay, Suffolk: Printed by R.
        Clay and Sons, Limited, 1933.

192.   NEWBIGIN, MARION ISABEL AND SIR JOHN SMITH FLETT.
        James Geikie: The Man and the Geologist. Edin-
        burgh: Oliver and Boyd, 1917.

193.   NORTH, F. J. "De la Beche and His Activities, As
        Revealed by His Diaries and Correspondence."
        Abstracts of the Proceedings of the Geological
        Society of London No. 1314 (June 1936), 104-106.

194.   NORTH, F. J. "Dean Conybeare, Geologist." Report
        and Transactions of the Cardiff Naturalists' Society
        66 (1933), 15-68.

195.   NORTH, F. J. "Geology's Debt to Henry Thomas de la
        Beche." Endeavour 3, No. 9 (Jan. 1944), 15-19.

196.   NORTH, F. J. "H. T. de la Beche: Geologist and
        Business Man." Nature 143 (1939), 254-255.

197.   NORTH, F. J.   "W. D. Conybeare, His Geological Con-
       temporaries and Bristol Associations."   Proceedings
       of the Bristol Naturalists' Society 29 (1954-58),
       133-146.

198.   NORTH, JOHN D.   The Measure of the Universe: A His-
       tory of Modern Cosmology.   New York: Oxford Univer-
       sity Press, 1965.

199.   NORTH, JOHN D. (Ed.)   Mid-Nineteenth-Century Scien-
       tists.   Oxford: Pergamon Press, 1969.   (Includes C.
       Babbage, Darwin, J. P. Joule, J. Lister, W. H. Perkin)

200.   OLBY, ROBERT CECIL. (Ed.)   Early Nineteenth-Century
       European Scientists.   New York: Pergamon Press,
       1967.   (H. Davy, J. Berzelius, T. Young, L. J. M.
       Daguerre, W. H. F. Talbot, C. Lyell, A. Quetelet)

201.   OLBY, ROBERT CECIL.   Origins of Mendelism.   New
       York: Schocken Books, 1966.

202.   OLIVER, FRANCIS WALL. (Ed.)   Makers of British Bot-
       any.   Cambridge: Cambridge University Press, 1913.
       (Includes R. Brown, W. Hooker, J. S. Henslow, J.
       Lindley, W. Griffith, A. Henfrey, W. H. Harvey, M.
       J. Berkeley, W. C. Williamson, J. D. Hooker, Pro-
       fessors of Botany in Edinburgh from 1670-1887)

203.   OPPENHEIMER, JANE MARION.   Essays in the History of
       Embryology and Biology.   Cambridge, Mass.:   M.I.T.
       Press, 1967.

204.   ORANGE, A. D.   "The British Association for the
       Advancement of Science: The Provincial Background."
       Science Studies (London) 1 (1971), 315-329.   (Deals
       with years 1831-1885)

205.   ORANGE, A. D.   "The Idols of the Theatre: The
       British Association and Its Early Critics."   Annals
       of Science 32 (1975), 277-294.

206.   ORANGE, A. D.   "The Origins of the British Associa-
       tion for the Advancement of Science."   British Jour-
       nal for the History of Science 6 (1972-73), 152-176.

207. PAYNE, L. G. "The Story of Our Society." London
Naturalist 27 (1948), 3-21. (London Natural History
Society since 1858)

208. PEACH, B. N. AND J. HORNE. "The Scientific Career
of Sir Archibald Geikie." Proceedings of the Royal
Society of Edinburgh 45 (1926), 346-361.

209. PICKFORD, RONALD F. Charles Moore, 1815-1881: A
Brief History of the Man and His Geological Collection. Bath: Bath Municipal Libraries, 1971.

210. PLEDGE, HUMPHREY T. Science since 1500: A Short
History of Mathematics, Physics, Chemistry, and
Biology. London: H. M. Stationery Office, 1939.

211. PORTER, ROY. "The Industrial Revolution and the
Rise of the Science of Geology." In Mikulás Teich
and Robert Young (Eds.), Changing Perspectives in
the History of Science: Essays in Honour of Joseph
Needham. London: Heinemann, 1973, 320-343.

212. PROVINE, WILLIAM B. The Origins of Theoretical
Population Genetics. Chicago: University of Chicago
Press, 1971. (T. H. Huxley, F. Galton, K. Pearson)

213. PUMPHREY, R. J. "The Forgotten Man: Sir John
Lubbock (1834-1913): His Contributions to Zoology
and His Liberal Record as a Member of Parliament
Ought to Be Remembered." Science 129 (1959),
1087-92.

214. RAYMOND, WILLIAM ODBER. "Philip Henry Gosse and
'The Canadian Naturalist'." Transactions of the
Royal Society of Canada, Series 3, 45, Section 2
(June 1951), 43-58.

215. REED, HOWARD SPRAGUE. A Short History of the Plant
Sciences. Waltham, Mass.: Chronica Botanica Company, 1942.

216. RICHARDSON, EDMUND WILLIAM. A Veteran Naturalist:
Being, the Life and Work of W. B. Tegetmeier.
London: Witherby and Co., 1916.

217. RICHARDSON, L.   The Worcestershire Naturalists' Club,
      1847-1947. Gloucester: Bellows, 1947.
218. ROBERTS, HERBERT FULLER. Plant Hybridization before
      Mendel. Princeton: Princeton University Press,
      1929.
219. RODERICK, GORDON W. The Emergence of a Scientific
      Society in England 1800-1965. New York: St.
      Martin's Press, 1967.
220. ROOK, ARTHUR J. (Ed.) The Origins and Growth of
      Biology. Baltimore: Penguin Books, 1964, 235-328.
221. RUDWICK, MARTIN J. S. "The Foundation of the Geo-
      logical Society of London: Its Scheme for Co-opera-
      tive Research and Its Struggle for Independence."
      British Journal for the History of Science 1 (1962-
      63), 325-355.
222. RUDWICK, MARTIN J. S. The Meaning of Fossils: Epi-
      sodes in the History of Palaeontology. London:
      Macdonald & Co.; New York: American Elsevier, 1972.
223. RUNCORN, STANLEY KEITH. (Ed.) Earth Sciences:
      Being the Friday Evening Discourses in Physical Sci-
      ences at the Royal Institution: 1851-1939. The
      Royal Institution Library of Science. 3 vols.
      Barking, England: Applied Science Publishers Ltd.,
      1971.
224. RUSSELL, SIR A. "John Hawkins, F.G.S., F.R.H.S.,
      F.R.S., 1761-1841: A Distinguished Cornishman and
      Early Mining Geologist." Journal of the Royal In-
      stitution of Cornwall New Series 2, Pt. 2 (1954),
      98-106.
225. SANDEMAN, C. "Richard Spruce, Portrait of a Great
      Englishman." Journal of the Royal Horticultural
      Society 74 (1949), 531-544.
226. SCHERREN, HENRY. The Zoological Society of London:
      A Sketch of Its Foundation and Development, and the
      Story of Its Farm, Museum, Gardens, Menagerie, and
      Library. London: Cassell and Company, Limited,
      1905.

227. SCHULTES, R. E. "Richard Spruce Still Lives."
      Northern Gardener 7 (1953), 20-27, 55-61, 87-93,
      121-125.
228. SCHUSTER, ARTHUR AND ARTHUR E. SHIPLEY. Britain's
      Heritage of Science. London: Constable and Co.,
      Ltd., 1917.
229. SCLATER, P. L. A Record of the Progress of the
      Zoological Society of London during the Nineteenth
      Century. London: William Clowes and Sons, 1901.
230. SHAPIN, STEVEN AND ARNOLD THACKRAY. "Prosopography
      as a Research Tool in the History of Science: The
      British Scientific Community, 1700-1900." History
      of Science 12 (1974), 1-28.
231. SHARLIN, HAROLD I. The Convergent Century: The Uni-
      fication of Science in the Nineteenth Century.
      London, New York: Abelard-Schuman, 1966.
232. SHARLIN, HAROLD I. "On Being Scientific: A Criti-
      que of Evolutionary Geology and Biology in the 19th
      Century." Annals of Science 29 (1972), 271-285.
233. SHARPEY-SCHAFER, SIR EDWARD ALBERT. History of the
      Physiological Society during Its First Fifty Years,
      1876-1926. London: Cambridge University Press,
      1927.
234. SHEPPARD, T. "John Phillips." Proceedings of the
      Yorkshire Geological Society 22 (1933), 153-187.
235. SHULL, C. A. AND J. F. STANFIELD. "Thomas Andrew
      Knight: In Memoriam." Plant Physiology 14 (1939),
      1-8.
236. SINGER, CHARLES JOSEPH. From Magic to Science:
      Essays on the Scientific Twilight. New York:
      Dover, 1957.
237. SINGER, CHARLES JOSEPH. A Short History of Biology:
      A General Introduction to the Study of Living Things.
      Oxford: Clarendon Press, 1931. (Revised edition, A
      History of Biology. New York: H. Schuman, 1950)

238.   SINGER, CHARLES JOSEPH.   A Short History of Science
       in the 19th Century.   Oxford: The Clarendon Press,
       1941.

239.   SINGER, CHARLES JOSEPH.   A Short History of Scienti-
       fic Ideas to 1900.   Oxford: Clarendon Press, 1959.

240.   SIRKS, M. J. AND CONWAY ZIRKLE.   The Evolution of
       Biology.   New York: Ronald Press, 1964.

241.   SMART, R. N. AND ELSPETH J. HILL. (Eds.)   An Index
       to the Correspondence and Papers of James David
       Forbes (1809-1868).   St. Andrews, Fife: University
       Library, 1968.

242.   SMITH, EDWARD.   Life of Sir Joseph Banks.   London:
       John Lane; New York: John Lane Company, 1911.

243.   SNELL, WILLIAM E.   "Frank Buckland - Medical Natural-
       ist."   Proceedings of the Royal Society of Medicine
       60 (1967), 291-296.

244.   SOWERBY, ARTHUR DE CARLE, ALICE MURIEL SOWERBY AND
       JOAN EVELYN STONE.   The Sowerby Saga.   Washington:
       1952.

245.   SPOKES, SIDNEY.   Gideon Algernon Mantell.   London:
       J. Bale and Co., 1927.

246.   STAGEMAN, PETER.   A Bibliography of the First Edi-
       tions of Philip Henry Gosse, F.R.S. with Introduc-
       tory Essays by Sacheverell Sitwell and Geoffrey
       Lapage.   Cambridge: The Golden Head Press, Ltd.,
       1955.

247.   STEARN, WILLIAM T.   "Bentham and Hooker's 'Genera
       Plantarum':   Its History and Dates of Publication."
       Journal of the Society of Bibliography and Natural
       History 3 (1956), 127-132.

248.   STEARN, WILLIAM T.   "The Self-Taught Botanists Who
       Saved the Kew Botanic Garden."   Taxon 14 (Dec.
       1965), 293-298.   (John Lindley)

249.   STEARN, WILLIAM T.   "William Herbert's 'Appendix'
       and 'Amaryllidaceae'."   Journal of the Society for
       the Bibliography of Natural History 2 (Nov. 1952),
       375-377.

250.   STIMSON, DOROTHY.   Scientists and Amateurs: A History of the Royal Society. New York: Schuman, 1948.

251.   SWAN, R. G.   "Naturalists and Beachcombers: The Victorian Mania for the Seashore." Country Life 143 (May 2, 1968), 1157, 1159-1160.

252.   SWEETING, GEORGE SCOTLAND. (Ed.)   The Geologists' Association, 1858-1958:   A History of the First Hundred Years. Colchester: Benham and Co., 1958.

253.   SWINTON, W. E.   "Gideon Mantell and the Maidstone Iguanodon." Notes and Records of the Royal Society of London 8 (1951), 261-276.

254.   SWINTON, W. E.   "Historical Interrelations of Geology and Other Sciences." Journal of the History of Ideas 36 (1975), 729-738.

255.   SYNGE, PATRICK MILLINGTON.   "The Botanical Magazine." Royal Horticultural Society Journal 73, Pt. 1 (Jan. 1948), 5-11.

256.   TATON, RENE. (Ed.)   A General History of the Sciences. 4 vols. Trans. Arnold Julius Pomerans. London: Thames & Hudson, 1963-65.

257.   TAYLOR, F. SHERWOOD.   The Century of Science, 1840-1940. London: Heinemann, 1941.

258.   THACKRAY, ARNOLD.   "Natural Knowledge in Cultural Context: The Manchester Model." American Historical Review 79 (1974), 672-709.

259.   THACKRAY, ARNOLD AND EVERETT MENDELSOHN. (Eds.) Science and Values: Patterns of Tradition and Change. New York: Humanities Press, 1974. (Includes articles on the image of science and the struggle between J. D. Hooker and Acton Ayrton for control over the Royal Gardens at Kew)

260.   THEODORIDES, JEAN.   "Humboldt and England." British Journal for the History of Science 3 (1966-67), 39-55.

261. THOMPSON, RUTH D'ARCY. D'Arcy Wentworth Thompson:
The Scholar Naturalist, 1860-1948. London, New
York: Oxford University Press, 1958.

262. THOMSON, JOHN ARTHUR. Progress of Science in the
Century. London, Philadelphia: The Linscott Publishing Co., 1903.

263. TURRILL, WILLIAM BERTRAM. Joseph Dalton Hooker.
London: Thomas Nelson, 1963.

264. TURRILL, WILLIAM BERTRAM. "Joseph Dalton Hooker,
F.R.S. (1817-1911)." Notes and Records of the Royal
Society of London 14, No. 1 (1959), 109-120.

265. TURRILL, WILLIAM BERTRAM. Pioneer Plant Geography:
The Phytogeographical Researches of Sir Joseph Dalton
Hooker. The Hague: Martinus Nijhoff, 1953.

266. TURRILL, WILLIAM BERTRAM. The Royal Botanic Gardens,
Kew: Past and Present. London: H. Jenkins, 1959.

267. VANDERVLIET, WILLIAM GLENN. Microbiology and the
Spontaneous Generation Debate during the 1870's.
Lawrence, Kansas: Coronado Press, 1971.

268. VAUGHAN, R. E. "A Forgotten Work by John Vaughan
Thompson." Proceedings of the Royal Society of Arts
and Sciences of Mauritius 1 (1953), 241-248.

269. WATTS, WILLIAM WHITEHEAD. "The Geological Society:
Its Works and Workers." Geological Society of London
Quarterly Journal 101, Pts. 3-4 (1946), liii-lxxi.

270. WEBB, DAVID A. "William Henry Harvey, 1811-1866,
and the Tradition of Systematic Botany." Hermathena
103 (Autumn 1966), 32-45.

271. WELSH, ALEXANDER. "Theories of Science and Romance,
1870-1920." Victorian Studies 17 (1973), 135-154.

272. WHETZEL, HERBERT HICE. An Outline of the History of
Phytopathology. Philadelphia and London: W. B.
Saunders Company, 1918.

273. WILCOCKSON, W. H. "The Geological Work of Henry
Clifton Sorby." Proceedings of the Yorkshire Geological Society 27 (1947), 1-22.

274. WILLIAMS, HENRY SMITH. The Story of Nineteenth-Century Science. New York and London: Harper and Brothers, 1900.
275. WILLIAMS, L. PEARCE. "The Royal Society and the Founding of the British Association for the Advancement of Science." Notes and Records of the Royal Society of London 16 (1961), 221-233.
276. WOODCOCK, GEORGE. Henry Walter Bates, Naturalist of the Amazons. London: Faber and Faber, 1969.
277. WOODWARD, HORACE BOLINGBROKE. History of Geology. London: Watts and Co., 1911.
278. WOODWARD, HORACE BOLINGBROKE. The History of the Geological Society of London. London: Geological Society, 1907.
279. WRIGHT, H. G. S. "Philip Henry Gosse's Microscope." The Microscope 9 (Jan.-Feb. 1953), 113-115.
280. YONGE, C. M. "Victorians by the Sea Shore." History Today 25 (Sept. 1975), 602-609. (P. H. Gosse)
281. YOUNG, ROBERT MAXWELL. "The Historiographic and Ideological Contexts of the Nineteenth-Century Debate on Man's Place in Nature." In Mikuláš Teich and Robert M. Young, (Eds.), Changing Perspectives in the History of Science: Essays in Honour of Joseph Needham. London: Heinemann Educational, 1973, 344-438.
282. ZITTEL, KARL ALFRED VON. History of Geology and Palaeontology to the End of the Nineteenth Century. Trans. M. M. Ogilvie-Gordon. London: W. Scott; New York: C. Scribner's Sons, 1901.

2. Method and Philosophy

This section concentrates on studies of Charles
Babbage, J. F. W. Herschel, J. S. Mill, and William
Whewell. It also includes a few entries on Darwin, W. S.
Jevons, Baden Powell, and Karl Pearson. See I. Science,
1. General; VI. Geology . . .; VII. Evolution; XIV.
Church of England, 4. Broad Church (Baden Powell); XIX.
Varieties of Belief . . ., 14. Mill, John Stuart.

283. AGASSI, JOSEPH. "Sir John Herschel's Philosophy of
      Success." Historical Studies in the Physical Sci-
      ences 1 (1969), 1-36. (On Herschel's Preliminary
      Discourse on the Study of Natural Philosophy, 1831)
284. ANSCHUTZ, RICHARD PAUL. "The Logic of John Stuart
      Mill." Mind 58 (1949), 277-305.
285. BECHER, HARVEY W. "William Whewell and Cambridge
      Mathematics." Dissertation Abstracts International
      32 (1971), 1430A (University of Missouri, Columbia,
      1971).
286. BECHER, SIEGFRIED. Erkenntnistheoretische Unter-
      suchungen zu Stuart Mills Theorie der Kausalität.
      Halle: M. Niemeyer, 1906.
287. BELL, WALTER LYLE. "Charles Babbage, Philosopher,
      Reformer, Inventor: A History of His Contributions
      to Science." Dissertation Abstracts International
      36 (1975), 1763A (Oregon State University, 1975).
288. BELSEY, ANDREW. "Interpreting Whewell." Studies in
      History and Philosophy of Science 5 (1974), 49-58.
289. BLACKWELL, RICHARD J. "The Inductivist Model of
      Science: A Study in Nineteenth-Century Philosophy
      of Science." Modern Schoolman 51 (Mar. 1974),
      197-212. (J. F. W. Herschel, J. S. Mill and W. S.
      Jevons)

290.   BLAKE, RALPH M., CURT JOHN DUCASSE AND EDWARD H.
       MADDEN. Theories of Scientific Method: The Renais-
       sance Through the Nineteenth-Century. Seattle:
       University of Washington Press, 1960.

291.   BLANCHE, ROBERT. Le Rationalisme de Whewell.
       Paris: F. Alcan, 1935.

292.   BRODY, THOMAS A. "Babbage and the History of Sci-
       ence." Modern Quarterly New Series 2, No. 4 (Autumn
       1947), 308-323.

293.   BROWN, WESLEY MILLER. "Rules and Norms in John
       Stuart Mill's Philosophy of Science." Ph.D. Dis-
       sertation, Harvard University, 1970.

294.   BUCHDAHL, GERD. "Inductivist vs. Deductivist Ap-
       proaches in the Philosophy of Science as Illustrated
       by Some Controversies between Whewell and Mill."
       Monist 55 (1971), 343-367.

295.   BUTTMAN, GUNTHER. The Shadow of the Telescope: A
       Biography of John Herschel. Trans. Bernard Dagel.
       Guildford: Lutterworth, 1974.

296.   BUTTS, ROBERT E. "Necessary Truth in Whewell's
       Theory of Science." American Philosophical Quarter-
       ly 2 (1965a), 1-21.

297.   BUTTS, ROBERT E. "On Walsh's Reading of Whewell's
       View of Necessity." Philosophy of Science 32
       (1965b), 175-181.

298.   BUTTS, ROBERT E. "Professor Marcucci on Whewell's
       Idealism." Philosophy of Science 34 (1967), 175-
       183.

299.   BUTTS, ROBERT E. "Whewell on Newton's Rules of
       Philosophizing." In Robert E. Butts and John W.
       Davis (Eds.), The Methodological Heritage of Newton.
       Toronto: University of Toronto Press, 1970, 132-
       149.

300.   BUTTS, ROBERT E. (Ed.) William Whewell's Theory of
       Scientific Method. Pittsburgh: University of
       Pittsburgh Press, 1968.

301. CANNON, WALTER F.   "John Herschel and the Idea of
     Science."   Journal of the History of Ideas 22 (1961),
     215-239.
302. CANNON, WALTER F.   "William Whewell, F.R.S. (1794-
     1866):   Contribution to Science and Learning."
     Notes and Records of the Royal Society of London 19
     (1964), 176-191.
303. CANTOR, G. N.   "Henry Brougham and the Scottish
     Methodological Tradition."   Studies in the History
     and Philosophy of Science 2 (1971), 69-89.
304. CASTELL, ALBUREY.   "Mill's Logic of the Moral Sci-
     ences: A Study of the Impact of Newtonism on Early
     Nineteenth Century Social Thought."   University of
     Chicago Abstract of Theses: Humanistic Series, 9
     (1930-32), 35.
305. COHEN, MORRIS RAPHAEL AND ERNEST NAGEL.   An Intro-
     duction to Logic and Scientific Method.   New York:
     Harcourt, Brace and Company, 1934.   (Includes J. S.
     Mill)
306. COPI, IRVING M.   "Causal Connections: Mill's
     Method of Experimental Inquiry."   In Introduction
     to Logic, 2nd ed. New York: Macmillan, 1961, 355-
     415.
307. CRAWFORD, JOHN FORSYTH.   The Relation of Inference
     to Fact in Mill's Logic.   Chicago: The University
     of Chicago Press, 1916.
308. DAY, JOHN PATRICK.   "Mill on Matter."   Philosophy 38
     (1963), 52-60.
309. DINGLE, HERBERT.   "Philosophy of Physics, 1850-
     1950."   Nature 168 (1951), 630-636.
310. DINGLE, HERBERT.   "The Scientific Outlook in 1851
     and 1951."   British Journal for the Philosophy of
     Science 2 (1951-52), 85-104.   (Newton and the Vic-
     torians)

311. DUCASSE, CURT JOHN. "John Herschel's Philosophy of
     Science." In Percy W. Long (Ed.), Studies in the
     History of Culture: The Disciplines of the Humani-
     ties. Menasha, Wisconsin: Conference of Secretar-
     ies of the American Council of Learned Societies,
     1942, 279-309. (Reprinted in Edward H. Madden, Ed.,
     Theories of Scientific Method. Seattle: University
     of Washington Press, 1960, 153-182)
312. DUCASSE, CURT JOHN. "Whewell's Philosophy of Sci-
     entific Discovery." Philosophical Review 60 (1951),
     56-69, 213-234.
313. ELLEGÅRD, ALVAR. "Darwinian Theory and Nineteenth-
     Century Philosophies of Science." Journal of the
     History of Ideas 18 (June 1957), 362-393.
314. FREUNDLICH, ELSA. John Stuart Mills Kausaltheorie.
     Düsseldorf: Schwann, 1914.
315. FYVIE, JOHN. "The Calculating Philosopher." In
     Some Literary Eccentrics. London: A. Constable and
     Company, Limited, 1906, 179-209. (C. Babbage)
316. GARWIG, PAUL L. "Charles Babbage (1792-1871)."
     American Documentation 20 (1969), 320-324.
317. GHISELIN, MICHAEL T. The Triumph of the Darwinian
     Method. Berkeley: University of California Press,
     1969.
318. GIERE, RONALD N. AND RICHARD S. WESTFALL. (Eds.)
     Foundations of Scientific Method: The Nineteenth
     Century. Bloomington: Indiana University Press,
     1973. (Includes Darwin, W. Whewell, and F. Galton)
319. HANSCHMANN, ALEXANDER BRUNO. Bernard Palissy der
     Künstler, Naturforscher und Schriftsteller, als
     Vater der induktiven Wissenschaftsmethode des Bacon
     von Verulam: Mit der Darstellung der Induktions-
     theorie Francis Bacons und John Stuart Mills.
     Leipzig: Weicher, 1903.
320. HARRE, ROMANO AND JOHN D. NORTH. "William Whewell
     and the History and Philosophy of Science." British
     Journal for the History of Science 4 (1969), 399-402.

321. HEATHCOTE, A. W. "Whewell's Philosophy of Science."
      M.Sc. Thesis, University of London, 1953.

322. HEATHCOTE, A. W. "William Whewell's Philosophy of
      Science." British Journal for the Philosophy of
      Science 4 (Feb. 1954), 302-314.

323. HENRY, WILBERT CAMERON. "The Analysis of Knowledge
      in John Stuart Mill and William Whewell." Ph.D.
      Dissertation, University of Toronto, 1958.

324. HESSE, MARY. "Whewell's Consilience of Inductions
      and Predictions." The Monist 55 (1971), 520-524.

325. JACKSON, REGINALD. An Examination of the Deductive
      Logic of J. S. Mill. London: Oxford University
      Press, 1941.

326. JAIN, CHAMAN LAL. "Methodology and Epistemology:
      An Examination of Sir John Frederick William
      Herschel's Philosophy of Science with Reference to
      His Theory of Knowledge." Dissertation Abstracts
      International 36 (1976), 5502-03A (Indiana Uni-
      versity, 1975).

327. KANNWISCHER, ARTHUR. "Psychology and Ethics in John
      Stuart Mill's Logic." University of Pittsburgh
      Abstracts of Dissertations 49 (1953), 25-30.

328. KARNS, C. FRANKLIN. "Causal Analysis and Rhetoric:
      A Survey of the Major Philosophical Conceptions of
      Cause Prior to John Stuart Mill." Speech Monographs
      32 (1965), 36-48.

329. KAVALOSKI, VINCENT CARL. "The 'Vera Causa' Princi-
      ple: A Historico-Philosophical Study of a Meta-
      theoretical Concept from Newton Through Darwin."
      Ph.D. Dissertation, University of Chicago, 1974.

330. KEENE, J. "Mill's Method of Hypothesis." Filosofia
      13 (1962), 595-598.

331. KENNEDY, GAIL. "The Psychological Empiricism of
      John Stuart Mill." Ph.D. Dissertation, Columbia
      University, 1928.

332. KNIGHT, DAVID M. "Professor Baden Powell and the
Inductive Philosophy." Durham University Journal 60
(1968), 81-87.

333. KOCKELMANS, JOSEPH J. (Ed.) Philosophy of Science:
The Historical Background. New York: The Free Press,
1968. (Includes J. F. W. Herschel, W. Whewell, J. S.
Mill, H. L. F. von Helmholtz, W. S. Jevons, K. Pearson)

334. KRAJEWSKI, WLADYSLAW. "The Idea of Statistical Law
in 19th-Century Science." Boston Studies in the
Philosophy of Science 14 (1974), 397-405. (Refers
to natural selection)

335. KUBITZ, OSKAR ALFRED. The Development of John
Stuart Mill's System of Logic. Urbana: The University of Illinois, 1932.

336. LAND, BEREL AND GARY STAHL. "Mill's 'Howlers' and
the Logic of Naturalism." Philosophy and Phenomeno-
logical Research 29 (1969), 562-574.

337. LAUDAN, LAURENS. "William Whewell on the Consilience of Inductions." The Monist 55 (1971A), 368-391.
(See also "Reply to Mary Hesse." The Monist 55,
1971b, 525)

338. LEVIN, DAVID MICHAEL. "Some Remarks on Mill's
Naturalism." Journal of Value Inquiry 3 (Winter
1969), 291-297.

339. LOSEE, JOHN. A Historical Introduction to the
Philosophy of Science. London: Oxford University
Press, 1972.

340. LUCAS, GERALD MORTON. "Whewell's Philosophy of the
Inductive Sciences." Dissertation Abstracts 16
(1956), 1468-69 (Columbia University, 1956).

341. MACFARLANE, ALEXANDER. Lectures on Ten British
Physicists of the Nineteenth Century. New York:
John Wiley and Sons, Inc., 1919. (Includes C.
Babbage, W. Whewell, J. F. W. Herschel)

342. MACLENNAN, BARBARA. "Jevons's Philosophy of
Science." Manchester School of Economic and Social
Studies 40 (Mar. 1972), 53-71.

343. MCRAE, ROBERT F.   "Phenomenalism and J. S. Mill's
      Theory of Causation."   Philosophy and Phenomeno-
      logical Research 9 (1948), 237-250.

344. MCRAE, ROBERT F.   "The Relation of John Stuart
      Mill's Logic to His Metaphysics and Epistemology."
      Ph.D. Dissertation, Johns Hopkins University, 1946.

345. MADDEN, EDWARD H.   Theories of Scientific Method:
      The Renaissance Through the Nineteenth Century.
      Seattle: University of Washington Press, 1960.
      (Includes J. F. W. Herschel, W. Whewell, J. S. Mill,
      W. S. Jevons)

346. MAKINDE, MOSES AKINOLA.   "John Stuart Mill's Theory
      of Logic and Scientific Method as a Rejection of
      Hume's Scepticism with Regard to the Validity of
      Inductive Reasoning."   Ph.D. Dissertation, Uni-
      versity of Toronto, 1975.

347. MARCUCCI, SILVESTRO.   "Di alcuni contributi di
      William Whewell alla nomenclatura scientifica."
      Physis 5 (1963), 373-382.

348. MARCUCCI, SILVESTRO.   L'"idealismo" scientifico di
      William Whewell.   Pisa: Istituto di filosofia,
      1963.

349. MARCUCCI, SILVESTRO.   "La teoria del metodo scien-
      tifico nell'epistemologia di William Whewell."
      Physis 11 (1969), 379-389.

350. MARCUCCI, SILVESTRO.   "William Whewell: Kantianism
      or Platonism?"   Physis 12 (1970), 69-72.

351. MARLIES, MICHAEL W.   "A Re-examination of Mill's
      'Utilitarianism' in the Context of His Philosophy of
      Science."   Dissertation Abstracts International 34
      (1974), 4328-29A (Brandeis University, 1973).

352. MAYS, W.   "Jevons's Conception of Scientific Method."
      Manchester School of Economics and Social Studies 30
      (1962), 223-249.

353. MEDAWAR, PETER BRIAN.   "Is the Scientific Paper
      Fraudulent?"   Saturday Review of Literature (Aug. 1,
      1964), 42-43.   (J. S. Mill's method)

354. MENDELSOHN, EVERETT. "Revolution and Reduction:
      The Sociology of Methodological and Philosophical
      Concerns in 19th-Century Biology." In Y. Elkana
      (Ed.), The Interaction Between Science and Philos-
      ophy. Atlantic Highlands: Humanities, 1974, 407-
      426.

355. MORRISON, PHILIP AND EMILY. (Eds.) Charles Babbage
      and His Calculating Engines. New York: Dover
      Publications, 1961.

356. MORRISON, PHILIP AND EMILY. (Eds.) "The Strange
      Life of Charles Babbage." Scientific American 186
      (Apr. 1952), 66-73.

357. MOSELY, MABOTH. Irascible Genius: A Life of Charles
      Babbage, Inventor. London: Hutchinson, 1964.

358. MULLETT, CHARLES FREDERIC. "Charles Babbage (1792-
      1871): A Scientific Gadfly." Scientific Monthly 67
      (Nov. 1948), 361-371.

359. MUNSON, JAMES RONALD. "The Science of Science: A
      Critical Examination of John Stuart Mill's Philos-
      ophy of Science." Dissertation Abstracts 28 (1967),
      1851A (Columbia University, 1967).

360. NAGEL, ERNEST. (Ed.) John Stuart Mill's Philosophy
      of Scientific Method. New York: Hafner, 1950.

361. OLSON, RICHARD. Scottish Philosophy and British
      Physics, 1750-1880: A Study in the Foundations of
      Victorian Scientific Style. Princeton, N.J.:
      Princeton University Press, 1975.

362. PEARL, PHILIP DAVID. "William Whewell's Conception
      of the Philosophy of Science." Dissertation
      Abstracts 27 (1967), 3490-91A (New School for Social
      Research, 1966).

363. PRICE, HENRY HABBERLEY. "Mill's View of the Ex-
      ternal World." Proceedings of the Aristotelian
      Society. New Series 27 (1926-27), 109-140.

364. RANDALL, JOHN HERMAN, JR. "J. S. Mill and the
      Working-out of Empiricism." Journal of the History
      of Ideas 16 (1965), 59-88.

365. REICHEL, H. "Darstellung und Kritik von J. St.
      Mills Theorie der induktiven Methode." Zeitschrift
      für Philosophie und philosophische Kritik 122 (1903),
      176-197; 123 (1904), 33-46, 121-151.

366. REINGOLD, NATHAN. "Babbage and Moll on the State of
      Science in Great Britain." British Journal for the
      History of Science 4 (1968-69), 58-64.

367. RIDDLE, CHAUNCEY CAZIER. "Karl Pearson's Philosophy
      of Science." Dissertation Abstracts 19 (1959), 3326
      (Columbia University, 1958).

368. ROBSON, ROBERT. "William Whewell, F.R.S. (1794-
      1866): Academic Life." Notes and Records of the
      Royal Society of London 19 (1964), 168-176.

369. ROGERSON, ALAN T. "A Study of John Stuart Mill's
      View of Scientific Inference with Particular Refer-
      ence to Its Philosophical Context." M.Sc. Disserta-
      tion, University of London, 1970.

370. ROSENBERG, A. "Mill and Some Contemporary Critics
      on 'Cause'." Personalist 54 (1973), 123-129.

371. RYAN, ALAN. John Stuart Mill. New York: Pantheon
      Books, 1970.

372. RYAN, ALAN. "Mill and the Naturalistic Fallacy."
      Mind 75 (July 1966), 422-425.

373. SCHAGRIN, MORTON L. "Whewell's Theory of Scientific
      Language." Studies in History and Philosophy of
      Science 4 (1973), 231-240.

374. SCHAGRIN, MORTON L. "William Whewell: Philosopher
      of Science." Dissertation Abstracts 27 (1967),
      3491A (University of California, Berkeley, 1966).

375. SCHMID, J. VON. "John Stuart Mill's Logica der
      Geesteswetenschappen." Algemeen Nederlands
      Tijdschrift voor Wijsbegeerte en Psychologie 37
      (1944), 11-20.

376. SELEM, ALESSANDRO. "La teoria dell'induzione di J.
      S. Mill." Studia patavina 6 (1959), 524-539.

377.   SEWARD, GEORGES C.   "Die theoretische Philosophie
       William Whewells und der kantische Einfluss."
       Dissertation, Universität Tübingen, 1938.

378.   SHORLAND, EILEEN.   "The Last of the Philosophers:
       Sir John Herschel, Bart., 1792-1871." Journal of
       the British Astronomical Association 83 (1973),
       335-340.

379.   SILVERS, STUART.   "The Evolutionary Development of
       Scientific Method in England from Bacon to Mill:
       Being an Historical Analysis of the Methods of Ex-
       perimental Investigation." Dissertation Abstracts
       25 (1965), 7318 (University of Pittsburgh, 1963).

380.   SMOKLER, HOWARD EDWARD.   "Scientific Concepts and
       Philosophical Theory: An Essay in the Philosophy of
       W. K. Clifford." Dissertation Abstracts 20 (1959),
       1394 (Columbia University, 1959).

381.   STEWART, HERBERT LESLIE.   "J. S. Mill's Logic: A
       Post Centenary Appraisal." University of Toronto
       Quarterly 17 (1948), 361-371.

382.   STOCKS, JOHN LEOFRIC.   "The Empiricism of John
       Stuart Mill." In Reason and Intuition, and Other
       Essays. London: Oxford University Press, 1939,
       208-217.

383.   STOLERMAN, HAROLD.   "Francis Bacon and the Vic-
       torians, 1830-1885." Dissertation Abstracts Inter-
       national 31 (1971), 6023A (New York University,
       1969).

384.   STOLL, MARION RUSH.   Whewell's Philosophy of Induc-
       tion. Lancaster, Pa.: Lancaster Press, Inc., 1929.

385.   STRONG, E. W.   "William Whewell and John Stuart
       Mill: Their Controversy about Scientific Knowledge."
       Journal of the History of Ideas 16 (1955), 209-231.

386.   TATARKIEWICZ, WLADYSLAW.   "John Stuart Mill and
       Empiricism." In Nineteenth Century Philosophy.
       Belmont: Wadsworth, 1974, 32-44.

387. TAWNEY, GUY ALLEN.   John Stuart Mill's Theory of
      Inductive Logic, in Two Parts.   2 vols.   Cincinnati,
      Ohio: University Press, 1909.

388. WALLACE, WILLIAM A.   Classical and Contemporary
      Science.   Ann Arbor: University of Michigan Press,
      1974.   (Volume 2 of Causality and Scientific Explana-
      tion)

389. WALSH, HAROLD TRUEMAN.   "The Philosophy of Science
      of William Whewell."   Dissertation Abstracts 21
      (1961), 2329 (University of Michigan, 1960).

390. WALSH, HAROLD TRUEMAN.   "Whewell and Mill on Induc-
      tion."   Philosophy of Science 29 (July 1962), 279-
      284.

391. WALSH, HAROLD TRUEMAN.   "Whewell on Necessity."
      Philosophy of Science 29 (1962), 139-145.

392. WENTSCHER, ELSE.   Das Problem des Empirismus
      dargestellt an John Stuart Mill.   Bonn: Marcus and
      Weber, 1922.

393. WEXLER, P. J.   "The Great Nomenclator: Whewell's
      Contributions to Scientific Terminology."   Notes and
      Queries 206 (1961), 27-29, 32.

394. WHITMORE, C. E.   "Mill and Mathematics: An Histori-
      cal Note."   Journal of the History of Ideas 6 (1945)
      109-112.

395. WILLIAMS, L. PEARCE.   "Epistemology and Experiment:
      The Case of Michael Faraday."   In Imre Lakatos and
      Alan Musgrave (Eds.), Problems in the Philosophy of
      Science.   Amsterdam: North-Holland Publishing
      Company, 1968, 231-239.

396. WILSON, CURTIS.   "Newton and Some Philosophers on
      Kepler's 'Laws'."   Journal of the History of Ideas
      35 (Apr.-June 1974), 231-258.   (Includes J. S. Mill,
      W. Whewell)

397. WILSON, DAVID B.   "Butts on Whewell's View of True
      Causes."   Philosophy of Science 40 (1973), 121-124.
      (Butts' reply, 125-128)

398.   WILSON, DAVID B.   "Herschel and Whewell's Version of
       Newtonianism."   Journal of the History of Ideas 35
       (1974), 79-97.
399.   WISNIEWSKI, JOSEPH.   Etude historique et critique de
       la théorie de la perception extérieure chez John
       Stuart Mill et Taine.   Paris: Jouve, 1925.
400.   WOLFE, JULIAN.   "Mill on Causality."   Personalist 57
       (Winter 1976), 96-97.

II.   Religion

1.   General

This section contains general works on religious
movements and organizations and on the history of religion,
as well as more specific studies on religion and politics
and on religion and the structure of society. Subjects
which do not fit easily into the various other categories
on religion find their place here, including the sociology
of religion, religion in poetry and fiction, the relationship of one Church to another, religion at Oxford and
Cambridge, the religious census of 1851, church-related
controversies, unification movements, the blasphemy laws,
toleration, church design, worship, and histories of the
Old and New Testaments. This bibliography concentrates
almost entirely on England, but in this section (as in
several others) there are a number of works on Scotland,
Ireland, and Wales. A few studies on non-Christian
religions (Buddhism and Judaism) are also cited here.
See II. Religion, 2. Theology; III. Ideas . . .; IV.
Education; V. Natural Theology; XIV. Church of England;
XV. Nonconformity; XVI. Catholicism; XIX. Varieties of
Belief . . . .

401.   ALCOCK, RICHARD A. "The Victorian Historical Novel:
        A Record of Religious Unrest." Ph.D. Dissertation,
        New York University, 1949.
402.   ALINGTON, CYRIL ARGENTINE. Christianity in England:
        An Historical Sketch. London: Oxford University
        Press, 1942.
403.   ALTHOLZ, JOSEF LEWIS. The Churches in the Nine-
        teenth Century. New York: Bobbs-Merrill, 1967.

404.   AMBLER, R. W.   "The 1851 Census of Religious Wor-
       ship." Local Historian 11 (1975), 375-381.

405.   ANDERSON, OLIVE.   "The Growth of Christian Militar-
       ism in Mid-Victorian Britain." English Historical
       Review 86 (Jan. 1971), 46-72.

406.   ANDERSON, OLIVE.   "The Reaction of Church and Dis-
       sent towards the Crimean War." Journal of Eccles-
       iastical History 16 (1965), 209-220.

407.   ANDERSON, OLIVE.   "Women Preachers in Mid-Victorian
       Britain: Some Reflections on Feminism, Popular
       Religion and Social Change." Historical Journal 12
       (1969), 467-484.

408.   ANDREWS, JAMES R.   "The Rationale of Nineteenth-
       Century Pacifism: Religious and Political Arguments
       in the Early British Peace Movement." Quaker History
       57 (Spring 1968), 17-27.

409.   ANDREWS, JOHN S.   "German Influence on English Reli-
       gious Life in the Victorian Era." Evangelical Quar-
       terly 44 (Oct.-Dec. 1972), 218-233.

410.   ANON.   "The Church and Dissent in Wales during the
       Nineteenth Century." Church Quarterly Review 58
       (1904), 51-73.

411.   ANSON, HAROLD.   "The Church in 19th Century Fiction."
       Listener 21 (1939), 945-946, 998-999, 1118-19. (J.
       H. Shorthouse, A. Trollope, George Eliot)

412.   ARNSTEIN, WALTER L.   "The Religious Issue in Mid-
       Victorian Politics: A Note on a Neglected Source."
       Albion 6 (Summer 1974), 114-175. (Source is Dod's
       Parliamentary Companion)

413.   BAKER, WILLIAM J. "The Attitude of English Churchmen,
       1800-1850, to the Reformation." Ph.D. Dissertation,
       University of Cambridge, 1966.

414.   BARING-GOULD, SABINE.   The Evangelical Revival.
       London: Methuen and Co., Ltd., 1920.

415.   BARNES, GEORGE NICOLL, EINAR LI, A. HENDERSON et al.
       The Religion in the Labour Movement: Speeches at the
       International Conference on Labour and Religion Held
       in London, 1919. London: Holborn Press, 1919.

416.   BENNETT, GARETH VAUGHAN AND J. D. WALSH. (Eds.)
       Essays in Modern Church History: In Memory of
       Norman Sykes. London: Black, 1966. (Includes
       Methodism, H. P. Hughes, the Nonconformist Conscience)

417.   BENZIGER, JAMES. Images of Eternity: Studies in
       the Poetry of Religious Vision from Wordsworth to
       T. S. Eliot. Carbondale: Southern Illinois Uni-
       versity Press, 1963.

418.   BIVORT DE LA SAUDEE, JACQUES DE.   Anglicans et
       Catholiques. 2 vols. Bruxelles: A. Goemaere;
       Paris: Plon, 1949. (Volume 1: Le Problème de
       l'union anglo-romaine, 1833-1933)

419.   BONNER, HYPATIA BRADLAUGH. Penalties upon Opinion:
       Or, Some Records of the Laws of Heresy and Blasphemy.
       London: ' Watts and Co., 1912.

420.   BRANTON, CLARENCE L. "The Church in the English
       Novel, 1800-1850." Ph.D. Dissertation, Harvard
       University, 1951.

421.   BREADY, JOHN WESLEY. England before and after
       Wesley: The Evangelical Revival and Social Reform.
       London: Hodder and Stoughton, Limited, 1938.

422.   BRETT, RAYMOND L. (Ed.) Poems of Faith and Doubt:
       The Victorian Age. London: Edward Arnold, 1965.

423.   BRIGGS, ASA. "Religion and Capitalism." Listener
       71 (Feb. 27, 1964), 339-341. (R. H. Tawney and
       Victorian England)

424.   BROWN, ALAN WILLARD. The Metaphysical Society:
       Victorian Minds in Crisis, 1869-1880. New York:
       Columbia University Press, 1947.

425.   BRUCE, FREDERICK FYVIE. The English Bible: A
       History of Translations. London: Lutterworth
       Press, 1961.

426. BULLOCK, FREDERICK WILLIAM BAGSHAWE. Evangelical
     Conversion in Great Britain, 1696-1845. St.
     Leonard's-on-Sea: Budd and Gilliat, 1959. (Both
     Nonconformists and Anglicans, including H. Venn,
     Rowland Hill, T. Scott, C. Simeon, W. Wilberforce,
     Legh Richmond, H. Bourne, T. Chalmers, Cornelius
     Neale, Robert McCheyne, Catherine Mumford)
427. BUSSEY, O. "The Religious Awakening of 1858-1860 in
     Great Britain and Ireland." Ph.D. Dissertation,
     University of Edinburgh, 1947.
428. CALDECOTT, ALFRED. The Philosophy of Religion in
     England and America. London: Methuen & Co., 1901.
429. CARPENTER, JOSEPH ESTLIN. The Bible in the Nine-
     teenth Century. London, New York and Bombay:
     Longmans, Green, and Co., 1903.
430. CARTER, CHARLES. "The Pre-Raphaelites as Religious
     Painters." Quarterly Review 286 (1948), 248-261.
431. CHADWICK, OWEN. The Victorian Church. 2 parts.
     Volumes 7 and 8 of J. C. Dickinson (Ed.), An Eccles-
     iastical History of England. London: Adam & Charles
     Black; New York: Oxford University Press, 1966-70.
     (Second ed. part 1, Adam & Charles Black, 1970;
     Third ed. Black, 1971)
432. CHAPMAN, EDWARD MORTIMER. English Literature in
     Account with Religion, 1800-1900. Boston and New
     York: Houghton Mifflin Co., 1910.
433. CHILD, PHILIP A. "Evangelicalism and English Litera-
     ture, 1798-1830: A Study in Literary, Religious,
     and Social Interrelations." Ph.D. Dissertation,
     Harvard University, 1928.
434. CLARK, BRUCE B. "The Spectrum of Faith in Victorian
     Literature." Brigham Young University Studies 4
     (1962), 183-207.
435. CLARKE, BASIL FULFORD LOWTHER. Church Builders of
     the Nineteenth Century: A Study of the Gothic
     Revival in England. London: Society for the Promo-
     tion of Christian Knowledge, 1938.

436. CLARKE, GEORGE HERBERT. "Christ and the English
      Poets." Queen's Quarterly 55 (1948), 292-307.
437. CLAUSEN, C. "Victorian Buddhism and the Origins of
      Comparative Religion." Religion: A Journal of
      Religion and Religions 5 (1975), 1-15.
438. COCKSHUT, ANTHONY O. J. (Ed.) Religious Contro-
      versies of the Nineteenth Century: Selected Docu-
      ments. Lincoln: University of Nebraska Press, 1966.
439. COLVILLE, DEREK. Victorian Poetry and the Romantic
      Religion. Albany: S.U.N.Y. Press, 1970.
440. COWIE, LEONARD W. "Exeter Hall." History Today 18
      (1968), 390-397.
441. CURZON, GORDON ANTHONY. "Paradise Sought: A Study
      of the Religious Motivation in Representative British
      and American Literary Utopias, 1850-1950." Disserta-
      tion Abstracts International 30 (1970), 4405A (Uni-
      versity of California, Riverside, 1969).
442. DAVIES, EBENEZER THOMAS. Religion in the Industrial
      Revolution in South Wales. Cardiff: University of
      Wales Press, 1965.
443. DAVIES, TREVOR H. Spiritual Voices in Modern Lit-
      erature. New York: George H. Doran, c.1919.
      (Includes F. Thompson, J. Ruskin, W. Wordsworth, J.
      Morley, R. Browning, A. Tennyson)
444. DAVIS, V. D. A History of Manchester College.
      London: Allen and Unwin, 1932.
445. DAWSON, CHRISTOPHER HENRY. Religion and the Modern
      State. London: Sheed and Ward, 1935.
446. DENNING, SIR ALFRED THOMPSON. "The Meaning of
      'Ecclesiastical Law'." Law Quarterly Review 60, No.
      239 (July 1944), 235-241. (Following the Irish
      Church Act, 1869, and the Welsh Church Act, 1914)
447. DIXON, H. N. "Religious Life in Cambridge in the
      80's of the Last Century." Congregational Quarterly
      20 (1942), 313-317.

448. DRUMMOND, ANDREW LANDALE. The Churches in English
     Fiction. Leicester: Backus, 1950. (Historical
     study of more than 250 novels)

449. DUNNING, ROBERT. "Nineteenth-Century Parochial
     Sources." In Derek Baker (Ed.), The Materials,
     Sources and Methods of Ecclesiastical History:
     Papers Read at the Twelfth Summer Meeting and the
     Thirteenth Winter Meeting of the Ecclesiastical
     History Society. Oxford: Published for the
     Ecclesiastical History Society by Blackwell, 1975,
     301-308.

450. EDWARDS, MALDWYN LLOYD. "The Church and the Rise of
     Socialism." London Quarterly and Holborn Review 20
     (1951), 201-208.

451. ELLIOTT-BINNS, LEONARD ELLIOTT. Religion in the Vic-
     torian Era. London: Lutterworth Press, 1936.

452. EVANS, ERIC J. "Some Reasons for the Growth of
     English Rural Anti-Clericalism c.1750-c.1830."
     Past and Present 66 (Feb. 1975), 84-109.

453. EVANS, HILARY A. "Religion and the Working Class in
     Mid Nineteenth Century England." M. Phil. Disserta-
     tion, University of London, 1970.

454. FAIRCHILD, HOXIE NEALE. Christianity and Roman-
     ticism in the Victorian Era 1830-1880. Volume 4 of
     Religious Trends in English Poetry. New York:
     Columbia University Press, 1957. (Includes A.
     Tennyson, M. Arnold, A. Clough)

455. FAIRCHILD, HOXIE NEALE. Gods of a Changing Poetry
     1880-1920. Volume 5 of Religious Trends in English
     Poetry. New York and London: Columbia University
     Press, 1962.

456. FAIRCHILD, HOXIE NEALE. "Religious Trends in Vic-
     torian Poetry." Victorian Newsletter No. 5 (Apr.
     1954), 3.

457.   FAIRCHILD, HOXIE NEALE.   Romantic Faith 1789-1830.
       Volume 3 of Religious Trends in English Poetry. New
       York: Columbia University Press, 1949. (Includes
       W. Wordsworth, S. T. Coleridge, P. B. Shelley, Lord
       Byron, J. Keats)

458.   FALLIS, JEAN THOMSON. "The Sacred and the Profane:
       Transvaluation of Religious Symbol in Hopkins,
       Rossetti, and Swinburne." Dissertation Abstracts
       International 36 (1975), 1524A (Princeton Uni-
       versity, 1974).

459.   FAULKNER, HAROLD UNDERWOOD.   Chartism and the
       Churches: A Study in Democracy. New York: The
       Columbia University Press, 1916.

460.   FAWKES, ALFRED. Studies in Modernism. London:
       Smith, Elder and Co., 1913. (In both the Catholic
       Church and the Church of England; includes G. Tyrrell,
       J. H. Newman, evolution, M. Arnold, Mrs. Humphry
       Ward)

461.   FLYNN, JOHN STEPHEN. The Influence of Puritanism on
       the Political and Religious Thought of the English.
       London: J. Murray, 1920.

462.   FOSTER, CHARLES I. An Errand of Mercy: The Evan-
       gelical United Front, 1790-1837. Chapel Hill:
       University of North Carolina Press, 1961.

463.   FRIETZSCHE, ARTHUR H. Disraeli's Religion: The
       Treatment of Religion in Disraeli's Novels. Logan:
       Utah State University Press, 1961.

464.   GARDNER, PERCY. Modernism in the English Church.
       London: Methuen and Co. Ltd., 1926. (In the Roman
       and English churches)

465.   GAY, JOHN D. The Geography of Religion in England.
       London: Duckworth, 1971. (Religious demography of
       England based on the Census of 1851)

466.   GILL, JOHN CLIFFORD. The Ten Hours Parson:
       Christian Social Action in the Eighteen-thirties.
       London: S.P.C.K., 1959. (George Stringer Bull)

467.   GILLEY, SHERIDAN. "Evangelical and Roman Catholic
       Missions to the Irish in London, 1830-1870." Ph.D.
       Dissertation, University of Cambridge, 1971.

468.   GREAVES, R. W. "'Church' and 'Chapel': The Histori-
       cal Background of Home Reunion, 1559-1952." Church
       Quarterly Review 153 (1952), 452-469. (The Church
       of England and Nonconformity)

469.   GREEN, VIVIAN HUBERT HOWARD. Religion at Oxford and
       Cambridge. London: SCM Press, 1964.

470.   GREENSLADE, STANLEY LAWRENCE. (Ed.) The Cambridge
       History of the Bible: The West from the Reformation
       to the Present Day. Cambridge: Cambridge Uni-
       versity Press, 1963.

471.   GREENSLADE, STANLEY LAWRENCE. "The Eighteenth and
       Nineteenth Centuries in England." In The Church and
       the Social Order: A Historical Sketch. London:
       S.C.M. Press, 1948, 98-121.

472.   GUPPY, HENRY. A Brief Sketch of the History of the
       Transmission of the Bible Down to the Revised English
       Version of 1881-1895. Manchester: Manchester
       University Press, 1926.

473.   HALEVY, ELIE. A History of the English People in
       the Nineteenth Century. 4 vols. Trans. E. I.
       Watkin and D. A. Barker. London: E. Benn, 1912-47.
       (Especially volume 1, England in 1815, Book 3,
       "Religion and Culture")

474.   HALL, THOMAS CUMMING. The Social Meaning of Modern
       Religious Movements in England. New York: C.
       Scribner's Sons, 1900. (Includes the Evangelical
       party and social reform, the Broad Church movement,
       the High Church reaction)

475.   HAMILTON-HOARE, HENRY WILLIAM. The Evolution of the
       English Bible: An Historical Sketch of the Succes-
       sive Versions from 1382 to 1885. New York: E. P.
       Dutton and Co., 1901.

476. HARRINGTON, HENRY RANDOLPH. "Muscular Christianity:
      A Study of the Development of a Victorian Idea."
      Dissertation Abstracts International 32 (1972),
      5738-39A (Stanford University, 1971). (Includes C.
      Kingsley, T. Hughes)

477. HARRISON, ARCHIBALD HAROLD WALTER. The Evangelical
      Revival and Christian Reunion. London: Epworth
      Press, 1942. (Evangelicalism both in and outside
      the Church of England, the unification of Protestant
      Christianity)

478. HARRISON, BRIAN. "Religion and Recreation in Nine-
      teenth Century England." Past and Present No. 38
      (1967), 98-125.

479. HARRISON, FREDERICK. The Bible in Britain. London,
      New York: T. Nelson, 1949.

480. HAWKS, EDWARD F. "The Anglican Reunion Movement and
      the Catholic Church." Catholic Historical Review
      24, No. 2 (July 1938), 129-140. (After the Oxford
      Movement)

481. HEENEY, BRIAN. "On Being a Mid-Victorian Clergy-
      man." Journal of Religious History 7 (June 1973),
      208-224.

482. HEMPHILL, SAMUEL. A History of the Revised Version
      of the New Testament. London: E. Stock, 1906.

483. HENKIN, LEO JUSTIN. "Problems and Digressions in
      the Victorian Novel." Bulletin of Bibliography
      and Dramatic Index 18 (1943-44), 40-43, 56-60,
      83-86. (Bibliographies consisting of a list of
      contemporary novels on: religious doubt and dis-
      belief; ritualism and the High Church; Protestantism
      versus Catholicism; Nonconformity and the Low Church)

484. HENNELL, MICHAEL. "Evangelicalism and Worldliness,
      1770-1870." In G. J. Cuming and Derek Baker (Eds.),
      Popular Belief and Practice: Papers Read at the
      Ninth Summer Meeting and the Tenth Winter Meeting
      of the Ecclesiastical History Society. Cambridge:
      Cambridge University Press, 1972, 229-236.

485.   HENRIQUES, URSULA. Religious Toleration in England,
       1787-1833. London: Routledge and Kegan Paul, 1961.

486.   HERBERT, ARTHUR SUMNER. Historical Catalogue of
       Printed Editions of the English Bible: 1525-1961;
       Revised and Expanded from the Edition of T. H.
       Darlow and H. F. Moule, 1903. London: British and
       Foreign Bible Society; New York: The American Bible
       Society, 1968.

487.   HOLTON, OSCAR DILE, JR. "The Victorian Sermon as
       Literature." Dissertation Abstracts 28 (1968),
       3144A (Texas Technological College, 1967).

488.   HOUGHTON, WALTER EDWARDS. "Victorian Anti-
       Intellectualism." Journal of the History of Ideas
       13 (1952), 291-313. (Puritanism's tendency to exalt
       conscience at the expense of intellect)

489.   HOYT, ARTHUR STEPHEN. The Spiritual Message of
       Modern English Poetry. New York: The Macmillan
       Company, 1924. (Includes R. Browning, M. Arnold, A.
       Tennyson)

490.   HUDSON, CYRIL E. AND MAURICE BENINGTON RECKITT.
       The Church and the World: Being Materials for
       the Historical Study of Christian Sociology. 3
       vols. London: G. Allen and Unwin Ltd., 1938-40.
       (Volume 3: Church and Society in England from 1800)

491.   HUMPHREYS, CHRISTMAS. The Development of Buddhism
       in England: Being a History of the Buddhist Move-
       ment in London and the Provinces. London: The
       Buddhist Lodge, 1937.

492.   HYAMSON, ALBERT MONTEFIORE. A History of the Jews
       in England. London: Macmillan, 1907.

493.   INGLIS, KENNETH STANLEY. Churches and the Working
       Classes in Victorian England. London: Routledge
       and Kegan Paul, 1963.

494.   INGLIS, KENNETH STANLEY. "English Churches and the
       Working Classes, 1880-1900, with an Introductory
       Survey of Tendencies Earlier in the Century." Ph.D.
       Dissertation, University of Oxford, 1956.

495. INGLIS, KENNETH STANLEY. "The Labour Church Move-
      ment." International Review of Social History 3
      (1958), 445-460.

496. INGLIS, KENNETH STANLEY. "Patterns of Religious
      Worship in 1851." Journal of Ecclesiastical History
      11 (Apr. 1960), 74-86.

497. JAMES, DAVID GWILYM. The Romantic Comedy. London,
      New York: Oxford University Press, 1948. (Includes
      Christianity in literature, similarities between J.
      H. Newman and S. T. Coleridge)

498. JAMES, EDWIN OLIVER. A History of Christianity in
      England. London, New York: Hutchinson's University
      Library, 1949.

499. JAMES, WALTER. The Christian in Politics. London:
      Oxford University Press, 1963.

500. JOHNSON, HENRY. Stories of Great Revivals. London:
      Religious Tract Society, 1906.

501. KELLETT, ERNEST EDWARD. Religion and Life in the
      Early Victorian Age. London: Epworth Press, 1938.

502. KELLETT, ERNEST EDWARD. "The Religious Biography."
      Life and Letters 9 (1933-34), 233-243.

503. KENT, JOHN HENRY SOMERSET. "The Role of Religion in
      the Cultural Structure of the Later Victorian City."
      Transactions of the Royal Historical Society 23
      (1973), 153-173.

504. KENT, JOHN HENRY SOMERSET. "The Victorian Resis-
      tance: Comments on Religious Life and Culture,
      1840-1880." Victorian Studies 12 (1968), 145-154.
      (Resistance to secularization)

505. KITSON CLARK, GEORGE SIDNEY ROBERTS. "The Religion
      of the People." In The Making of Victorian England.
      London: Methuen, 1962, 147-205.

506. KNOTT, JAMES P. "Evangelicalism and Its Influence
      on English Social Reform during Part of the Eigh-
      teenth and Nineteenth Centuries." Ph.D. Disserta-
      tion, University of Southern California, 1939.

507.   KRUMP, JACQUELINE.   "The Clergyman in the Victorian
       Novel."   Summaries of Doctoral Dissertations . . .
       Northwestern University 19 (1951), 25-30.   (Includes
       Tractarians, Old High Churchmen, Evangelicals, Broad
       Churchmen, Nonconformists, Roman Catholics)

508.   LAND, W. L. "A Study of the Influence of the Vic-
       torian Home on the Religious Development of the
       Child in the Last Three Decades of the Nineteenth
       Century." B.Litt. Dissertation, University of
       Oxford, 1948.

509.   LATOURETTE, KENNETH SCOTT. A History of the Expan-
       sion of Christianity. New York and London: Harper
       and Brothers, 1937-45. (Volume 4: The Great Cen-
       tury, A.D. 1800 - A.D. 1914)

510.   LATOURETTE, KENNETH SCOTT. The Nineteenth Century
       in Europe: The Protestant and Eastern Churches.
       New York: Harper, c.1959.

511.   LESOURD, J. A. "La Déchristianisation en Angleterre
       vers le milieu du 19e siècle." Cahiers d'histoire 9
       (1964), 279-294.

512.   LILLY, WILLIAM SAMUEL. Studies in Religion and Lit-
       erature. London: Chapman and Hall, 1904. (In-
       cludes N. P. S. Wiseman, A. Tennyson, Tractarianism)

513.   LONGFORD, ELIZABETH (HARMAN) PAKENHAM. Piety in
       Queen Victoria's Reign. London: Dr. Williams's
       Trust, 1973.

514.   MACAN, REGINALD WALTER. Religious Changes in Oxford
       during the Last Fifty Years. London, New York: H.
       Milford, Oxford University Press, 1917.

515.   MACLAREN, A. ALLAN. Religion and Social Class: The
       Disruption Years in Aberdeen. London: Routledge
       and Kegan Paul, 1974.

516.   MCLEOD, HUGH. Class and Religion in the Late Vic-
       torian City. London: Croom Helm, 1974.

517.   MCLEOD, HUGH. "Class, Community and Region: The
       Religious Geography of XIXth-Century England."
       A Sociological Yearbook of Religion in Britain 6
       (1973), 29-72.

518. MACLEOD, J. "Religious Movements of the Past Cen-
      tury in England, 1830-1930." Bibliotheca Sacra 87
      (Oct. 1930), 405-422.

519. MACLEOD, R. D. "Church Statistics for England."
      Hibbert Journal 46 (1948), 351-357. (Changes in
      adherents, 1851-1946)

520. MCLOUGHLIN, WILLIAM GERALD. Modern Revivalism:
      Charles Grandison Finney to Billy Graham. New York:
      Ronald Press Co., 1959. (Victorian religious re-
      vivals)

521. MADELEVA, MARY, SISTER. Chaucer's Nuns, and Other
      Essays. New York, London: D. Appleton and Company,
      1925. (Includes F. Thompson and the religious
      poetry of the nineteenth century)

522. MAISON, MARGARET M. The Victorian Vision: Studies
      in the Religious Novel. New York: Sheed and Ward,
      1961.

523. MARLOWE, JOHN. The Puritan Tradition in English
      Life. London: Cresset Press, 1956.

524. MARSH, PETER T. "The Other Victorian Christians."
      Victorian Studies 15 (Mar. 1972), 357-368.

525. MARTIN, DAVID A. A Sociology of English Religion.
      London: S.C.M. Press, 1967. (Includes a historical
      chapter)

526. MATTHEWS, WALTER ROBERT. "Religious Movements in
      the Lifetime of Charles Dickens." Dickensian 52
      (1956), 52-59.

527. MAYOR, STEPHEN H. The Churches and the Labour Move-
      ment. London: Independent Press, 1967.

528. MAYOR, STEPHEN H. "The Relations between Organised
      Religion and English Working-Class Movements, 1850-
      1914." Ph.D. Dissertation, University of Manchester,
      1960.

529.   MELLONE, SYDNEY HERBERT.   Leaders of Religious
       Thought in the Nineteenth Century: Newman,
       Martineau, Comte, Spencer, Browning.   Edinburgh and
       London: W. Blackwood and Sons, 1902.   (Includes
       Agnosticism and Positivism)

530.   MEWS, STUART.   "Religion and Emotion in Working-
       Class Religion, 1794-1824." In Derek Baker (Ed.),
       Schism, Heresy and Religious Protest: Papers Read
       at the Tenth Summer Meeting and the Eleventh Winter
       Meeting of the Ecclesiastical History Society.
       Cambridge: Cambridge University Press, 1972, 364-
       382.

531.   MILLER, CHARLES J.   "British and American Influences
       on the Religious Revival in French Europe, 1816-
       1848."   Summaries of Doctoral Dissertations . . .
       Northwestern University 15 (1948), 173-178.

532.   MODDER, MONTAGU F.   "An Aspect of Jewish Emancipa-
       tion in England."   London Quarterly and Holborn
       Review 158 (1933), 453-463.

533.   MOORE, ROBERT H.   "Victorian Religious Liberalism
       Reflected in Autobiography."   Ph.D. Dissertation,
       University of Illinois at Urbana-Champaign, 1948.

534.   MORGAN, JOHN VYRNWY. (Ed.) Welsh Religious Leaders
       in the Victorian Era.   London: James Nisbet and
       Co., 1905.

535.   MORGAN, P. B. "A Study of the Work of Revivalist
       Movements in Great Britain and Ireland from 1870-
       1914 and of Their Effect upon Organised Christianity
       There."   B.Litt. Dissertation, University of Oxford,
       1961.

536.   MORLEY, JOHN.   Death, Heaven, and the Victorians.
       London: Studio Vista, 1971.

537.   MORRIS, R. J. "Religion and Medicine: The Cholera
       Pamphlets of Oxford, 1832, 1849 and 1854."   Medical
       History 19 (1975), 256-270.

538.   MUDIE-SMITH, RICHARD. (Ed.) The Religious Life of
       London.   London: Hodder and Stoughton, 1904.

539. MUNBY, L. M.   "Religious Reaction in the Epoch of
      Imperialism." _Modern Quarterly_ 5, No. 4 (1950),
      328-341.

540. NETTEL, REGINALD. "Folk Elements in Nineteenth
      Century Puritanism." _Folklore_ 80 (Winter 1969),
      272-285. (Religion and popular culture)

541. NEVILLE, MARGARET M. "The Relationship of Aesthe-
      ticism to Christian and Neo-Pagan Currents in
      English Poetry, 1850-1900." Ph.D. Dissertation,
      Loyola University of Chicago, 1950.

542. NEWBOULD, J. "The Significance of Conflict between
      Selected Religious Thinkers and Religious Orthodoxy
      in the Nineteenth Century." M.A. Dissertation,
      University of Nottingham, 1971.

543. NEWMAN, ERNST. _Evangeliska alliansen: en studie i_
      protestantisk enhets-och frihetssträvan. Lund:
      Gleerup, 1937. ("The Evangelical Alliance: A Study
      in Protestant Struggles for Unity and Independence")

544. NICHOLS, JAMES HASTINGS. _History of Christianity,_
      1650-1950: Secularization of the West. New York:
      Ronald Press Co., 1956.

545. ORR, JAMES EDWIN. _The Second Evangelical Awakening:_
      An Account of the Second Worldwide Evangelical
      Revival Beginning in the Mid-Nineteenth Century.
      London: Marshall, Morgan and Scott, 1949.

546. PARSONS, E. "Religious and Philosophical Ideas
      Reflected in the Novel, 1870-1900." Ph.D. Dis-
      sertation, University of London, 1936.

547. PATERSON, JOHN. _The Novel as Faith: The Gospel_
      According to James, Hardy, Conrad, Joyce, Lawrence,
      and Virginia Woolf. New York: Gambit, Houghton,
      1973.

548. PELLING, HENRY MATHISON. "The Labour Church Move-
      ment." _International Review of Social History_ 4
      (1959), 111-112. (Reply by Kenneth Stanley Inglis,
      112-113)

549.   PELLING, HENRY MATHISON. Popular Politics and
        Society in Late Victorian Britain. London, Mel-
        bourne: Macmillan; New York: St. Martin's Press,
        1968. (Popular attitudes to religion)

550.   PELLING, HENRY MATHISON. "Religion and the Nine-
        teenth Century British Working Class." Past and
        Present 27 (1964), 128-133.

551.   PERKIN, HAROLD JAMES. The Origins of Modern English
        Society 1780-1880. London: Routledge and Kegan
        Paul, 1969. (Religion as the "midwife" of class)

552.   PHILLIPS, PAUL THOMAS. "The Sectarian Spirit: A
        Study of Sectarianism, Society and Politics in the
        North and West of England, 1832-1870." Disserta-
        tion Abstracts International 33 (1973), 6285-86A
        (University of Toronto, 1971).

553.   PICKERING, W. S. F. "The 1851 Religious Census - A
        Useless Experiment?" British Journal of Sociology
        18 (Dec. 1967), 382-407.

554.   PIERSON, STANLEY ARTHUR. "John Trevor and the
        Labour Church Movement in England, 1891-1900."
        Church History 29 (1960), 463-478.

555.   PIERSON, STANLEY ARTHUR. "Socialism and Religion:
        A Study of Their Interaction in Great Britain,
        1889-1911." Ph.D. Dissertation, Harvard University,
        1957.

556.   PINNINGTON, JOHN E. "Denominational Loyalty and
        Loyalty to Christ." Canadian Journal of Theology 14
        (1968), 125-130. (The early years of the Evan-
        gelical Alliance)

557.   PITTENGER, W. NORMAN. Reconceptions in Christian
        Thinking, 1817-1967. New York: Seabury, 1968.

558.   POLLOCK, JOHN CHARLES. A Cambridge Movement.
        London: Murray, 1953. (Cambridge Intercollegiate
        Christian Union)

559.   POOLE-CONNER, EDWARD JOSHUA. Evangelicalism in
        England. London: Fellowship of Independent Evan-
        gelical Churches, 1951.

560. POPE, HUGH. "A Brief History of the English Version
      of the New Testament First Published at Rheims in
      1582, Continued Down to the Present Day." Library
      20 (1940), 351-376; 21 (1940), 44-77.

561. PRINGLE, JOHN CHRISTIAN. Social Work of the London
      Churches: Being Some Account of the Metropolitan
      Visiting and Relief Association, 1843-1937. London:
      Oxford University Press, 1937.

562. RACK, HENRY D. "Domestic Visitation: A Chapter in
      Early Nineteenth-Century Evangelism." Journal of
      Ecclesiastical History 24 (1973), 357-376.

563. REARDON, BERNARD M. G. "Religion and the Romantic
      Movement." Theology 76 (Aug. 1973), 403-416.

564. REARDON, BERNARD M. G. Religious Thought in the
      Nineteenth Century, Illustrated from the Writers of
      the Period. Cambridge: Cambridge University Press,
      1968. (Introductions and selections, including S.
      T. Coleridge, F. D. Maurice, J. H. Newman, H. L.
      Mansel, J. S. Mill, B. Jowett, M. Arnold, S. Holland,
      J. Caird, E. Caird, F. H. Bradley)

565. REID, F. "Socialist Sunday Schools in Britain,
      1892-1939." International Review of Social History
      11, No. 1 (1966), 18-47.

566. RENNIE, I. S. "Evangelicalism and English Public
      Life, 1823-1850." Ph.D. Dissertation, University of
      Toronto, 1963.

567. RICHARDSON, CYRIL C. The Church through the Cen-
      turies. New York: Scribner's, 1938. (Includes
      Catholicism, Anglicanism, Methodism)

568. ROACH, JOHN P. C. "'The Rudiments of Faith and
      Religion': Religious Controversy at Oxford, 1860-
      1865." Journal of Ecclesiastical History 22 (1971),
      333-353. (Comparison of Oxford and Cambridge)

569. ROGAN, JOHN. "The Religious Census of 1851."
      Theology 66 (Jan. 1963), 11-15.

570.   ROGERS, ALAN.   "The 1851 Religious Census Returns
       for the City of Nottingham."   Transactions of the
       Thoroton Society 76 (1972), 74-87.

571.   ROTH, CECIL.   A History of the Jews in England.
       London: Oxford University Press, 1941.

572.   ROUSE, RUTH AND STEPHEN CHARLES NEILL.   (Eds.)
       A History of the Ecumenical Movement, 1517-1948.
       Second Edition.   Philadelphia: Westminster Press,
       1953.   (Includes the ecumenical ideals of English
       Evangelicals, the Broad Church, the Oxford Movement,
       Nonconformists)

573.   RUSSELL, A. J.   "A Sociological Analysis of the
       Clergyman's Role, with Special Reference to its·
       Development in the Early Nineteenth Century."   Ph.D.
       Dissertation, University of Oxford, 1970.

574.   RUSSELL, ARTHUR JAMES.   Their Religion.   London:
       Hodder and Stoughton, 1934.   (Includes W. E. Glad-
       stone, B. Disraeli, C. Dickens, Darwin)

575.   ST. JOHN-STEVAS, NORMAN.   "The Victorian Conscience:
       An Assessment and Explanation."   Wiseman Review 236
       (Autumn 1962), 247-259.

576.   SCHARPFF, PAULUS.   History of Evangelism: Three
       Hundred Years of Evangelism in Germany, Great
       Britain and the United States of America.   Trans.
       Helga Bender Henry.   Grand Rapids: Eerdmans, 1966.

577.   SCOTT, PATRICK GREIG.   "The Business of Belief: The
       Emergence of 'Religious' Publishing."   In Derek
       Baker (Ed.), Sanctity and Secularity: The Church
       and the World.   Papers Read at the Eleventh Summer
       Meeting and the Twelfth Winter Meeting of the
       Ecclesiastical History Society.   Oxford: Published
       for the Ecclesiastical History Society by Blackwell,
       1973, 213-224.

578.   SCOTT, PATRICK GREIG.   "Cricket and the Religious
       World in the Victorian Period."   Church Quarterly 3
       (Oct. 1970), 134-144.   (Muscular Christianity)

579. SCOTT, PATRICK GREIG. "Listing Victorian Religious
      Periodicals." Victorian Periodicals Newsletter No.
      10 (1970), 29-33.

580. SCOTT, PATRICK GREIG. "Richard Cope Morgan, Re-
      ligious Periodicals, and the Pontifex Factor."
      Victorian Periodicals Newsletter Nos. 1-6 (June
      1972), 1-14.

581. SCOTT, PATRICK GREIG. "Victorian Religious Peri-
      odicals: Fragments That Remain." In Derek Baker
      (Ed.), The Materials, Sources and Methods of Eccles-
      iastical History: Papers Read at the Twelfth Summer
      Meeting and the Thirteenth Winter Meeting of the
      Ecclesiastical History Society. Oxford: Published
      for the Ecclesiastical History Society by Blackwell,
      1975, 325-339.

582. SHAROT, STEPHEN. "Religious Change in Native Or-
      thodoxy in London, 1870-1914: The Synagogue Ser-
      vice." Jewish Journal of Sociology 15 (June 1973),
      57-78.

583. SHEA, F. X., S.J. "Religion and the Romantic Move-
      ment." Studies in Romanticism 9 (1970), 285-296.

584. SIEKER, EGON. "Christliche Religion und Fabian-
      ismus." Ph.D. Dissertation, Universität Marburg,
      1969.

585. SMITH, ALAN W. "Popular Religion." Past and Present
      No. 40 (July 1968), 181-186.

586. SMITH, C. F. "The Attitude of the Clergy to the
      Industrial Revolution as Reflected in the First and
      Second Statistical Accounts." Ph.D. Dissertation,
      University of Glasgow, 1953.

587. SOLOWAY, RICHARD ALLEN. "Church and Society:
      Recent Trends in Nineteenth Century Religious
      History." Journal of British Studies 11 (1972),
      142-159.

588. SPRING, DAVID. "Aristocracy, Social Structure, and
      Religion in the Early Victorian Period." Victorian
      Studies 6 (1963), 263-280.

589.   STARZYK, LAWRENCE J.   "'That Promised Land':   Poetry
       and Religion in the Early Victorian Period."   Vic-
       torian Studies 16 (Mar. 1973), 269-290.
590.   STEVENSON, LLOYD G.   "Religious Elements in the
       Background of the British Anti-Vivisection Move-
       ment."   Yale Journal of Biology and Medicine 29
       (1956), 125-157.
591.   STOUGHTON, JOHN.   The History of Religion in England
       from the Opening of the Long Parliament to 1850.   8
       vols.   London: Hodder and Stoughton, 1901.   (Vol-
       umes 7 and 8:   The Church of the First Half of
       the Nineteenth Century)
592.   SUMMERS, D. E. F.   "The Labour Church and Allied
       Movements of the Late Nineteenth and Early Twentieth
       Centuries."   Ph.D. Dissertation, University of
       Edinburgh, 1961.
593.   SYKES, NORMAN.   The English Religious Tradition:
       Sketches of Its Influence on Church, State, and
       Society.   London: S.C.M. Press, 1953.   (Includes
       the Oxford Movement, the Roman Catholic revival,
       science, history)
594.   TAMKE, SUSAN SMITH.   "Nineteenth-Century English
       Hymns as a Reflection of Victorian Society."   Dis-
       sertation Abstracts International 35 (1975), 6082A
       (University of Delaware, 1975).
595.   TATLOW, TISSINGTON.   The Story of the Student Chris-
       tian Movement of Great Britain and Ireland.   London:
       Student Christian Movement Press, 1933.
596.   THOLFSEN, TRYGVE R.   "The Chartist Crisis in Bir-
       mingham."   International Review of Social History 3
       (1958), 461-480.   (Case study of an unusual Chartist
       church)
597.   THOLFSEN, TRYGVE R.   "The Intellectual Origins of
       Mid-Victorian Stability."   Political Science
       Quarterly 86 (1971), 57-91.   (The Evangelical Re-
       vival)

598. THOMPSON, DAVID MICHAEL. "The Churches and Society
      in Nineteenth-Century England: A Rural Perspec-
      tive." In G. J. Cuming and Derek Baker (Eds.),
      Popular Belief and Practice: Papers Read at the
      Ninth Summer Meeting and the Tenth Winter Meeting
      of the Ecclesiastical History Society. Cambridge:
      Cambridge University Press, 1972, 267-276.
599. THOMPSON, DAVID MICHAEL. "The 1851 Religious Census:
      Problems and Possibilities." Victorian Studies 11
      (Sept. 1967), 87-97.
600. THOMPSON, EDWARD PALMER. The Making of the English
      Working Class. London: Gollancz, 1963. Revised,
      Harmondsworth, Middlesex: Penguin Books, 1968.
      (Includes Methodism)
601. THURMANN, ERICH. "Der Niederschlag der evangelischen
      Bewegung in der englischen Literatur." Ph.D. Dis-
      sertation, Universität Münster, 1936.
602. TURNELL, MARTIN. Modern Literature and the Chris-
      tian Faith. London: Darton, Longman, and Todd,
      1961. (Includes C. Patmore, G. M. Hopkins)
603. TWAMLEY, W. B. "The Influence of Religious Bodies
      on Political and Social Life between the Evangelical
      Revival and the Oxford Movement." Ph.D. Disserta-
      tion, University of Cambridge, 1925.
604. WALKER, R. B. "Religious Change in Cheshire, 1750-
      1850." Journal of Ecclesiastical History 17 (1966),
      77-94. (The impact of the Industrial Revolution on
      religion)
605. WALKER, R. B. "Religious Change in Liverpool in the
      Nineteenth Century." Journal of Ecclesiastical
      History 19 (1968), 195-211.
606. WARD, WILLIAM REGINALD. Religion and Society in
      England, 1790-1850. London: Batsford, 1972.
607. WEALES, GERALD. Religion in Modern English Drama.
      Philadelphia: University of Pennsylvania Press,
      1961. (Includes Henry Arthur Jones)

608.   WEIGLE, LUTHER ALLEN.   The English New Testament:
       From Tyndale to the Revised Standard Version.   New
       York: Abingdon-Cokesbury Press, 1949.
609.   WHITE, JAMES FLOYD.   Protestant Worship and Church
       Architecture: Theological and Historical Considera-
       tions.   New York: Oxford University Press, 1964.
610.   WILLMER, HADDON.   "Dechristianisation in England in
       the XIXth and XXth Centuries."   Miscellanea Historiae
       Ecclesiasticae 3 (1970), 315-327.
611.   WILLMER, HADDON.   "'Holy Worldliness' in Nineteenth-
       Century England."   In Derek Baker (Ed.), Sanctity
       and Secularity: The Church and the World: Papers
       Read at the Eleventh Summer Meeting and the Twelfth
       Winter Meeting of the Ecclesiastical History Society.
       Oxford: Published for the Ecclesiastical History
       Society by B. Blackwell, 1973, 193-211.
612.   WILSON, BRYAN R.   Religion in Secular Society: A
       Sociological Comment.   London: Watts, 1966.
613.   WILSON, ROBERT SYDNEY.   "A House Divided: British
       Evangelical Parliamentary Influence in the Latter
       Nineteenth-Century, 1860-1902."   Dissertation
       Abstracts International 35 (1974), 2923A (University
       of Guelph, 1973).
614.   WOLFF, ROBERT LEE.   "Some Erring Children in
       Children's Literature: The World of Victorian
       Religious Strife in Miniature."   In Jerome Hamilton
       Buckley (Ed.), The Worlds of Victorian Fiction.
       Cambridge, Mass.:   Harvard University Press, 1975,
       295-318.
615.   WOLFRUM, HELGA.   "Christentum und Griechentum in der
       viktorianischen Prosa."   Ph.D. Dissertation, Uni-
       versität München, 1943.
616.   WOODHOUSE, A. S. P.   The Poet and His Faith: Re-
       ligion and Poetry in England from Spenser to Eliot
       and Auden.   Chicago: University of Chicago Press,
       1965.

2.   Theology

This section includes both general and specialized
studies of Protestant and Catholic theology, as well as
some writings on the impact of science, particularly
Darwin's work, on theological speculation. See II.
Religion, 1. General; V. Natural Theology; XIV. Church
of England; XV. Nonconformity; XVI. Catholicism; XIX.
Varieties of Belief. . . .

617.   ALLCHIN, ARTHUR MACDONALD.   The Spirit and the Word:
Two Studies in 19th-Century Anglican Theology.
London: Faith Press, 1963. (Richard Meux Benson
and Thomas Hancock)

618.   ANON. "The Five-Fold Growth of Theology: Catholicism and Liberalism (1837-1937)." Times Literary
Supplement No. 1839 (May 1, 1937), 330-331.

619.   BAKER, RONALD DUNCAN. "The Influence of Charles
Darwin on the Development of Theological Thought."
B.D. Thesis, Trinity College (Dublin), 1962.

620.   BARTH, KARL. Protestant Theology in the Nineteenth
Century: Its Background and History. New York:
Judson/Thomas Nelson, 1973.

621.   BERTOCCI, PETER ANTHONY. The Empirical Argument for
God in Late British Thought. Cambridge: Harvard
University Press, 1938. (J. Martineau, A. Pringle-
Pattison, J. Ward, W. Sorley, F. Tennant)

622.   CHADWICK, OWEN. From Bossuet to Newman: The Idea
of Doctrinal Development. Cambridge, England:
Cambridge University Press, 1957.

623.   COLEMAN, ARTHUR MAINWARING AND FREDERICK HENRY
AMPHLETT MICKLEWRIGHT. "The Bampton Lectures."
Notes and Queries 190 (1946), 250-251; 191 (1946),
20, 35-36, 174. (List of lectures, 1780-1940)

624.   CRATCHLEY, W. J.   "Influence of the Theory of Evolu-
        tion on the Christian Doctrine of the Atonement."
        M.A. Thesis, University of Bristol, 1933.

625.   CUPITT, DON.   "Darwinism and English Religious
        Thought."   Theology 78 (Mar. 1975), 125-130.

626.   DAVIES, HORTON.   From Newman to Martineau, 1850-1900.
        Volume 4 of Worship and Theology in England.   London:
        Oxford University Press, 1962.

627.   DAVIES, HORTON.   From Watts and Wesley to Maurice,
        1690-1850.   Volume 3 of Worship and Theology in
        England.   London: Oxford University Press, 1961.

628.   DICKIE, JOHN.   Fifty Years of British Theology: A
        Personal Retrospect.   Being the Gunning Lectures
        Delivered in the Martin Hall, New College, Edin-
        burgh University, in December 1936.   Edinburgh: T.
        & T. Clark, 1937.

629.   ELLIOTT-BINNS, LEONARD ELLIOTT.   The Development of
        English Theology in the Later Nineteenth Century.
        London: Longmans, 1952.

630.   ELLIOTT-BINNS, LEONARD ELLIOTT.   English Thought,
        1860-1900: The Theological Aspect.   London: Long-
        mans, Green and Co., Ltd., 1956.

631.   FARRIS, W. J. S.   "The Concept of Divine Immanence
        in the Theology of the Nineteenth Century."   Ph.D.
        Dissertation, University of Edinburgh, 1954.

632.   GARVIE, ALFRED ERNEST.   "The Theology of Dr. Andrew
        Martin Fairbairn."   London Quarterly and Holborn
        Review 164 (Jan. 1939), 28-39.

633.   GILKEY, LANGDON BROWN.   "Maker of Heaven and Earth:
        A Thesis on the Relation between Metaphysics and
        Christian Theology with Special Reference to the
        Problem of Creation As That Problem Appears in the
        Philosophies of F. H. Bradley and A. N. Whitehead
        and in the Historic Leaders of Christian Thought."
        Dissertation Abstracts 14 (1954), 2141 (Columbia
        University, 1954).

634. GREEN, PETER. "Religious Controversy: I. The Knowable and the Unknowable." Twentieth Century 151 (1952), 225-231.

635. HUNT, REV. JOHN. Religious Thought in England in the Nineteenth Century. London: Gibbings and Co., Limited, 1896.

636. INGE, WILLIAM R. The Platonic Tradition in English Religious Thought. London: Longmans, Green and Co., 1926.

637. JOHNSON, ROBERT CLYDE. Authority in Protestant Theology. Philadelphia: Westminster Press, 1959.

638. LAWTON, JOHN STEWART. Conflict in Christology: A Study of British and American Christology, from 1889-1914. London: S.P.C.K.; New York: Macmillan, 1947.

639. LIDGETT, JOHN SCOTT. The Victorian Transformation of Theology. London: Epworth Press, 1934. (Includes F. D. Maurice)

640. MCDONALD, HUGH DERMOT. Ideas of Revelation: An Historical Study, A.D. 1700 to A.D. 1860. London: Macmillan, 1959. (Includes Evangelicalism, F. D. Maurice, S. T. Coleridge, J. Wesley)

641. MCDONALD, HUGH DERMOT. Theories of Revelation: An Historical Study, 1860-1890. London: G. Allen and Unwin, 1963.

642. MCPHEETERS, CHILTON C. "The Changing Apologetic Emphasis of Anglican Theology as Represented in the 'Bampton Lectures', 1780-1940." Ph.D. Dissertation, Drew University, 1948.

643. MELAND, BERNARD EUGENE. "From Darwin to Whitehead: A Study in the Shift in Ethos and Perspective Underlying Religious Thought." Journal of Religion 40 (1960), 229-245.

644. MOZLEY, JOHN KENNETH. Some Tendencies in British Theology from the Publication of Lux Mundi to the Present Day. London: S.P.C.K., 1951.

645. MURPHY, HOWARD R. "The Origins of the Humanitarian
      Ethos in England: With Special Reference to the
      History of Ethical and Theological Ideas, 1700-
      1870." Ph.D. Dissertation, Harvard University,
      1952.

646. NEDONCELLE, MAURICE GUSTAVE. La Philosophie
      religieuse en Grande-Bretagne de 1850 à nos jours.
      Paris: Bloud and Gay, 1934.

647. REARDON, BERNARD M. G. From Coleridge to Gore: A
      Century of Religious Thought in Britain. London:
      Longman, 1971.

648. REARDON, BERNARD M. G. Religious Thought in the
      Nineteenth Century: Illustrated from Writers of
      the Period. Cambridge: The University Press, 1966.
      (Introductions and selections including S. T.
      Coleridge, F. D. Maurice, J. H. Newman, H. L. Mansel,
      J. S. Mill, B. Jowett, M. Arnold, S. Holland, J.
      Caird, E. Caird, F. H. Bradley)

649. ROE, WILLIAM GORDON. Lamennais and England: The
      Reception of Lamennais's Religious Ideas in England
      in the Nineteenth Century. Oxford: Oxford Uni-
      versity Press, 1966.

650. ROWELL, GEOFFREY. Hell and the Victorians: A Study
      of the Nineteenth-Century Theological Controversies
      concerning Eternal Punishment and the Future Life.
      Oxford: Clarendon Press, 1974.

651. SCHIRMER, WALTER FRANZ. "German Literature, His-
      toriography and Theology in Nineteenth-Century
      England." German Life and Letters New Series 1, No.
      3 (Apr. 1948), 165-174.

652. SELWYN, EDWARD GORDON. "Christ in the Thought of
      the Nineteenth Century." Theology 34, No. 199
      (Jan. 1937), 7-19. (German thought and English
      Protestant theology)

653. STORR, VERNON FAITHFUL. The Development of English
      Theology in the Nineteenth Century. London, New
      York: Longmans, Green and Co., 1913.

654. SWANSTON, HAMISH F. G. Ideas of Order: Anglicans and the Renewal of Theological Method in the Middle Years of the Nineteenth Century. Assen: Gorcum, 1974. (Includes R. D. Hampden, H. L. Mansel, F. D. Maurice, B. Jowett)

655. TENNANT, F. R. "Influence of Darwinism upon Theology." Quarterly Review 211 (Oct. 1909), 418-440.

656. WEATHERHEAD, LESLIE DIXON. The Afterworld of the Poets: The Contribution of Victorian Poets to the Development of the Idea of Immortality. London: Epworth Press, J. A. Sharp, 1929.

657. WEBB, CLEMENT CHARLES JULIAN. A Century of Anglican Theology, and Other Lectures. Oxford: B. Blackwell, 1923.

658. WEBB, CLEMENT CHARLES JULIAN. A Study of Religious Thought in England from 1850. Oxford: Clarendon Press, 1933.

659. WELCH, CLAUDE. "The Problem of a History of Nineteenth-Century Theology." Journal of Religion 52 (1972), 1-21.

660. WELCH, CLAUDE. Protestant Thought in the Nineteenth Century. New Haven: Yale University Press, 1972- . (Volume 1: 1799-1870)

661. WHITE, ANTONIA. "Religious Controversy: III. The Soul and Future Life." The Twentieth Century 151 (1952), 241-245.

III. Ideas - Trends, Conflicts, Comparisons

Listed under this broad heading are general works on philosophy, political thought, belief, and unbelief, comprehensive studies of intellectual themes or movements, comparative studies of believers and unbelievers, and writings dealing directly with the relationship between science and religion. A variety of specific subjects are included as well, e.g., histories of newspapers and journals, and portraits of clubs and intellectual networks. There is clearly a considerable degree of overlap between this category and others preceding and following it. See particularly II. Religion; IV. Education; XIX. Varieties of Belief. . .; XXI. Atheism . . . .

662. AARSLEFF, HANS. The Study of Language in England, 1780-1860. Princeton: Princeton University Press, 1967.

663. AIMEE, SISTER. "The Religious Beliefs of Three Victorian Poets: Tennyson, Browning, and Arnold, and Their Influence on English Literature." Ph.D. Dissertation, University of Ottawa, 1942.

664. ALIOTTA, ANTONIO. The Idealistic Reaction against Science. Trans. Agnes McCaskill. London: Macmillan and Co., 1914. (Includes positivism, H. Spencer, T. H. Green, F. H. Bradley)

665. ALIOTTA, ANTONIO. "Science and Religion in the Nineteenth Century." In Joseph Needham (Ed.), Science, Religion and Reality. London: The Sheldon Press, 1926, 149-186.

666. ALTICK, RICHARD D. The English Common Reader: A Social History of the Mass Reading Public, 1800-1900. Chicago: The University of Chicago Press, 1957.

667. ALTICK, RICHARD D. Victorian People and Ideas. New
      York: Norton, 1973.

668. ANNAN, NOEL GILROY. "The Intellectual Aristocracy."
      In John Harold Plumb (Ed.), Studies in Social His-
      tory: A Tribute to G. M. Trevelyan. London, New
      York: Longmans, Green, 1955, 243-287.

669. APPLEMAN, PHILIP, WILLIAM ANTHONY MADDEN AND MICHAEL
      WOLFF. (Eds.) 1859: Entering an Age of Crisis.
      Bloomington: Indiana University Press, 1959.
      (Includes Darwin, H. L. Mansel, F. D. Maurice, W.
      Pater)

670. BADGER, KINGSBURY. "Christianity and Victorian
      Religious Confessions." Modern Language Quarterly
      25 (1964), 86-101. (Liberal confessional litera-
      ture, F. Newman, W. R. Greg, J. A. Froude)

671. BARBOUR, IAN G. Issues in Science and Religion.
      Englewood Cliffs, N. J.: Prentice-Hall, 1966.

672. BAUMER, FRANKLIN L. Religion and the Rise of
      Scepticism. New York: Harcourt, Brace and World,
      Inc., 1960.

673. BENN, ALFRED WILLIAM. Modern England: A Record of
      Opinion and Action, from the Time of the French
      Revolution to Present Day. 2 vols. London: Watts
      and Co., 1908.

674. BENSON, ARTHUR CHRISTOPHER. The Leaves of the Tree:
      Studies in Biography. London: Smith, Elder, and
      Co., 1911. (Bishop Westcott, H. Sidgwick, J. K.
      Stephen, Bishop Wilkinson, Professor Newton, F.
      Myers, Bishop Lightfoot, H. Bradshaw, C. Kingsley,
      Bishop Wordsworth of Lincoln, M. Arnold)

675. BEVINGTON, MERLE MOWBRAY. The Saturday Review,
      1855-1868. New York: Columbia University Press,
      1941.

676. BLYTON, W. J. "Altered Atmosphere: Victorians and
      the Lack of Faith." Month 179 (1943), 187-195.

677. BRETT, RAYMOND L. (Ed.) Poems of Faith and Doubt:
      The Victorian Age. London: Edward Arnold, 1965.

678. BRIGGS, ASA. *Victorian People: Some Reassessments
      of People, Institutions, Ideas, and Events, 1851-
      1867*. London: Odhams Press, 1954. Revised.
      Chicago: University of Chicago Press, 1970. (In-
      cludes W. Bagehot, T. Hughes)

679. BRINTON, CRANE. *English Political Thought in the
      Nineteenth Century*. Cambridge: Harvard University
      Press, 1949.

680. BROCK, WILLIAM H. "The Fortieth Article of Religion
      and the F.R.S. Who Fairly Represents Science. The
      Declaration of the Students of the Natural and
      Physical Sciences, 1865." *Clio* (University of
      Leicester History Society), No. 6 (1974), 15-21.
      (Later version by Brock and R. M. Macleod. "The
      Scientists' Declaration . . . ." *The British
      Journal for the History of Science* 9, No. 31,
      Mar. 1976, 39-66)

681. BRODY, BORUCH A. "Reid and Hamilton on Perception."
      Monist 55 (1971), 423-441.

682. [BROOKE, JOHN HEDLEY AND ALAN RICHARDSON.] *The
      Crisis of Evolution*. Milton Keynes: Open Uni-
      versity Press, 1974. (Includes science and belief)

683. BROOKFIELD, FRANCES M. *The Cambridge "Apostles"*.
      New York: Scribner's, 1906. (Includes A. H. Hallam,
      F. D. Maurice, R. M. Milnes, J. Sterling, A. Tenny-
      son, R. C. Trench)

684. BROWN, ALAN WILLARD. *The Metaphysical Society:
      Victorian Minds in Crisis, 1869-1880*. New York:
      Columbia University Press, 1947.

685. BRUSH, STEPHEN G. "The Prayer Test." *American
      Scientist* 62 (1974), 561-563. (The power of prayer)

686. BRYCE, JAMES. *Studies in Contemporary Biography*.
      New York: The Macmillan Company; London: Macmillan
      and Co., Ltd., 1903. (Includes A. P. Stanley, T. H.
      Green, A. C. Tait, J. R. Green, J. Fraser, Bishop of
      Manchester, H. Manning, W. R. Smith, H. Sidgwick,
      Lord Acton)

687. BUCKLER, WILLIAM EARL. "The Diffusion of Religious
      Doubt in Late Nineteenth-Century Fiction." Micro-
      film Abstracts 9, No. 3 (1950), 140-141 (University
      of Illinois, 1949).

688. BURCHFIELD, JOE D. Lord Kelvin and the Age of the
      Earth. London: Macmillan, 1975.

689. BUTLER, HENRY MONTAGU. Ten Great and Good Men.
      London: E. Arnold, 1909. (Includes W. Wilberforce,
      Lord Shaftesbury, T. Arnold)

690. CACOULLOS, ANN R. Thomas Hill Green: Philosopher
      of Rights. Boston: Twayne, 1974.

691. CADMAN, SAMUEL PARKES. Charles Darwin and Other
      English Thinkers: With Reference to Their Religious
      and Ethical Value. Boston, New York: The Pilgrim
      Press, 1911. (Darwin, T. H. Huxley, J. S. Mill,
      J. Martineau, M. Arnold)

692. CANNON, WALTER F. "History in Depth: The Early
      Victorian Period." History of Science 3 (1964),
      20-38.

693. CANNON, WALTER F. "The Normative Role of Science in
      Early Victorian Thought." Journal of the History of
      Ideas 25 (1964), 487-502.

694. CANNON, WALTER F. "The Problem of Miracles in the
      1830's." Victorian Studies 4 (1960-61), 6-32. (W.
      Whewell, R. Chambers, W. Buckland, C. Babbage)

695. CANNON, WALTER F. "Scientists and Broad Churchmen:
      An Early Victorian Intellectual Network." Journal
      of British Studies 4 (1964), 65-86.

696. CARRE, MEYRICK HEATH. "Earlier Conflict with Sci-
      ence." Church Quarterly Review 167 (July-Sept.
      1966), 347-355.

697. CARRE, MEYRICK HEATH. Phases of Thought in England.
      Oxford: The Clarendon Press, 1949.

698. CASSIRER, ERNST. The Problem of Knowledge: Philo-
      sophy, Science, and History Since Hegel. Trans.
      William H. Woglom and Charles W. Hendel. New Haven:
      Yale University Press, 1950.

699.   CECIL, ALGERNON. Six Oxford Thinkers: Edward
       Gibbon, John Henry Newman, R. W. Church, James
       Anthony Froude, Walter Pater, Lord Morley of Black-
       burn. London: J. Murray, 1909.
700.   CHORLEY, LADY KATHERINE. "Victorian Agnostics and
       Wordsworth." Month 11 (Apr. 1954), 207-221.
701.   CLIVE, GEOFFREY H. "The Connection between Ethics
       and Religion in Kant, Kierkegaard, and F. H. Brad-
       ley." Ph.D. Dissertation, Harvard University, 1953.
702.   COCKSHUT, ANTHONY O. J. The Unbelievers: English
       Agnostic Thought 1840-1890. London: Collins, 1964.
703.   COLLINI, STEFAN. "Idealism and 'Cambridge Ideal-
       ism'." Historical Journal 18 (1975), 171-177.
704.   COLLINS, JAMES DANIEL. "Darwin's Impact on Philo-
       sophy." Thought 34 (1959), 185-248. (Reprinted in
       Walter J. Ong, Ed., Darwin's Vision and Christian
       Perspectives. New York: Macmillan and Co., 1960,
       33-103). (The impact of Darwin on Natural Theology,
       agnosticism, theories of progress)
705.   COLSON, PERCY. Victorian Portraits. London: Rich-
       and Cowan, Ltd., 1932. (Includes H. Martineau,
       S. Wilberforce)
706.   COURTNEY, MRS. JANET ELIZABETH. Freethinkers of the
       Nineteenth Century. London: Chapman and Hall,
       1920.
707.   CRUSE, AMY. The Englishman and His Books in the
       Early Nineteenth Century. London: G. G. Harrap and
       Company, Ltd., 1930.
708.   CRUSE, AMY. The Victorians and Their Books. London:
       George Allen and Unwin, Ltd., 1936.
709.   DALGLISH, DORIS N. "Faith and Doubt in Victorian
       Literature." Contemporary Review 173 (Jan.-June
       1948), 106-112.
710.   DAMPIER, SIR WILLIAM CECIL. A History of Science
       and Its Relations with Philosophy and Religion.
       Cambridge: The University Press, 1929.

711.   DAVIES, HUGH SYKES AND GEORGE WATSON. (Eds.) The
       English Mind: Studies in the English Moralists
       Presented to Basil Willey. Cambridge: Cambridge
       University Press, 1964. (Includes J. S. Mill, S. T.
       Coleridge, J. H. Newman, M. Arnold)

712.   DAVIES, RUPERT ERIC. Religious Authority in an Age
       of Doubt. London: Epworth Press, 1968. (J. H.
       Newman, W. Palmer, Essays and Reviews, C. Gore)

713.   DAVISON, W. T. "Poetic Agnosticism: Meredith and
       Swinburne." London Quarterly Review 112 (July
       1909), 127-130.

714.   DAWSON, MARSHALL. Nineteenth Century Evolution and
       After: A Study of Personal Forces Affecting the
       Social Process in the Light of the Life-Sciences
       and Religion. New York: The Macmillan Co., 1923.

715.   DILLENBERGER, JOHN. Protestant Thought and Natural
       Science: A Historical Interpretation. Garden City,
       New York: Doubleday and Co. Inc., 1960.

716.   DOCKRILL, D. W. "The Origin and Development of
       Nineteenth Century English Agnosticism." Historical
       Journal (University of Newcastle, New South Wales)
       1, No. 4 (1971), 3-31.

717.   DOWNES, DAVID ANTHONY. The Temper of Victorian
       Belief: Studies in the Religious Novels of Pater,
       Kingsley, and Newman. New York: Twayne Publishers,
       1972.

718.   DOWNES, DAVID ANTHONY. Victorian Portraits: Hopkins
       and Pater. New York: Bookman Associates, 1965.

719.   DRAPER, JOHN W. History of the Conflict between
       Religion and Science. London: H. S. King, 1875.
       (New York: D. Appleton and Co., [1874])

720.   ELLEGÅRD, ALVAR. "The Readership of the Periodical
       Press in Mid-Victorian Britain." Göteborgs Un-
       iversitets Årsskrift 63, No. 3 (1957), 3-41. (Part
       II, a directory, is reprinted in Victorian Periodi-
       cals Newsletter No. 13, Sept. 1971, 3-22) (Includes
       religious and scientific journals)

721. EMDEN, CECIL STUART. Oriel Papers. Oxford: Clarendon Press, 1948. (Includes J. H. Newman, M.
Pattison, J. A. Froude)

722. EROS, JOHN. "The Rise of Organised Freethought in
Mid-Victorian England." Sociological Review New
Series 2 (July 1954), 98-120.

723. EVERETT, EDWIN MALLARD. The Party of Humanity: The
Fortnightly Review and Its Contributors, 1865-1874.
Chapel Hill: University of North Carolina Press,
1939. (Includes J. Morley, J. S. Mill, G. H. Lewes)

724. FEILING, KEITH GRAHAME. Sketches in Nineteenth
Century Biography. London, New York: Longmans,
Green and Co., 1930. (Includes S. T. Coleridge, J.
H. Newman, W. Bagehot)

725. FISKE, JOHN. Essays Historical and Literary. 2
vols. New York: The Macmillan Company; London:
Macmillan and Co., Ltd., 1902. (Includes T. H.
Huxley, J. Tyndall, H. Spencer)

726. FLEMING, DONALD HARNISH. John William Draper and
the Religion of Science. Philadelphia: University
of Pennsylvania Press, 1950.

727. FLINT, R. Agnosticism. Edinburgh and London: Wm.
Blackwood and Sons, 1903.

728. FOTHERGILL, PHILIP G. Evolution and Christians.
London: Longmans, Green, and Co., 1961.

729. GERSTNER, JOHN HENRY, JR. "Scotch Realism: Kant
and Darwin in the Philosophy of James McCosh."
Ph.D. Dissertation, Harvard University, 1945.

730. GINSBERG, MORRIS. The Idea of Progress: A
Revaluation. London: Methuen and Co., Ltd., 1953.

731. GLÜCKSMANN, HEDWIG L. "Die Gegenüberstellung von
Antike-Christentum in der englischen Literatur des
19. Jahrhunderts." Ph.D. Dissertation, Universität
Freiburg, 1932.

732. GRAHAM, WALLER. English Literary Periodicals. New
     York: Thomas Nelson and Sons, 1930.

733. GRANT DUFF, SIR MOUNTSTUART ELPHINSTONE. Out of the
     Past: Some Biographical Essays. 2 vols. London:
     J. Murray, 1903. (Includes H. Manning, A. P. Stan-
     ley, W. Bagehot, M. Arnold, Lord Acton, The Duke of
     Argyll)

734. GREEN, PETER. "Religious Controversy: I. The
     Knowable and the Unknowable." Twentieth Century
     151 (1952), 225-231.

735. GRISEWOOD, HARMAN. (Ed.) Ideas and Beliefs of the
     Victorians. London: The Sylvan Press, 1949.

736. GRØNBECH, VILHELM PETER. Religious Currents in the
     Nineteenth Century. Trans. Phillip M. Mitchell and
     William D. Paden. Lawrence: University of Kansas
     Press, 1964. (A. Tennyson, Darwin, J. H. Newman)

737. HALLAM, GEORGE W. "Source of the Word 'Agnostic'."
     Modern Language Notes 70 (Apr. 1955), 265-269.

738. HARKNESS, GEORGIA ELMA. "The Philosophy of Thomas
     Hill Green, with Special Reference to the Relations
     between Ethics and the Philosophy of Religion."
     Ph.D. Dissertation, Boston University, 1923.

739. HEARNSHAW, F. J. C. (Ed.) The Social and Political
     Ideas of Some Representative Thinkers of the Vic-
     torian Age. London: G. G. Harrap and Co., 1933.
     (Includes T. Carlyle, H. Spencer, H. Maine, T. H.
     Green, M. Arnold, W. Bagehot)

740. HEGNER, ANNA. "Die Evolutionsidee bei Tennyson und
     Browning." Ph.D. Dissertation, Universität Frei-
     burg, 1931.

741. HEIMANN, P. M. "The 'Unseen Universe': Physics and
     the Philosophy of Nature in Victorian England."
     British Journal for the History of Science 6 (1972-
     73), 73-79. (Balfour Stewart and P. G. Tait's
     Unseen Universe)

742.   HESS, MARY WHITCOMB. "Two Evolutionists: Teilhard
       and Browning." Contemporary Review 207 (Nov. 1965),
       261-264.

743.   HICKS, GRANVILLE. Figures in Transition: A Study
       of British Literature at the End of the Nineteenth
       Century. New York: Macmillan Company, 1939.
       (Includes T. Hardy, S. Butler)

744.   HIEBERT, ERWIN N. "The Use and Abuse of Thermo-
       dynamics in Religion." Daedalus 95 (1966), 1046-80.

745.   HIMMELFARB, GERTRUDE. Victorian Minds. New York:
       Alfred A. Knopf, 1968. (J. S. Mill, Lord Acton, L.
       Stephen, W. Bagehot, J. A. Froude, Social Darwinism)

746.   HOLLOWAY, JOHN. The Victorian Sage: Studies in
       Argument. London: Macmillan, 1953. (M. Arnold, J.
       H. Newman, George Eliot)

747.   HOUGHTON, WALTER EDWARDS. "Victorian Anti-
       Intellectualism." Journal of the History of Ideas
       13 (1952), 291-313. (Puritanism's tendency to exalt
       conscience over intellect)

748.   HOUGHTON, WALTER EDWARDS. The Victorian Frame of
       Mind, 1830-1870. New Haven: Yale University Press,
       1957.

749.   HOUGHTON, WALTER EDWARDS. (Ed.) The Wellesley Index
       to Victorian Periodicals, 1824-1900. 3 vols.
       Toronto: University of Toronto Press; London:
       Routledge and Kegan Paul, 1966-79.

750.   HOYT, ARTHUR STEPHEN. The Spiritual Message of
       Modern English Poetry. New York: The Macmillan
       Company, 1924. (Includes R. Browning, M. Arnold, A.
       Tennyson)

751.   JACOBS, LEO. Three Types of Practical Ethical Move-
       ments of the Past Half Century. New York: Mac-
       millan Company, 1922.

752.   JENSEN, JOHN VERNON. "Interrelationships within the
       Victorian X-Club." Dalhousie Review 51 (1971),
       539-552.

753. JENSEN, JOHN VERNON. "The X-Club: Fraternity of
     Victorian Scientists." British Journal for the
     History of Science 5 (June 1970), 63-72.
754. KENT, JOHN HENRY SOMERSET. From Darwin to Blatch-
     ford: The Role of Darwinism in Christian Apolo-
     getics, 1875-1910. London: Dr. Williams's Trust,
     1966.
755. KEYSER, C. J. "A Story of Whewell and Hamilton."
     Scientific American Supplement 65 (June 6, 1908),
     354.
756. KNIGHT, MARGARET. Humanist Anthology. London:
     Barrie and Rockliff, 1961. (Selections from C.
     Bradlaugh, T. H. Huxley, H. Spencer, L. Stephen, J.
     S. Mill, W. Reade, W. E. H. Lecky, J. Morley, W. K.
     Clifford)
757. KNIGHT, WILLIAM ANGUS. Retrospects: First Series.
     London: Smith, Elder, 1904. (Includes F. D.
     Maurice, A. P. Stanley, T. Carlyle, A. Tennyson, R.
     Browning, M. Arnold)
758. KNOEPFLMACHER, U. C. Religious Humanism and the
     Victorian Novel: George Eliot, Walter Pater, and
     Samuel Butler. Princeton, N.J.: Princeton Uni-
     versity Press, 1965.
759. LACK, DAVID L. Evolutionary Theory and Christian
     Belief: The Unresolved Conflict. London: Methuen,
     1957.
760. LASKI, HAROLD J. Studies in the Problem of Sov-
     ereignty. New Haven: Yale University Press, 1917.
     (Includes the political theory of the Oxford Move-
     ment and the Catholic revival)
761. LETWIN, SHIRLEY R. The Pursuit of Certainty: David
     Hume, Jeremy Bentham, John Stuart Mill and Beatrice
     Webb. London: Cambridge University Press, 1965.
762. LEVINE, RICHARD A. (Ed.) Backgrounds to Victorian
     Literature. San Francisco: Chandler Publishing
     Co., 1967. (Includes religion and science)

763. LIVINGSTON, JAMES C. The Ethics of Belief: An
      Essay on the Victorian Religious Conscience. Talla-
      hassee, Fla.: American Academy of Religion, 1974.
764. LOVEJOY, ARTHUR O. "Kant and the English Platon-
      ists." In Essays Philosophical and Psychological in
      Honor of William James, by His Colleagues at
      Columbia University. London: Longmans, Green, and
      Co., 1908, 265-302. (Includes T. H. Green and F. H.
      Bradley)
765. LUNN, SIR ARNOLD HENRY MOORE AND J. B. S. HALDANE.
      Science and the Supernatural: A Correspondence
      Between A. Lunn and J. B. S. Haldane. London: Eyre
      and Spottiswoode, 1935.
766. MACLEOD, ROY M. "The X-Club: A Social Network of
      Science in Late-Victorian England." Notes and
      Records of the Royal Society of London 24 (1970),
      305-322.
767. MANDELBAUM, MAURICE. History, Man, and Reason: A
      Study in Nineteenth-Century Thought. Baltimore:
      Johns Hopkins Press, 1971.
768. MANSFIELD, BRUCE E. "Erasmus in the Nineteenth
      Century: The Liberal Tradition." Studies in the
      Renaissance 15 (1968), 193-219. (H. H. Milman, M.
      Pattison, R. B. Drummond, J. A. Froude)
769. MARCHAND, LESLIE ALEXIS. The Athenaeum: A Mirror
      of Victorian Culture. Chapel Hill: University of
      North Carolina Press, 1941.
770. MASSINGHAM, H. J. AND HUGH H. MASSINGHAM. (Eds.)
      The Great Victorians. London: Nicolson and Watson,
      1932. (Includes M. Arnold, S. Butler, T. Carlyle,
      Darwin, George Eliot, T. Hardy, T. H. Huxley, J. S.
      Mill, J. H. Newman, W. Pater, C. Patmore, J. Ruskin,
      A. Tennyson)
771. MASUR, GERHARD. Prophets of Yesterday: Studies in
      European Culture, 1890-1914. New York: Macmillan,
      1961. (Includes A. Comte, Darwin)

772. MEAD, GEORGE HERBERT.   Movements of Thought in the
      Nineteenth Century.   Chicago: University of Chicago
      Press, 1936.

773. MEADOWS, ARTHUR JACK.   The High Firmament: A Survey
      of Astronomy in English Literature.   London: Mac-
      millan, 1972.

774. MERZ, JOHN THEODORE.   A History of European Thought
      in the Nineteenth Century.   4 vols. Edinburgh: W.
      Blackwood and Sons, 1912-28; New York: Dover Pub-
      lications, Inc., 1965.

775. METZ, RUDOLF.   A Hundred Years of British Philoso-
      phy.   London: G. Allen and Unwin, 1938.

776. MEYERSON, EMILE.   "Hegel, Hamilton, Hamelin et le
      concept de cause."   Revue Philosophique 96 (July
      1923), 33-55.

777. MILLER, JOSEPH HILLIS.   The Disappearance of God:
      Five Nineteenth-Century Writers.   Cambridge, Mass.:
      Belknap Press of Harvard University Press, 1963.
      (T. de Quincey, R. Browning, E. Brontë, M. Arnold,
      G. M. Hopkins)

778. MILLER, JOSEPH HILLIS.   "The Theme of the Disappear-
      ance of God in Victorian Poetry."   Victorian Studies
      6 (1962-63), 207-228.

779. MOORE, JAMES R.   "Evolutionary Theory and Christian
      Faith: A Bibliographical Guide to the Post-Darwin
      Controversies."   Christian Scholar's Review 4, No. 3
      (1975), 211-230.   (Includes Neo-Lamarckism, Neo-
      Darwinism)

780. MORE, PAUL ELMER.   The Drift of Romanticism: Shel-
      burne Essays.   Eighth Series. Boston and New York:
      Houghton Mifflin Company, 1913. (Includes J. H.
      Newman, W. Pater, T. H. Huxley)

781. MUIRHEAD, JOHN HENRY. (Ed.)   Nine Famous Birmingham
      Men.   Birmingham: Cornish Brothers, Ltd., 1909.
      (Includes J. H. Newman, R. W. Dale, Bishop Westcott)

782. MUIRHEAD, JOHN HENRY. The Platonic Tradition in
     Anglo-Saxon Philosophy. London: George Allen and
     Unwin, 1931. (Includes T. Carlyle, British Hegelians,
     F. H. Bradley)

783. MURPHY, HOWARD R. "The Ethical Revolt against
     Christian Orthodoxy in Early Victorian England."
     American Historical Review 60 (1954-55), 800-817.
     (F. W. Newman, J. A. Froude, George Eliot)

784. MURRAY, REV. ROBERT HENRY. Studies in the English
     Social and Political Thinkers of the Nineteenth Cen-
     tury. Cambridge: W. Heffer and Sons, 1929. (In-
     cludes T. R. Malthus, J. Bentham, James Mill, S. T.
     Coleridge, Oxford Movement, T. Carlyle, J. S. Mill,
     C. Kingsley, H. Spencer, H. Maine, J. Ruskin, M.
     Arnold, J. R. Seeley, W. Bagehot, T. H. Green)

785. NEEDHAM, JOSEPH. (Ed.) Science, Religion and Reality.
     New York and Toronto: The Macmillan Company, 1925.

786. NESBITT, GEORGE LYMAN. Benthamite Reviewing: The
     First Twelve Years of the Westminster Review, 1824-
     1836. New York: Columbia University Press, 1934.

787. NEUBURG, VICTOR E. "The Reading of the Victorian
     Freethinkers." Library 28 (Sept. 1973), 191-214.

788. ONIONS, C. T. "Agnostic." Times Literary Supple-
     ment (May 10, 1947), 225; (Sept. 6, 1947), 451.
     (Origin of the term)

789. ORANGE, A. D. Philosophers and Provincials: The
     Yorkshire Philosophical Society from 1822 to 1844.
     York: Yorkshire Philosophical Society, 1973.

790. OSMOND, PERCY HERBERT. The Mystical Poets of the
     English Church. London: S.P.C.K.; New York: The
     Macmillan Co., 1919. (Includes S. T. Coleridge, A.
     Tennyson, F. W. H. Myers, George MacDonald, F.
     Thompson, J. Keble, I. Williams, R. C. Trench, C.
     Kingsley, E. Brontë)

791. PASSMORE, JOHN A. A Hundred Years of English Philos-
     ophy. London: Duckworth, 1959.

792. PECKHAM, MORSE. _Victorian Revolutionaries: Specula-
      tions on Some Heroes of a Culture Crisis_. New York:
      Braziller, 1970. (A. Tennyson, T. Carlyle, R.
      Browning, E. B. Tylor, J. Lubbock)

793. PREYER, ROBERT OTTO. "The Benthamite and Colerid-
      gean Versions of History." _Dissertation Abstracts_
      14 (1954), 1727 (Columbia University, 1953).

794. PUCELLE, JEAN. _L'Idéalisme en Angleterre, de Col-
      eridge à Bradley_. Neuchâtel: La Baconnière, 1955.

795. QUINTON, ANTHONY M. "The Neglect of Victorian
      Philosophy." _Victorian Studies_ 1 (1957-58), 245-254.

796. RAMSEY, I. T. (Ed.) _Biology and Personality: Fron-
      tier Problems in Science, Philosophy and Religion_.
      New York: Barnes and Noble, 1966.

797. RANDALL, JOHN HERMAN, JR. "T. H. Green: The Devel-
      opment of English Thought from J. S. Mill to F. H.
      Bradley." _Journal of the History of Ideas_ 27 (1966),
      217-244.

798. RASMUSSEN, S. V. _The Philosophy of Sir William
      Hamilton_. Copenhagen: Levin and Munksgaard, 1925.

799. RAVEN, CHARLES EARLE. _Science, Religion and the
      Future: A Course of Eight Lectures_. Cambridge:
      Cambridge University Press, 1943.

800. RICHTER, MELVIN. _The Politics of Conscience: T. H.
      Green and His Age_. Cambridge: Harvard University
      Press, 1964.

801. RICHTER, MELVIN. "T. H. Green and His Audience:
      Liberalism as a Surrogate Faith." _Review of Poli-
      tics_ 18 (1956), 444-472.

802. ROBERTS, RICHARD. _The Jesus of the Poets and Pro-
      phets_. London: Student Christian Movement, 1919.
      (Includes A. Tennyson, R. Browning, J. Ruskin, F.
      Thompson)

803. ROBERTSON, JOHN MACKINNON. _A History of Freethought
      in the Nineteenth Century_. 2 vols. London: Watts
      and Co., 1929.

804. ROBERTSON, JOHN MACKINNON. Modern Humanists Recon-
      sidered. London: Watts and Co., 1927. (H. Spencer,
      J. S. Mill, T. Carlyle)

805. ROBERTSON, JOHN MACKINNON. Modern Humanists: Socio-
      logical Studies of Carlyle, Mill, Emerson, Arnold,
      Ruskin and Spencer. London: S. Sonnenschein and
      Co., 1891.

806. ROBERTSON SCOTT, JOHN WILLIAM. The Life and Death
      of a Newspaper: An Account of the Temperaments,
      Perturbations and Achievements of John Morley, W. T.
      Stead, E. T. Cooke, Harry Cust, J. L. Garvin, and
      Three Other Editors of the Pall Mall Gazette.
      London: Methuen, 1952.

807. ROBSON, ROBERT. (Ed.) Ideas and Institutions of
      Victorian Britain: Essays in Honour of George
      Kitson Clark. London: G. Bell and Sons, 1967.
      (Includes J. R. Seeley, philology, popular Protes-
      tantism)

808. ROGERS, ARTHUR KENYON. English and American Philoso-
      phy Since 1800: A Critical Survey. New York: The
      Macmillan Company, 1922.

809. ROLL-HANSEN, DIDERIK. The Academy 1869-1879: Vic-
      torian Intellectuals in Revolt. Copenhagen: Rosen-
      kilde and Bagger, 1957.

810. ROUTH, HAROLD VICTOR. Towards the Twentieth Cen-
      tury: Essays in the Spiritual History of the Nine-
      teenth. Cambridge: The University Press, 1937.
      (Includes Oxford Movement, J. H. Newman, A. Tenny-
      son, R. Browning, T. Carlyle, J. A. Froude, A.
      Clough, M. Arnold, J. S. Mill, H. Spencer, George
      Eliot, Darwin, G. Romanes)

811. RUSE, MICHAEL. "The Relationship between Science
      and Religion in Britain, 1830-1870." Church History
      44 (1975), 505-522.

812. RUSSELL, C. A. (Ed.) Science and Religious Belief:
      A Selection of Recent Historical Studies. London:
      University of London Press Ltd., 1973.

813. ST. JOHN-STEVAS, NORMAN. "Science and Faith."
      Wiseman Review No. 500 (Summer 1964), 138-146.

814. SHAFER, ROBERT. Christianity and Naturalism:
      Essays in Criticism. New Haven: Yale University
      Press, 1926. (Includes S. T. Coleridge, J. H.
      Newman, T. H. Huxley, M. Arnold, S. Butler, T.
      Hardy)

815. SHAPLEY, HARLOW. (Ed.) Science Ponders Religion.
      New York: Appleton-Century-Crofts, 1960.

816. SHELDON, HENRY C. Unbelief in the Nineteenth Cen-
      tury: A Critical History. London: Robert Culley,
      1907.

817. SIMPSON, JAMES YOUNG. Landmarks in the Struggle
      between Science and Religion. New York: G. H.
      Doran, 1925.

818. SMITH, WARREN SYLVESTER. The London Heretics, 1870-
      1914. London: Constable, 1967. (Includes spiritual-
      ism, Catholic Modernism, religious Positivism,
      secularism)

819. SOMERVELL, D. C. English Thought in the Nineteenth
      Century. New York: Longmans, 1929.

820. SOPER, DON LAURENCE. "An English Liberal." Down-
      side Review 88 (Jan. 1970), 27-35. (T. H. Green and
      religion)

821. SORLEY, WILLIAM RITCHIE. A History of British Phil-
      osophy to 1900. Cambridge: Cambridge University
      Press, 1920.

822. SPILLER, GUSTAV. The Ethical Movement in Great
      Britain: A Documentary History. London: Printed
      for the Author at the Farleigh Press, 1934.

823. STENSON, STEN HAROLD. "A History of Scottish
      Empiricism from 1730 to 1856." Dissertation Ab-
      stracts 12 (1952), 642 (Columbia University, 1952).
      (Includes W. Hamilton)

824. STEPHEN, SIR LESLIE. The English Utilitarians. 3
      vols. London: Duckworth and Co., 1900.

825. STRACHEY, GILES LYTTON. Eminent Victorians. Garden
City, N.Y.: Garden City Publishing Co., Inc., 1918.
(Includes H. Manning, T. Arnold)

826. SYMONDSON, ANTHONY. (Ed.) The Victorian Crisis of
Faith: Six Lectures by Robert M. Young, Geoffrey
Best, Max Warren, David Newsome, Owen Chadwick, R. C.
D. Jasper. London: Society for Promoting Christian
Knowledge, 1970. (Includes Darwin, evangelicalism,
missionaries, J. H. Newman, the Prayer Book)

827. TATARKIEWICZ, WLADYSLAW. Nineteenth-Century Philosophy. Trans. Chester A. Kisiel. Belmont, California: Wadsworth, 1973.

828. TENER, ROBERT H. "Agnostic." Times Literary Supplement No. 3415 (Aug. 10, 1967), 732. (R. H. Hutton)

829. TENER, ROBERT H. "R. H. Hutton and 'Agnostic'."
Notes and Queries 209 (Nov. 1964), 429-431.

830. THOMSON, ARTHUR. "The Philosophy of J. F. Ferrier."
Philosophy 39 (1964), 46-62.

831. TOULMIN, STEPHEN EDELSTON AND JUNE GOODFIELD. The
Discovery of Time. London: Hutchinson, 1965.
(Includes Uniformitarian-Catastrophist Debate,
Darwin)

832. TURNER, FRANK MILLER. Between Science and Religion:
The Reaction to Scientific Naturalism in Late Victorian England. New Haven and London: Yale University Press, 1974. (H. Sidgwick, A. R. Wallace,
F. W. H. Myers, G. Romanes, S. Butler, J. Ward)

833. TURNER, FRANK MILLER. "Lucretius among the Victorians." Victorian Studies 16 (1973), 329-348.

834. TURNER, FRANK MILLER. "Rainfall, Plagues, and the
Prince of Wales: A Chapter in the Conflict of
Religion and Science." Journal of British Studies
13 (1974), 46-65.

835. VANDER WAAL, JOHN ANTHONY. "The Religious Philosophy of Andrew Seth Pringle-Pattison." Dissertation Abstracts 14 (1954), 198-199 (Columbia University, 1953).

836. VAUGHAN, HERBERT MILLINGCHAMP. From Anne to Vic-
      toria: Fourteen Biographical Studies between 1702
      and 1901. London: Methuen and Co., Ltd., 1931.
      (Includes S. Butler, F. Thompson)

837. WADDINGTON, M. M. The Development of British Thought
      from 1820-1890. Toronto: J. M. Dent and Sons,
      1919.

838. WALKER, FRANK FISH. "British Liberalism: Some
      Philosophic Origins. The Contributions of Adam
      Smith, Thomas Robert Malthus, Jeremy Bentham, and
      Herbert Spencer." Dissertation Abstracts 17 (1957),
      2996-97 (Stanford University, 1957).

839. WARD, WILFRID PHILIP. Men and Matters. New York,
      London: Longmans, Green and Co., 1914. (Includes
      J. S. Mill, Cardinal Vaughan, A. Tennyson, J. H.
      Newman)

840. WARD, WILFRID PHILIP. Problems and Persons. Lon-
      don, New York: Longmans, Green, and Co., 1903.
      (Includes A. Tennyson, T. H. Huxley, J. H. Newman
      and E. Renan, N. P. S. Wiseman)

841. WARD, WILFRID PHILIP. Ten Personal Studies. Lon-
      don, New York: Longmans, Green, and Co., 1908.
      (Includes A. J. Balfour, H. Sidgwick, J. H. Newman
      and H. E. Manning, Father I. Ryder, N. P. S. Wise-
      man)

842. WAUGH, ARTHUR. "The Fortnightly's Seventieth Birth-
      day." The Fortnightly 143 (May 1935), 627-629.
      (Includes J. S. Mill, G. H. Lewes, W. K. Clifford,
      M. Arnold)

843. WEBB, ROBERT KIEFER. The British Working-Class
      Reader, 1790-1848: Literary and Social Tension.
      London: Allen and Unwin, 1955.

844. WEDGWOOD, JULIA. Nineteenth Century Teachers and
      Other Essays. London: Hodder and Stoughton, 1909.
      (S. T. Coleridge, F. D. Maurice, C. Kingsley, A. P.
      Stanley, R. H. Hutton, T. Carlyle, James Fitzjames
      Stephen, George Eliot)

845. WERTHEIMER, DOUGLAS L. "The Victoria Institute, 1865-1919: A Study in Collective Biography Meant as an Introduction to the Conflict of Science and Religion after Darwin." M.A. Dissertation, University of Toronto, 1971.

846. WHITE, ANDREW DICKSON. A History of the Warfare of Science with Theology in Christendom. 2 vols. New York: D. Appleton and Co., 1896; In one volume, New York: Dover, 1960.

847. WIENER, PHILIP PAUL. Evolution and the Founders of Pragmatism. Cambridge: Harvard University Press, 1949.

848. WILLEY, BASIL. "Honest Doubt." In Christianity Past and Present. Cambridge: Cambridge University Press, 1952, 107-130.

849. WILLEY, BASIL. More Nineteenth Century Studies: A Group of Honest Doubters. London: Chatto and Windus, 1956. (F. W. Newman, A. Tennyson, J. A. Froude, Essays and Reviews, W. H. White, J. Morley)

850. WILLEY, BASIL. Nineteenth Century Studies. London: Chatto and Windus, 1949. (S. T. Coleridge, T. Arnold, J. H. Newman, T. Carlyle, J. S. Mill, A. Comte, George Eliot, M. Arnold)

851. WILLIAMS, RAYMOND. Culture and Society, 1780-1950. London: Chatto and Windus, 1958. (Includes J. H. Newman, M. Arnold, J. S. Mill)

852. WILLIAMS, RAYMOND. The Long Revolution. London: Chatto and Windus, 1961. (Includes education, the press, the reading public)

853. WILSON, DAVID B. "Kelvin's Scientific Realism: The Theological Context." Philosophical Journal 11 (1974), 41-60.

854. WILSON, SAMUEL LAW. The Theology of Modern Literature. Edinburgh: T. and T. Clark, 1899. (T. Carlyle, R. Browning, George Eliot, George MacDonald, Mrs. Humphry Ward, T. Hardy, G. Meredith)

855. WOOD, HERBERT GEORGE.   Belief and Unbelief since
      1850. Cambridge: Cambridge University Press, 1955.
856. YOUNG, GEORGE MALCOLM.   Victorian England: Portrait
      of an Age. London, New York, Toronto: Oxford Uni-
      versity Press, 1936; Annotated Edition (notes and
      Introduction by G. Kitson Clark), London, Toronto:
      Oxford University Press, 1977.

IV. Education

This category focuses on education as it is related
to science, religion, and technology. It includes broad
works on education as well as studies dealing with specific
institutions or events where these pertain directly to
developments in science or religion. See II. Religion,
1. General; III. Ideas. . .; VIII. Evolution and Social
and Political Thought, 6. Spencer, Herbert; XIV. Church
of England, 1. General.

857. ADAMS, FRANCIS. History of the Elementary School
      Contest in England. London: Chapman & Hall,
      Limited, 1882.

858. ADAMSON, JOHN WILLIAM. English Education, 1782-
      1902. Cambridge, England: The University Press,
      1930.

859. ALWALL, ELLEN. The Religious Trend in Secular Scot-
      tish School-Books, 1850-1861 and 1873-1882: With a
      Survey of the Debate on Education in Scotland in the
      Middle and Late Nineteenth Century. Lund: Gleerup,
      1970.

860. ANON. "Henry Edward Armstrong." The Central 35
      (1938), 1-94.

861. ARGLES, MICHAEL. "English Education for Technology
      and Science: The Formative Years, 1880-1902."
      History of Education Quarterly 2 (Sept. 1962),
      182-191.

862. ARGLES, MICHAEL. "The Royal Commission on Technical
      Education, 1881-4: Its Inception and Composition."
      Vocational Aspect of Secondary and Further Education
      11 (Autumn 1959), 97-104.

863. ARGLES, MICHAEL. South Kensington to Robbins: An
      Account of English Scientific and Technical Educa-
      tion Since 1851. London: Longmans, 1964.

864. ARMSTRONG, E. F. "The Royal College of Chemistry."
      Nature 156, No. 3966 (Nov. 3, 1945), 524-527.

865.   ARMYTAGE, WALTER H. G.   "The 1870 Education Act."
       British Journal of Educational Studies 18 (June
       1970), 121-133.

866.   ARMYTAGE, WALTER H. G.   Four Hundred Years of Eng-
       lish Education. Cambridge: University Press, 1964;
       Second Edition, London: Cambridge University Press,
       1970.

867.   ARMYTAGE, WALTER H. G.   "Lyon Playfair and Technical
       Education in Britain." Nature 161 (1948), 752-753.

868.   ASHBY, ERIC. Technology and the Academics: An
       Essay on Universities and the Scientific Revolution.
       London: Macmillan, 1958.

869.   BARNARD, H. C. A History of English Education from
       1760. London: University of London Press, 1961.

870.   BARNES, ARTHUR STAPYLTON. The Catholic Schools of
       England. London: Williams and Norgate, Ltd., 1926.

871.   BELL, ENID HESTER CHATAWAY MOBERLY. A History of
       the Church Schools Company, 1883-1958. London:
       S.P.C.K., 1958.

872.   BEST, GEOFFREY FRANCIS ANDREW. "The Religious Dif-
       ficulties of National Education in England, 1800-
       70." Cambridge Historical Journal 12, No. 2 (1956),
       155-173.

873.   BETTERIDGE, D. "The Teaching of Analytical
       Chemistry in the United Kingdom before 1914."
       Talanta 16 (1969), 995-1022.

874.   BISHOP, A. S. The Rise of a Central Authority for
       English Education. Cambridge: Cambridge University
       Press, 1971. (A history of the Department of Educa-
       tion and Science)

875.   BISHOP, GEORGE DANIEL. Physics Teaching in England
       from Early Times Up to 1850. London: P.R.M. Pub-
       lishers, 1961.

876.   BRADDOCK, A. P. "The Activities of Catholics in
       Matters of Education in England." M.A. Disserta-
       tion, University of London, 1917.

877.   BREMNER, JEAN P.   "Some Aspects of Botany Teaching
       in English Schools in the Second Half of the Nine-
       teenth Century." School Science Review 38 (1957),
       376-383.

878.   BREMNER, JEAN P.   "Some Developments in Teaching
       Zoology in Schools in the Nineteenth Century."
       School Science Review 39 (1957), 70-77.

879.   BROAD, E. G.   "The Christian Churches and the Forma-
       tion and Maintenance of a System of State-Aided
       Elementary Education in England and Wales." M.A.
       Dissertation, University of Bristol, 1938.

880.   BROCK, WILLIAM H. (Ed.) H. E. Armstrong and the
       Teaching of Science, 1880-1930. Cambridge and New
       York: Cambridge University Press, 1973.

881.   BROCK, WILLIAM H.   "Prologue to Heurism." In The
       Changing Curriculum. London: Methuen, 1971, 71-85.
       (H. E. Armstrong)

882.   BROWN, CHARLES KENNETH FRANCIS.   "The Church of
       England's Contribution to Popular Education in
       England after 1833." B.Litt. Dissertation, Uni-
       versity of Oxford, 1941.

883.   BROWN, CHARLES KENNETH FRANCIS. The Church's Part
       in Education, 1833-1941: With Special Reference to
       the Work of the National Society. London: National
       Society, Society for Promoting Christian Knowledge,
       1942.

884.   BURGESS, H. J. Enterprise in Education: The Story
       of the Work of the Established Church in the Educa-
       tion of the People Prior to 1870. London: National
       Society, S.P.C.K., 1958.

885.   BURSTYN, JOAN N.   "Education and Sex: The Medical
       Case Against Higher Education for Women in England,
       1870-1900." Proceedings of the American Philo-
       sophical Society 117 (1973), 79-89.

886.   BURSTYN, JOAN N.   "Religious Arguments against
       Higher Education for Women in England 1840-1890."
       Women's Studies 1 (1972), 111-132.

887.   BUTTERWORTH, H.   "The Development and Influence of the Department of Science and Art, 1853-1899." Ph.D. Dissertation, University of Sheffield, 1968.

888.   BYRNE, M. B.   "Studies in the History of Science Teaching, 1799-1875." M.A. Dissertation, University of Durham, 1968.

889.   CLOKE, H.   "Wesleyan Methodism's Contribution to National Education (1739-1902)." M.A. Dissertation, University of London, 1936.

890.   CORBISHLEY, J.   "Catholic Secondary Education during the Victorian Era." Ushaw Magazine 11 (1901), 105-119.

891.   COTGROVE, STEPHEN F.   Technical Education and Social Change. London: Allen and Unwin, 1958.

892.   CRUICKSHANK, MARJORIE.   Church and State in English Education: 1870 to the Present Day. New York: St. Martin's Press, 1963.

893.   CULLEN, MISS M. M.   "The Growth of Roman Catholic Training Colleges for Women in England during the Nineteenth and Twentieth Centuries." M.Ed. Dissertation, University of Durham, 1964.

894.   CURTIS, STANLEY JAMES.   The History of Education in Great Britain. London: University Tutorial Press, 1948.

895.   CURTIS, STANLEY JAMES AND M. E. A. BOULTWOOD. An Introductory History of English Education since 1800. London: University Tutorial Press, 1960.

896.   DASGUPTA, DEBENDRA CHANDRA.   "A Comparative Study of the Board and the Voluntary Schools of England and Wales from 1870 to 1926." Calcutta Review 77 (1940), 71-82, 197-210, 275-288; 78 (1941), 113-124.

897.   DAVIS, V. D.   A History of Manchester College. London: Allen and Unwin, 1932.

898.   DEWEY, J. A.   "An Examination of the Role of Church and State in the Development of Elementary Education in North Staffordshire between 1870 and 1903." Ph.D. Dissertation, University of Keele, 1971.

899. DOCKING, JAMES W. Victorian Schools and Scholars:
      Church of England Elementary Schools in 19th Century
      Coventry. Coventry, England: Coventry Branch of
      the Historical Association, 1967.

900. DUKE, C. "The Department of Science and Art:
      Politics and Administration to 1864." Ph.D. Dis-
      sertation, University of London, 1966.

901. EDMONDS, J. M. "The First Geological Lecture Course
      at the University of London, 1831." Annals of
      Science 32 (1975), 257-275.

902. EDMONDSON, ERNEST MORTON. "Church-State Relation-
      ship in England 1800-1840 and Its Implications for
      Public Education." Dissertation Abstracts 12 (1952),
      288 (New York University, 1952).

903. ELLINGHAM, H. J. T. "The Imperial College of Sci-
      ence and Technology." Endeavour 5, No. 19 (July
      1946), 90-95.

904. ENGEL, ARTHUR JASON. "From Clergyman to Don: The
      Rise of the Academic Profession in Nineteenth-
      Century Oxford." Dissertation Abstracts Interna-
      tional 36 (1976), 6241A (Princeton University,
      1975).

905. EVANS, LESLIE W. "School Boards and the Work Schools
      System after the Education Act of 1870." National
      Library of Wales Journal 15 (1967), 89-100. (The
      decline of the voluntary system)

906. EVENNETT, H. O. The Catholic Schools of England and
      Wales. Cambridge: Cambridge University Press,
      1944.

907. EYRE, JOHN VARGAS. Henry Edward Armstrong, 1848-
      1937: The Doyen of British Chemists and Pioneer of
      Technical Education. London: Butterworths Sci-
      entific Publications, 1958.

908. FAIRHURST, J. R. "Some Aspects of the Relationship
      between Education, Politics and Religion from 1895
      to 1906." Ph.D. Dissertation, University of Oxford,
      1974.

909.   FINDLAY, ALEXANDER.   The Teaching of Chemistry in
       the Universities of Aberdeen.   Aberdeen: Aberdeen
       University Press, 1935.
910.   FOREMAN, H.   "Nonconformity and Education in England
       and Wales, 1870-1902."   M.A. Dissertation, Uni-
       versity of London, 1967.
911.   FOSTER, CHARLES.   "One Hundred Years of Science
       Teaching in Great Britain."   Annals of Science 2
       (July 15, 1937), 335-344.
912.   GARLAND, MARTHA MCMACKIN.   "A Liberal Education:
       The Development of an Ideal at the University of
       Cambridge, 1800-1860."   Dissertation Abstracts
       International 36 (1976), 7582A (Ohio State Uni-
       versity, 1975).   (Includes Natural Theology, W.
       Whewell, A. Sedgwick, J. F. W. Herschel, J. Hare, G.
       B. Airy)
913.   GILBERT, EDMUND W.   "The RGS and Geographical Educa-
       tion in 1871."   Geographical Journal 137 (1971),
       200-202.   (The Royal Geographical Society)
914.   GRABAR, TERRY H.   "'Scientific' Education and Richard
       Feverel."   Victorian Studies 14 (1970), 129-141.
       (On George Meredith's The Ordeal of Richard Feverel)
915.   GRANT, A. CAMERON.   "Note on Secular Education in
       the Nineteenth Century."   British Journal of Educa-
       tional Studies 16 (Oct. 1968), 308-317.
916.   GREEN, T. B.   "Physical Science Teaching in English
       Schools from 1850 to 1970."   M.Sc. Dissertation,
       University of Sheffield, 1971.
917.   GREGORY, J. M.   "Physics Teaching in the Late 19th
       Century: A Case History."   Physics Education 8
       (1973), 368-373.   (Winchester College)
918.   HAINES, GEORGE.   Essays on German Influence upon
       English Education and Science, 1850-1919.   New
       London: Connecticut College, 1969.
919.   HAINES, GEORGE.   German Influence upon English Edu-
       cation and Science, 1800-1866.   New London: Con-
       necticut College, 1957.

920. HAINES, GEORGE. "German Influence upon Scientific Instruction in England, 1867-1887." Victorian Studies 1 (1957-58), 215-244.

921. HARRISON, J. G. "The Relationship of Matthew Arnold as Inspector of Schools (1851-1886) with the Nonconformist Educationalists of the Time." Ph.D. Dissertation, Queen's University of Belfast, 1956.

922. HEENEY, BRIAN. Mission to the Middle Classes: The Woodard Schools, 1848-1891. London: S.P.C.K., 1969. (Nathaniel Woodard)

923. HILLIARD, F. H. et al. Christianity in Education. London: Allen and Unwin, 1966.

924. HOLLAND, MARY G. "The British Catholic Press and the Educational Controversy, 1847-1865." Dissertation Abstracts International 36 (1975), 1723A (Catholic University of America, 1975)

925. HOLLOWAY, S. W. F. "Medical Education in England, 1830-1858: A Sociological Analysis." History 49 (1964), 299-324.

926. HOLMES, R. F. "An Historical Note on the Natural Science Tripos." Cambridge Review (Jan. 23, 1965), 199-241.

927. HURT, JOHN. Education in Evolution: Church, State, Society and Popular Education, 1800-1870. London: Rupert Hart-Davis, 1971.

928. IKIN, A. E. "The State and Religious Education in Great Britain and Ireland." Year Book of Education (1940), 229-263. (From the Norman Conquest to the twentieth century)

929. INKSTER, IAN. "A Note on Itinerant Science Lecturers, 1790-1850." Annals of Science 28 (1972), 235-236.

930. INKSTER, IAN. "Science and the Mechanics' Institutes, 1820-1850: The Case of Sheffield." Annals of Science 32 (1975), 451-474.

931.   INKSTER, IAN.   "Science Instruction for Youth in the
       Industrial Revolution: The Informal Network in
       Sheffield." Vocational Aspect of Education 25 (Aug.
       1973), 91-98.

932.   JEFFERY, FREDERICK.   "The Centenary of the Methodist
       Education Committee." Proceedings of the Wesley
       Historical Society 21, Pt. 5 (Mar. 1938), 116-119.

933.   JONES, D. K.   "Lancashire, The American Common
       School, and the Religious Problem in British Educa-
       tion in the Nineteenth Century." British Journal of
       Educational Studies 15, No. 3 (1967), 292-306.

934.   JONES, L. S.   "Church and Chapel as Sources and
       Centres of Education in Wales during the Second Half
       of the Nineteenth Century." M.A. Dissertation,
       University of Liverpool, 1940.

935.   JUDGES, A. V. (Ed.)   Pioneers of English Education.
       London: Faber and Faber Ltd., 1952. (Includes J.
       Bentham, J. Kay-Shuttleworth, J. H. Newman, H.
       Spencer, M. Arnold, W. E. Forster)

936.   KINGSLAND, J. P.   "Lancashire College Sixty-Five
       Years Ago." Transactions of the Congregational
       Historical Society 14, No. 3 (Apr. 1943), 173-180.

937.   KINLOCH, TOM FLEMING.   Pioneers of Religious Educa-
       tion. London, New York: Oxford University Press,
       1939. (Includes T. Arnold, F. D. E. Schleiermacher)

938.   KIRK, KENNETH ESCOTT.   The Story of the Woodard
       Schools. London: Hodder and Stoughton, Limited,
       1937.

939.   LAYTON, DAVID.   Science for the People: The Origins
       of the School Science Curriculum in England. Lon-
       don: Allen and Unwin; New York: Science History,
       1973.

940.   LAYTON, DAVID.   "Science in General Education: The
       Rise and Fall of the First Movement, 1851-1857."
       Journal of Educational Administration and History 5
       (1973), 7-20.

941. LEVINE, ARNOLD SIDNEY. "The Politics of Taste: The
     Science and Art Department of Great Britain, 1852-
     1873." Dissertation Abstracts International 32
     (1972), 6898A (University of Wisconsin, 1972).

942. MCCANN, W. P. "Trade Unionists, Artisans and the
     1870 Education Act." British Journal of Educational
     Studies 18 (1970), 134-150.

943. MCCLELLAN, A. "Congregationalism and the Education
     of the People, 1760-1914." M.A. Dissertation, Uni-
     versity of Birmingham, 1964.

944. MCCLELLAND, VINCENT ALAN. English Roman Catholics
     and Higher Education. Oxford: Clarendon Press; New
     York: Oxford University Press, 1973.

945. MCCLELLAND, VINCENT ALAN. "The Protestant Alliance
     and Roman Catholic Schools, 1872-1874." Victorian
     Studies 8 (1964), 173-182.

946. MACDONALD, R. M. "The History of the Teaching of
     the Biological Sciences in English Grammar Schools,
     1850-1952." M.Ed. Dissertation, University of
     Durham, 1953.

947. MCLACHLAN, HERBERT. English Education under the
     Test Acts: Being the History of the Nonconformist
     Academies, 1662-1820. Manchester: Manchester
     University Press, 1931.

948. MCLEISH, JOHN. Evangelical Religion and Popular
     Education: A Modern Interpretation. London:
     Methuen, 1969. (Includes Hannah More and Griffith
     Jones)

949. MCPHERSON, ROBERT G. Theory of Higher Education in
     Nineteenth-Century England. Athens, Ga.: Uni-
     versity of Georgia Press, 1959.

950. MALONE, J. S. "The Content and Quality of Secondary
     Education in an English Catholic Seminary in the
     Nineteenth Century - Ushaw, 1808-1863." M.A. Dis-
     sertation, University of London, 1969.

951.   MARCHAM, A. J.   "Question of Conscience: The Church
       and the 'Conscience Clause', 1860-70."   Journal of
       Ecclesiastical History 22 (1971), 237-249.

952.   MARSHALL, B. R.   "The Theology of Church and State
       in Relation to the Concern for Popular Education in
       England, 1800-70."   Ph.D. Dissertation, University
       of Oxford, 1956.

953.   MATHEWS, HORACE FREDERICK.   Methodism and the Educa-
       tion of the People, 1791-1851.   London: Epworth
       Press, 1949.

954.   MATHEWS, HORACE FREDERICK.   "Methodism and the
       Education of the People (since 1851)."   Ph.D. Dis-
       sertation, University of London, 1954.

955.   MEIR, J. K.   "The Development of the Sunday School
       Movement in England from 1780 to 1880 in Relation to
       the State Provision of Education."   Ph.D. Disserta-
       tion, University of Edinburgh, 1954.

956.   MIDWINTER, ERIC.   Nineteenth Century Education.
       London: Longman Group Limited, 1970.

957.   MORRELL, J. B.   "The Chemist Breeders: The Research
       Schools of Liebig and Thomson."   Ambix 19 (1972),
       1-46.

958.   MORRELL, J. B.   "The Patronage of Mid-Victorian
       Science in the University of Edinburgh."   Science
       Studies (London) 3 (1973), 353-388.

959.   MORRELL, J. B.   "Science and Scottish University
       Reform: Edinburgh in 1826."   British Journal for
       the History of Science 6 (1972), 39-56.

960.   MUNSON, J. E. B.   "The London School Board Election
       of 1894:   A Study in Victorian Religious Contro-
       versy."   British Journal of Educational Studies 23
       (Feb. 1975), 7-23.

961.   MURPHY, JAMES M.   Church, State, and Schools in
       Britain, 1800-1970.   London: Routledge and Kegan
       Paul, 1971.

962.   MURPHY, JAMES M.   "Religion, the State, and Educa-
       tion in England."   History of Education Quarterly 8
       (Spring 1968), 3-34.

963.   MURPHY, JAMES M.   The Religious Problem in English
       Education.   Liverpool: Liverpool University Press,
       1959.   (The Liverpool Corporation School's attempt
       to meet Church education requirements in a non-sect-
       arian school, Dec. 1835-Dec. 1841)

964.   MUSGRAVE, P. W.   "Constant Factors in the Demand for
       Technical Education, 1860-1960."   British Journal of
       Educational Studies 14 (May 1966), 173-187.

965.   MUSGRAVE, P. W.   "The Definition of Technical Educa-
       tion, 1860-1910."   Vocational Aspect of Secondary
       and Further Education 34 (May 1964), 105-110.

966.   MUSGRAVE, P. W.   Society and Education in England
       since 1800.   London: Methuen and Co. Ltd., 1968.

967.   NEWSOME, DAVID H.   Godliness and Good Learning:
       Four Studies on a Victorian Ideal.   London: Murray,
       1961.   (Includes Bishop Westcott, A. P. Stanley, J.
       P. Lee, M. W. Benson, C. Kingsley, T. Hughes)

968.   O'BRIEN, M. A.   "Studies from Roman Catholic Govern-
       ment Inspected Elementary Schools in Bristol, 1847-
       1902."   M.Litt. Dissertation, University of Bristol,
       1971.

969.   OSBORNE, F. M.   "The Work of Religious Societies in
       English Education, 1660-1870."   M.A. Dissertation,
       University of London, 1925.

970.   OTTER, SIR JOHN LONSDALE.   Nathaniel Woodard: A
       Memoir of His Life.   London: John Lane, 1925.

971.   OWEN, R. M.   "The Methodist Contribution to Educa-
       tion in North-West England from 1850."   M.Ed. Dis-
       sertation, University of Liverpool, 1965.

972.   PARKER, IRENE.   Dissenting Academies in England:
       Their Rise and Progress and Their Place among the
       Education Systems of the Country.   Cambridge:
       Cambridge University Press, 1914.

973.   PAYNE, ERNEST ALEXANDER.   "The Development of Non-
       conformist Theological Education in the Nineteenth
       Century, with Special Reference to Regent's Park
       College."   In Ernest Alexander Payne (Ed.), <u>Studies
       in History and Religion Presented to Dr. H. Wheeler
       Robinson</u>.   London: Lutterworth Press, 1942, 229-253.

974.   PERRY, R.   "The Life and Work of Nathaniel Woodard,
       with Special Reference to the Influence of the
       Oxford Movement on English Education in the 19th
       Century."   M.A. Dissertation, University of Bristol,
       1932.

975.   PETERS, A. J.   "The Changing Idea of Technical
       Education."   <u>British Journal of Educational Studies</u>
       11 (May 1963), 142-166.

976.   PLATTEN, S. G.   "Conflict over the Control of Ele-
       mentary Education 1870-1902 and Its Effect upon the
       Life and Influence of the Church."   <u>British Journal
       of Educational Studies</u> 23 (Oct. 1975), 276-302.

977.   POOLE, M. D. W.   "The English Bishops and Education,
       c.1830-c.1870."   Ph.D. Dissertation, University of
       Sheffield, 1972.

978.   PRITCHARD, FRANK CYRIL.   <u>Methodist Secondary Educa-
       tion: A History of the Contribution of Methodism
       to Secondary Education in the United Kingdom</u>.
       London: The Epworth Press, 1949.

979.   RICH, EDWIN ERNEST.   "Education and the Dissenters:
       A Sidelight on Nineteenth-Century Political
       Thought."   <u>Economica</u> 10 (1930), 188-199.

980.   RICH, ERIC E.   <u>The Education Act 1870:   A Study of
       Public Opinion</u>.   London: Longmans, 1970.

981.   RICHARDS, NOEL JUDD.   "Religious Controversy and the
       School Boards, 1870-1902."   <u>British Journal of Educa-
       tional Studies</u> 18 (June 1970), 180-196.

982.   RICHMOND, W. KENNETH.   <u>Education in England</u>.   Har-
       mondsworth, Middlesex: Penguin Books, 1945.

983.   RICKETT, BERNARD A. A.   "The Cowper-Temple Clause."
       <u>Clergy Review</u> 33 (Apr. 1950), 232-240.

984. ROACH, JOHN P. C. "Victorian Universities and the National Intelligencia." Victorian Studies 3 (1959-60), 131-150.

985. ROBERTS, GERRYLYNN KUSZEN. "The Royal College of Chemistry (1845-1853): A Social History of Chemistry in Early-Victorian England." Dissertation Abstracts International 34 (1974), 5074-75A (Johns Hopkins University, 1973).

986. RODERICK, GORDON W. AND MICHAEL D. STEPHENS. "Private Enterprise and Chemical Training in 19th-Century Liverpool." Annals of Science 27 (1971), 85-93.

987. RODERICK, GORDON W. AND MICHAEL D. STEPHENS. "Science and Secondary Education in 19th-Century Liverpool." Annals of Science 31 (1974), 131-163.

988. RODERICK, GORDON W. "Science and Technology at English Universities and Colleges and the Economic Development during the 19th Century." Technikgeschichte 40 (1973), 226-250.

989. RODERICK, GORDON W. AND MICHAEL D. STEPHENS. Scientific and Technical Education in Nineteenth-Century England. Newton Abbot: David and Charles, 1972.

990. RODERICK, GORDON W. "Scientific and Technical Training in Liverpool and Its Relevance to Industrial Merseyside, 1870-1914." M.A. Dissertation, University of Liverpool, 1972.

991. RODERICK, GORDON W. AND MICHAEL D. STEPHENS. "Scientific Studies and Scientific Manpower in the English Civic Universities, 1870-1914." Science Studies (London), 4 (1974), 41-63.

992. RODERICK, GORDON W. AND MICHAEL D. STEPHENS. "Scientific Studies in the Public Schools and Endowed Grammar Schools in the 19th Century: The Evidence of the Royal Commissions." Vocational Aspect of Education 23 (Aug. 1971), 97-105.

993.   ROPER, HENRY. "Toward an Elementary Education Act
       for England and Wales, 1865-1868." British Journal
       of Educational Studies 23 (June 1975), 181-208.

994.   ROTHBLATT, SHELDON. The Revolution of the Dons:
       Cambridge and Society in Victorian England. London:
       Faber, 1968.

995.   ROTHBLATT, SHELDON. "A Victorian University Re-
       former." Minerva 8 (1970), 470-473. (A. Sedgwick)

996.   SAFFIN, N. W. Science, Religion and Education in
       Britain 1804-1904. Kilmore, Australia: Lowden
       Publishing Co., 1973.

997.   SELBY, D. E. "Henry Edward Manning and the Educa-
       tion Bill of 1870." British Journal of Educational
       Studies 18 (June 1970), 197-212.

998.   SELLECK, R. J. W. "The Scientific Educationist,
       1870-1914." British Journal of Educational Studies
       15 (June 1967), 148-165.

999.   SHEPHERD, T. B. "Liberal Education: The Views of
       the Rev. John Scott Compared with Those of Other
       Educationalists in the Mid-19th Century." London
       Quarterly and Holborn Review 164 (Oct. 1939), 459-
       470.

1000. SIMON, BRIAN. Studies in the History of Education
       1780-1870. London: Lawrence and Wishart, 1960.

1001. SMYTHE, B. H. "Church Attitudes to Education at
       the End of the 19th Century." M.Ed. Dissertation,
       University of Durham, 1969.

1002. STAUFFER, ROBERT C. "The Introduction of Modern
       Laboratory Science into the English University
       System." Ph.D. Dissertation, Harvard University,
       1948.

1003. STEPHENS, MICHAEL D. AND GORDON W. RODERICK.
       "American and English Attitudes to Scientific
       Education during the 19th Century." Annals of
       Science 30 (1973), 435-456.

1004.   STEPHENS, MICHAEL D.   "British Artisan, Scientific
        and Technical Education in the Early 19th Century."
        Annals of Science 29 (1972), 87-98.

1005.   STEPHENS, MICHAEL D.   "Changing Attitudes to Educa-
        tion in England and Wales, 1833-1902: The Govern-
        mental Reports, with Particular Reference to Sci-
        ence and Technical Studies." Annals of Science 30
        (1973), 149-164.

1006.   STEPHENS, MICHAEL D. AND GORDON W. RODERICK. "The
        Late Victorians and Scientific and Technical Educa-
        tion." Annals of Science 28 (1972), 385-400.

1007.   STEPHENS, MICHAEL D. AND GORDON W. RODERICK.
        "National Attitudes towards Scientific Education in
        Early Nineteenth-Century England." Vocational
        Aspect of Education 26 (Dec. 1974), 115-120.

1008.   STEPHENS, MICHAEL D. AND GORDON W. RODERICK.
        "Nineteenth-Century Educational Finance: The
        Literary and Philosophical Societies." Annals of
        Science 31 (1974), 335-349.

1009.   STEPHENS, MICHAEL D. AND GORDON W. RODERICK.
        "19th-Century Ventures in Liverpool's Scientific
        Education." Annals of Science 28 (1972), 61-86.

1010.   STEPHENS, MICHAEL D. AND GORDON W. RODERICK.
        "Science, the Working Classes and Mechanics' Insti-
        tutes." Annals of Science 29 (1972), 349-360.

1011.   STEPHENS, MICHAEL D. AND GORDON W. RODERICK.
        "Science Training for the Nineteenth Century Eng-
        lish Amateur: The Penzance Natural History and
        Antiquarian Society." Annals of Science 27 (July
        1971), 135-141.

1012.   STEPHENS, MICHAEL D. AND GORDON W. RODERICK.
        "Supplementary Education in a 19th-Century British
        Mining Area." Annals of Science 29 (1972), 59-79.
        (Cornwall)

1013.   STEWART, WILLIAM ALEXANDER CAMPBELL. Quakers and
        Education, as Seen in Their Schools in England.
        London: Epworth Press, 1953.

1014.   STROUD, L. J.   "The History of Quaker Education in
        England, 1647-1903."   M.Ed. Dissertation, Uni-
        versity of Leeds, 1945.

1015.   STURT, MARY.   The Education of the People: A His-
        tory of Primary Education in England and Wales in
        the Nineteenth Century.   London: Routledge and
        Kegan Paul, 1967.

1016.   SVIEDRYS, ROMUALDAS.   "The Rise of Physical Science
        in Victorian Cambridge."   Historical Studies in the
        Physical Sciences 2 (1970), 127-151.

1017.   TAYLOR, F. SHERWOOD.   "The Teaching of Science at
        Oxford in the Nineteenth Century."   Annals of
        Science 8 (1952), 82-112.

1018.   THOMAS, D. H.   "An Enquiry into the Policy of the
        Free Churches with Reference to Religious Education
        in the Schools, 1870-1914."   M.A. Dissertation,
        University College of Swansea, 1955.

1019.   THOMPSON, D.   "The Influence of Sir H. E. Roscoe on
        the Development of Scientific and Technical Educa-
        tion during the Second Half of the 19th Century."
        M.Ed. Thesis, University of Leeds, 1957-58.

1020.   THOMPSON, D.   "Science Teaching in Schools during
        the Second Half of the Nineteenth Century."   School
        Science Review 37 (1956), 298-305.

1021.   TILLYARD, A. I.   A History of University Reform.
        Cambridge: Heffer, 1913.

1022.   TOMS, V. G.   "The Secular Education Movement in
        England and Wales, 1800-1870."   Ph.D. Disserta-
        tion, University of London, 1972.

1023.   TRISTRAM, HENRY.   "London University and Catholic
        Education."   Dublin Review 199 (1936), 269-282.

1024.   TURNER, D. A.   "1870, the State and the Infant
        School System."   British Journal of Educational
        Studies 18 (June 1970), 151-165.

1025.   TURNER, DOROTHY M.   History of Science Teaching
        in England.   London: Chapman and Hall Ltd., 1927.

1026. TURNER, DOROTHY M. "The Philosophical Aspect of
       Education in Science." Isis 9 (1927), 402-419.
       (Includes W. Whewell, H. Spencer, T. H. Huxley)

1027. VAN PRAAGH, GORDON. (Ed.) H. E. Armstrong and
       Science Education. London: John Murray, 1973.

1028. WALKER, W. B. "Medical Education in Nineteenth
       Century Britain." Journal of Medical Education 31
       (1956), 765-776.

1029. WARD, JOHN T. AND J. H. TREBLE. "Religion and
       Education in 1843: Reaction to the 'Factory Educa-
       tion Bill'." Journal of Ecclesiastical History 20
       (1969), 79-110.

1030. WARDLE, DAVID. English Popular Education 1780-1970.
       Cambridge: Cambridge University Press, 1970.

1031. WATSON, J. E. "The Educational Activities of
       Baptists in England during the Eighteenth and
       Nineteenth Centuries with Particular Reference to
       the North-West." M.A. Dissertation, University of
       Liverpool, 1947.

1032. WEBSTER, WENDY H. "The Introduction of Science
       into University Education in England, 1820-80."
       M.Litt. Dissertation, University of Cambridge,
       1975.

1033. WILLIAMS, D. G. "The Significance of Alexander
       Bain (1818-1903) as an Educationalist." Ph.D.
       Dissertation, University of Sheffield, 1973.

1034. WRIGHT, W. H. "The Voluntary Principle in Educa-
       tion: The Contribution to English Education Made
       by the Clapham Sect and Its Allies and the Con-
       tinuance of Evangelical Endeavour by Lord Shaftes-
       bury." M.Ed. Dissertation, University of Durham,
       1964.

PART B. NATURAL THEOLOGY, GEOLOGY, AND EVOLUTION

V.   Natural Theology

The juxtaposition of "Natural Theology," "Geology," and "Evolution" is intended to convey the intimacy between science and religion, particularly in the late eighteenth and early nineteenth centuries. The subject of Natural Theology is a broad one, and the emphasis here is on William Paley and on the period of The Bridgewater Treatises on the Power, Wisdom and Goodness of God as Manifested in the Creation (1833-1836). A few works on Michael Faraday, Samuel Butler, Robert Browning, and post-Darwinian thinking on Natural Theology and "design" are included as well. See I. Science, 2. Method and Philosophy (William Whewell); II. Religion, 2. Theology; III. Ideas . . . ; VI. Geology . . . .

1035.   BENFEY, OTTO THEODOR. "Prout's Hypothesis."
        Journal of Chemical Education 29 (1952), 78-81.
1036.   BOWLER, PETER J. "Sir Francis Palgrave on Natural
        Theology." Journal of the History of Ideas 35
        (1974), 144-147.
1037.   BOYLAN, PATRICK J. "An Unpublished Portrait of
        Dean William Buckland, 1784-1856." Journal of the
        Society for the Bibliography of Natural History 5
        (1970), 350-354.
1038.   BREUER, HANS-PETER. "Samuel Butler's 'The Book of
        the Machines' and the Argument from Design."
        Modern Philology 72 (1975), 365-383.
1039.   BROCK, WILLIAM H. "The Chemical Career of William
        Prout." Ph.D. Dissertation, University of Leicester,
        1966.
1040.   BROCK, WILLIAM H. "The Life and Work of William
        Prout." Medical History 9 (1965), 101-126.

1041. BROCK, WILLIAM H.   "Prout's Chemical Bridgewater
       Treatise." Journal of Chemical Education 40 (1963),
       652-655.
1042. BROCK, WILLIAM H.   "The Selection of the Authors of
       the Bridgewater Treatises." Notes and Records of
       the Royal Society of London 21 (1966), 162-179.
1043. BROCK, WILLIAM H.   "Studies in the History of
       Prout's Hypothesis." Annals of Science 25 (1969),
       49-80, 127-137.
1044. BROCK, WILLIAM H.   "William Prout and Barometry."
       Notes and Records of the Royal Society of London 24
       (1970), 281-294. (Includes his Bridgewater
       Treatise)
1045. CAIRNS, D.   "Thomas Chalmers' Astronomical Dis-
       courses: A Study in Natural Theology?" Scottish
       Journal of Theology 9 (Dec. 1956), 410-421.
1046. CANNON, WALTER F.   "The Problem of Miracles in the
       1830's." Victorian Studies 4 (1960-61), 6-32. (W.
       Whewell, T. Chalmers, W. Buckland, C. Babbage)
1047. CANNON, WALTER F.   "The Role of the Cambridge
       Movement in Early Nineteenth Century Science."
       Proceedings of the Tenth International Congress of
       the History of Science (Ithaca, 1962), 1 (1964),
       317-320.
1048. CANNON, WALTER F.   "Scientists and Broad Churchmen:
       An Early Victorian Intellectual Network." Journal
       of British Studies 4 (1964), 65-86.
1049. CANNON, WALTER F.   "William Whewell, F.R.S. (1794-
       1866) Contributions to Science and Learning."
       Notes and Records of the Royal Society of London 19
       (1964), 176-191.
1050. CLARK, ROBERT E. D.   "Michael Faraday on Science
       and Religion." Hibbert Journal 65 (1967), 144-147.
1051. CLARKE, MARTIN LOWTHER. Paley: Evidences for the
       Man. London: S.P.C.K., 1974.

1052. COPEMAN, WILLIAM S. C.   "William Prout, M.D.,
       F.R.S., Physician and Chemist."   Notes and Records
       of the Royal Society of London 24 (1969), 273-280.

1053. EDMONDS, J. M.   "William Buckland (1784-1856)."
       Nature 178 (1956), 290-291.

1054. GILLISPIE, CHARLES COULSTON.   Genesis and Geology:
       A Study in the Relations of Scientific Thought,
       Natural Theology, and Social Opinion in Great
       Britain, 1790-1850.   Cambridge: Harvard University
       Press, 1951.

1055. GORDON-TAYLOR, SIR GORDON AND ELDRED WRIGHT WALLS.
       Sir Charles Bell, His Life and Times.   Edinburgh:
       E. & S. Livingstone, 1958.

1056. GUNDRY, D. W.   "Bicentenary of an English Divine."
       Theology 47 (July 1943), 145-150.   (William Paley)

1057. GUNDRY, D. W.   "The Bridgewater Treatises and Their
       Authors."   History New Series 31 (1946), 140-152.

1058. GUNDRY, D. W.   "Paleyan Argument from Design."
       Church Quarterly Review 151 (Jan.-Mar. 1951),
       181-198.

1059. HOUGH, GRAHAM.   "Books in General."   New States-
       man and Nation 32 (Aug. 10, 1946), 101.   (William
       Paley)

1060. HUNT, A. R.   "On Kent's Cavern with Reference to
       Buckland and His Detractors."   Geological Magazine
       9 (1902), 114-118.

1061. LAMMERS, JOHN HUNTER.   "Browning's Treatment of
       Natural Theology."   Dissertation Abstracts Inter-
       national 34 (1974), 4210A (Auburn University, 1973).

1062. LARDER, DAVID F.   "Prout's Hypothesis: A Recon-
       sideration."   Centaurus 15 (1970), 44-50.

1063. LEMAHIEU, DAN LLOYD.   "The Mind of William Paley."
       Ph.D. Dissertation, Harvard University, 1973.

1064. LEVERE, T. H.   "Faraday, Matter, and Natural
       Theology - Reflections on an Unpublished Manuscript."
       British Journal for the History of Science 4 (1968-
       69), 95-107.

1065. NORTH, F. J. "Centenary of the Glacial Theory."
       Proceedings of the Geologists' Association 54
       (1943), 1-28.

1066. NORTH, F. J. "Paviland Cave, the 'Red Lady', the
       Deluge, and William Buckland." Annals of Science 5
       (1942), 91-128. ("Red Lady" - name given to a
       skeleton in the cave)

1067. ORANGE, A. D. "Hyaenas in Yorkshire: William
       Buckland and the Cave at Kirkdale." History Today
       22 (1972), 777-785.

1068. RAVEN, CHARLES EARLE. Organic Design: Scientific
       Thought from Ray to Paley. London: Oxford Uni-
       versity Press, 1954.

1069. RICE, DANIEL FREDERICK. "Natural Theology and the
       Scottish Philosophy in the Thought of Thomas
       Chalmers." Scottish Journal of Theology 24 (Feb.
       1971), 23-46.

1070. SCHNEIDER, HOWARD A. "And Then There Were Nine."
       Perspectives in Biology and Medicine 12 (1969),
       514-528. (C. Babbage's self-styled Ninth Bridge-
       water Treatise)

1071. SIEGFRIED, ROBERT. "The Chemical Basis for Prout's
       Hypothesis." Journal of Chemical Education 33
       (1956), 263-266.

1072. SPECTOR, BENJAMIN. "Sir Charles Bell and the
       Bridgewater Treatises." Bulletin of the History of
       Medicine 12 (1942), 314-322.

1073. SWINBURNE, R. G. "The Argument from Design."
       Philosophy 43 (1968), 199-212.

1074. TIMKO, MICHAEL. "Browning upon Butler: Or Natural
       Theology in the English Isle." Criticism 7 (1965),
       141-150.

1075. WEBB, CLEMENT CHARLES JULIAN. Studies in the
       History of Natural Theology. Oxford: The Claren-
       don Press, 1915.

1076.   WHITE, GEORGE W.   "Announcement of Glaciation in
         Scotland: William Buckland (1784-1856)."   Journal
         of Glaciology 9 (1970), 143-145.
1077.   WILSON, LEONARD GILCHRIST.   "Paley and Natural
         Theology: A Response to M. J. S. Hodge."   Isis 63
         (1972), 296.
1078.   YOKOYAMA, TOSHIAKI.   "The Influence of Theological
         Thought on Charles Darwin: Consideration of the
         Relation between William Paley and Charles Darwin."
         Kagakusi Kenkyu 10 (1971), 49-59.   (In Japanese
         with an English summary)
1079.   YOUNG, ROBERT MAXWELL.   "Malthus and the Evolu-
         tionists: The Common Context of Biological and
         Social Theory."   Past and Present No. 43 (May
         1969), 109-145.

VI. Geology and the Development of the Earth

1. Lyell, Charles

See I. Science, 1. General; V. Natural Theology; VI.
Geology . . .; 2. Uniformitarian-Catastrophist Debate;
VII. Evolution.

1080. ADAMS, FRANK DAWSON. "Sir C. Lyell, His Place in
Geological Science and His Contributions to the
Geology of North America." Science 78 (Sept. 1,
1933), 177-183.

1081. ANON. "Darwin's Letters to Lyell." Notes and
Records of the Royal Society of London 8 (1950),
122-124.

1082. BAILEY, SIR EDWARD BATTERSBY. Charles Lyell.
London: Thomas Nelson and Sons, 1962.

1083. BAILEY, SIR EDWARD BATTERSBY. "Charles Lyell
F.R.S." Notes and Records of the Royal Society
of London 14 (1959), 121-138.

1084. BARTHOLOMEW, MICHAEL J. "The Intellectual Background to the Work of Charles Lyell and Charles
Darwin." Ph.D. Dissertation, University of Lan-
caster, 1975.

1085. BARTHOLOMEW, MICHAEL J. "Lyell and Evolution: An
Account of Lyell's Response to the Prospect of an
Evolutionary Ancestry for Man." British Journal
for the History of Science 6 (1972-73), 261-303.

1086. CANNON, WALTER F. "The Impact of Uniformitarian-
ism: Two Letters from John Herschel to Charles
Lyell, 1836-1837." Proceedings of the American
Philosophical Society 105 (1961), 301-314.

1087. COLEMAN, WILLIAM R. "Lyell and the Reality of
Species: 1830-1833." Isis 53 (1962), 325-338.

1088. COLLARD, EDGAR ANDREW. "Lyell and Dawson: A
Centenary." Dalhousie Review 22 (1942), 133-144.

1089.   CONKLIN, EDWIN GRANT.   "Letters of Charles Darwin
        and Other Scientists and Philosophers to Sir
        Charles Lyell, Bart."   Proceedings of the American
        Philosophical Society 95, No. 3 (1951), 220-222.

1090.   EISELEY, LOREN COREY.   "Charles Lyell."   Scientific
        American 201 (1959), 98-101.

1091.   EYLES, VICTOR A.   "The History of Geology: Sug-
        gestions for Further Research."   History of Science
        5 (1966), 77-86.

1092.   EYLES, VICTOR A.   "James Hutton (1726-1797) and Sir
        Charles Lyell (1797-1875)."   Nature 160 (1947),
        694-695.

1093.   GOULD, STEPHEN JAY.   "Catastrophes and Steady State
        Earth."   Natural History 84 (Feb. 1975), 14-18.

1094.   JUDD, JOHN WESLEY.   The Coming of Evolution: The
        Story of a Great Revolution in Science.   Cambridge,
        England: University Press, 1910.   (Includes Darwin)

1095.   LINGELBACH, W. E.   "Notable Letters and Papers:
        the Darwin-Lyell Letters."   Proceedings of the
        American Philosophical Society 95, No. 3 (1951),
        216-217.

1096.   MOORE, JAMES R.   "Charles Lyell and the Noachian
        Deluge."   Evangelical Quarterly 45 (July-Sept.
        1973), 141-160.

1097.   NORTH, F. J.   Sir Charles Lyell, Interpreter of
        the Principles of Geology.   London: Arthur Baker,
        1965.

1098.   RUDWICK, MARTIN J. S.   "Caricature as a Source for
        the History of Science: De la Beche's Anti-Lyellian
        Sketches of 1831."   Isis 66 (1975), 534-560.

1099.   RUDWICK, MARTIN J. S.   "Charles Lyell, F.R.S.
        (1797-1875) and His London Lectures on Geology,
        1832-33."   Notes and Records of the Royal Society
        of London 29 (1975), 231-263.

1100.   RUDWICK, MARTIN J. S.   "A Critique of Uniformi-
        tarian Geology: A Letter from W. D. Conybeare to

Charles Lyell, 1841." Proceedings of the American
Philosophical Society 3 (1967), 272-287.

1101. RUDWICK, MARTIN J. S. "The Strategy of Lyell's
'Principles of Geology'." Isis 61 (1970), 5-33.

1102. SNYDER, EMILY EVELETH. "Sir Charles Lyell's Nuggets
of American History: Visit to the United States
1841-42." Scientific Monthly 53 (1941), 303-308.

1103. TASCH, PAUL. "A Quantitative Estimate of Geological Time by Lyell." Isis 66 (1975), 406.

1104. TOMKEIEFF, S. I. "Letters from Charles Lyell and
Adam Sedgwick to H. B. Tristram." British Journal
for the History of Science 2 (1965), 251-253.

1105. WIELAND, G. R. "Lyell's Geological Texts." Nature
145 (Feb. 10, 1940), 227.

1106. WILSON, LEONARD GILCHRIST. Charles Lyell, the
Years to 1841: The Revolution in Geology. New
Haven and London: Yale University Press, 1972.

1107. WILSON, LEONARD GILCHRIST. "The Development of the
Concept of Uniformitarianism in the Mind of Charles
Lyell." Proceedings of Xth International Congress
of the History of Science (Ithaca, 1962), 2 (1964),
992-996.

1108. WILSON, LEONARD GILCHRIST. "The Origins of Charles
Lyell's Uniformitarianism." In Claude Carrol
Albritton (Ed.), Uniformity and Simplicity: A
Symposium on the Principle of the Uniformity of
Nature. New York: Geological Society of America,
1967, 35-62.

1109. WILSON, LEONARD GILCHRIST. "Sir Charles Lyell and
the Species Question." American Scientist 59
(1971), 43-55.

1110. WILSON, LEONARD GILCHRIST. (Ed.) Sir Charles
Lyell's Scientific Journals on the Species Question. New Haven: Yale University Press, 1970.

2.   Uniformitarian-Catastrophist Debate

The age and development of the earth and the clash of opinion on these issues, particularly in the late eighteenth and early nineteenth centuries, are the subjects of this section. Of the entries on individual geologists, the majority are concerned with the work of James Hutton and of William Smith. While there are a number of references to the conflict between the Huttonians and Wernerians (Abraham Gottlob Werner), the emphasis is on the debate between the Uniformitarians and Catastrophists. See V. Natural Theology; VI. Geology . . ., 1. Lyell, Charles; VII. Evolution.

1111.   ANON. "James Hutton: Geologist and Agricultur-
        ist." Nature 160 (Nov. 22, 1947), 727.
1112.   ANON. "James Hutton, 1726-97." Discovery 8 (Dec.
        1947), 357.
1113.   ARMYTAGE, WALTER H. G. "John Pye Smith, F.R.S.
        (1775-1851): A Synthesis of Geology and Genesis."
        London Quarterly and Holborn Review 180 (Apr.
        1955), 143-147.
1114.   BADASH, LAWRENCE. "Rutherford, Boltwood, and the
        Age of the Earth: The Origin of Radioactive Tech-
        niques." Proceedings of the American Philosophical
        Society 112 (1968), 157-169. (Includes Lord Kelvin)
1115.   BAILEY, SIR EDWARD BATTERSBY. "James Hutton:
        Father of Modern Geology, 1726-1797." Nature 119
        (1927), 582.
1116.   BAILEY, SIR EDWARD BATTERSBY. James Hutton: The
        Founder of Modern Geology. With a Foreword by J.
        E. Richey. Amsterdam, Barking (Essex), New York:
        Elsevier, 1967.
1117.   BASSETT, D. A. "James Hutton, the Founder of
        Modern Geology: An Anthology." Geology: The
        Journal of The Association of Teachers of Geology 2
        (1970), 55-76.

1118.   BASSETT, D. A.   "William Smith, the Father of
        English Geology and Stratigraphy: An Anthology."
        Geology: Journal of the Association of Teachers of
        Geology 1 (1969), 38-51.

1119.   BATHER, FRANCIS ARTHUR.   Address Delivered on
        July 10th, 1926, on William Smith, the Founder of
        English Geology.   Bath: Royal Literary and
        Scientific Institution, 1926.

1120.   BURCHFIELD, JOE D.   Lord Kelvin and the Age of
        the Earth.   London: Macmillan, 1975.

1121.   CANNON, WALTER F.   "On Uniformity and Progression
        in Early Victorian Cosmography."   Ph.D. Disserta-
        tion, Harvard University, 1956.

1122.   CANNON, WALTER F.   "The Uniformitarian-Catastro-
        phist Debate."   Isis 51 (1960), 38-55.

1123.   CAROZZI, ALBERT V.   "Une Nouvelle Interprétation du
        soi-disant catastrophisme de Cuvier."   Archives
        des Sciences 24 (1971), 367-377.

1124.   CHALLINOR, JOHN.   "The Early Progress of British
        Geology. III. From Hutton to Playfair, 1788-1802."
        Annals of Science 10 (June 1954), 107-148.

1125.   CHITNIS, ANAND C.   "The University of Edinburgh's
        Natural History Museum and the Huttonian-Wernerian
        Debate."   Annals of Science 26 (1970), 85-94.

1126.   COX, L. R.   "New Light on William Smith and His
        Work."   Proceedings of the Yorkshire Geological
        Society 25 (1942), 1-99.

1127.   DANIEL, GLYN EDMUND.   The Idea of Prehistory.
        Cleveland: World Publishing Co., 1963.

1128.   DAVIES, GORDON L.   The Earth in Decay: A History
        of British Geomorphology 1578 to 1878.   London:
        Macdonald and Co., 1969.   (Includes J. Hutton,
        glaciation)

1129.   DAVIES, GORDON L.   "The Eighteenth Century Denuda-
        tion Dilemma and the Huttonian Theory of the Earth."
        Annals of Science 22 (1966), 129-138.

1130. DAVIES, GORDON L. "George Hoggart Toulmin and the
       Huttonian Theory of the Earth." Bulletin of the
       Geological Society of America 78 (1967), 121-124.
1131. DAVIES, GORDON L. "The University of Dublin and
       Two Pioneers of English Geology: William Smith and
       John Phillips." Hermathena 109 (1969), 24-36.
1132. DAVIS, ARTHUR G. "William Smith, Civil Engineer,
       Geologist (1769-1839)." Transactions of the New-
       comen Society 23 (1942-43), 93-98.
1133. DAVIS, ARTHUR G. "William Smith's 'Geological
       Atlas' and the Later History of the Plates."
       Journal of the Society for the Bibliography of
       Natural History 2 (1952), 388-395.
1134. DEAN, DENNIS RICHARD. "Geology and English Litera-
       ture, 1770-1830." Dissertation Abstracts 29 (1968),
       1864A (University of Wisconsin, 1968).
1135. DEAN, DENNIS RICHARD. "James Hutton on Religion
       and Geology: An Unpublished Preface to His 'Theory
       of the Earth' (1788)." Annals of Science 32 (May
       1975), 187-193.
1136. DEAN, DENNIS RICHARD. "Scott and Mackenzie: New
       Poems." Philological Quarterly 52 (1973), 265-273.
       (Poems by Sir Walter Scott and Henry Mackenzie
       relating to the controversy between the Huttonians
       and Wernerians)
1137. DELAIR, J. B. AND W. A. S. SARJEANT. "Earliest
       Discoveries of Dinosaurs." Isis 66 (Mar. 1975),
       5-25. (Includes William Buckland)
1138. DOUGLAS, J. A. AND L. R. COX. "An Early List of
       Strata by William Smith." Geological Magazine 86
       (1949), 180-188.
1139. EDMONDS, J. M. "The Geological Lecture-Courses
       Given in Yorkshire by William Smith and John
       Phillips, 1824-1825." Proceedings of the York-
       shire Geological Society 40 (1975), 373-412.

1140. EDWARDS, NICHOLAS. "One of the Last Letters of Adam Sedgwick, Geologist, (1785-1873)." Annals of Science 28 (1972), 109-112.

1141. EYLES, JOAN M. "William Smith (1769-1839): A Bibliography of His Published Writings, Maps and Geological Sections, Printed and Lithographed." Journal of the Society for the Bibliography of Natural History 5 (1969), 87-109.

1142. EYLES, JOAN M. "William Smith: The Sale of His Geological Collection to The British Museum." Annals of Science 23 (1967), 177-212.

1143. EYLES, VICTOR A. "A Bibliographical Note on the Earliest Printed Version of James Hutton's 'Theory of the Earth', Its Form and Date of Publication." Journal of the Society for the Bibliography of Natural History 3 (1955), 105-108.

1144. EYLES, VICTOR A. "Note on the Original Publication of Hutton's 'Theory of the Earth', and on the Subsequent Forms in Which It Was Issued." Proceedings of the Royal Society of Edinburgh 63B (1950), 377-386.

1145. EYLES, VICTOR A. "Roderick Murchison, Geologist and Promoter of Science." Nature 234 (1971), 287-289.

1146. EYLES, VICTOR A. AND JOAN M. EYLES. "Some Geological Correspondence of James Hutton." Annals of Science 7 (1951), 316-339.

1147. FULLER, G. C. M. "The Industrial Basis of Stratigraphy: John Strachey, 1671-1743, and William Smith, 1769-1839." American Association of Petroleum Geologists Bulletin 53 (1969), 2256-73.

1148. GALBRAITH, WINSLOW HACKLEY. "James Hutton: An Analytic and Historical Study." Dissertation Abstracts International 36 (1975), 1054A (University of Pittsburgh, 1974).

1149. GERSTNER, PATSY A.   "James Hutton's Theory of the
       Earth and His Theory of Matter."   Isis 59 (1968),
       26-31.

1150. GERSTNER, PATSY A.   "The Reaction to James Hutton's
       Use of Heat as a Geological Agent."   British
       Journal of the History of Science 5 (Dec. 1971),
       353-362.

1151. GILBERT, EDMUND W. AND ANDREW GOUDIE.   "Sir Roderick
       Impey Murchison, Bart, KCB, 1792-1871."   Geograph-
       ical Journal 137 (Dec. 1971), 505-511.

1152. GOULD, STEPHEN JAY.   "Is Uniformitarianism Neces-
       sary?"   American Journal of Science 263 (1965),
       223-228.

1153. GRUBER, JACOB W. A.   "Brixham Cave and the Anti-
       quity of Man."   In Melford E. Spiro (Ed.), Context
       and Meaning in Cultural Anthropology: In Honor of
       A. Irving Hallowell.   New York: Free Press, 1965,
       373-402.

1154. HABER, FRANCIS C.   The Age of the World, Moses
       to Darwin.   Baltimore: Johns Hopkins Press, 1959.

1155. HABER, FRANCIS C.   "The Darwinian Revolution in the
       Concept of Time."   Studium Generale: Zeitschrift
       für interdisziplinäre Studien 24 (1971), 289-307.

1156. HANSEN, BERT.   "The Early History of Glacial Theory
       in British Geology."   Journal of Glaciology 9
       (1970), 135-141.

1157. HEMINGWAY, J. E. AND J. S. OWEN.   "William Smith
       and the Jurassic Coals of Yorkshire."   Proceedings
       of the Yorkshire Geological Society 40 (1975),
       297-308.

1158. HOBBS, WILLIAM H.   "James Hutton: The Pioneer of
       Modern Geology."   Science 64 (1926), 261-265.

1159. HOOYKAAS, REIJER.   Catastrophism in Geology: Its
       Scientific Character in Relation to Actualism and
       Uniformitarianism.   Amsterdam: North-Holland
       Publishing Co., 1970.

1160. HOOYKAAS, REIJER. "Geological Uniformitarianism
       and Evolution." Archives Internationales
       d'Histoire des Sciences 19 (1966), 3-19.
1161. HOOYKAAS, REIJER. "James Hutton und die Ewigkeit
       der Welt." Gesnerus 23 (1966), 55-66.
1162. HOOYKAAS, REIJER. Natural Law and Divine Miracle:
       A Historical-Critical Study of the Principle of
       Uniformity in Geology, Biology and Theology.
       Leiden: Brill, 1959.
1163. HOOYKAAS, REIJER. "Nature and History." Organon
       (Warsaw) 2 (1965), 5-16.
1164. HOOYKAAS, REIJER. "The Parallel between the His-
       tory of the Earth and the History of the Animal
       World." Archives Internationales d'Histoire des
       Sciences 10 (1957), 3-18.
1165. HOOYKAAS, REIJER. "The Principle of Uniformity in
       Geology, Biology and Theology." Journal of the
       Transactions of the Victoria Institute 88 (1956),
       101-116.
1166. HORNE, J. "The Influence of James Geikie's
       Researches on the Development of Glacial Geology."
       Proceedings of the Royal Society of Edinburgh 36
       (1917), 1-25.
1167. HUGHES, ARTHUR. "Science in English Encyclopaedias,
       1704-1875. IV. Theories of the Earth." Annals of
       Science 11 (1955), 74-92. (Theories of the deluge
       before Lyell)
1168. MACGREGOR, MURRAY, SIR EDWARD BATTERSBY BAILEY, G.
       W. TYRRELL, VICTOR A. EYLES AND S. I. TOMKEIEFF.
       "James Hutton 1726-1797: Commemoration of the
       150th Anniversary of His Death." Proceedings of
       the Royal Society of Edinburgh (Section B: Biology)
       63, Pt. 4 (1950), 347-400.
1169. MACGREGOR, MURRAY. "James Hutton, the Founder of
       Modern Geology: 1726-97." Endeavour 6 (1947),
       109-111.

1170.   MCINTYRE, D. B.   "James Hutton and the Philosophy
        of Geology."   In Claude Carrol Albritton, Jr. (Ed.),
        The Fabric of Geology.   Reading, Mass.:   Addison-
        Wesley Publishing Co., 1963, 1-11.

1171.   MILLHAUSER, MILTON.   "The Scriptural Geologists:
        An Episode in the History of Opinion."   Osiris 11
        (1954), 65-86.

1172.   NORTH, F. J.   "Verses about Buckland: With Ode to
        a Professor's Hammer, by Conybeare."   Nature 142
        (Dec. 10, 1938), 1040-41.

1173.   PAGE, LEROY EARL.   "John Playfair and Huttonian
        Catastrophism."   Actes du XIe Congrès International
        d'Histoire des Sciences (Warsaw, 1965), 4 (1968),
        221-225.

1174.   PAGE, LEROY EARL.   "The Rise of Diluvial Theory in
        British Geological Thought."   Dissertation Abstracts
        24 (1963), 2450 (University of Oklahoma, 1963).

1175.   PICKFORD, RONALD F.   William Smith, Father of
        English Geology, 1769-1839:   A Brief Memoir of His
        Work in the Bath District.   Bath, Somerset: Bath
        Municipal Libraries, 1969.

1176.   RAVIKOVICH, A. I.   "Ideas of Catastrophism and
        Uniformitarianism in Geology."   Actes du XIe Con-
        grès International d'Histoire des Sciences (Warsaw,
        1965), 4 (1968), 231-236.

1177.   RITCHIE, J.   "A Double Centenary: Robert Jameson
        and Edward Forbes."   Proceedings of the Royal
        Society of Edinburgh 66B (1956), 29-58.

1178.   RUDWICK, MARTIN J. S.   "The Devonian System, 1834-
        1840:   A Study in Scientific Controversy."   Actes
        du XIIe Congrès International d'Histoire des
        Sciences (Paris, 1968), 7 (1971), 39-43.   (Includes
        Henry de la Beche)

1179.   RUDWICK, MARTIN J. S.   "Hutton and Werner Compared:
        George Greenough's Geological Tour of Scotland."
        British Journal for the History of Science 1 (1962-
        63), 117-135.

1180.   RUDWICK, MARTIN J. S.   "Poulett Scrope on the
        Volcanoes of Auvergne: Lyellian Time and Political
        Economy."   British Journal for the History of
        Science 7 (1974), 205-242.

1181.   RUDWICK, MARTIN J. S.   "The Principle of Uniform-
        ity."   History of Science 1 (1962), 82-86.

1182.   RUDWICK, MARTIN J. S.   "Uniformity and Progression:
        Reflections on the Structure of Geological Theory
        in the Age of Lyell."   In Duane H. D. Roller (Ed.),
        Perspectives in the History of Science and Tech-
        nology.   Norman: University of Oklahoma Press,
        1971, 209-227.

1183.   SCHNEER, CECIL J. (Ed.)   Toward a History of
        Geology.   Cambridge: M.I.T. Press, 1969.   (In-
        cludes E. Geoffroy Saint-Hilaire, G. Cuvier,
        William Smith, C. Lyell, J. Hutton)

1184.   SEYLAZ, LOUIS.   "A Forgotten Pioneer of the Glacial
        Theory: John Playfair."   Journal of Glaciology 4
        (1961), 124-126.

1185.   SHEPPARD, T.   "William Smith: His Maps and
        Memoirs."   Proceedings of the Yorkshire Geological
        and Polytechnic Society 19 (1917), 75-253.   (Re-
        printed as a book; Hull: A. Brown and Sons, Ltd.,
        1920)

1186.   SWEET, J. M.   "Abraham Gottlob Werner Gedenkschrift."
        Freiberger Forschungshefte 223C (1967), 205-218.
        (The Wernerian Natural History Society in Edinburgh)

1187.   SWEET, J. M. AND C. D. WATERSTON.   "Robert Jameson's
        Approach to the Wernerian Theory of the Earth,
        1796."   Annals of Science 23 (1967), 81-95.

1188.   THACKRAY, JOHN C.   "Essential Source-Material of
        Roderick Murchison."   Journal of the Society for
        the Bibliography of Natural History 6 (1972),
        162-170.

1189.   TOMKEIEFF, S. I.   "Hutton's Uniformity, Isle of
        Arran." Geological Magazine 90 (Nov. 1953), 404-
        408. (See reply by J. G. C. Anderson in 91, Jan.
        1954, 85)

1190.   TOMKEIEFF, S. I.   "James Hutton and the Philosophy
        of Geology." Transactions of the Edinburgh Geolo-
        gical Society 14 (1948), 253-276. (Also in Pro-
        ceedings of the Royal Society of Edinburgh 63B,
        1950, 387-400)

1191.   TOMKEIEFF, S. I.   "Unconformity - An Historical
        Study." Proceedings of the Geologists' Association
        73 (1962), 383-401. (J. Hutton)

VII.   Evolution

1.   General

This section includes general works on the meaning, history, and impact of evolution. It also lists a few studies of Gregor Mendel and of contemporary and later opposition to Darwin's theory. See I. Science; III. Ideas . . .; VII. Evolution (other sections).

1192.   APPLEMAN, PHILIP. "The Logic of Evolution: Some Reconsiderations." Victorian Studies 3 (Sept. 1959), 115-125.

1193.   BOWLER, PETER J. "The Changing Meaning of 'Evolution'." Journal of the History of Ideas 36 (1975), 95-114.

1194.   CAMERON, THOMAS W. M. (Ed.) Evolution, Its Science and Doctrine: Symposium Presented to the Royal Society of Canada in 1959. Toronto: University of Toronto Press, 1960.

1195.   CANGUILHEM, GEORGES, G. LAPASSADE, J. PIGUEMAL AND J. ULMANN. "Histoire de la biologie: du développement à l'évolution au XIXe siècle." Thalès 11 (1960), 1-63.

1196.   CAWS, PETER. "Evidence and Testimony: Philip Henry Gosse and the 'Omphalos' Theory." In Harold Orel and George J. Worth (Eds.), Six Studies in Nineteenth Century Literature and Thought. Lawrence: University of Kansas, 1962, 69-90.

1197.   CENTORE, FLOYD F. "Evolution after Darwin." Thomist 33 (Oct. 1969), 718-736.

1198.   COCK, A. G. "William Bateson, Mendelism and Biometry." Journal of the History of Biology 6 (1973), 1-36.

1199.   CRATCHLEY, W. J. "Influence of the Theory of Evolution on the Christian Doctrine of the Atonement." M.A. Thesis, University of Bristol, 1933.

1200. DAWSON, MARSHALL. Nineteenth Century Evolution and
       After: A Study of Personal Forces Affecting the
       Social Process, in the Light of the Life-Sciences
       and Religion. New York: The Macmillan Co., 1923.
1201. DE BEER, SIR GAVIN RYLANDS. A Handbook on Evo-
       lution. London: Printed by Order of the Trustees
       of the British Museum (Natural History), 1958.
1202. DEWAR, DOUGLAS. Difficulties of the Evolution
       Theory. London: Edward Arnold, 1931.
1203. DOBZHANSKY, THEODOSIUS. "Evolutionism and Man's
       Hope." Sewanee Review 68 (1960), 274-288.
1204. DOBZHANSKY, THEODOSIUS. "Mendelism, Darwinism, and
       Evolutionism." Proceedings of the American Philo-
       sophical Society 109 (1965), 205-215.
1205. DRACHMAN, JULIAN MOSES. Studies in the Litera-
       ture of Natural Science and Religion. New York:
       The Macmillan Company, 1923. (Natural history and
       evolution)
1206. EISELEY, LOREN COREY. "Adventures of the Mind:
       1. An Evolutionist Looks at Modern Man." Saturday
       Evening Post 230 (Apr. 26, 1958), 28-29.
1207. FLEW, ANTONY G. N. "The Concept of Evolution: A
       Comment." Philosophy 41 (1966), 70-75.
1208. FOTHERGILL, PHILIP G. Historical Aspects of
       Organic Evolution. London: Hollis and Carter,
       1952.
1209. FULTON, JAMES STREET. "Philosophical Adventures of
       the Idea of Evolution: 1859-1959." Rice Institute
       Pamphlets 46, No. 1 (1959), 1-31.
1210. GASKING, E. B. "Why Was Mendel's Work Ignored?"
       Journal of the History of Ideas 20 (1959), 60-84.
1211. GLASS, BENTLEY. "Evolution and Heredity in the
       Nineteenth Century." In Lloyd G. Stevenson and
       R. Multhauf (Eds.), Medicine, Science and Culture.
       Baltimore: Johns Hopkins Press, 1968, 209-246.

1212.   GORDON, MARK A.   "The Social History of Evolution
in Britain."   American Antiquity 39 (1974), 194-
204.

1213.   GOUDGE, T. A.   "Some Philosophical Aspects of the
Theory of Evolution."   University of Toronto
Quarterly 23 (1954), 386-401.

1214.   GRASSE, PIERRE-P.   "Les Deux Phases de l'évolu-
tionnisme."   Nouvelles Littéraires (Oct. 1, 1959),
5.

1215.   GREENE, JOHN C.   The Death of Adam: Evolution
and Its Impact on Western Thought.   Iowa: Iowa
State University, 1959.

1216.   GROSSER, OTTO.   "Contemporary Views on the Oper-
ation of the Nineteenth Century Theories of Evolution."   Research and Progress 5 (1939), 39-50.

1217.   HULL, DAVID L.   "The Metaphysics of Evolution."
British Journal for the History of Science 3 (1966-
67), 309-337.

1218.   HUXLEY, JULIAN SORELL.   The Story of Evolution:
The Wonderful World of Life.   London: Rathbone
Books, 1958.

1219.   JOHNSON, R. H.   "Malthusian Principle and Natural
Selection."   American Naturalist 48 (Jan. 1914),
63-64.

1220.   LESCH, JOHN E.   "The Role of Isolation in Evolution: George J. Romanes and John T. Gulick."
Isis 66 (1975), 483-503.

1221.   LURIE, EDWARD.   "Louis Agassiz and the Idea of
Evolution."   Victorian Studies 3 (Sept. 1959),
87-108.

1222.   MACBRIDE, ERNEST WILLIAM.   "The Theory of Evolution
since Darwin."   Nature 115 (Jan. 10-17, 1925),
52-55, 89-92.

1223. MCKINNEY, HENRY LEWIS. (Ed.) Lamarck to Darwin:
       Contributions to Evolutionary Biology, 1809-1859.
       Lawrence, Kansas: Coronado Press, 1971. (Selec-
       tions from J.-B. de Lamarck, W. C. Wells, P. Matthew,
       E. Blyth, R. Chambers, A. R. Wallace, Darwin)

1224. MACLEOD, ROY M. "Evolutionism and Richard Owen,
       1830-1868: An Episode in Darwin's Century." Isis
       56 (1965), 259-280.

1225. MANDELBAUM, MAURICE. "The Scientific Background of
       Evolutionary Theory in Biology." In Philip Paul
       Wiener and Aaron Noland (Eds.), Roots of Scientific
       Thought: A Cultural Perspective. New York: Basic
       Books, 1957, 517-536. (Also in Journal of the
       History of Ideas 18, 1957, 342-361)

1226. MANN, GUNTER. (Ed.) Biologismus im 19. Jahr-
       hundert: Vorträge eines Symposiums vom 30. bis 31.
       Oktober 1970 in Frankfurt am Main. Stuttgart:
       Enke, 1973. (Includes Darwin)

1227. MANSER, A. R. "The Concept of Evolution."
       Philosophy 40 (1965), 18-34.

1228. MEAD, GEORGE HERBERT. "Evolution Becomes a General
       Idea." In Merritt Hadden Moore (Ed.), Movements
       of Thought in the Nineteenth Century. Chicago:
       University of Chicago Press, 1936, 153-168. (Re-
       printed in Anselm Strauss, Ed., George Herbert
       Mead on Social Psychology. Chicago: Phoenix
       Paperback, 1964, 3-18)

1229. MOORE, RUTH. Man, Time, and Fossils: The Story
       of Evolution. New York: Knopf Press, 1953.

1230. MORE, LOUIS TRENCHARD. The Dogma of Evolution.
       Princeton: Princeton University Press, 1925.

1231. MORGAN, THOMAS HUNT. "The Bearing of Mendelism on
       the Origin of Species." Scientific Monthly 16
       (1923), 237-247.

1232. MORTON, HAROLD CHRISTOPHERSON. The Bankruptcy
       of Evolution. London: Marshall Bros., 1925.

1233. NEWELL, NORMAN D.   "Special Creation and Organic
      Evolution." Proceedings of the American Philo-
      sophical Society 117 (1973), 323-331.

1234. OLBY, ROBERT CECIL. Origins of Mendelism. New
      York: Schocken Books, 1966. (Includes Darwin and
      F. Galton)

1235. OSTOYA, PAUL. Les Théories de l'évolution. Paris:
      Payot, 1951.

1236. POULTON, SIR EDWARD B.   "The History of Evolu-
      tionary Thought: As Recorded in Meetings of the
      British Association." Science 86 (1937), 203-214.
      (Also in Nature 140, 1937, 395-407 and British
      Association Annual Meeting Report, 1937, 1-21)

1237. POULTON, SIR EDWARD B.   "A Hundred Years of Evolu-
      tion." Report of the British Association for
      the Advancement of Science, London 1931 (1932),
      71-95. (Also in Science 74, 1931, 345-360)

1238. SIMPSON, GEORGE GAYLORD.   "Anatomy and Morphology:
      Classification and Evolution: 1859 and 1959."
      Proceedings of the American Philosophical Society
      103 (1959), 286-306.

1239. SIMPSON, GEORGE GAYLORD. The Meaning of Evolu-
      tion: A Study of the History of Life and of Its
      Significance for Man. New Haven: Yale University
      Press, 1949.

1240. SIMPSON, GEORGE GAYLORD.   "The Study of Evolution:
      Methods and Present Status of Theory." In Ann Roe
      and George Gaylord Simpson (Eds.), Behaviour and
      Evolution. New Haven: Yale University Press,
      1958, 7-26.

1241. WIENER, PHILIP PAUL. Evolution and the Founders
      of Pragmatism. Cambridge: Harvard University
      Press, 1949.

1242. YOUNG, ROBERT MAXWELL.   "Evolutionary Biology and
      Ideology: Then and Now." In Watson Fuller (Ed.),
      The Biological Revolution. Garden City, N.Y.:
      Doubleday, 1972, 241-282.

2.   Ideas of Organic Development Prior to Darwin

Writings dealing with the formation of species are included in this section. While studies of Jean-Baptiste de Lamarck predominate, there are a number of works on Georges Leopold Cuvier, Georges Louis Leclerc de Buffon, Etienne Geoffroy Saint-Hilaire, and Erasmus Darwin, as well as those naturalists who anticipated some of Darwin's specific conclusions. See I. Science; III. Ideas . . .; V. Natural Theology; VI. Geology . . .; VII. Evolution (other sections).

1243.   ANON. "Lamarck Bicentenary Celebrations." Nature 158 (1946), 35.

1244.   ANON. "Lamarck, Darwin and Weismann." Living Age 235 (Nov. 29, 1902), 517-529.

1245.   ANON. "Remarkable Anticipation of Darwin." Nature 92 (Jan. 22, 1914), 588-589.

1246.   ARDOUIN, PAUL ALPHONSE. Georges Cuvier: promoteur de l'idée évolutioniste et créateur de la biologie moderne. Paris: Expansion Scientifique Française, 1970.

1247.   ARON, JEAN-PAUL. "Les Circonstances et le plan de la nature chez Lamarck." Revue Générale des Sciences Pures et Appliquées 64 (1957), 243-250.

1248.   BARLOW, NORA. "Erasmus Darwin." Notes and Records of the Royal Society of London 14, No. 1 (1959), 85-98.

1249.   BELTRAN, ENRIQUE. Lamarck: intérprete de la naturaleza. México, D. F.: Sociedad Mexicana de Historia Natural, 1945.

1250.   BELTRAN, ENRIQUE. "Lamarck y Geoffroy Saint-Hilaire: su obra y su tiempo." Revista de la Sociedad Mexicana de Historia Natural 5 (1944), 155-166.

1251. BERTHELOT, RENE. "Lamarck et Goethe: l'évolu-
tionnisme de la continuité au début du XIXe siècle."
Revue de Métaphysique et de Morale 36 (1929),
285-341.

1252. BOURDIER, FRANCK. "Poprzednicy Darwina w Latach
1550-1859." Kwartalnik Historii Nauki i Techniki 6
(1961), 431-456, 607-643. (Darwin's predecessors,
in Polish with Russian summary)

1253. BOURDIER, FRANCK. "Trois siècles d'hypothèses sur
l'origine et la transformation des êtres vivants
(1550-1859)." Revue d'Histoire des Sciences 13
(1960), 1-44.

1254. BREMOND, J. AND J. LESSERTISSEUR. "Lamarck et
l'entomologie." Revue d'Histoire des Sciences et
de leurs Applications 26 (1973), 231-250.

1255. BREWSTER, EDWIN TENNEY. Creation: A History
of Non-Evolution Theories. Indianapolis: The
Bobbs-Merrill Company, 1927.

1256. BRUNET, PIERRE. "La Notion d'évolution dans la
science moderne avant Lamarck." Archeion 19
(1937), 21-43.

1257. BURKHARDT, RICHARD W., JR. "The Evolutionary
Thought of Jean-Baptiste Lamarck." Ph.D. Dissertation, Harvard University, 1972.

1258. BURKHARDT, RICHARD W., JR. "The Inspiration of
Lamarck's Belief in Evolution." Journal of the
History of Biology 5 (1972), 413-438.

1259. BURKHARDT, RICHARD W., JR. "Lamarck, Evolution,
and the Politics of Science." Journal of the
History of Biology 3 (1970), 275-298.

1260. BURLINGAME, LESLIE JEAN. "The Importance of
Lamarck's Chemistry for His Theories of Nature and
Evolution or Transformism." Proceedings of XIIIth
International Congress of the History of Science
(Moscow, 1971), 8 (1974), 92-97.

1261. BURLINGAME, LESLIE JEAN. "Lamarck's Theory of
      Transformism in the Context of His Views of Nature
      from 1776 to 1809." Dissertation Abstracts Inter-
      national 34 (1974), 6557A (Cornell University,
      1973).

1262. CAHN, THEOPHILE. La Vie et l'oeuvre d'Etienne
      Geoffroy Saint-Hilaire. Paris: Presses Universi-
      taires de France, 1962.

1263. CALDER, GRACE J. "Erasmus Darwin, Friend of Thomas
      and Jane Carlyle." Modern Language Quarterly 20
      (1959), 36-48.

1264. CANGUILHEM, GEORGES. (Ed.) "Georges Cuvier: jour-
      nées d'études organisées par l'Institut d'Histoire
      des Sciences de l'Université de Paris les 30 et 31
      mai 1969 pour le bicentenaire de la naissance de G.
      Cuvier." Revue d'Histoire des Sciences et de leurs
      Applications 23 (1970), 7-92.

1265. CANNON, HERBERT GRAHAM. Lamarck and Modern
      Genetics. Manchester: Manchester University
      Press, 1959.

1266. CAROZZI, ALBERT V. "Lamarck's Theory of the Earth:
      'Hydrogéologie'." Isis 55 (1964), 293-307.

1267. CAROZZI, ALBERT V. "Lamarck's Theory of the Earth:
      Reply." Isis 56 (1956), 356-357.

1268. CHAINE, J. "La Grande Epoque de l'anatomie compara-
      tive." Scientia 50 (1931), 365-374. (G. Cuvier
      versus E. Geoffroy Saint-Hilaire)

1269. COLEMAN, WILLIAM R. "Georges Cuvier: Biological
      Variation and the Fixity of Species." Archives
      Internationales d'Histoire des Sciences 15 (1962),
      315-331.

1270. COLEMAN, WILLIAM R. Georges Cuvier, Zoologist:
      A Study in the History of Evolution Theory. Cam-
      bridge, Mass.: Harvard University Press, 1964.

1271. DALL, W. H. "Lamarck: The Founder of Evolution."
      Popular Science Monthly 60 (Jan. 1902), 263-264.

1272. DAUDIN, HENRI. Cuvier et Lamarck: les classes
      zoologiques et l'idée de série animale (1790-1830).
      2 vols. Paris: F. Alcan, 1926.
1273. DEHAUT, EMILE GEORGES. Les Doctrines de Georges
      Cuvier dans leurs rapports avec le transformisme.
      Paris: P. Lechevalier, 1945.
1274. EGERTON, FRANK N. "Studies of Animal Populations
      from Lamarck to Darwin." Journal of the History of
      Biology 1 (1968), 225-259.
1275. EMERY, CLARK. "Scientific Theory in Erasmus
      Darwin's 'The Botanic Garden' (1789-91)." Isis 33
      (1941), 315-325.
1276. FARBER, PAUL LAWRENCE. "Buffon and Daubenton:
      Divergent Traditions within the 'Histoire natur-
      elle'." Isis 66 (1975), 63-74.
1277. FARBER, PAUL LAWRENCE. "Buffon and the Concept of
      Species." Journal of the History of Biology 5
      (1972), 259-284.
1278. FARBER, PAUL LAWRENCE. "Buffon's Concept of
      Species." Dissertation Abstracts International 31`
      (1971), 4656A (Indiana University, 1970).
1279. FAURE, J.-P. "Actualité de Lamarck." Raison
      Présente 30 (1974), 97-104.
1280. FISCHER, JEAN-LOUIS et al. "Etienne Geoffroy
      Saint-Hilaire." Revue d'Histoire des Sciences et
      de leurs Applications 25 (1972), 293-390.
1281. FOTHERGILL, PHILIP G. "Modern Evidence in Support
      of Lamarckism." In Historical Aspects of Organic
      Evolution. London: Hollis & Carter, 1952, 253-
      274.
1282. GEIKIE, SIR ARCHIBALD. "Lamarck and Playfair."
      Geological Magazine 43 (1906), 145-152, 193-202.
1283. GIENAPP, JOHN CHARLES. "Animal Hybridization and
      the Species Question from Aristotle to Darwin."
      Dissertation Abstracts International 31 (1971),
      5967A (University of Kansas, 1970).

1284.   GILLISPIE, CHARLES COULSTON.   "The Formation of
        Lamarck's Evolutionary Theory."   Archives Inter-
        nationales d'Histoire des Sciences 9 (1956), 323-
        338.

1285.   GILLISPIE, CHARLES COULSTON.   "Lamarck and Darwin
        in the History of Science."   The American Scientist
        46 (1958), 388-409.

1286.   GILLISPIE, CHARLES COULSTON.   "The Origin of
        Lamarck's Evolutionary Views."   Actes du VIIIe
        Congrès International d'Histoire des Sciences
        (Florence, 1956), 2 (1958), 544-548.

1287.   GLASS, BENTLEY, OWSEI TEMKIN AND WILLIAM L. STRAUS
        JR. (Eds.)   Forerunners of Darwin, 1745-1859.
        Baltimore: Johns Hopkins Press, 1959.

1288.   GOODFIELD-TOULMIN, JUNE.   "Blasphemy and Biology."
        Rockefeller University Review 4 (1966), 9-18.
        (William Lawrence)

1289.   GOODFIELD-TOULMIN, JUNE.   "Some Aspects of English
        Physiology: 1780-1840."   Journal of the History
        of Biology 2 (1969), 283-320.   (The controversy
        over the theory of "Vital Principle" and William
        Lawrence)

1290.   GRASSE, PIERRE-P.   "Lamarck, Wallace et Darwin."
        Revue d'Histoire des Sciences et de leurs Applica-
        tions 13 (1960), 73-79.

1291.   HARRISON, JAMES ERNEST.   "Erasmus Darwin's View of
        Evolution."   Journal of the History of Ideas 32
        (Apr.-June 1971), 247-264.

1292.   HASSLER, DONALD M.   Erasmus Darwin.   New York:
        Twayne Publishers, 1973.

1293.   HILDEBRANDT, KURT.   "Goethe und Darwin."   Archiv
        für Geschichte der Philosophie 41 (1932), 57-79.

1294.   HO, WING MENG.   "Methodological Issues in Evo-
        lutionary Theory, with Special Reference to Dar-
        winism and Lamarckism."   D.Phil. Thesis, University
        of Oxford, 1966.

1295.   HODGE, M. J. S.   "Lamarck's Science of Living
        Bodies."   British Journal for the History of
        Science 5 (1971), 323-352.

1296.   HOFSTEN, NILS VON.   "From Cuvier to Darwin."   Isis
        24 (1935-36), 361-366.   (Comparative anatomy)

1297.   HOFSTEN, NILS VON.   "Ideas of Creation and Spon-
        taneous Generation Prior to Darwin."   Isis 25
        (1936), 80-94.

1298.   HOLMES, SAMUEL JACKSON.   "K. E. von Baer's Per-
        plexities over Evolution."   Isis 37 (1947), 7-14.

1299.   KING-HELE, DESMOND G.   "Dr. Erasmus Darwin and the
        Theory of Evolution."   Nature 200 (1963), 304-306.

1300.   KING-HELE, DESMOND G.   Erasmus Darwin, 1731-1802.
        New York: Scribner, 1963.

1301.   KOFOID, CHARLES.   "An American Pioneer in Science:
        Dr. William Charles Wells, 1757-1817."   Scientific
        Monthly 57 (1943), 77-80.

1302.   KRUMBHAAR, E. B.   "The Bicentenary of Erasmus
        Darwin and His Relation to the Doctrine of Evolu-
        tion."   Annals of Medical History 3 (1931), 487-
        500.

1303.   LANDRIEU, MARCEL.   "Lamarck et ses précurseurs."
        Revue de l'Ecole d'Anthropologie de Paris 16
        (1906), 152-169.

1304.   LANDRIEU, MARCEL.   Lamarck, le fondateur du trans-
        formisme: sa vie, son oeuvre.   Paris: La Société
        Zoologique de France, 1909.

1305.   LANESSAN, J. L. DE.   "L'Attitude de Darwin à l'égard
        de ses prédécesseurs au sujet de l'origine des
        espèces."   Revue Anthropologique 24 (1914), 33-45.

1306.   LEFANU, WILLIAM RICHARD.   "Past Presidents: Sir
        William Lawrence, Bart."   Annals of the Royal
        College of Surgeons of England 25 (1959), 201-202.

1307.   LEGEE, GEORGETTE.   "Cuvier (1769-1832), Geoffroy
        Saint-Hilaire (1772-1844) et Flourens (1794-1867)."
        Histoire et Biologie 2 (1969), 10-34.

1308. LENOIR, RAYMOND. "Lamarck." Monist 34 (1924),
      187-235.

1309. LENOIR, RAYMOND. "Lamarck." Revue Philosophique
      90 (1920), 351-392.

1310. LILLEY, SAMUEL. "The Origin and Fate of Erasmus
      Darwin's Theory of Organic Evolution." Actes du
      XIe Congrès International d'Histoire des Sciences
      (Warsaw, 1965), 5 (1968), 70-75.

1311. LOVEJOY, ARTHUR O. "The Argument for Organic
      Evolution before 'The Origin of Species'." Popular
      Science Monthly 75 (1909), 499-514, 537-549.

1312. LOVEJOY, ARTHUR O. "Buffon and the Problem of
      Species." Popular Science Monthly 79 (1911),
      464-473, 554-567.

1313. LOVEJOY, ARTHUR O. "Some Eighteenth Century Evolu-
      tionists." Popular Science Monthly 65 (1904),
      238-251, 323-340. (Reprinted in Scientific Monthly
      71, Sept. 1950, 162-178)

1314. MANTOY, BERNARD. (Ed.) Lamarck. Paris: Seghers,
      1968.

1315. MAYR, ERNST. "Lamarck Revisited." Journal of
      the History of Biology 5 (1972), 55-94.

1316. MONY, P. "Lamarck, créateur du transformisme."
      Bulletin de la Société Linnéenne du Nord de la
      France 24 (1929), 6-21.

1317. MUDFORD, PETER G. "William Lawrence and 'The
      Natural History of Man'." Journal of the History
      of Ideas 29 (1968), 430-436.

1318. OLBY, ROBERT CECIL. (Ed.) Late Eighteenth Century
      European Scientists. New York: Pergamon Press,
      1966. (Includes J.-B. de Lamarck)

1319. OMODEO, PIETRO. "Cuvier e l'evoluzionismo."
      Episteme 7 (1973), 3-14.

1320. PACKARD, ALPHEUS SPRING (THE YOUNGER). Lamarck,
      the Founder of Evolution: His Life and Work. With
      translations of his writings on organic evolution.
      New York: Longmans & Co., 1901.

1321.   PANCALDI, GIULIANO. "'L'economia della natura' da
        Cuvier a Darwin." Rivista di filosofia 66 (Feb.
        1975), 77-111.

1322.   PEARSON, HESKETH. Dr. Darwin. London: Dent,
        1930. (Erasmus Darwin)

1323.   PELSENEER, PAUL. "Les Premiers Temps de l'idée
        évolutionniste: Lamarck, Geoffroy Saint-Hilaire,
        et Cuvier." Annales de la Société Zoologique et
        Malacologique de Belgique 50 (1914-19), 53-89.

1324.   PERRIER, EDMOND. Lamarck. Paris: Payot, 1925.

1325.   PIVETEAU, JEAN. "Le Débat entre Cuvier et Geoffroy
        Saint-Hilaire sur l'unité de plan et de composi-
        tion." Revue d'Histoire des Sciences et de leurs
        Applications 3 (1950), 343-363.

1326.   POLIAKOV, I. M. Zh. B. Lamark i uchenie ob
        evoliutsii organicheskogo mira. Moscow: Gosizdat
        "Vysshaia Shkola", 1962. (Lamarck and organic evo-
        lution)

1327.   PRIMER, IRWIN. "Erasmus Darwin's 'Temple of Nature':
        Progress, Evolution, and the Eleusinian Mysteries."
        Journal of the History of Ideas 25 (1964), 58-76.

1328.   RABEL, GABRIELE. "Lamarck: In Honour of the 200th
        Anniversary of His Birth." Nineteenth Century 137
        (1945), 258-264.

1329.   ROSTAND, JEAN. "Un Grand Révolutionnaire de
        l'esprit: Jean Lamarck (1744-1829)." Les Lettres
        Françaises No. 111 (7 juin, 1946), 1-2.

1330.   ROSTAND, JEAN. "Les Précurseurs français de Charles
        Darwin." Revue d'Histoire des Sciences et de leurs
        Applications 13 (1960), 45-58.

1331.   ROUCHE, MAX. Herder précurseur de Darwin? Histoire
        d'un mythe. Paris: Société d'Edition, Les Belles-
        lettres, 1940.

1332.   ROUSSEAU, GEORGES. "Lamarck et Darwin." Bulletin
        du Muséum National d'Histoire Naturelle 5 (1969),
        1029-41.

1333. RUSSELL, C. A. (Ed.) Science and Religious Belief:
       A Selection of Recent Historical Studies. London:
       University of London Press Ltd., 1973. (Includes
       T. Chalmers, Uniformity, Diluvialism, G. Cuvier,
       G. L. L. de Buffon, J.-B. de Lamarck and Darwin,
       evolution and the churches)

1334. SCHILLER, JOSEPH. (Ed.) Colloque international
       "Lamarck", tenu au Muséum National d'Histoire
       Naturelle, Paris, les 1-2 et 3 juillet 1971.
       Paris: A. Blanchard, 1971.

1335. SCHILLER, JOSEPH. "Physiologie et classification
       dans l'oeuvre de Lamarck." Histoire et Biologie 2
       (1969), 35-57.

1336. SCHOFIELD, M. "Lamarck: Precursor of Darwin."
       Contemporary Review 166 (Aug. 1944), 107-110.

1337. SHANER, R. F. "Lamarck and the Evolution Theory."
       Scientific Monthly 24 (Mar. 1927), 251-255.

1338. SHRYOCK, RICHARD HARRISON. "The Strange Case of
       Wells' Theory of Natural Selection (1813): Some
       Comments on the Dissemination of Scientific Ideas."
       In M. F. Ashley Montagu (Ed.), Studies and Essays
       in the History of Science and Learning Offered in
       Homage to George Sarton. New York: Schuman, 1947,
       197-207.

1339. SIMPSON, GEORGE GAYLORD. "Lamarck, Darwin and
       Butler." American Scholar 30 (1961), 238-249.

1340. SNELDERS, H. A. M. "Lamarck's Theory of the Earth:
       Comments." Isis 56 (1965), 356-357.

1341. SOBOL', S. L. "Printsip estestvennogo otbora v
       rabotakh nekotorykh angliiskikh biologov 10-30-kh
       godov XIXv." Trudy Institut Istorii Estestvoznaniia
       i Tekhniki 40 (1962), 17-117. (Natural selection
       in the works of some English biologists in the
       first third of the 19th century)

1342. STAFLEU, FRANS A. "Lamarck: The Birth of Bio-
       logy." Taxon 20 (1971), 397-442.

1343.   TASKER, J. G.   "Goethe and Darwin."   London
        Quarterly Review 113 (Jan. 1910), 130-133.
1344.   TEMKIN, OWSEI.   "Basic Science, Medicine, and the
        Romantic Era."   Bulletin of the History of Medicine
        37 (1963), 97-129.   (William Lawrence)
1345.   UNGERER, EMIL.   Lamarck-Darwin: Die Entwicklung
        des Lebens. Stuttgart: Frommann, 1923.
1346.   VANDEL, ALBERT.   "Lamarck et Darwin."   Revue
        d'Histoire des Sciences et de leurs Applications 13
        (1960), 59-72.
1347.   VICKERS, H. M.   "An Apparently Hitherto Unnoticed
        'Anticipation' of the Theory of Natural Selection."
        Nature 85 (Feb. 16, 1911), 510-511.   (Edward Blyth)
1348.   WEIDENREICH, FRANZ.   "Lamarck: Seine Persönlich-
        keit und sein Werk: Zur 100. Wiederkehr seines
        Todestages."   Natur und Museum 60 (1930), 326-333,
        363-369.
1349.   WELLS, GEORGE A.   "Goethe and Evolution."   Journal
        of the History of Ideas 28 (1967), 537-550.   (In-
        cludes J.-B. de Lamarck and G. Cuvier)
1350.   WELLS, KENTWOOD D.   "The Historical Context of
        Natural Selection: The Case of Patrick Matthew."
        Journal of the History of Biology 6 (1973), 225-
        258.
1351.   WELLS, KENTWOOD D.   "Sir William Lawrence (1783-
        1867): A Study of Pre-Darwinian Ideas on Heredity
        and Variation."   Journal of the History of Biology
        4 (1971), 319-361.
1352.   WELLS, KENTWOOD D.   "William Charles Wells and the
        Races of Man."   Isis 64 (1973), 215-225.   (See
        reply by P. A. Erickson 66, Mar. 1975, 96-97)
1353.   WHEELER, WILLIAM MORTON AND THOMAS BARBOUR. (Eds.)
        The Lamarck Manuscripts at Harvard. Cambridge,
        Mass.: Harvard University Press, 1933.
1354.   WILKIE, J. S.   "The Idea of Evolution in the Writ-
        ings of Buffon."   Annals of Science 12 (1956),
        48-62, 212-227, 255-266.

1355. ZIMMERMANN, WALTER. Evolution: Die Geschichte
       ihrer Probleme und Erkenntnisse. Freiburg-München:
       Karl Alber, 1953.
1356. ZIRKLE, CONWAY. "The Early History of the Idea of
       the Inheritance of Acquired Characters and of
       Pangenesis." Transactions of the American Philo-
       sophical Society 35 (1946), 91-151.
1357. ZIRKLE, CONWAY. "Natural Selection before the
       'Origin of Species'." Proceedings of the American
       Philosophical Society 84 (1941), 71-123.
1358. ZIRKLE, CONWAY. "Species before Darwin." Proceed-
       ings of the American Philosophical Society 103
       (1959), 636-644.

3.   Chambers, Robert

See V. Natural Theology; VI. Geology . . .;
VII. Evolution (other sections); XVIII. Spiritualism . . .;
XIX. Varieties of Belief . . ., 20. Tennyson, Alfred.

1359. ANON. "Dr. Robert Chambers and Spiritualism."
       Light 25 (July 15, 1905), 331-333; (July 22, 1905),
       343-345.
1360. BRACKEN, HARRY M. "Berkeley and Chambers."
       Journal of the History of Ideas 17 (1956), 120-126.
1361. BUSH, DOUGLAS. "Evolution and the Victorian
       Poets." In Science and English Poetry: A His-
       torical Sketch, 1590-1950. New York: Oxford
       University Press, 1950, 109-138. (Includes
       A. Tennyson, M. Arnold)
1362. EGERTON, FRANK N. "Refutation and Conjecture:
       Darwin's Response to Sedgwick's Attack on
       Chambers." Studies in the History and Philo-
       sophy of Science 1 (1970), 176-183.

1363.   FLEMING, DONALD HARNISH.   "The Centenary of the
        Origin of Species."   Journal of the History of
        Ideas 20 (1959), 437-446.   (Includes Darwin)

1364.   FRAZER, PERSIFOR.   "Was the Development Theory
        Influenced by 'The Vestiges of the Natural History
        of Creation'?"   The American Geologist 30 (Oct.
        1902), 262-263.

1365.   HODGE, M. J. S.   "The Universal Gestation of Nature:
        Chambers' 'Vestiges' and 'Explanations'."   Journal
        of the History of Biology 5 (1972), 127-151.

1366.   MILLHAUSER, MILTON.   Just Before Darwin: Robert
        Chambers and "Vestiges".   Middletown: Wesleyan
        University Press, 1959.

1367.   MILLHAUSER, MILTON.   "The Literary Impact of
        'Vestiges of Creation'."   Modern Language Quarterly
        17 (1956), 213-226.

1368.   MILLHAUSER, MILTON.   "Robert Browning, Robert
        Chambers, and Mr. Home, the Medium."   Victorian
        Newsletter No. 39 (Spring 1971), 15-19.

1369.   MILLHAUSER, MILTON.   "Robert Chambers and the
        'Supernatural'."   Journal of the American Society
        for Psychical Research 47 (1953), 104-118.

1370.   OGILVIE, MARILYN BAILEY.   "Robert Chambers and the
        Nebular Hypothesis."   British Journal for the
        History of Science 8 (Nov. 1975), 214-232.

1371.   OGILVIE, MARILYN BAILEY.   "Robert Chambers and the
        Successive Revisions of the 'Vestiges of the
        Natural History of Creation'."   Dissertation
        Abstracts International 34 (1973), 1833-34A (Uni-
        versity of Oklahoma, 1973)

1372.   ROSTAND, JEAN.   "Une Oeuvre méconnue: les
        'Vestiges of the natural history of creation'."
        Revue d'Histoire des Sciences et de leurs Applica-
        tions 9 (1956), 62-73.

4.   DARWIN, CHARLES ROBERT

This large section includes works on Darwin's
family, health, and personality, and his education, intellectual development, and scientific achievements. It also
deals with his relationship with contemporary naturalists,
the comparison of his ideas with those thinkers who
preceded and followed him, his impact on science and
thought in general and on religion in particular, and
the reception of his writings in his own day as well as
later celebrations of the anniversaries of the publication
of the Origin. Though our aim has been to be more
exhaustive here than in other sections, so great is the
volume of literature on Darwin, that very brief notices
or articles have, for the most part, been omitted. See
I. Science; III. Ideas . . .; V. Natural Theology;
VI. Geology . . .; VII. Evolution (other sections);
categories IX-XIII below.

1373.   ABBOTT, LAWRENCE FRASER. "Charles R. Darwin, the
        Saint." In Twelve Great Modernists. New York:
        Doubleday, Page, 1927, 225-251.
1374.   ADLER, SAUL. "Darwin's Illness." British Medical
        Journal No. 5444 (May 8, 1965), 1249-50.
1375.   ADLER, SAUL. "Darwin's Illness." Nature 184
        (1959), 1102-03.
1376.   ALEKSEEV, VALERII ANDREEVICH. Osnovy darvinizma:
        istoricheskoe i teoreticheskoe vvedenie. Moscow:
        1964. (Historical and theoretical aspects of
        Darwinism)
1377.   ALTNER, GÜNTER. Charles Darwin und Ernst Haeckel:
        Ein Vergleich nach theologischen Aspekten mit einem
        Geleitwort von Wulf Emmo Ankel. Zürich: EVZ-
        Verlag, 1966.
1378.   ALVAREZ, WALTER C. "The Nature of Charles Darwin's
        Lifelong Ill-Health." New England Journal of
        Medicine 261 (1959), 1109-12.

1379. ANON. "Achievement of Darwin." Living Age 260
      (Mar. 20, 1909), 761-764.

1380. ANON. "Charles Darwin: 1809-1859-1909." London
      Quarterly Review 112 (Oct. 1909), 323-327.

1381. ANON. "Darwin and Evolution." Scientific American
      130 (Mar. 1924), 155.

1382. ANON. "Darwin's Evolution." New York Times Maga-
      zine (Nov. 22, 1959), 92-93.

1383. ANON. "Difference between Darwin and Wallace."
      Current Literature 40 (Feb. 1906), 181-182.

1384. ANON. Handlist of Darwin Papers at the University
      Library, Cambridge. Cambridge: University Press,
      for the University Library, 1960.

1385. ANON. "Lamarck, Darwin and Weismann." Living Age
      235 (Nov. 29, 1902), 517-529.

1386. ANON. "A Letter of Charles Darwin." Cambridge
      Review 51 (1930), 410. (To Professor Owen, Dec.
      19, 1836)

1387. ANON. "Mistakes of Darwin and His Would-be Fol-
      lowers." Bibliotheca Sacra 66 (Apr. 1909), 332-
      343.

1388. APPLEMAN, PHILIP. (Ed.) Darwin. New York: W. W.
      Norton & Company Inc., 1970. Second Edition, 1979.
      (Selections from primary and secondary sources,
      with an Epilogue, 1969, and a Postscript, 1979)

1389. ASHWORTH, J. H. "Charles Darwin as a Student in
      Edinburgh, 1825-27." Nature 136 (1935), 1011-14.

1390. ASIMOV, I. "Darwin and Wallace." Senior Scholastic
      74 (Mar. 20, 1959), 22-23.

1391. ATKINS, HEDLEY. Down, the Home of the Darwins:
      The Story of a House and the People Who Lived
      There. London: Royal College of Surgeons of Eng-
      land, 1974.

1392. AULIE, RICHARD P. "An American Contribution to
      Darwin's 'Origin of Species'." American Biology
      Teacher 32 (1970), 85-87. (A. Gray)

1393. AULIE, RICHARD P. "Darwin and Spontaneous Genera-
       tion." Journal of the American Scientific Affilia-
       tion 22 (1970), 31-34.

1394. AULIE, RICHARD P. "Darwin, Immutability, and
       Creation." American Biology Teacher 22 (Oct.
       1960), 420-425.

1395. AVERY, G. S., JR. et al. "Darwin and Early Dis-
       coveries in Connection with Plant Hormones."
       Science 87 (Jan. 21, 1938), 66.

1396. AVINERI, SHLOMO. "From Hoax to Dogma: A Footnote
       on Marx and Darwin." Encounter 28, No. 3 (1967),
       30-32.

1397. BAEHNI, CHARLES. "Correspondance de Charles Darwin
       et d'Alphonse de Candolle." Gesnerus 12 (1955),
       109-156.

1398. BAILLAUD, LUCIEN. "Le Mémoire de Charles Darwin
       sur les plantes grimpantes." Archives Internation-
       ales d'Histoire des Sciences 19 (1966), 235-246.

1399. BAIRD, T. "Darwin and the Tangled Bank." American
       Scholar 15, No. 4 (1946), 477-486.

1400. BAKER, H. G. "Darwin and after Darwin." Evolution
       14 (1960), 272-274.

1401. BAKER, RONALD DUNCAN. "The Influence of Charles
       Darwin on the Development of Theological Thought."
       B.D. Thesis, Trinity College (Dublin), 1962.

1402. BALDWIN, JAMES M. "The Influence of Darwin on
       Theory of Knowledge and Philosophy." Psychological
       Review 16 (May 1909), 207-218.

1403. BARLOW, NORA. "Charles Darwin and the Galapagos
       Islands." Nature 136 (Sept. 7, 1935), 391, 534-
       535.

1404. BARLOW, NORA. Charles Darwin and the Voyage of
       the Beagle. New York: Philosophical Library,
       1946.

1405. BARLOW, NORA. (Ed.) Darwin and Henslow: The
       Growth of an Idea: Letters, 1831-1860. London:
       Murray, [for] Bentham-Moxon Trust, 1967.

1406. BARLOW, NORA. "Darwin's Ornithological Notes."
       Bulletin of the British Museum (Natural History).
       Historical Series 2 (1963), 201-278.

1407. BARLOW, NORA. "Robert FitzRoy and Charles Darwin:
       With Five Unpublished Letters from Darwin." Corn-
       hill Magazine 72 (Apr. 1932), 493-510.

1408. BARLOW, NORA. "The Voyage of the 'Beagle'." Nature
       129 (1932), 439.

1409. BARNETT, LINCOLN. "Darwin's World of Nature." Life
       45 (Sept. 8, 1958), 56-69, 73, 75, 76; (Nov. 3,
       1958), 54-68, 70; 46 (Jan. 26, 1959), 46-59, 63,
       64; (Mar. 16, 1959), 56-67, 71, 72; (June 1, 1959),
       68-78, 80, 82, 87; 47 (July 20, 1959), 54-65,
       67-68; (Oct. 19, 1959), 96-101, 103, 104, 107, 108,
       113.

1410. BARNETT, SAMUEL ANTHONY. (Ed.) A Century of
       Darwin. London: Heinemann, 1958.

1411. BARRETT, PAUL H. "Darwin's 'Gigantic Blunder'."
       Journal of Geological Education 21 (1973), 19-28.

1412. BARRETT, PAUL H. "From Darwin's Unpublished Note-
       books." The Centennial Review of Arts and Sciences
       3 (1959), 398-406.

1413. BARRETT, PAUL H. AND ALAIN F. CORCOS. "A Letter
       from Alexander Humboldt to Charles Darwin."
       Journal of the History of Medicine and Allied
       Sciences 27 (1972), 159-172.

1414. BARRETT, PAUL H. "The Sedgwick-Darwin Geologic
       Tour of North Wales." Proceedings of the American
       Philosophical Society 118 (1974), 146-164.

1415. BARRY, DAVID G. "The Darwinian Synthesis." Mid-
       west Quarterly 1 (1960), 363-376.

1416. BARRY, DAVID G. "The Mosaic Heritage of Charles
       Darwin." Midwest Quarterly 1 (1960), 209-225.

1417. BARTHOLOMEW, MICHAEL J. "The Intellectual Back-
       ground to the Work of Charles Lyell and Charles
       Darwin." Ph.D. Dissertation, University of
       Lancaster, 1975.

1418.   BARZUN, JACQUES. Darwin, Marx, Wagner: Critique
        of a Heritage. Boston: Little, Brown and Company,
        1941.

1419.   BASALLA, GEORGE. "Darwin's Orchid Book." Pro-
        ceedings of the 10th International Congress of
        the History of Science (Ithaca, 1962), 2 (1964),
        971-974.

1420.   BASALLA, GEORGE. "The Voyage of the Beagle without
        Darwin." Mariner's Mirror 49 (1963), 42-48.

1421.   BEDDALL, BARBARA G. "'Notes for Mr. Darwin':
        Letters to Charles Darwin from Edward Blyth at
        Calcutta: A Study in the Process of Discovery."
        Journal of the History of Biology 6 (1973), 69-95.

1422.   BELL, PETER ROBERT. (Ed.) Darwin's Biological
        Work: Some Aspects Reconsidered. Cambridge:
        Cambridge University Press, 1959.

1423.   BERRILL, N. J. "Darwin and the Islands of Evolu-
        tion." Atlantic Monthly 190 (Mar. 1952), 54-59.

1424.   BIRCH, L. C. "Darwin and His Successors." Aus-
        tralian Journal of Science 22 (1959), 17-22.

1425.   BOLLER, PAUL F., JR. "Darwin's American Champion."
        Southwest Review 45 (1960), 156-164. (A. Gray)

1426.   BOWLER, PETER J. "Darwin's Concepts of Variation."
        Journal of the History of Medicine and Allied
        Sciences 29 (1974), 196-212.

1427.   BOWNE, B. P. "Darwin and Darwinism." Hibbert
        Journal 8 (Oct. 1909), 122-138.

1428.   BRADFORD, GAMALIEL. Darwin. Boston and New York:
        Houghton Mifflin Co., 1926.

1429.   BRETT, RAYMOND L. "The Influence of Darwin on His
        Contemporaries." South Atlantic Quarterly 59
        (1960), 69-81.

1430.   BRIEN, PAUL. "A l'occasion d'un glorieux cen-
        tenaire: 'L'Origine des espèces' de Charles Darwin
        et le problème de l'évolution." Scientia 95 (1960),
        156-160.

1431. BRITTON, N. L. "Darwin and Botany." Popular
       Science Monthly 74 (Apr. 1909), 355-360.

1432. BRONOWSKI, J. "The Ladder of Creation." Listener
       90 (July 5, 1973), 10-16. (Includes A. R. Wallace)

1433. BROWNE, B. "Darwin's Health." Nature 151 (1943),
       14-15.

1434. BRUSSEL, J. A. "The Nature of the Naturalist's
       Unnatural Illness: A Study of Charles Robert
       Darwin." Psychiatric Quarterly Supplement 40
       (1966), 315-331.

1435. BRYCE, JAMES. "Personal Reminiscences of Charles
       Darwin and of the Reception of the 'Origin of
       Species'." Proceedings of the American Philo-
       sophical Society 48 (1909), iii-xiv.

1436. BUCHSBAUM, RALPH. (Ed.) A Book that Shook the
       World: Anniversary Essays on Charles Darwin's
       "Origin of Species". Pittsburgh, Pa.: University
       of Pittsburgh Press, 1958.

1437. BUMPUS, H. C. "Darwin and Zoology." Popular
       Science Monthly 74 (Apr. 1909), 361-366.

1438. BURCHFIELD, JOE D. "Darwin and the Dilemma of
       Geological Time." Isis 65 (1974), 301-321.

1439. BURLA, HANS. "Darwin und sein Werk." Gesnerus 15
       (1958), 164-175.

1440. BURLA, HANS. Darwin und sein Werk. Zürich:
       Kommission-Verlag Gebr. Fretz, 1959.

1441. BURNABY, JOHN. Darwin and the Human Situation.
       Cambridge: Heffer, 1960.

1442. BURROUGHS, JOHN. "Critical Glance into Darwin."
       Atlantic Monthly 126 (Aug. 1920), 237-247.

1443. BURROW, JOHN W. "Charles Darwin." Horizon 8,
       No. 4 (1966), 41-47.

1444. BURROW, JOHN W. "Charles Darwin." Listener 87
       (1972), 173-174.

1445. BURSTYN, HAROLD L. "If Darwin Wasn't the 'Beagle's'
       Naturalist, Why Was He on Board?" British Journal
       for the History of Science 8 (1975), 62-69.

1446. BYL, SIMON. "Le Jugement de Darwin sur Aristote."
       Antiquité Classique 42 (1973), 519-521.
1447. CAMPBELL, JOHN ANGUS. "Charles Darwin and the
       Crisis of Ecology: A Rhetorical Perspective."
       Quarterly Journal of Speech 60 (1974), 442-449.
1448. CAMPBELL, JOHN ANGUS. "Darwin and 'The Origin of
       Species': The Rhetorical Ancestry of an Idea."
       Speech Monographs 37 (1970), 1-14.
1449. CAMPBELL, JOHN ANGUS. "Nature, Religion and Emo-
       tional Response: A Reconsideration of Darwin's
       Affective Decline." Victorian Studies 18 (1974),
       159-174.
1450. CAMPBELL, JOHN ANGUS. "The Polemical Mr. Darwin."
       Quarterly Journal of Speech 61 (1975), 375-390.
1451. CANGUILHEM, GEORGES. Les Concepts de "lutte pour
       l'existence" et de "sélection naturelle" en 1858:
       Charles Darwin et Alfred Russel Wallace. Paris:
       En vente à la Librairie du Palais de la Découverte,
       1959.
1452. CANGUILHEM, GEORGES. "L'Homme et l'animal du point
       de vue psychologique selon Charles Darwin." Revue
       d'Histoire des Sciences et de leurs Applications 13
       (1960), 81-94.
1453. CANNON, WALTER F. "The Basis of Darwin's Achieve-
       ment: A Revaluation." Victorian Studies 5 (1961-
       62), 109-134.
1454. CANNON, WALTER F. "Darwin's Vision in 'On the
       Origin of Species'." In George Levine and William
       Anthony Madden (Eds.), The Art of Victorian Prose.
       New York: Oxford University Press, 1968, 154-176.
1455. CARTER, GEORGE STUART. "The Centenary of 'Darwin-
       ism'." Nature 182 (1958), 1351.
1456. CARTER, GEORGE STUART. A Hundred Years of Evolu-
       tion. London: Sidgwick and Jackson, 1957.
1457. CAVERNO, CHARLES. "Louis Agassiz and Charles
       Darwin: A Synthesis." Bibliotheca Sacra 73 (Jan.
       1916), 137-140.

1458. CENTORE, FLOYD F. "Darwin on Evolution: A Re-estimation." Thomist 33 (July 1969), 456-496.

1459. CENTORE, FLOYD F. "Neo-Darwinian Reactions to the Social Consequences of Darwin's Nominalism." Thomist 35 (1971), 113-142.

1460. CHANCELLOR, JOHN. Charles Darwin. London: Weidenfeld and Nicolson, 1974.

1461. CHAPIN, E. A. AND C. C. CHAPIN. "Darwin and the Galapagos Islands." Bulletin of the Pan American Union 69 (Sept. 1935), 655-666.

1462. CHEESMAN, EVELYN. Charles Darwin and His Problems. London: Bell, 1953.

1463. CHESTERTON, G. K. "The Case of Darwin." G K's Weekly 19 (May 17, 1934), 168-170.

1464. CHOLODNY, N. "Charles Darwin and the Modern Theory of Tropisms." Science 86 (Nov. 19, 1937), 468.

1465. CLARK, ROBERT E. D. Darwin: Before and After. The Story of Evolution. London: Paternoster Press, 1948.

1466. CLEMENTS, FREDERICK E. "Darwin's Influence upon Plant Geography and Ecology." American Naturalist 43 (Mar. 1909), 143-151.

1467. COLBERT, E. H. "Darwin-Wallace Centennial." Science 129 (Jan. 16, 1959), 154.

1468. COLEMAN, BRIAN. "Samuel Butler, Darwin and Darwinism." Journal of the Society for the Bibliography of Natural History 7 (1974), 93-105.

1469. COLP, RALPH, JR. "The Contacts between Karl Marx and Charles Darwin." Journal of the History of Ideas 35 (1974), 329-338.

1470. COMAS, J. "Darwin y la evolución biológica." Revista Colombiana de Antropología 8 (1959), 127-158.

1471. CONKLIN, EDWIN GRANT. "Letters of Charles Darwin and Other Scientists and Philosophers to Sir Charles Lyell, Bart." Proceedings of the American Philosophical Society 95, No. 3 (1951), 220-222.

1472.   CONKLIN, EDWIN GRANT.   "The World's Debt to Darwin."
        Proceedings of the American Philosophical Society
        48 (1909), xxxviii-lvii.   (Also in University of
        Chicago Magazine 1, Mar.-Apr. 1909, 184-192, 251-
        259; and in Outlook 91, Feb. 20, 1909, 378-379)

1473.   CONRY, YVETTE.   "Broca et Darwin."   Actes du XIIe
        Congrès International d'Histoire des Sciences
        (Paris, 1968), 8 (1971), 25-29

1474.   CONRY, YVETTE.   Correspondance entre Charles
        Darwin et Gaston de Saporta.   Paris: Presses
        Universitaires de France, 1972.

1475.   COUSINS, FRANK.   "'The Ascent of Man'."   Listener
        90 (1973), 56-57.

1476.   COWLES, T.   "Malthus, Darwin and Bagehot: A Study
        in the Transference of a Concept."   Isis 26 (1937),
        341-348.

1477.   COX, CHARLES F.   "Charles Darwin and the Mutation
        Theory."   American Naturalist 43 (Feb. 1909),
        65-91.

1478.   COX, CHARLES F.   "Individuality of Darwin."
        Popular Science Monthly 74 (Apr. 1909), 344-348.

1479.   CRAMPTON, H. E.   "Darwin and after Darwin."
        Scientific American Supplement 66 (Aug. 22, 1908),
        122-123.

1480.   CREIGHTON, J. E.   "Darwin and Logic."   Psycho-
        logical Review 16 (May 1909), 170-187.   (See
        James M. Baldwin.   "Darwinism and Logic: A Reply
        to Professor Creighton."   Psychological Review 16,
        Nov. 1909, 431-436)

1481.   CRELLIN, J. K. (Comp.)   Darwin and Evolution.
        London: Cape, 1968.

1482.   CROMBIE, A. C.   "Darwin's Scientific Method."
        Actes du IXe Congrès International d'Histoire des
        Sciences (Barcelona, 1959), 1 (1960), 354-362.

1483.   CROMBIE, D. L.   "Back to Darwin."   Journal of
        the Royal College of General Practitioners 13
        (1967), 22-29.

1484. CROWTHER, JAMES GERALD. Charles Darwin. London: Methuen, 1972.

1485. CUPITT, DON. "Darwinism and English Religious Thought." Theology 78 (Mar. 1975), 125-130.

1486. DARLING, L. "Beagle, A Search for a Lost Ship." Natural History 69 (May 1960), 48-59.

1487. DARLINGTON, C. D. Darwin's Place in History. Oxford: Basil Blackwell, 1959.

1488. DARLINGTON, C. D. "The Origin of Darwinism." Scientific American 200 (May 1959), 60-66.

1489. DARLINGTON, P. J. "Darwin and Zoogeography." Proceedings of the American Philosophical Society 103 (1959), 307-319.

1490. DARMSTAEDTER, ERNST. "Erinnerung an Darwin: Antritt seiner Weltreise am 27. Dezember 1831." Münchener medizinische Wochenschrift 50 (1931), 2126-29.

1491. DARWIN, SIR CHARLES GALTON. "Charles Darwin: Man and Scientist." American Biology Teacher 20 (Nov. 1958), 235-236.

1492. DARWIN, SIR CHARLES GALTON. "Darwin as a Traveller." Geographical Journal 126 (June 1960), 129-136.

1493. DARWIN, SIR CHARLES GALTON. "Some Episodes in Darwin's Voyage on the 'Beagle'." Proceedings of the Royal Institution of Great Britain 38, No. 171 (1960), 198-210.

1494. DARWIN, SIR CHARLES GALTON. "Some Episodes in the Life of Charles Darwin." Proceedings of the American Philosophical Society 103 (1959), 609-615.

1495. DARWIN, FRANCIS. "FitzRoy and Darwin, 1831-1836." Nature 88 (1912), 547-548.

1496. DATTA, NARESH CHANDRA. "One Hundred Years of Darwinism." Science and Culture 29 (1963), 280-287.

1497. DAVIS, WILLIAM M.   "Wharton's and Darwin's Theories
       of Coral Reefs." Science Progress 24 (July 1929),
       42-56.

1498. DAVISON, C. "Education of Charles Darwin." Educa-
       tional Review 43 (Feb. 1912), 125-133.

1499. DE BEER, SIR GAVIN RYLANDS. "Charles Darwin."
       Proceedings of the British Academy 44 (1958),
       163-183.

1500. DE BEER, SIR GAVIN RYLANDS. Charles Darwin:
       Evolution by Natural Selection. London, New York:
       T. Nelson, 1963.

1501. DE BEER, SIR GAVIN RYLANDS. Charles Darwin:
       Lecture on a Master Mind. Master Mind Lectures,
       1958. London: Oxford University Press, 1958.

1502. DE BEER, SIR GAVIN RYLANDS. "The Darwin Letters at
       Shrewsbury School." Notes and Records of the
       Royal Society of London 23 (1968), 68-85.

1503. DE BEER, SIR GAVIN RYLANDS. "Darwin's Evolution."
       Listener 68 (Sept. 6, 1962), 347-348; (Sept. 13,
       1962), 387-388.

1504. DE BEER, SIR GAVIN RYLANDS. "Darwin's Journal."
       Bulletin of the British Museum (Natural History).
       Historical Series 2 (1959), 1-21.

1505. DE BEER, SIR GAVIN RYLANDS. "Darwin's Notebooks on
       Transmutation of Species. Part I. First Notebook
       (July 1837-Feb. 1838)." Bulletin of the British
       Museum (Natural History). Historical Series, 2
       (1960), 23-73.

1506. DE BEER, SIR GAVIN RYLANDS. "Darwin's Notebooks on
       Transmutation of Species. Part II. Second Note-
       book (Feb. to July 1838)." Bulletin of the British
       Museum (Natural History). Historical Series, 2
       (1960), 75-117.

1507. DE BEER, SIR GAVIN RYLANDS. "Darwin's Notebooks on
       Transmutation of Species. Part III. Third Note-
       book (July 15th 1838-Oct. 2nd 1838)." Bulletin of
       the British Museum (Natural History). Historical
       Series, 2 (1960), 119-150.

1508.   DE BEER, SIR GAVIN RYLANDS.   "Darwin's Notebooks on
        Transmutation of Species.   Part IV.   Fourth Note-
        book (Oct. 1838-10 July 1839)."   Bulletin of the
        British Museum (Natural History).   Historical
        Series, 2 (1960), 151-183.

1509.   DE BEER, SIR GAVIN RYLANDS AND M. J. ROWLANDS.
        "Darwin's Notebooks on Transmutation of Species.
        [Part V.]   Addenda and Corrigenda."   Bulletin of
        the British Museum (Natural History).   Historical
        Series, 2 (1961), 187-200.

1510.   DE BEER, SIR GAVIN RYLANDS, M. J. ROWLANDS AND
        B. M. SKRAMOVSKY.   "Darwin's Notebooks on Trans-
        mutation of Species.   Part VI.   Pages Excised by
        Darwin."   Bulletin of the British Museum (Natural
        History).   Historical Series, 3 (1967), 129-176.

1511.   DE BEER, SIR GAVIN RYLANDS.   "Darwin's Views on the
        Relations between Embryology and Evolution."
        Journal of the Linnean Society of London (Zoology)
        44 (1958), 15-23.

1512.   DE BEER, SIR GAVIN RYLANDS.   "The Darwin-Wallace
        Centenary."   Endeavour 17 (1958), 65.

1513.   DE BEER, SIR GAVIN RYLANDS.   "Further Unpublished
        Letters of Charles Darwin."   Annals of Science 14
        (1958), 83-115.

1514.   DE BEER, SIR GAVIN RYLANDS.   "Makers of Modern
        Science: Charles Darwin."   Times Educational
        Supplement 2131 (Mar. 23, 1956), 368.

1515.   DE BEER, SIR GAVIN RYLANDS.   "Mendel, Darwin, and
        Fisher."   Notes and Records of the Royal Society of
        London 19 (1964), 192-226; 21 (1966), 64-71.

1516.   DE BEER, SIR GAVIN RYLANDS.   "Mendel, Darwin, and
        the Centre of Science."   Listener 73 (Mar. 11,
        1965), 364-366.

1517.   DE BEER, SIR GAVIN RYLANDS.   "The Origins of Dar-
        win's Ideas on Evolution and Natural Selection."
        Proceedings of the Royal Society of London.   Sec-
        tion B, 155 (1961), 321-338.

1518.   DE BEER, SIR GAVIN RYLANDS.   "Some Unpublished
        Letters of Charles Darwin."   Notes and Records of
        the Royal Society of London 14 (1959), 12-66.

1519.   DEELY, JOHN N.   "The Philosophical Dimensions of
        the 'Origin of Species'."   Thomist 33 (1969),
        75-149, 251-342.

1520.   DELAGE, YVES AND MARIE GOLDSMITH.   The Theories
        of Evolution.   Translated by André Tridon.   London:
        Frank Palmer, 1912.

1521.   DELANY, SELDEN PEABODY.   "Charles Darwin and the
        Church."   American Church Monthly 21 (1927), 134-
        139.

1522.   DEWEY, JOHN.   The Influence of Darwin on Philo-
        sophy.   New York: H. Holt and Company, 1910.

1523.   DIBNER, BERN.   Darwin of the Beagle.   Norwalk,
        Conn.:   Burndy Library, 1960.

1524.   DICK, WILLIAM E.   "Darwin."   Discovery 20 (1959),
        482-487.

1525.   DICKINSON, ALICE.   Charles Darwin and Natural
        Selection.   New York: Watts, 1964.

1526.   DOBZHANSKY, THEODOSIUS.   "Darwin and Our Intel-
        lectual Heritage."   Science 95 (Mar. 20, 1942),
        303-304.

1527.   DODSON, EDWARD O.   "Some Problems of Evolution and
        Religion: A Darwin Centennial Address."   Revue
        de l'Université d'Ottawa 31 (1961), 380-395.

1528.   DOELLO-JURADO, M. (Ed.)   "Letter of Ch. Darwin in
        Argentina to Dr. F. J. Muñiz."   Nature 99 (June 14,
        1917), 305-306.

1529.   DORLODOT, HENRI DE.   Darwinism and Catholic
        Thought.   Trans. the Rev. Ernest Wessenger.
        London: Burns, Oates and Washbourne, Ltd., 1922.

1530. DORRANCE, ANNE. "The Darwin Centennial Address: Charles Darwin, 1809-1909." Proceedings and Collections of the Wyoming Historical and Geological Society 11 (1910), 45-64.

1531. DORSEY, GEORGE A. The Evolution of Charles Darwin. Garden City, New York: Doubleday, Page and Company, 1927.

1532. DOWDESWELL, WILFRID HOGARTH. The Mechanism of Evolution. London: Heinemann, 1955.

1533. DÜRKEN, BERNHARD. Allgemeine Abstammungslehre, zugleich eine gemeinverständliche Kritik des Darwinismus und des Lamarckismus. Berlin: Gebrüder Borntraeger, 1923.

1534. DUPREE, A. HUNTER. "Some Letters from Charles Darwin to Jeffries Wyman." Isis 42 (1951), 104-110.

1535. DUROUX, PAUL EMILE. Les Darwin. Paris: Editions Universitaires, 1972. (Erasmus, Charles, and Sir George Darwin)

1536. EGAMI, FUYUKO AND RYUICHI YASUGI. "The Influence of the Writings on the Philosophy of Science on the Works of Charles Darwin." Kagakusi Kenkyu 11 (1972), 57-64. (In Japanese; includes W. Whewell, J. F. W. Herschel)

1537. EGERTON, FRANK N. "Darwin's Method or Methods?" Studies in History and Philosophy of Science 2 (1971), 281-286.

1538. EGERTON, FRANK N. "Humboldt, Darwin, and Population." Journal of the History of Biology 3 (1970), 325-360.

1539. EGERTON, FRANK N. "Refutation and Conjecture: Darwin's Response to Sedgwick's Attack on Chambers." Studies in the History and Philosophy of Science 1 (1970), 176-183.

1540. EGERTON, FRANK N. "Studies of Animal Populations from Lamarck to Darwin." Journal of the History of Biology 1 (1968), 225-259.

1541. EHRLE, E. B. "Notes on the Teaching of Evolution:
       Paperback Literature on Charles Darwin, His Life
       and Thought." American Biology Teacher 27 (Dec.
       1965), 769-770.
1542. EISELEY, LOREN COREY. "Charles Darwin." Scientific
       American 194 (Feb. 1956), 62-72.
1543. EISELEY, LOREN COREY. "Charles Darwin, Edward
       Blyth, and the Theory of Natural Selection."
       Proceedings of the American Philosophical Society
       103 (1959), 94-158.
1544. EISELEY, LOREN COREY. "Darwin, Coleridge, and the
       Theory of Unconscious Creation." Library Chronicle,
       University of Pennsylvania 31 (1965), 7-22. (Re-
       printed in Daedalus 94, Summer 1965, 588-602)
1545. EISELEY, LOREN COREY. Darwin's Century: Evolu-
       tion and the Men Who Discovered It. Garden City,
       New York: Doubleday, 1958.
1546. EISELEY, LOREN COREY. "The Intellectual Ante-
       cedents of 'The Descent of Man'." In Bernard
       Campbell (Ed.), Sexual Selection and the Descent
       of Man, 1871-1971. London: Heinemann, 1972, 1-16.
1547. EISELEY, LOREN COREY. "The Reception of the First
       Missing Links." Proceedings of the American Philo-
       sophical Society 98 (1954), 453-465.
1548. EISELEY, LOREN COREY. "Was Darwin Wrong about the
       Human Brain?" Harper's Magazine 211 (Nov. 1955),
       66-70.
1549. EKMAN, PAUL. (Ed.) Darwin and Facial Expression:
       A Century of Research in Review. New York:
       Academic, 1973.
1550. ELLEGÅRD, ALVAR. Darwin and the General Reader:
       The Reception of Darwin's Theory of Evolution in
       the British Periodical Press, 1859-1872. Göteborg:
       [University of Gothenburg,] 1958.
1551. ELLEGÅRD, ALVAR. "The Darwinian Revolution: A
       Review Article." Lychnos (1960-61), 55-85.

1552. ELLEGÅRD, ALVAR. "The Darwinian Theory and the Argument from Design." Lychnos (1956), 173-192.

1553. ELLEGÅRD, ALVAR. "Darwin's Theory and Nineteenth-Century Philosophies of Science." In Philip Paul Wiener and Aaron Noland (Eds.), Roots of Scientific Thought: A Cultural Perspective. New York: Basic Books, 1957, 537-568. (Reprinted from Journal of the History of Ideas 18, June 1957, 362-393)

1554. ELLEGÅRD, ALVAR. "Public Opinion and the Press: Reactions to Darwinism." Journal of the History of Ideas 19 (1958), 379-387.

1555. ENGEL, H. AND M. S. J. ENGEL. "Charles Robert Darwin." Janus 49 (1960), 53-66.

1556. ERIKSON, ERIK H. "First Psychoanalyst." Yale Review 46, No. 1 (1956), 40-62. (S. Freud and Darwin)

1557. EVANS, MARY ALICE. "Mimicry and the Darwinian Heritage." Journal of the History of Ideas 26 (1965), 211-220. (Darwin's debt to H. W. Bates)

1558. EWING, J. FRANKLIN. "Darwin Up to Date." Catholic World 189 (1959), 344-349.

1559. FARRINGTON, BENJAMIN. What Darwin Really Said. London: Macdonald & Co., 1966.

1560. FAUCHER, EUGENE. "Fontane et Darwin." Etudes Germaniques 25 (1970), 7-24, 141-154.

1561. FEIBLEMAN, JAMES K. "Darwin and Scientific Method." Tulane Studies in Philosophy 8 (1959), 3-14.

1562. FEUER, LEWIS. "Is the 'Darwin-Marx Correspondence' Authentic?" Annals of Science 32 (Jan. 1975), 1-12.

1563. FISCHER, JEAN-LOUIS. "Lettre inédite de Charles Darwin à Dareste." Archives Internationales d'Histoire des Sciences 23 (1970), 81-86. (To Camille Dareste dated May 23, 1867)

1564.   FITCH, ROBERT E.   "Charles Darwin: Science and the
        Saintly Sentiments." Columbia University Forum 2
        (1959), 7-12.
1565.   FITCH, ROBERT E.   "Darwinism and Christianity."
        Antioch Review 19 (1959), 20-32.
1566.   FITCH, ROBERT E.   "Darwin's Gift of Positive Think-
        ing." New Republic 140 (Feb. 9, 1959), 12-14.
1567.   FLEMING, DONALD HARNISH. "The Centenary of the
        Origin of Species." Journal of the History of
        Ideas 20 (1959), 437-446.
1568.   FLEMING, DONALD HARNISH. "Charles Darwin, the
        Anaesthetic Man." Victorian Studies 4 (1961),
        219-236.
1569.   FLEMING, DONALD HARNISH. John William Draper
        and the Religion of Science. Philadelphia: Uni-
        versity of Pennsylvania Press, 1950. (Includes
        debate on Darwin's Origin at the meeting of the
        British Association for the Advancement of Science,
        1860)
1570.   FLEW, ANTONY G. N. "The Concept of Evolution: A
        Comment." Philosophy 41 (1966), 70-75.
1571.   FLEW, ANTONY G. N. "The Structure of Darwinism."
        New Biology No. 28 (1959), 25-44.
1572.   FLEWELLING, R. T. "From Darwin to Du Noüy."
        Personalist 29 (July 1948), 229-241.
1573.   FOSTER, W. D. "A Contribution to the Problem of
        Darwin's Ill-Health." Bulletin of the History of
        Medicine 39 (1965), 476-478.
1574.   FOTHERGILL, PHILIP G. "Charles Darwin and Darwin-
        ism." In Historical Aspects of Organic Evolution.
        London: Hollis & Carter, 1952, 104-134.
1575.   FOTHERGILL, PHILIP G. "Darwinian Theory and Its
        Effects." London Quarterly and Holborn Review 28
        (Oct. 1959), 289-294. (Darwinism and religion)
1576.   FRANKLIN, H. BRUCE. "The Island Worlds of Darwin
        and Melville." Centennial Review 11 (1967), 353-
        370.

1577. FREEMAN, DEREK. "The Evolutionary Theories of
Charles Darwin and Herbert Spencer." Current
Anthropology 15 (1974), 211-237. (Includes com-
ments by various scholars and Freeman's reply)

1578. FREEMAN, R. B. AND PETER J. GAUTREY. "Charles
Darwin's 'Queries about Expression'." Bulletin of
the British Museum (Natural History) 4 (1972),
207-219.

1579. FREEMAN, R. B. AND PETER J. GAUTREY. "Darwin's
'Questions about the Breeding of Animals', with a
Note on 'Queries about Expression'." Journal of
the Society for the Bibliography of Natural History
5 (1969), 220-225.

1580. FREEMAN, R. B. "Note 358: Issues and States of
Charles Darwin's 'The Expression of the Emotions in
Man and Animals'." Book Collector 21 (1972),
557-558.

1581. FREEMAN, R. B. "On 'The Origin of Species' 1859."
Book Collector 16 (1967), 340-344.

1582. FREEMAN, R. B. The Works of Charles Darwin: An
Annotated Bibliographical Handlist. London:
Dawson's, 1965.

1583. FRENCH, RICHARD D. "Darwin and the Physiologists,
or the Medusa and Modern Cardiology." Journal of
the History of Biology 3 (1970), 253-274.

1584. GAGER, C. S. "Darwin's Contribution to Evolution."
Open Court 23 (Oct. 1909), 577-587.

1585. GAISINOVICH, A. E. "Vzgliady Ch. Darvina na
izmenchivost' i nasledstvennost'." Iz Istorii Bio-
logicheskikh Nauk 2 (1970), 33-59. (Darwin's views
on variability and heredity)

1586. GALE, BARRY G. "Darwin and the Concept of a Struggle for Existence: A Study in the Extrascientific
Origins of Scientific Ideas." Isis 63 (1972),
321-344.

1587. GALSTON, ARTHUR W. "Botanist Charles Darwin."
Natural History 82, No. 10 (Dec. 1973), 85-86,
91-93.

1588. GATES, EUNICE JOINER. "Charles Darwin and Benito
Lynch's 'El inglés de los güesos'." Hispania 44
(May 1961), 250-253.

1589. GEIKIE, SIR ARCHIBALD. Charles Darwin as Geologist. Cambridge: University Press, 1909.

1590. GEISON, GERALD L. "Darwin and Heredity: The
Evolution of His Hypothesis of Pangenesis."
Journal of the History of Medicine 24 (1969),
375-411.

1591. GENSCHEL, R. Charles Darwin: Mensch zwischen
Glauben und Wissen. Göttingen: Arbeitskreis für
angewandte Anthropologie, 1959.

1592. GERRATANA, VALENTINO. "Marx and Darwin." New
Left Review 82 (Nov.-Dec. 1973), 60-82.

1593. GERSON, ELLIOT F. "Natural Selection and Late
19th-Century Paleontologists." Synthesis (Cambridge) 1, No. 2 (1973), 14-27.

1594. GHISELIN, MICHAEL T. "Darwin and Evolutionary
Psychology." Science 179 (1973), 964-968.

1595. GHISELIN, MICHAEL T. "The Individual in the Darwinian Revolution." New Literary History 3 (1971),
113-134.

1596. GHISELIN, MICHAEL T. "Mr. Darwin's Critics, Old
and New." Journal of the History of Biology 6
(1973), 155-165.

1597. GHISELIN, MICHAEL T. The Triumph of the Darwinian
Method. Berkeley: University of California Press,
1969.

1598. GILKEY, LANGDON BROWN. "Darwin and Christian
Thought." Christian Century 77 (Jan. 6, 1960),
7-11.

1599. GILLISPIE, CHARLES COULSTON. "Lamarck and Darwin
in the History of Science." The American Scientist
46 (1958), 388-409.

1600. GILSON, ETIENNE. "Darwin sans l'évolution."
       Revue des Deux Mondes (Apr.-June 1970), 264-282.

1601. GIRVETZ, HARRY K. "Philosophical Implications of
       Darwinism." Antioch Review 19 (1959), 9-19.

1602. GLICK, THOMAS F. (Ed.) The Comparative Reception
       of Darwinism. Austin and London: University of
       Texas Press, 1972. (Includes the Catholic reaction,
       and the reception of Darwinism in England, Germany,
       France, the U.S.A., and Russia)

1603. GOLDSCHMIDT, R. B. "Mimetic Polymorphism, a
       Controversial Chapter of Darwinism." Quarterly
       Review of Biology 20 (1945), 147-164, 205-230.

1604. GOOD, R. "The Origin of 'The Origin': A Psycho-
       logical Approach." Biology and Human Affairs 20,
       No. 1 (1954B), 10-16.

1605. GOODALE, GEORGE LINCOLN. "The Influence of Darwin
       on the Natural Sciences." Proceedings of the
       American Philosophical Society 48 (1909), xv-xxiv.

1606. GOTTLIEB, L. D. "Uses of Place: Darwin and Mel-
       ville on the Galapagos." BioScience 25 (Mar. 1975),
       172-175.

1607. GOULD, STEPHEN JAY. "Darwin and the Captain."
       Natural History 85 (Jan. 1976), 32-34.

1608. GOULD, STEPHEN JAY. "Darwin's Delay." Natural
       History 83 (Dec. 1974), 68-70.

1609. GOULD, STEPHEN JAY. "Darwin's Dilemma." Natural
       History 83 (June 1974), 16, 20, 22.

1610. GREENACRE, PHYLLIS. The Quest for the Father:
       A Study of the Darwin-Butler Controversy, as a
       Contribution to the Understanding of the Creative
       Individual. New York: International Universities
       Press, 1963.

1611. GREENE, JOHN C. "Darwin and Religion." Pro-
       ceedings of the American Philosophical Society 103
       (1959), 716-725. (Also in Harlow Shapley, Ed.,
       Science Ponders Religion. New York: Appleton-
       Century Crofts Inc., 1960, 254-276 and in W. Warren
       Wagar, Ed., European Intellectual History since
       Darwin and Marx. New York, Evanston, and London:
       Harper and Row Publishers, 1967, 12-34)

1612. GREENE, JOHN C. Darwin and the Modern World View.
       Baton Rouge: Louisiana State University Press,
       1961.

1613. GREENE, JOHN C. "The Kuhnian Paradigm and the
       Darwinian Revolution." In Duane H. D. Roller
       (Ed.), Perspectives in the History of Science and
       Technology. Norman: University of Oklahoma Press,
       1971, 3-25.

1614. GRELL, KARL. "Darwins Pangenesis-Theorie." Natur-
       wissenschaftliche Rundschau 13 (1960), 239-240,
       247-248, 250-252.

1615. GRENE, MARJORIE. "The Faith of Darwinism."
       Encounter 13, No. 5 (1959), 48-56.

1616. GRINNELL, GEORGE J. "The Darwin Case, a Computer
       Analysis of Scientific Creativity." Dissertation
       Abstracts International 31 (1970), 1725A (Uni-
       versity of California, Berkeley, 1969).

1617. GRINNELL, GEORGE J. "The Rise and Fall of Darwin's
       First Theory of Transmutation." Journal of the
       History of Biology 7 (1974), 259-273.

1618. GROVE, RICHARD STANLEY. "A Re-examination of
       Darwin's Argument in 'On the Origin of Species'."
       Dissertation Abstracts International 30 (1970),
       4411-12A (University of Missouri, Columbia, 1969).

1619. GRUBER, HOWARD E. "Darwin and 'Das Kapital'."
       Isis 52 (1961), 582-583.

1620.   GRUBER, HOWARD E.   Darwin on Man: A Psychological
        Study of Scientific Creativity.   Together with
        Darwin's Early and Unpublished Notebooks.   Tran-
        scribed and annotated by Paul H. Barrett.   New
        York: E. P. Dutton and Co., Inc., 1974.

1621.   GRUBER, HOWARD E. AND VALMAI GRUBER.   "The Eye of
        Reason: Darwin's Development during the Beagle
        Voyage."   Isis 53 (1962), 186-200.

1622.   GRUBER, HOWARD E.   "Pensée créatrice et vitesse du
        changement adaptif: le développement de la pensée
        de Darwin."   In [François Bresson and Maurice de
        Montmollin (Eds.)]   Psychologie et épistémologie
        génétiques: thèmes piagétiens.   Paris: Dunod,
        1966, 407-421.

1623.   GRUBER, JACOB W. A.   "Who Was the 'Beagle's'
        Naturalist?"   British Journal for the History of
        Science 4 (1968-69), 266-282.

1624.   GULICK, ADDISON.   "Charles Darwin, the Man."
        Scientific Monthly 15 (Aug. 1922), 132-143.

1625.   GUNTHER, ALBERT E.   "The Darwin Letters at Shrews-
        bury School."   Notes and Records of the Royal
        Society of London 30 (July 1975), 439-453.

1626.   HABGOOD, JOHN.   "They Changed Our Thinking: Darwin
        (1809-82) and After."   Expository Times 84 (Jan.
        1973), 100-105.

1627.   HAMILTON, R.   "Darwin and Newman."   Pax 42 (1952),
        28-33.

1628.   HARDIN, GARRETT JAMES.   "Darwin and the Heterotroph
        Hypothesis."   Scientific Monthly 70 (1950), 178-179.

1629.   HARDIN, GARRETT JAMES.   Nature and Man's Fate.
        London: Cape, 1960.

1630.   HATTIANGADI, J. N.   "Alternatives and Incommen-
        surables: The Case of Darwin and Kelvin."
        Philosophy of Science 38 (1971), 502-507.

1631.   HEBERER, GERHARD.   Charles Darwin: Sein Leben
        und sein Werk.   Stuttgart: Kosmos, 1959.

1632. HEBERER, GERHARD AND FRANZ SCHWANITZ. (Eds.)
Hundert Jahre Evolutionsforschung: Das wissen-
schaftliche Vermächtnis Charles Darwins. Stuttgart:
Gustav Fischer, 1960.

1633. HEGENBARTH, HANS. Darwin, die Bibel und die
Tatsachen. Graz: Steiermärkische Landesregierung,
Steiermärkische Landesbibliothek, 1972.

1634. HELLMANN, R. A. "Natural Selection: The Impact
of Darwin's 'The Origin of Species' on Biological
Science." Science Teacher 31 (Apr. 1964), 18-20.

1635. HEMLEBEN, JOHANNES. Charles Darwin in Selbst-
zeugnissen und Bilddokumenten. Reinbeck bei Ham-
burg: Rowohlt, 1968.

1636. HERBERT, SANDRA. "Darwin, Malthus, and Selection."
Journal of the History of Biology 4 (1971), 209-
217.

1637. HERBERT, SANDRA. "The Logic of Darwin's Discovery."
Dissertation Abstracts 29 (1969), 3547-48A (Brandeis
University, 1968).

1638. HERBERT, SANDRA. "The Place of Man in the Development of Darwin's Theory of Transmutation, Part I:
To July 1837." Journal of the History of Biology 7
(1974), 217-258.

1639. HEYER, PAUL. "Marx and Darwin: A Related Legacy
on Man, Nature, and Society." Dissertation
Abstracts International 36 (1975), 2941A (Rutgers
University, 1975).

1640. HILLENIUS, D. Inleiding tot het denken van Darwin.
Assen: Born, 1956. (Introduction to Darwin's
thought)

1641. HILTNER, S. "Darwin and Religious Development."
Journal of Religion 40 (Oct. 1960), 282-295.

1642. HIMMELFARB, GERTRUDE. Darwin and the Darwinian
Revolution. London: Chatto and Windus, 1959; New
York: W. W. Norton & Company, 1968.

1643. HIMMELFARB, GERTRUDE. "Sanctification of Darwin."
Nation 192 (Jan. 7, 1961), 14-15.

1644. HINGSTON, RICHARD WILLIAMS G. _Darwin_. London:
       Duckworth, 1934.

1645. HO, WING MENG. "Methodological Issues in Evo-
       lutionary Theory, with Special Reference to Dar-
       winism and Lamarckism." D.Phil. Thesis, University
       of Oxford, 1966.

1646. HODGE, M. J. S. "Origins and Species: A Study of
       the Historical Sources of Darwinism and of the
       Contexts of Some Other Accounts of Organic Diver-
       sity from Plato and Aristotle On." Ph.D. Disserta-
       tion, Harvard University, 1970.

1647. HOLIFIELD, E. BROOKS. "English Methodist Response
       to Darwin." _Methodist History_ 10 (Jan. 1972),
       196-217.

1648. HOLMES, SAMUEL JACKSON. "What Is Natural Selec-
       tion?" _Scientific Monthly_ 67 (1948), 324-330.

1649. HOOKER, SIR JOSEPH DALTON. "First Presentation of
       the Theory of Natural Selection." _Popular Science
       Monthly_ 74 (Apr. 1909), 401-406.

1650. HOPKINS, LOUIS J. "Darwin and His Interpreters."
       Personalist 18 (1937), 134-151.

1651. HOPKINS, ROBERT S. _Darwin's South America_. New
       York: John Day, 1969.

1652. HOSKIN, MICHAEL. _The Mind of the Scientist:
       Imaginary Conversations with Galileo, Newton,
       Herschel, Darwin, and Pasteur_. New York: Tap-
       linger, 1972.

1653. HOUSE, HUMPHRY. "Charles Darwin and the Voyage of
       the Beagle: Critical Essay." _New Statesman and
       Nation_ 31 (Apr. 6, 1946), 249.

1654. HUBBLE, DOUGLAS. "Charles Darwin and Psycho-
       therapy." _Lancet_ 244 (Jan. 30, 1943), 129-133.

1655. HUBBLE, DOUGLAS. "The Evolution of Charles Dar-
       win." _Horizon_ 14 (Aug. 1946), 74-85.

1656. HUGHESDON, P. J. "Spencer, Darwin and the Evolu-
       tion-Hypothesis." _Sociological Review_ 17 (Jan.
       1925), 31-44.

1657.   HULL, DAVID L.   Darwin and His Critics: The Recep-
        tion of Darwin's Theory of Evolution by the
        Scientific Community.   Cambridge, Mass.:   Harvard
        University Press, 1973.   (Introduction and con-
        temporary reviews)

1658.   HUNTLEY, WILLIAM B.   "David Hume and Charles Dar-
        win."   Journal of the History of Ideas 33 (1972),
        457-470.

1659.   HUXLEY, FRANCIS.   "Charles Darwin: Life and Habit."
        American Scholar 28 (Fall 1959), 489-499; 29 (Winter
        1959-60), 85-93.

1660.   HUXLEY, JULIAN SORELL et al.   A Book That Shook the
        World: Anniversary Essays on Charles Darwin's
        Origin of Species.   By Julian S. Huxley, Theodosius
        Dobzhansky, Reinhold Niebuhr, Oliver L. Reiser,
        Swami Nikhilananda.   Pittsburgh: University of
        Pittsburgh Press, 1958.

1661.   HUXLEY, JULIAN SORELL.   "Charles Darwin."   Con-
        temporary Review 142 (Oct. 1932), 424-429.

1662.   HUXLEY, JULIAN SORELL.   "Charles Darwin."   Con-
        temporary Review 209 (Sept. 1966), 145-149.

1663.   HUXLEY, JULIAN SORELL AND H. B. D. KETTLEWELL.
        Charles Darwin and His World.   New York: Viking
        Press, 1965.

1664.   HUXLEY, JULIAN SORELL.   "Charles Darwin: The
        Decisive Years."   Nature 157 (1946), 536-538.

1665.   HUXLEY, JULIAN SORELL AND JAMES FISHER. (Eds.)
        Darwin.   New York: Longmans, Green, 1939.

1666.   HUXLEY, JULIAN SORELL.   "Darwin and the Idea of
        Evolution."   Hibbert Journal 58 (1959), 1-12.

1667.   HUXLEY, JULIAN SORELL.   "Darwin Discovers Nature's
        Plan."   Life 44 (June 30, 1958), 64-75, 77, 78, 80,
        83, 86.

1668.   HUXLEY, JULIAN SORELL.   "Darwin's Theory of Sexual
        Selection and the Data Subsumed by It, in the Light
        of Recent Research."   American Naturalist 72 (1938),
        416-433.

1669. HUXLEY, JULIAN SORELL. "The Emergence of Darwin-
      ism." Journal of the Linnaean Society of London 44
      (1958), 1-14.
1670. HUXLEY, JULIAN SORELL. Evolution: The Modern
      Synthesis. New York: Harper and Brothers, Pub-
      lishers, 1942. (Includes natural selection)
1671. HUXLEY, JULIAN SORELL AND THOMAS FISHER. (Eds.)
      The Living Thoughts of Darwin. Greenwich, Conn.:
      Fawcett Publications, 1963.
1672. HUXLEY, JULIAN SORELL. "The Three Types of Evolu-
      tionary Process." Nature 180 (1957), 454-455.
1673. HUXLEY, LEONARD. "Charles Darwin: A Centenary
      Sketch." Cornhill Magazine 26 (1909), 376-389.
1674. HYMAN, STANLEY EDGAR. "After the Great Metaphors."
      American Scholar 31 (Spring 1962), 236-258.
      (Includes K. Marx, J. G. Frazer, S. Freud)
1675. HYMAN, STANLEY EDGAR. "Darwin the Dramatist."
      Centennial Review of Arts and Science 3 (1959),
      364-375.
1676. HYMAN, STANLEY EDGAR. "Descent, Fall and Sex:
      Darwin's Victorianism." Carleton Miscellany 2
      (Fall 1961), 11-25.
1677. HYMAN, STANLEY EDGAR. "The 'Origin' as Scripture."
      Virginia Quarterly Review 35 (1959), 540-552.
1678. HYMAN, STANLEY EDGAR. The Tangled Bank: Darwin,
      Marx, Frazer and Freud as Imaginative Writers. New
      York: Atheneum, 1962.
1679. IHMELS, CARL HEINRICH. Die Entstehung der
      organischen Natur nach Schelling, Darwin und Wundt:
      Eine Untersuchung über den Entwicklungsgedanken.
      Naumburg a.d.S.: G. Pätz'sche Buchdruckerei
      Lippert & Co., G.m.b.H., 1916.
1680. IRVINE, WILLIAM. Apes, Angels and Victorians:
      Darwin, Huxley and Evolution. New York: McGraw-
      Hill Book Co. Inc., 1955. (Reprinted, Cleveland:
      The World Publishing Company, 1959)

1681. JANSSENS, EMILE. "Le Voyage du 'Beagle'." Annales de la Société Zoologique de Belgique 90 (1959-60), 5-17.

1682. JESPERSEN, P. HELVEG. "Charles Darwin and Dr. Grant." Lychnos (1948-49), 159-167.

1683. JOHNSTON, W. W. "The Ill Health of Charles Darwin: Its Nature and Its Relation to His Work." American Anthropologist New Series 3 (1901), 139-158.

1684. JONES, FRANK MORTON. "The Most Wonderful Plant in the World: With Some Unpublished Correspondence of Charles Darwin." Natural History 23 (1923), 589-596. (Venus's-flytrap)

1685. JONES, HENRY FESTING. Charles Darwin and Samuel A. Butler: A Step towards Reconciliation. London: A. C. Fifield, 1911.

1686. JUDD, JOHN WESLEY. "Charles Darwin's Earliest Doubts Concerning the Immutability of Species." Nature 88 (Nov. 2, 1911), 8-12.

1687. JUDD, JOHN WESLEY. The Coming of Evolution: The Story of a Great Revolution in Science. Cambridge, England: University Press, 1910.

1688. KANNOWSKI, PAUL B. "The Newton of Biology." North Dakota Quarterly 36, No. 3 (Summer 1968), 5-16. (Darwin's theory - reception and later refinements)

1689. KARP, WALTER. Charles Darwin and the Origin of Species. By the Editors of Horizon Magazine. New York: American Heritage Pub. Co.; Distributed by Harper & Row, 1968.

1690. KEITH, SIR ARTHUR. Darwin Revalued. London: Watts, 1955.

1691. KEITH, SIR ARTHUR. "Darwinian Exhibition at Moscow." Nature 150 (1942), 393-395.

1692. KEITH, SIR ARTHUR. "A Postscript to Darwin's 'Formation of Vegetable Mould through the Action of Worms'." Nature 149 (June 27, 1942), 716-720.

1693.   KELLER, A. G.   "What Did Darwin Really Say?"
Saturday Review of Literature 28 (Oct. 6, 1945),
5-8.

1694.   KELLOGG, VERNON LYMAN.   Darwinism To-day: A
Discussion of Present-day Scientific Criticism
of the Darwinian Selection Theories.   London:
George Bell & Sons, 1907.

1695.   KELLY, MICHAEL.   "Darwin Really Was Sick."   Journal
of Chronic Diseases 20 (1967), 341.

1696.   KELLY, MICHAEL.   "Darwin's Illness."   British
Medical Journal No. 5470 (Nov. 6, 1965), 1128.

1697.   KEMPF, EDWARD J.   "Charles Darwin: The Affective
Sources of His Inspiration and Anxiety Neurosis."
Psychoanalytic Review (Washington), 5 (1918),
151-192.

1698.   KETTLEWELL, H. B. D.   "Darwin's Missing Evidence."
Scientific American 200 (Mar. 1959), 48-53.

1699.   KHOD'KOV, L. E.   "Charles Darwin, créateur de
l'association entre la méthode historique et la
méthode expérimentale."   Vestnik Leningradskogo
Gosudarstvennogo Universiteta 15, No. 9 (1960),
51-63.   (In Russian)

1700.   KNAPP, GUNTRAM.   Der antimetaphysische Mensch:
Darwin, Marx, Freud.   Stuttgart: Klett, 1973.

1701.   KNAPP, GUNTRAM.   "Darwin-Marx-Freud: Ein philoso-
phisch-anthropologischer Vergleich."   Stimmen der
Zeit 192 (July 1974), 463-473.

1702.   KNEALE, ELMER J.   "Darwin, Science and the Church:
A Symposium."   Forum 51 (June 1914), 821-845.

1703.   KOHLBRUGGE, J. H. F.   "War Darwin ein originelles
Genie?"   Biologisches Centralblatt 35 (1915),
93-111.

1704.   KOHN, LAWRENCE A.   "Charles Darwin's Chronic Ill
Health."   Bulletin of the History of Medicine 37
(1963), 239-256.

1705. KOLB, HAVEN. "Essay in Commemoration: On the
       Occasion of the Centennial of Darwin's Theory."
       Nature Magazine 51 (Oct. 1958), 415-416, 445.

1706. KOLMAN, E. "Marx and Darwin." Labour Monthly 13
       (Nov. 1931), 702-705.

1707. KONDRATIOUK, T. "Charles Darwin: la formation de
       la théorie matérialiste du développement du monde
       organique." Ukrayins'kyi botanichnyi zhurnal 16,
       No. 5 (1959), 3-14. (In Ukrainian)

1708. KONINCK, CHARLES DE. "Darwin's Dilemma: Theory of
       Natural Selection." Thomist 24 (Oct. 1961), 367-
       382.

1709. KORSUNSKAIA, VERA M. Charles Darvin. Moscow:
       Prosveshchenie, 1969. (In Russian)

1710. KRAGLIEVICH, LUCAS. "Darwin: algo sobre su labor
       científica en nuestro país." Anales de la Sociedad
       Científica Argentina 109 (1930), 353-376.

1711. KRAMER, HERBERT J. "The Intellectual Background
       and Immediate Reception of Darwin's 'Origin of
       Species'." Ph.D. Dissertation, Harvard University,
       1949.

1712. KRITSKY, G. "Mendel, Darwin, and Evolution."
       American Biology Teacher 35 (Nov. 1973), 477-479.
       (Reply with rejoinder, R. R. Hedtke, 36, May 1974,
       310-311)

1713. LACK, DAVID L. Darwin's Finches. Cambridge,
       England: University Press, 1947.

1714. LACK, DAVID L. "Darwin's Finches: With Biograph-
       ical Sketch." Scientific American 188 (Apr. 1953),
       66-68, 70, 72.

1715. LANESSAN, J. L. DE. "L'Attitude de Darwin à
       l'égard de ses prédécesseurs au sujet de l'origine
       des espèces." Revue Anthropologique 24 (1914),
       33-45.

1716. LANKESTER, E. R. "Increase of Knowledge in the
       Several Branches of Science." Nature 74 (Aug. 2,
       1906), 321-335. (Excerpt in Science 24, Aug. 24,

1906, 225-229)   (On Darwin's theory of the origin
of species)

1717.   LEAR, JOHN.   "The Book That Made Man Timeless:
'Origin of Species'."   Saturday Review 42 (Nov. 14,
1959), 51-64.

1718.   LEE, K. K.   "Popper's Falsifiability and Darwin's
Natural Selection."   Philosophy 44 (1969), 291-302.

1719.   LERNER, I. MICHAEL.   "The Concept of Natural Selection: A Centennial View."   Proceedings of the
American Philosophical Society 103 (1959), 173-182.

1720.   LEROY, JEAN F.   Charles Darwin et la théorie
moderne de l'évolution.   Paris: Seghers, 1966.

1721.   LEROY, JEAN F.   "Naudin, Spencer et Darwin dans
l'histoire des théories de l'hérédité."   Actes du
XIe Congrès International d'Histoire des Sciences
(Warsaw, 1965), 5 (1968), 64-69.

1722.   LEWONTIN, R. C.   "Darwin and Mendel: The Materialist Revolution."   In J. Neyman (Ed.), The Heritage
of Copernicus: Theories "Pleasing to the Mind."
Cambridge: M.I.T. Press, 1974, 166-183.

1723.   LIMOGES, CAMILLE.   "Darwin, Milne-Edwards et le
principe de divergence."   Actes du XIIe Congrès
International d'Histoire des Sciences (Paris,
1968), 8 (1971), 111-115.

1724.   LIMOGES, CAMILLE.   "Darwinisme et adaptation."
Revue des Questions Scientifiques 31 (1970), 353-
374.

1725.   LIMOGES, CAMILLE.   "Une Lecture nouvelle de
Darwin."   Sciences 58-59 (1969), 70-73.

1726.   LIMOGES, CAMILLE.   La Sélection naturelle: étude
sur la première constitution d'un concept (1837-
1859).   Paris: Presses Universitaires de France,
1970.

1727.   LINDROTH, STEN.   Charles Darwin.   Stockholm: Lind-
fors, 1946.

1728.   LINGELBACH, W. E.   "Notable Letters and Papers:
        The Darwin-Lyell Letters." Proceedings of the
        American Philosophical Society 95, No. 3 (1951),
        216-217.
1729.   LINTON, E.   "Examination of Darwin's 'Origin of
        Species' in the Light of Recent Observations and
        Experiments." American Naturalist 43 (Mar. 1909),
        163-172.
1730.   LIOUBINSKI, T.   "Apport de Darwin dans la physio-
        logie des plantes." Ukrayins'kÿi botanichnÿi
        zhurnal 16, No. 5 (1959), 28-40. (In Ukrainian)
1731.   LOEWENBERG, BERT JAMES. (Ed.) Charles Darwin:
        Evolution and Natural Selection. Boston: Beacon
        Press, 1959. (Introduction and selections)
1732.   LOEWENBERG, BERT JAMES. "Darwin and Darwin
        Studies, 1959-63." History of Science 4 (1965),
        15-54.
1733.   LOEWENBERG, BERT JAMES. Darwin, Wallace and the
        Theory of Natural Selection Including the Linnean
        Society Papers. New Haven: G. E. Cinamon, 1957.
1734.   LOEWENBERG, BERT JAMES. "The Mosaic of Darwinian
        Thought." Victorian Studies 3 (1959-60), 3-18.
1735.   LONGRIGG, JAMES. "Darwinism and Pre-Socratic
        Philosophy." Durham University Journal 65 (1973),
        307-315.
1736.   LUNN, SIR ARNOLD HENRY MOORE. "Enigma of Charles
        Darwin." Month 20 (Aug. 1958), 95-102.
1737.   LYON, JOHN JOSEPH. "Immediate Reactions to Darwin:
        The English Catholic Press' First Reviews of the
        'Origin of the [sic] Species'." Church History 41
        (Mar. 1972), 78-93.
1738.   MCATEE, W. L. "The Cats to Clover Chain."
        Scientific Monthly 65 (1947), 241-242. (A criticism
        of Darwin on the balance of nature)
1739.   MACBETH, NORMAN. Darwin Retried: An Appeal to
        Reason. Boston: Gambit Press, 1972.

1740. MCCLURG, JACK. "A Philosophical Study of the
       Biological Theories of Aristotle, Darwin, and
       Weismann." Ph.D. Dissertation, University of
       Chicago, 1962.

1741. MACHIN, ALFRED. Darwin's Theory Applied to Man-
       kind. London: Longmans, 1937.

1742. MACLEOD, ROY M. "Evolutionism and Richard Owen,
       1830-1868: An Episode in Darwin's Century." Isis
       56 (1965), 259-280.

1743. MAGOUN, H. W. "Darwin and Concepts of Brain Func-
       tion." In A. Fessard, R. W. Gerard, J. Konorski
       (Eds.), The Council for International Organiza-
       tions of Medical Sciences, Brain Mechanisms and
       Learning. Springfield, Illinois: Charles C.
       Thomas, 1961, 1-20.

1744. MANDELBAUM, MAURICE. "Darwin's Religious Views."
       Journal of the History of Ideas 19 (1958), 363-378.

1745. MARSHALL, ALAN J. Darwin and Huxley in Australia.
       London: Hodder and Stoughton, 1970.

1746. MARZA, VASILE D. AND ION T. TARNAVSCHI. "Darwin's
       Theory of Unity of Reacting Mechanisms in Plants
       and Animals: Its Present Day Importance." Indian
       Journal of History of Science 9 (1974), 185-220.

1747. MARZA, VASILE D. AND ION T. TARNAVSCHI. "The
       Problem of the Fertilization and Evolution of
       Phanerograms in Darwin's Work: A Critical Study."
       Indian Journal of History of Science 2 (1967),
       71-104.

1748. MAYR, ERNST. "Agassiz, Darwin and Evolution."
       Harvard Library Bulletin 13 (1959), 165-194.

1749. MAYR, ERNST. "Darwin and the Evolutionary Theory
       in Biology." In Betty J. Meggers (Ed.), Evolution
       and Anthropology: A Centennial Appraisal. Wash-
       ington, D.C.: The Anthropological Society of
       Washington, 1959, 1-10.

1750.   MAYR, ERNST.   "The Nature of the Darwinian Revolu-
         tion." Science 176 (1972), 981-989.
1751.   MAYR, ERNST.   "Open Problems of Darwin Research."
         Studies in History and Philosophy of Science 2
         (1971), 273-280.
1752.   MAZLISH, BRUCE.   "Darwin and the Benchuca."
         Horizon 17 (Summer 1975), 102-105. (Insect which
         bit Darwin, 1835)
1753.   MEDAWAR, PETER BRIAN.   "Darwin's Illness." New
         Statesman 67 (Apr. 3, 1964), 527-528. (Also
         Douglas Hubble's letter 67, Apr. 10, 1964, 561)
1754.   MELDOLA, RAPHAEL.   Evolution, Darwinian and
         Spencerian: The Herbert Spencer Lecture Delivered
         at the Museum, 8 December 1910. Oxford: Clarendon
         Press, 1910.
1755.   MELLERSH, HAROLD EDWARD LESLIE.   Charles Darwin:
         Pioneer of the Theory of Evolution. London:
         Arthur Barker, 1964.
1756.   MELLERSH, HAROLD EDWARD LESLIE.   FitzRoy of the
         Beagle. London: Hart-Davis, 1968.
1757.   MODILEWSKI, T.   "Les Problèmes de la fécondation
         dans les travaux de Darwin." Ukrayins'kȳi
         botanichnȳi zhurnal 16, No. 5 (1959), 41-47. (In
         Ukrainian)
1758.   MONTAGU, M. F. ASHLEY.   Darwin: Competition and
         Cooperation. New York: H. Schuman, 1952.
1759.   MONTAGU, M. F. ASHLEY.   "Theognis, Darwin and
         Social Selection." Isis 37 (1947), 24-26.
1760.   MOODY, J. W. T.   "The Reading of the Darwin and
         Wallace Papers: An Historical 'Non-event'."
         Journal of the Society for the Bibliography of
         Natural History 5 (1971), 474-476.
1761.   MOORE, JAMES R.   "Charles Darwin and the Doctrine
         of Man." Evangelical Quarterly 44 (Oct.-Dec.
         1972), 196-217.
1762.   MOORE, RUTH.   Charles Darwin: A Great Life in
         Brief. New York: Alfred A. Knopf, 1955.

1763. MOOREHEAD, ALAN. Darwin and the Beagle. New York:
       Harper & Row, Publishers, 1969.

1764. MORGAN, THOMAS HUNT. "For Darwin." Popular
       Science Monthly 74 (1909), 367-380.

1765. MORRISON, JOHN L. "Orestes Brownson and the
       Catholic Reaction to Darwinism." Duquesne Review 6
       (Spring 1961), 75-87.

1766. MUCKERMAN, H. Attitude of Catholics to Darwinism
       and Evolution. St. Louis, Mo.: B. Herder, 1928.

1767. MÜNTZING, ARNE. "Darwin's Views on Variation under
       Domestication." American Scientist 47 (1959),
       314-325.

1768. MÜNTZING, ARNE. "Darwin's Views on Variation under
       Domestication in the Light of Present-day Know-
       ledge." Proceedings of the American Philosophical
       Society 103 (1959), 190-220.

1769. MULLER, HERBERT J. "Darwinism and Modern Concep-
       tions of Natural Selection." Proceedings of the
       American Philosophical Society 93, No. 6 (1949),
       459-470.

1770. MULLER, HERBERT J. "Reflections on Re-reading
       Darwin." Bulletin of the Atomic Scientists 29, No.
       2 (1973), 5-8.

1771. MURPHY, GARDNER. "Evolution: Charles Darwin."
       Bulletin of the Menninger Clinic 32 (1968), 86-101.

1772. NACHTWEY, ROBERT. Der Irrweg des Darwinismus.
       Berlin: Morus-Verlag, 1959.

1773. NASH, J. V. "The Religious Evolution of Darwin."
       Open Court 42 (1928), 449-463.

1774. NESBITT, HERBERT HUGH JOHN. "Darwinism." In
       Herbert Hugh John Nesbitt (Ed.), Darwin in Retro-
       spect. Toronto: Ryerson Press, 1960, 1-16.

1775. NEWBERRY, J. S. "Darwinian Theory and Religion."
       Scientific Monthly 51 (Oct. 1940), 298.

1776. NURSALL, J. R. "The Consequences of Darwinism."
       Dalhousie Review 42 (Winter 1962-63), 472-480.

1777.   OLBY, ROBERT CECIL.   Charles Darwin.   London:
        Oxford University Press, 1967.

1778.   OLBY, ROBERT CECIL.   "Charles Darwin's Manuscript
        of 'Pangenesis'."   British Journal for the History
        of Science 1 (1962-63), 251-263.

1779.   OMODEO, PIETRO.   "Darwin e l'ereditarietà dei
        caratteri acquisiti."   Scientia 95 (1960), 22-31.

1780.   ONG, WALTER J. (Ed.)   Darwin's Vision and
        Christian Perspectives.   New York: Macmillan &
        Co., 1960.

1781.   ORESME, NICOLAS DE.   "Darwin y Teilhard de Chardin."
        Abside 29 (1965), 336-341.

1782.   OSBORN, HENRY FAIRFIELD.   "Darwin's Theory of
        Evolution by the Selection of Minor Saltations."
        American Naturalist 46 (Feb. 1912), 76-82.

1783.   OSBORN, HENRY FAIRFIELD.   "Life and Works of
        Darwin."   Popular Science Monthly 74 (Apr. 1909),
        314-343.

1784.   OSBORN, HENRY FAIRFIELD.   "A Priceless Darwin
        Letter."   Science 64 (1926), 476-477.   (To T. H.
        Huxley, on the publication of the Origin)

1785.   OSBORN, HENRY FAIRFIELD.   "The Problem of the
        Origin of the Species As It Appeared to Darwin in
        1859 and As It Appears Today."   Nature 118 (1926),
        270-273.

1786.   OSBORNE, E. A.   "The First Edition of 'On the
        Origin of Species'."   Book Collector 9 (1960),
        77-78.

1787.   PANCALDI, GIULIANO.   "'L'economia della natura' da
        Cuvier a Darwin."   Rivista di filosofia 66 (Feb.
        1975), 77-111.

1788.   PANCALDI, GIULIANO.   "Spazio e tempo nella teoria
        darwiniana."   Rivista di filosofia 64, No. 1
        (1973), 3-17.

1789.   PANTIN, C. F. A.   "Darwin's Theory and the Causes
        of Its Acceptance."   The School Science Review 32

(Oct. 1950), 75-83; (Mar. 1951), 197-205; (June
1951), 313-321.

1790.   PANTIN, C. F. A.   One Hundredth Anniversary of
the Publication of Charles Darwin's "Origin of
Species." Cambridge: Christ's College, 1959.

1791.   PANTIN, C. F. A.   Young Darwin and the "Origin
of the [sic] Species." Cambridge: Christ's
College, 1959.

1792.   PAS, PETER W. VAN DER.   "The Correspondence of Hugo
de Vries and Charles Darwin." Janus 57 (1970),
173-213.

1793.   PASSMORE, JOHN A.   "Darwin and the Climate of
Opinion." Australian Journal of Science 22 (1959),
8-15.

1794.   PASSMORE, JOHN A.   "Darwin's Impact on British
Metaphysics." Victorian Studies 3 (Sept. 1959),
41-54.

1795.   PEARSON, KARL.   Charles Darwin, 1809-1882. London:
Cambridge University Press, 1923.

1796.   PEATTIE, D. C.   "Evolution of Charles Darwin."
Reader's Digest 47 (Sept. 1945), 53-57.

1797.   PECKHAM, MORSE.   "Darwinism and Darwinisticism."
Victorian Studies 3 (Sept. 1959), 19-40.

1798.   PECKHAM, MORSE. (Ed.)   The Origin of Species by
Charles Darwin: A Variorum Text. Philadelphia:
University of Pennsylvania Press, 1959.

1799.   PETIT, GEORGES AND JEAN THEODORIDES.   "Quatre
lettres inédites de Darwin à des savants français."
Janus 48 (1959), 208-213.

1800.   PETRONIEVICS, BRANISLAV.   "Charles Darwin und
Alfred Russel Wallace: Beitrag zur höheren Psy-
chologie und zur Wissenschaftsgeschichte." Isis 7
(1925), 25-57.

1801.   PICKERING, GEORGE.   Creative Malady: Illness
in the Lives and Minds of Charles Darwin, Florence
Nightingale, Mary Baker Eddy, Sigmund Freud, Marcel

Proust, Elizabeth Barrett Browning. London: Allen and Unwin, 1974.

1802. PIKE, F. H. "Darwin and Wallace on Sexual Selection and Warning Coloration." Popular Science Monthly 84 (Apr. 1914), 403-410.

1803. PILKINGTON, R. "Darwin and the Christians." Cornhill Magazine 170 (Spring 1958), 52-68.

1804. PLAINE, HENRY L. (Ed.) Darwin, Marx, and Wagner: A Symposium. Columbus: Ohio State University Press, 1962.

1805. PLATONOV, GEORGII VASIL'EVICH. Darvin, darvinizm i filosofia. Moscow, 1959.

1806. PLATT, ROBERT. "Darwin, Mendel, and Galton." Medical History 3 (1959), 87-99.

1807. PLOCHMANN, GEORGE KIMBALL. "Darwin or Spencer?" Science 130 (Nov. 27, 1959), 1452-56.

1808. PORGES, A. "Darwin and Fabre: A Sidelight." School Science and Mathematics 61 (Nov. 1961), 573-578.

1809. PORTMAN, ARNE. (Ed.) "Three Unknown Darwin Letters." Lychnos (1948-49), 206-210.

1810. POYNTER, FREDERICK NOEL LAWRENCE. "Centenary of the Darwin-Wallace Paper on Natural Selection." Science and Culture 25 (1959), 125-129. (Reprinted from British Medical Journal 1, 1958, 1538-40)

1811. PRAEGER, WILLIAM E. (Ed.) "Six Unpublished Letters of Charles Darwin." Papers of the Michigan Academy of Sciences, Arts and Letters 20 (1935), 711-715.

1812. PRENANT, MARCEL. Darwin. Paris: Editions Sociales Internationales, 1938.

1813. QUERNER, HANS. "Ideologisch-weltanschauliche Konsequenzen der Lehre Darwins." Studium Generale: Zeitschrift für die Einheit der Wissenschaften im Zusammenhang ihrer Begriffsbildungen und Forschungsmethoden 24 (1971), 231-245.

1814. RANDALL, JOHN HERMAN, JR.   "The Changing Impact of
       Darwin on Philosophy."   Journal of the History
       of Ideas 22 (1961), 435-462.

1815. RAVEN, CHARLES EARLE.   "Charles Darwin: The Man
       and His Work."   South Atlantic Quarterly 58 (1959),
       421-426.

1816. RAVEN, CHARLES EARLE.   "Darwinism: Past and
       Present."   South Atlantic Quarterly 58 (1959),
       568-571.

1817. RAVEN, CHARLES EARLE.   The History of Science:
       Origins and Results of the Scientific Revolution.
       London: Cohen and West, 1951.

1818. RAVIKOVICH, A. I.   "Idei uniformizma v 'Proiskho-
       shdenii vidov' Charlza Darvina."   Ocherki po
       istorii geologischeskikh znanii 10 (1962), 46-64.
       (Uniformitarian ideas in Darwin's Origin)

1819. REID, LESLIE.   "Darwin and the Galapagos Islands."
       Contemporary Review 194 (Sept. 1958), 139-143.

1820. REID, LESLIE.   "Something Curious and Hitherto
       Unknown."   Quarterly Review 300 (1962), 206-217.
       (A. R. Wallace and Darwin compared)

1821. REZNECK, SAMUEL.   "Notes on a Correspondence be-
       tween Charles Darwin and James Dwight Dana, 1861-
       1863."   Yale University Library Gazette 36 (Apr.
       1962), 176-183.

1822. RICE, EDWARD L.   "Darwin and Bryan: A Study in
       Method."   Science 61 (1925), 243-250.

1823. RICHARDS, H. M.   "Darwin's Work on Movement in
       Plants."   American Naturalist 43 (Mar. 1909),
       152-162.

1824. RICHARDSON, ROBERT ALAN.   "The Development of the
       Theory of Geographical Race Formation: Buffon to
       Darwin."   Dissertation Abstracts 29 (1969), 3059-
       60A (University of Wisconsin, 1968).

1825. RITCHIE, J.   "Conversion of Charles Darwin: Review
       of His Diary of the Voyage of H.M.S. Beagle."
       Science Progress 28 (Apr. 1934), 736-742.

1826.   ROBERTS, H. J.   "Reflections on Darwin's Illness."
        Geriatrics 22, No. 9 (1967), 160-168.

1827.   ROBERTS, H. J.   "Reflections on Darwin's Illness."
        Journal of Chronic Diseases 19 (1966), 723-725.

1828.   ROBERTSON, G. M.   "Darwin's View of Heredity."
        Science 80 (Aug. 10, 1934), 140.

1829.   ROBINSON, GLORIA.   "Theories of a Material Sub-
        stance of Heredity: Darwin to Weismann."   Dis-
        sertation Abstracts International 30 (1970), 3505B
        (Yale University, 1969).

1830.   ROGERS, JAMES ALLEN.   "Darwinism, Scientism and
        Nihilism."   Russian Review 19 (Jan. 1960), 10-23.

1831.   ROMER, A. S.   "Darwin and the Fossil Record."
        Natural History 68 (Oct. 1959), 456-469.

1832.   ROSTAND, JEAN.   Charles Darwin.   Paris: Gallimard,
        1947.

1833.   ROSTAND, JEAN.   "Il y a cent ans Darwin inventait
        une nouvelle façon de voir et de comprendre la
        nature."   Nouvelles Littéraires (Oct. 1, 1959), 1,
        5.

1834.   ROUSSEAU, GEORGES.   "Lamarck et Darwin."   Bulletin
        du Muséum National d'Histoire Naturelle 5 (1969),
        1029-41.

1835.   ROUSSEAU, RICHARD W.   "Secular and Christian Images
        of Man."   Thought 47 (1972), 165-200.   (Includes
        Darwinism)

1836.   RUBAILOVA, N. G.   "Ch. Darvin i problema proisk-
        hozhdeniia cheloveka."   Iz istorii biologii 4
        (1973), 89-111.   (Darwin and the problem of man's
        origin)

1837.   RUBAILOVA, N. G.   "Istoriia sozdaniia i znachenie
        truda Ch. Darvina 'Izmeneniia domashnikh zhivotnykh
        i kul'turnykh rastenii'."   Iz istorii biologi-
        cheskikh nauk 2 (1970), 21-32.   (History and
        importance of Darwin's Variation of Animals and
        Plants under Domestication)

1838. RUDWICK, MARTIN J. S. "Darwin and Glen Roy: A
      'Great Failure' in Scientific Method?" Studies in
      History and Philosophy of Science 5 (1974), 99-185.
1839. RUNKLE, G. "Marxism and Charles Darwin." Journal
      of Politics 23 (Feb. 1961), 108-126.
1840. RUSE, MICHAEL. "Charles Darwin and Artificial
      Selection." Journal of the History of Ideas 36
      (Apr.-June 1975), 339-350.
1841. RUSE, MICHAEL. "Charles Darwin's Theory of Evolu-
      tion: An Analysis." Journal of the History of
      Biology 8 (1975), 219-241.
1842. RUSE, MICHAEL. "Darwin's Debt to Philosophy: An
      Examination of the Influence of the Philosophical
      Ideas of John F. W. Herschel and William Whewell on
      the Development of Charles Darwin's Theory of
      Evolution." Studies in History and Philosophy of
      Science 6 (June 1975), 159-181.
1843. RUSE, MICHAEL. "Natural Selection in the 'Origin
      of Species'." Studies in History and Philosophy of
      Science 1 (1970-71), 311-351.
1844. RUSE, MICHAEL. "The Nature of Scientific Models:
      Formal v. Material Analogy." Philosophy of the
      Social Sciences 3 (1973), 63-80. (Relationship
      between Darwin's Origin and Malthus's Essay)
1845. RUSE, MICHAEL. "The Value of Analogical Models in
      Science." Dialogue: Canadian Philosophical Review
      12 (June 1973), 246-253.
1846. RUSSELL, ARTHUR JAMES. "Darwin." In Their Reli-
      gion. London: Hodder & Stoughton, 1934, 273-297.
1847. RUTHERFORD, HENRY W. Catalog of the Library of
      Charles Darwin. Cambridge: University Press,
      1908.
1848. SANDOW, ALEXANDER. "Social Factors in the Origin
      of Darwinism." Quarterly Review of Biology 13
      (1938), 315-326.

1849.   SARTON, GEORGE.   "Darwin's Conception of the Theory
        of Natural Selection."   Isis 26 (1937), 336-340.

1850.   SARTON, GEORGE.   "Discovery of the Theory of
        Natural Selection."   Isis 14 (1930), 133-154.
        (Includes facsimile reproductions of Darwin's and
        A. R. Wallace's earliest publications on the sub-
        ject)

1851.   SARTON, GEORGE.   "Experiments with Truth by Fara-
        day, Darwin, and Gandhi."   Osiris 11 (1954), 87-
        107.

1852.   SAYRE, S. A.   "Adventures of Charles Darwin."
        Science and Children 12 (Sept. 1974), 19-20.

1853.   SCHEIK, WILLIAM J.   "Epic Traces in Darwin's
        'Origin of Species'."   South Atlantic Quarterly 72
        (1973), 270-279.

1854.   SCHIERBEEK, A.   Darwin's werk en persoonlijkheid.
        Amsterdam: Wereldbiblioteek, 1958.   (Darwin's work
        and personality)

1855.   SCHIERBEEK, A.   "De pangenesistheorie van Darwin."
        Bijdragen tot de Geschiedenis der Geneeskunde 23
        (1943), 29-35.   (Darwin's theory of pangenesis)

1856.   SCHWARTZ, JOEL S.   "Charles Darwin's Debt to
        Malthus and Edward Blyth."   Journal of the History
        of Biology 7 (1974), 301-318.

1857.   SCOON, ROBERT.   "Retrospect to Darwin."   Centennial
        Review of Arts and Science 3 (1959), 376-390.

1858.   SCRIVEN, MICHAEL.   "Explanation and Prediction in
        Evolutionary Theory."   Science 130 (1959), 477-482.

1859.   SEARS, PAUL BIGELOW.   Charles Darwin: The Natur-
        alist as a Cultural Force.   New York: Scribner,
        1950.

1860.   SELSAM, HOWARD.   "Charles Darwin and Karl Marx."
        Mainstream 12, No. 6 (1959), 23-26.

1861.   SEREJSKI, M. H.   "Charles Darwin's Views on His-
        tory."   Scientia 105 (1970), 757-761.

1862. SEWARD, ALBERT CHARLES. (Ed.) Darwin and Modern
       Science: Essays in Commemoration of the Centenary
       of the Birth of Charles Darwin and of the Fiftieth
       Anniversary of the Publication of the Origin of
       Species. Cambridge: University Press, 1909.

1863. SHEPPERSON, GEORGE. "The Intellectual Background
       of Charles Darwin's Student Years at Edinburgh."
       In Michael Banton (Ed.), Darwinism and the Study of
       Society. London: Tavistock Pub., 1961, 17-36.

1864. SHIDELER, EMERSON WAYNE. "Darwin and the Doctrine
       of Man." Journal of Religion 40 (1960), 198-211.

1865. SIMPSON, GEORGE E. "Darwin and 'Social Darwin-
       ism'." Antioch Review 19 (Spring 1959), 33-45.

1866. SIMPSON, GEORGE GAYLORD. "Charles Darwin in Search
       of Himself." Scientific American 199, No. 2
       (1958), 117-122.

1867. SIMPSON, GEORGE GAYLORD. "Lamarck, Darwin and
       Butler." American Scholar 30 (1961), 238-249.

1868. SIMPSON, GEORGE GAYLORD. "The World into Which
       Darwin Led Us." Science 131 (1960), 966-974.
       (Discussion: 131, June 17, 1960, 1820-22)

1869. SINETY, ROBERT DE. "Un Demi-Siècle de Darwinisme."
       Revue des Questions Scientifiques 67 (1910), 5-38,
       480-513.

1870. SMITH, DAVID S. "Inheritance." London Quarterly
       and Holborn Review 28 (1959), 299-305.

1871. SMITH, SYDNEY. "The Darwin Collection at Cambridge
       with One Example of Its Use: Charles Darwin and
       Cirripedes." Actes du XIe Congrès International
       d'Histoire des Sciences (Warsaw, 1965), 5 (1968),
       96-100.

1872. SMITH, SYDNEY. "The Origin of 'The Origin', as
       Discerned from Charles Darwin's Notebooks and His
       Annotations in the Books He Read between 1837 and
       1842." Advancement of Science 16 (1960), 391-401.

1873. SNODDY, ELMER ELLSWORTH. "Darwin and Calvin: Natural Selection and Supernatural Selection." Christian Century 40 (July 12, 1923), 874-876.

1874. SOBOL', S. L. "Ch. Darwin's Evolutionary Conception at the Early Stages of Its Formation." Annals of Biology, Moscow Society of Natural History, Section in the History of Natural Sciences 1 (1959), 13-34. (In Russian; English summary)

1875. SOBOL', S. L. "Darwin's Achievements." Voprosy istorii estestvoznaniia i tekhniki 8 (1959), 57-65. (In Russian)

1876. SOMKIN, FRED. "The Contributions of Sir John Lubbock, F.R.S. to the 'Origin of Species': Some Annotations to Darwin." Notes and Records of the Royal Society of London 17, No. 2 (Dec. 1962), 183-191.

1877. SPENCER, T. J. B. From Gibbon to Darwin. Birmingham: University of Birmingham Press, 1959.

1878. SPILLER, GUSTAV. "Charles Darwin and the Theory of Evolution: A Sociological Study." Sociological Review 18 (Apr. 1926), 110-130.

1879. SPILSBURY, RICHARD. Providence Lost: A Critique of Darwinism. London and New York: Oxford University Press, 1974.

1880. STACKHOUSE, REGINALD FRANCIS. "Darwin and a Century of Conflict." Christian Century 76 (Aug. 19, 1959), 944-946. (Discussion: 76, Sept. 30, 1959, 1121-22)

1881. STAUFFER, ROBERT C. (Ed.) Charles Darwin's 'Natural Selection': Being the Second Part of His Big Species Book Written from 1856 to 1858. London: Cambridge University Press, 1975.

1882. STAUFFER, ROBERT C. "Ecology in the Long Manuscript Version of Darwin's 'Origin of Species' and Linnaeus' 'Oeconomy of Nature'." Proceedings of the American Philosophical Society 114 (1960), 235-241.

1883.   STAUFFER, ROBERT C.   "Haeckel, Darwin, and Ecology."
        Quarterly Review of Biology 32 (1957), 138-144.

1884.   STAUFFER, ROBERT C.   "'On the Origin of Species':
        An Unpublished Version." Science 130 (1959),
        1449-52.

1885.   STECHER, ROBERT M.   "The Darwin-Bates Letters:
        Correspondence between Two Nineteenth-Century
        Travellers and Naturalists." Annals of Science 25
        (June 1969), 1-47, 95-125.

1886.   STECHER, ROBERT M.   "The Darwin-Innes Letters: The
        Correspondence of an Evolutionist with His Vicar,
        1848-1884." Annals of Science 17 (Dec. 1961),
        201-258.

1887.   STERN, BERNHARD JOSEPH.   "Darwin on Spencer."
        Scientific Monthly 26 (1928), 180-181.

1888.   STEVENSON, J. J.   "Darwin and Geology." Popular
        Science Monthly 74 (Apr. 1909), 349-354.

1889.   STEVENSON, LIONEL. Darwin among the Poets.
        Chicago: University of Chicago Press, 1932.

1890.   STEVENSON, ROBERT SCOTT. Famous Illnesses in
        History. London: Eyre and Spottiswoode, 1962.

1891.   STODDARD, D. R.   "Darwin's Impact on Geography." In
        Wayne K. D. Davies (Ed.), The Conceptual Revolution
        in Geography. London: University of London Press,
        1972, 52-76. (Also in Annals of the Association of
        American Geographers 56, Dec. 1966, 683-698)

1892.   STRAELEN, VICTOR VAN.   "L'Incidence de l'oeuvre de
        Darwin sur le progrès de la géologie." Annales
        de la Société Zoologique de Belgique 90 (1959-60),
        19-25.

1893.   STRUVE, O.   "Charles Darwin and the Problem of
        Stellar Evolution." Sky and Telescope 18 (Mar.
        1959), 240-243.

1894.   SWISHER, CHARLES N.   "Charles Darwin on the Origins
        of Behaviour." Bulletin of the History of Medicine
        41 (1967), 24-43.

1895.   SYMOENS, J. J. et al.   Actualité de Darwin.   Brux-
        elles: Les Naturalistes Belges, 1960.
1896.   TASCH, PAUL.   "Darwin and the Forgotten Mr. Lons-
        dale."   Geological Magazine 87 (1950), 292-296.
1897.   TAX, SOL. (Ed.)   Evolution after Darwin.   3 vols.
        Chicago: University of Chicago Press, 1960.
1898.   TEIDMAN, S. J.   "Darwin's Reverend Friend."   Modern
        Churchman 6 (July 1963), 286-290.   (Leonard Jenyns)
1899.   TENNANT, F. R.   "Influence of Darwinism upon Theo-
        logy."   Quarterly Review 211 (Oct. 1909), 418-440.
1900.   THEODORIDES, JEAN.   "Humboldt et Darwin."   Actes
        du XIe Congrès International d'Histoire des
        Sciences (Warsaw, 1965), 5 (1968), 87-92.
1901.   TINKLE, W. J.   "Darwin versus Experimental Bio-
        logy."   School Science and Mathematics 47 (Apr.
        1947), 369-372.
1902.   TODD, WILLIAM B.   "Variant Issues of 'On the Origin
        of Species', 1859."   Book Collector 9 (1960), 78.
1903.   TOLPIN, MARTHA.   "The Darwinian Influence on
        Psychological Definitions of Feminity in England,
        1871-1914."   Ph.D. Dissertation, Harvard University,
        1972.
1904.   TOWERS, BERNARD.   "The Impact of Darwin's 'Origin
        of Species' on Medicine and Biology."   In Medicine
        and Science in the 1860s: Proceedings of the Sixth
        British Congress on the History of Medicine, Uni-
        versity of Sussex, 6-9 September 1967.   London:
        Wellcome Institute of the History of Medicine,
        1968, 45-55.
1905.   TRELEASE, WILLIAM.   "Darwin's Work on Cross Pol-
        lination in Plants."   American Naturalist 43 (Mar.
        1909), 131-142.
1906.   TRENN, THADDEUS J.   "Charles Darwin, Fossil Cir-
        ripedes, and Robert Fitch: Presenting Sixteen
        Hitherto Unpublished Darwin Letters of 1849-1851."
        Proceedings of the American Philosophical Society
        118 (1974), 471-491.

1907.   TRIFILO, S.   "Darwin and the Second Beagle Expedi-
        tion in Tierra del Fuego."   Pacific Historical
        Review 29 (1960), 222-229.

1908.   TRIFILO, S.   "Naturalist's Delight: Darwin in
        South America."   Américas 14 (June 1962), 14-19.

1909.   TSCHULOCK, S.   "Über Darwins Selektionslehre:
        Historisch-kritische Betrachtungen."   Viertel-
        jahrsschrift der naturforschenden Gesellschaft in
        Zürich 81 (1936), 1-68.

1910.   TUTIN, T. G.   "Centenary of 'The Origin of Species':
        Some Botanical Aspects."   Nature 185 (1960), 216-
        217.

1911.   TUTTLE, RUSSELL.   "Darwin's Apes, Dental Apes, and
        the Descent of Man: Normal Science in Evolutionary
        Anthropology."   Current Anthropology 15 (1974),
        389-398.

1912.   TWIESSELMANN, FRANCOIS.   "Darwin et les causes de
        l'évolution de l'homme."   Annales de la Société
        Zoologique de Belgique 90 (1959-60), 27-35.

1913.   UEXKÜLL, J. VON.   "Darwin und die englische Moral."
        Deutsche Rundschau 173 (Nov. 1917), 215-242.

1914.   UNGERER, EMIL.   Lamarck-Darwin: Die Entwicklung
        des Lebens. Stuttgart: Frommann, 1923.

1915.   VALLAUX, CAMILLE.   "Deux précurseurs de la géo-
        graphie humaine: Volney et Charles Darwin."
        Revue de Synthèse 15 (1938), 81-93.

1916.   VANDERPOOL, HAROLD Y.   "Charles Darwin and Dar-
        winism."   In Roger A. Johnson et al.   Critical
        Issues in Modern Religion. Englewood Cliffs, New
        Jersey: Prentice-Hall, 1973, 77-113.

1917.   VANDERPOOL, HAROLD Y. (Ed.)   Darwin and Darwinism:
        Revolutionary Insights Concerning Man, Nature,
        Religion, and Society. Lexington, Mass.: Heath,
        1973.

1918.   VILLIERS, ALAN.   "In the Wake of Darwin's 'Beagle'."
        National Geographic Magazine 136, No. 4 (Oct.
        1969), 449-495.

1919. VON HAGEN, VICTOR WOLFGANG. South America Called
      Them: Explorations of the Great Naturalists.
      London: R. Hale, 1949. (Includes Richard Spruce)
1920. VORZIMMER, PETER J. "Charles Darwin and Blending
      Inheritance." Isis 54 (1963), 371-390.
1921. VORZIMMER, PETER J. Charles Darwin: The Years
      of Controversy. The "Origin of Species" and Its
      Critics, 1859-1882. Philadelphia: Temple Uni-
      versity Press, 1970.
1922. VORZIMMER, PETER J. "Darwin and Mendel: The
      Historical Connection." Isis 59 (1968), 77-82.
1923. VORZIMMER, PETER J. "Darwin, Malthus, and the
      Theory of Natural Selection." Journal of the
      History of Ideas 30 (1969), 527-542.
1924. VORZIMMER, PETER J. "Darwin's Ecology and Its
      Influence upon His Theory." Isis 56 (1965), 148-
      155.
1925. VORZIMMER, PETER J. "Darwin's 'Lamarckism' and the
      'Flat-Fish Controversy' (1863-1871)." Lychnos
      (1969-70), 121-170.
1926. VORZIMMER, PETER J. "Darwin's 'Questions about the
      Breeding of Animals' (1839)." Journal of the
      History of Biology 2 (1969), 269-281.
1927. VRIES, HUGO DE. "Darwin's bezoek aan de Galapagos-
      eilanden." Gids 73, No. 4 (1909), 386-394.
1928. WALLACE, R. W. "Charles Darwin." Journal of
      Education 69 (Feb. 11, 1909), 146-147.
1929. WARD, CHARLES HENSHAW. Charles Darwin: The Man
      and His Warfare. Indianapolis: Bobbs-Merrill,
      1927.
1930. WEISMANN, AUGUST. "Charles Darwin." Contemporary
      Review 96 (1909), 1-22.
1931. WELLS, GEOFFREY HARRY (GEOFFREY WEST). Charles
      Darwin: The Fragmentary Man. London: G. Rout-
      ledge and Sons, Ltd., 1937. (Also published as
      Charles Darwin: A Portrait. New Haven: Yale
      University Press, 1938)

1932. WERNHAM, J. C. S. "The Religious Controversy." In
       Herbert Hugh John Nesbitt (Ed.), Darwin in Retro-
       spect. Toronto: Ryerson Press, 1960, 17-34.

1933. WEST, ANTHONY. Principles and Persuasions: The
       Literary Essays of Anthony West. New York: Har-
       court and Brace, 1957. (Includes T. H. Huxley)

1934. WICHLER, GERHARD. Charles Darwin: The Founder
       of the Theory of Evolution and Natural Selection.
       New York: Pergamon Press, 1961.

1935. WICHLER, GERHARD. "Darwin als Botaniker."
       Sudhoffs Archiv für Geschichte der Medizin 44
       (1960), 289-313.

1936. WILBER, CHARLES G. "Charles Darwin - The Man As He
       Was." The Texas Quarterly 3 (1960), 104-112.

1937. WILLEY, BASIL. Darwin and Butler: Two Versions
       of Evolution. London: Chatto & Windus, 1960.

1938. WILSON, AMBROSE J. "What Charles Darwin Really
       Found." Princeton Theological Review 26 (Oct.
       1928), 515-530.

1939. WILSON, JOHN A. "Darwinian Natural Selection and
       Vertebrate Paleontology." Journal of the Washing-
       ton Academy of Sciences 49 (1959), 231-233.

1940. WILSON, JOHN B. "Darwin and the Transcendental-
       ists." Journal of the History of Ideas 26 (1965),
       286-290.

1941. WINDLE, SIR BERTRAM COGHILL ALAN. "Darwin and
       Darwinism and Certain Other Isms." Catholic World
       95 (Apr. 1912), 1-10.

1942. WINDLE, SIR BERTRAM COGHILL ALAN. "Darwin and the
       Theory of Natural Selection." Dublin Review 150
       (Apr. 1912), 307-324.

1943. WINSLOW, JOHN H. Darwin's Victorian Malady: Evi-
       dence for Its Medically Induced Origin. Phila-
       delphia: American Philosophical Society, 1971.

1944. WOLSKY, ALEXANDER. "A Hundred Years of Darwinism
       in Biology." Thought 24 (1959), 165-184.

1945. WOODFIELD, ANDREW. "Darwin, Teleology and Tax-
       onomy." Philosophy 48 (1973), 35-49.
1946. WOODRUFF, A. W. "Darwin's Health in Relation to
       His Voyage to South America." British Medical
       Journal No. 5437 (Mar. 20, 1965), 745-750.
1947. WOODRUFF, A. W. "The Impact of Darwin's Voyage to
       South America on His Work and Health." Bulletin
       of the New York Academy of Medicine 44 (1968),
       661-672.
1948. WYNN-TYSON, ESME. "Darwinism and Spiritual Evolu-
       tion." Contemporary Review 196 (1959), 234-236.
1949. WYSS, WALTER VON. Charles Darwin: Ein Forscher-
       leben. Zürich: Artemis-Verlag, 1958.
1950. YANCEY, P. H. "Darwin-Wallace: A Century After."
       America 99 (Sept. 20, 1958), 636.
1951. YOKOYAMA, TOSHIAKI. "The Influence of Theological
       Thought on Charles Darwin: Consideration of the
       Relation between William Paley and Charles Darwin."
       Kagakusi Kenkyu 10 (1971), 49-59. (In Japanese;
       English summary)
1952. YONGE, C. M. "Darwin's Achievement." Spectator
       155 (Sept. 6, 1935), 349-350.
1953. YOUNG, ROBERT MAXWELL. "Darwin's Metaphor: Does
       Nature Select?" The Monist 55 (1971), 442-503.
1954. YOUNG, ROBERT MAXWELL. "Malthus and the Evolu-
       tionists: The Common Context of Biological and
       Social Theory." Past and Present No. 43 (May
       1969), 109-145.
1955. YOUNG, ROBERT MAXWELL. "'Non-Scientific' Factors
       in the Darwinian Debate." Actes du XIIe Congrès
       International d'Histoire des Sciences (Paris,
       1968), 8 (1971), 221-226.
1956. ZAPPLER, G. "Darwin's Worms." Natural History 67
       (Nov. 1958), 488-495.
1957. ZIMMERMAN, PAUL A. (Ed.) Darwin, Evolution and
       Creation. St. Louis: Concordia Publishing House,
       1959. (By Lutheran scholars)

1958.   ZIRKLE, CONWAY.   "Further Notes on Pangenesis and
         the Inheritance of Acquired Characters."   American
         Naturalist 70 (1936), 529-546.

1959.   ZIRKLE, CONWAY.   "The Inheritance of Acquired
         Characters and the Provisional Hypothesis of
         Pangenesis."   American Naturalist 69 (1935), 417-
         445.

1960.   ZIRNSTEIN, GOTTFRIED.   Charles Darwin.   Leipzig:
         B. G. Teubner, 1974.

1961.   ZIRNSTEIN, GOTTFRIED.   "Das Problem der Artverän-
         derung von Darwin bis zur Begründung der Mutations-
         theorie."   Zeitschrift für Geschichte der Naturwis-
         senschaft, Technik und Medizin 11, No. 2 (1974),
         82-94.

1962.   ZUIDEMA, H. P. (Ed.)   "Discovery of Letters by
         Lyell and Darwin."   Journal of Geology 55 (Sept.
         1947), 439-445.   (Alternate glaciation in the
         northern and southern hemispheres)

5.   Wallace, Alfred Russel

See III.   Ideas . . .; VII.   Evolution (other
sections, especially 4.   Darwin, Charles Robert).

1963.   ANON.   "Alfred Russel Wallace."   Nature 92 (Nov. 20,
         1913), 347-349.

1964.   BAUMEL, H. B.   "Alfred Wallace: Man in a Shadow."
         Science Teacher 43 (Apr. 1976), 29-30.

1965.   BEDDALL, BARBARA G.   "Wallace, Darwin and Edward
         Blyth: Further Notes on the Development of Evolu-
         tion Theory."   Journal of the History of Biology 5
         (1972), 153-158.

1966.   BEDDALL, BARBARA G.   "Wallace, Darwin, and the
         Theory of Natural Selection."   Journal of the
         History of Biology 1 (1968), 261-323.

1967.   BROOKS, JOHN L.   "Re-assessment of A. R. Wallace's
        Contribution to the Theory of Organic Evolution."
        American Philosophical Society Yearbook (1968),
        534-535.
1968.   CARSE, ROBERT.   The Castaways: A Narrative History
        of Some Survivors from the Dangers of the Sea.
        Chicago: Rand McNally, 1966.
1969.   EISELEY, LOREN COREY.   "Alfred Russel Wallace."
        Scientific American 200 (Feb. 1959), 70-82.
1970.   GEORGE, WILMA.   Biologist Philosopher: A Study of
        the Life and Writings of Alfred Russel Wallace.
        London: Abelard-Schuman, 1964.
1971.   GRASSE, PIERRE-P.   "Lamarck, Wallace et Darwin."
        Revue d'Histoire des Sciences et de leurs Applica-
        tions 13 (1960), 73-79.
1972.   HOGBEN, LANCELOT THOMAS.   Alfred Russel Wallace:
        The Story of a Great Discoverer.   London: Society
        for Promoting Christian Knowledge, 1918.
1973.   KOTTLER, MALCOLM JAY.   "Alfred Russel Wallace, the
        Origin of Man, and Spiritualism."   Isis 65 (1974),
        145-192.
1974.   MCKINNEY, HENRY LEWIS.   "Alfred Russel Wallace and
        the Discovery of Natural Selection."   Journal of
        the History of Medicine 21 (1966), 333-357.
1975.   MCKINNEY, HENRY LEWIS.   Wallace and Natural Selec-
        tion.   New Haven and London: Yale University
        Press, 1972.
1976.   MCKINNEY, HENRY LEWIS.   "Wallace's Earliest
        Observations on Evolution: 28 December 1845."
        Isis 60 (1969), 370-373.
1977.   MOOG, F.   "Alfred Russel Wallace: Evolution's
        Forgotten Man."   American Biology Teacher 22 (Oct.
        1960), 414-418.
1978.   OSBORN, HENRY FAIRFIELD.   "Alfred Russel Wallace
        (1823-1913)."   Popular Science Monthly 83 (1913),
        523-537.

1979. PANTIN, C. F. A. "Alfred Russel Wallace, F.R.S.,
       and His Essays of 1858 and 1855." Notes and
       Records of the Royal Society of London 14 (1959),
       67-84.

1980. PANTIN, C. F. A. "Alfred Russel Wallace: His
       Pre-Darwinian Essay of 1855." Proceedings of the
       Linnean Society of London 171 (1960), 139-153.

1981. PICKENS, A. L. "Letters Concerning Wallace, Newton
       and Tristram." Auk 51 (July 1934), 404-406.

1982. POULTON, SIR EDWARD B. "Alfred Russel Wallace."
       Proceedings of the Royal Society of London, Series
       B, 95 (1924), i-xxv.

1983. REID, LESLIE. "Something Curious and Hitherto
       Unknown." Quarterly Review 300 (1962), 206-217.
       (Wallace compared with Darwin)

1984. RITCHIE, J. "Wallace's Line and the Distribution
       of Mammals." Nature 136 (Aug. 31, 1935), 325-326.

1985. SMITH, ROGER. "Alfred Russel Wallace: Philosophy
       of Nature and Man." British Journal for the History
       of Science 6 (1972-73), 177-199.

1986. SMITH, SUSY. "Alfred R. Wallace: Scientific
       Enthusiast." Tomorrow 8 (1960), 95-105.

1987. STERN, BERNHARD JOSEPH. "Letters of Alfred Russel
       Wallace to Lester F. Ward." Scientific Monthly 40
       (1935), 375-379.

1988. WICHLER, GERHARD. "Alfred Russel Wallace (1823-
       1913), sein Leben, seine Arbeiten, sein Wesen:
       Zugleich ein Beitrag zu dem Verhältnis von Wallace
       zu Darwin." Sudhoffs Archiv für Geschichte der
       Medizin 30 (1938), 364-400.

1989. WILLIAMS-ELLIS, AMABEL. Darwin's Moon: A Bio-
       graphy of Alfred Russel Wallace. London: Blackie,
       1966.

VIII. Evolution and Social and Political Thought

1. General

    This section contains material on the impact of
evolution on social attitudes and theories; many entries
are concerned with the meaning of Social Darwinism.
Selected studies of a number of social thinkers are also
included, e.g., L. T. Hobhouse and E. A. Westermarck. For
all sections under this category, see VII. Evolution, 1.
General; III. Ideas . . .; and categories IX-XIII. (For
an explanation of the grouping of works under individual
authors, see general Introduction.)

1990. ABRAMS, PHILIP. (Ed.) The Origins of British
        Sociology: 1834-1914. Chicago: University of
        Chicago Press, 1968. (Includes H. Spencer, F.
        Galton, L. T. Hobhouse)
1991. BANNISTER, ROBERT C. "'The Survival of the Fittest
        Is Our Doctrine': History or Histrionics." Journal
        of the History of Ideas 31 (1970), 377-398.
1992. BANTON, MICHAEL. (Ed.) Darwinism and the Study of
        Society: A Centenary Symposium. London: Tavi-
        stock Publications, 1961.
1993. BOCK, KENNETH ELLIOTT. "Darwin and Social Theory."
        Philosophy of Science 22 (Apr. 1955), 123-134.
1994. BRANFORD, VICTOR. "The Sociological Work of Leonard
        Hobhouse." Sociological Review 21 (1929), 273-280.
1995. BURNS, TOM. "Darwinism and the Study of Society."
        Nature 183 (1959), 1562-64.
1996. BURROW, JOHN W. Evolution and Society: A Study
        in Victorian Social Theory. Cambridge: The Cam-
        bridge University Press, 1966. (London: Cambridge
        University Press, 1970)

1997. CARTER, HUGH. The Social Theories of L. T. Hob-
       house. Chapel Hill: The University of North
       Carolina Press; London: H. Milford, Oxford Uni-
       versity Press, 1927.

1998. EDDY, L. K. "Education and Social Darwinism."
       Educational Theory 19 (Winter 1969), 76-87.

1999. ELLWOOD, CHARLES A. "Influence of Darwin on
       Sociology." Psychological Review 16 (May 1909),
       188-194.

2000. FARRALL, LYNDSAY ANDREW. "The Origins and Growth
       of the English Eugenics Movement, 1865-1925."
       Dissertation Abstracts International 31 (1970),
       1187A (Indiana University, 1970). (Includes K.
       Pearson, F. Galton)

2001. GINSBERG, MORRIS. Reason and Unreason in Society:
       Essays in Sociology and Social Philosophy. London:
       Heinemann, 1947. (Includes L. T. Hobhouse, E.
       Westermarck)

2002. GORDON, SCOTT. "Darwinism and Social Thought." In
       Herbert Hugh John Nesbitt (Ed.), Darwin in Retro-
       spect. Toronto: Ryerson Press, 1960, 49-66.

2003. GRACE, EMILY R., M. F. ASHLEY MONTAGU AND BERNHARD
       JOSEPH STERN. "More on Social Darwinism." Science
       and Society 6 (1942), 71-78.

2004. HALLIDAY, R. J. "Social Darwinism: A Definition."
       Victorian Studies 14 (June 1971), 389-405.

2005. HANS, N. "Historical Evolution of International-
       ism: The Contribution of Biology, Darwin and the
       Theory of Evolution." Year Book of Education
       (1964), 30-31.

2006. HOBSON, JOHN ATKINSON AND MORRIS GINSBERG. L. T.
       Hobhouse: His Life and Work. London: G. Allen
       and Unwin Ltd., 1931.

2007. JONES, GRETA J. "Darwinism and Social Thought: A
       Study of the Relationship between Science and the
       Development of Sociological Theory in Britain,

1860-1914." Ph.D Dissertation, London School of Economics, 1974.

2008.   KARDINER, ABRAM AND EDWARD PREBLE. They Studied Man. Cleveland: World Publishing Co., 1961. (Includes Darwin, H. Spencer, E. B. Tylor, J. G. Frazer)

2009.   KELLER, CHESTER Z. "An Examination of L. T. Hobhouse's Conception of Morals in Evolution with Special Reference to Cultural Anthropology." Ph.D. Dissertation, University of Southern California, 1958.

2010.   MCCONNAUGHEY, GLORIA. "Darwin and Social Darwinism." Osiris 9 (1950), 397-412.

2011.   MACRAE, DONALD G. "Darwinism and the Concept of Social Evolution." British Journal of Sociology 10 (June 1959), 105-113.

2012.   MARIZ, GEORGE ERIC. "L. T. Hobhouse as a Theoretical Sociologist." Albion 6 (1974), 307-319.

2013.   MARIZ, GEORGE ERIC. "The Life and Work of L. T. Hobhouse: A Study in the History of Ideas." Dissertation Abstracts International 31 (1971), 4092A (University of Missouri - Columbia, 1970).

2014.   MAZRUI, ALI A. "From Social Darwinism to Current Theories of Modernization: A Tradition of Analysis." World Politics 21 (Oct. 1968), 69-83.

2015.   MOORE, LEWIS DURWARD. "A Study of George Gissing and Social Darwinism with Special Emphasis on 'New Grub Street' and 'The Private Papers of Henry Ryecroft'." Dissertation Abstracts International 35 (1974), 2233A (American University, 1974).

2016.   NICHOLSON, JOHN ANGUS. Some Aspects of the Philosophy of L. T. Hobhouse: Logic and Social Theory. Urbana: University of Illinois, 1928.

2017.   OWEN, JOHN E. L. T. Hobhouse, Sociologist. London: Nelson, 1974.

2018.   ROGERS, JAMES ALLEN.   "Darwinism and Social Dar-
        winism."   Journal of the History of Ideas 33
        (1972), 265-280.

2019.   SIMPSON, GEORGE E.   "Darwin and 'Social Darwin-
        ism'."   Antioch Review 19 (Spring 1959), 33-45.

2020.   WELLS, DAVID COLLIN.   "Social Darwinism."   American
        Journal of Sociology 12 (Mar. 1907), 695-708.

2021.   WERNER, STEVEN EDWIN.   "Reason and Reform: The
        Social Thought of Leonard T. Hobhouse."   Disserta-
        tion Abstracts International 34 (1974), 5892A
        (University of Wisconsin, 1973).

2022.   WILLIAMS, RAYMOND.   "Social Darwinism."   Listener
        88 (1972), 696-700.

2023.   ZACHARIAH, M.   "The Impact of Darwin's Theory of
        Evolution on Theories of Society."   Social Studies
        62 (Feb. 1971), 69-77.

2. Bagehot, Walter

See introduction to section 1.

2024.   AMES, ROBERT JAMES.   "Walter Bagehot: A Study in
        Religious Compromise."   Dissertation Abstracts 12
        (1952), 419-420 (University of Minnesota, 1952).

2025.   BARNES, HARRY ELMER.   "Walter Bagehot (1826-77) and
        the Psychological Interpretation of Political
        Evolution."   American Journal of Sociology 27 (Mar.
        1922), 573-581.

2026.   BUCHAN, ALASTAIR.   The Spare Chancellor: The
        Life of Walter Bagehot.   London: Chatto & Windus,
        1959.

2027.   COWLES, T.   "Malthus, Darwin and Bagehot: A Study
        in the Transference of a Concept."   Isis 26 (1937),
        341-348.

2028. CROSSMAN, R. H. S. "Walter Bagehot." Encounter 20
       (Mar. 1963), 42-55; (Apr. 1963), 17-26.

2029. HALSTED, JOHN BURT. "The Idea of Agitation: A
       Study of the Social and Political Ideas of Walter
       Bagehot." Dissertation Abstracts 14 (1954), 1373-
       74 (Columbia University, 1954).

2030. HALSTED, JOHN BURT. "Walter Bagehot on Tolera-
       tion." Journal of the History of Ideas 19 (1958),
       119-128.

2031. IRVINE, WILLIAM. Walter Bagehot. London and New
       York: Longmans, Green and Co., 1939.

2032. NASR, MOHAMMED ABDULL-MUIZZ. Walter Bagehot: A
       Study in Victorian Ideas. Alexandria: Alexandria
       University Press, 1959.

2033. ST. JOHN-STEVAS, NORMAN. (Ed.) Bagehot's Histor-
       ical Essays. New York: New York University Press,
       1965.

2034. ST. JOHN-STEVAS, NORMAN. Walter Bagehot: A Study
       of His Life and Thought Together with a Selection
       from His Political Writings. Bloomington: Indiana
       University Press, 1959.

2035. SISSON, CHARLES HUBERT. The Case of Walter Bagehot.
       London: Faber & Faber Ltd., 1972.

2036. SMITH, ADRIAN. "Walter Bagehot." Dublin Review
       507 (Spring 1966), 48-59.

2037. SULLIVAN, HARRY R. Walter Bagehot. Boston:
       Twayne, 1975.

2038. YOUNG, GEORGE MALCOLM. "The Greatest Victorian."
       Spectator 158 (June 18, 1937), 1137-38; 159 (July
       2, 1937), 9-10.

3. Galton, Francis

See introduction to section 1; VII. Evolution,
1. General; XIII. Evolution and Psychology.

2039. ANON. "Milestone: Sir Francis Galton (1822-1911)."
Science Digest 28 (Nov. 1950), 89-92.

2040. ANON. "Subject Mind Explores Object Mind: An
Unfixated Psychology." Times Literary Supplement
No. 3054 (Sept. 9, 1960), lxv.

2041. BLACKER, CHARLES PATON. Eugenics, Galton and
After. London: Duckworth, 1952.

2042. BLACKER, CHARLES PATON. "Galton on Eugenics as
Science and Practice." Eugenics Review 38 (Jan.
1947), 240-241.

2043. BLACKER, CHARLES PATON. "Galton's Outlook on
Religion." Eugenics Review 38 (July 1946), 69-78.

2044. BLACKER, CHARLES PATON. "Galton's Views on Race."
Eugenics Review 43 (Apr. 1951), 19-22.

2045. BRAMWELL, B. S. "Galton's 'Hereditary Genius' and
the Three Following Generations since 1869."
Eugenics Review 39 (Jan. 1948), 146-153.

2046. BROMBERG, WALTER AND J. K. WINKLER. "Galton:
Student of Mankind." Science Digest 7 (Feb. 1940),
85-91.

2047. BURT, CYRIL. "Francis Galton and His Contributions
to Psychology." British Journal of Statistical
Psychology 15 (1962), 1-41.

2048. BURT, CYRIL. "Galton's Contribution to Psychology."
Bulletin of the British Psychological Society 45
(1961), 10-21.

2049. CASTLE, W. E. "The Laws of Heredity of Galton and
Mendel, and Some Laws Governing Race Improvement by
Selection." Proceedings of the American Academy of
Arts and Sciences 39 (1903), 223-242.

2050. CORNING, CONSTANCE HELLYER. "Francis Galton and
      Eugenics." History Today 23 (Oct. 1973), 724-732.
2051. COWAN, RUTH SCHWARTZ. "Francis Galton's Contribu-
      tion to Genetics." Journal of the History of
      Biology 5 (1972), 389-412.
2052. COWAN, RUTH SCHWARTZ. "Francis Galton's Statis-
      tical Ideas: The Influence of Eugenics." Isis 63
      (1972), 509-528.
2053. COWAN, RUTH SCHWARTZ. "Sir Francis Galton and the
      Continuity of Germ-Plasm: A Biological Idea with
      Political Roots." Actes du XIIe Congrès Interna-
      tional d'Histoire des Sciences (Paris, 1968), 8
      (1971), 181-186.
2054. COWAN, RUTH SCHWARTZ. "Sir Francis Galton and the
      Study of Heredity in the Nineteenth Century."
      Dissertation Abstracts International 32 (1972),
      6883A (Johns Hopkins, 1969).
2055. CRACKENTHORPE, MONTAGUE. "Sir Francis Galton,
      F.R.S.: A Memoir." Eugenics Review 3 (1912), 1-9.
2056. DARWIN, FRANCIS. "Francis Galton, 1822-1911."
      Eugenics Review 6 (1914-15), 1-17.
2057. DE MARRAIS, ROBERT. "The Double-edged Effect of
      Sir Francis Galton: A Search for the Motives in
      the Biometrician-Mendelian Debate." Journal of
      the History of Biology 7 (1974), 141-174.
2058. EAST, E. M. "Centenary of Gregor Mendel and of
      Francis Galton." Scientific Monthly 16 (Mar.
      1923), 225-268.
2059. FORREST, D. W. Francis Galton: The Life and Work
      of a Victorian Genius. London: Elek, 1974.
2060. FROGGATT, P. AND N. C. NEVIN. "Galton's 'Law of
      Ancestral Heredity': Its Influence on the Early
      Development of Human Genetics." History of Science
      10 (1971), 1-27.
2061. FROGGATT, P. AND N. C. NEVIN. "The 'Law of An-
      cestral Heredity' and the Mendelian-Ancestrian

Controversy in England, 1889-1906." _Journal of Medical Genetics_ 8 (1971), 1-36.

2062. GARRETT, HENRY EDWARD.   _Great Experiments in Psychology_.   New York, London: The Century Co., 1930.

2063. GATES, R. R.   "Galton and Discontinuity in Variation."   _American Naturalist_ 48 (Nov. 1914), 697-699.

2064. HILTS, VICTOR L.   _A Guide To Francis Galton's "English Men of Science_."   Philadelphia: American Philosophical Society, 1975.

2065. KANAEV, IVAN I.   "F. Gal'ton, i bliznetsovyi metod genetiki."   _Iz Istorii Biologii_ 3 (1971), 193-201.   (Galton and the Twin Genetic Method)

2066. KANAEV, IVAN I.   _Frensis Gal'ton, 1822-1911_.   Leningrad: Nauka, 1972.

2067. KEITH, SIR ARTHUR.   "Galton's Place among Anthropologists."   _Eugenics Review_ 12 (Apr. 1920), 14-28.

2068. MORSIER, G. DE.   "Correspondance inédite entre Alphonse de Candolle (1806-1893) et Francis Galton (1822-1911)."   _Gesnerus_ 29 (1972), 129-160.

2069. MÜLLER, D.   "Drei Briefe über reine Linien von Galton, de Vries und Yule an Wilhelm Johannsen in 1903 geschrieben."   _Centaurus_ 16 (1972), 316-319.

2070. NEWMAN, JAMES R.   "Francis Galton."   _Scientific American_ 190 (Jan. 1954), 72-76.

2071. NORTON, B. J.   "The Biometric Defense of Darwinism."   _Journal of the History of Biology_ 6 (1973), 283-316.   (Includes K. Pearson)

2072. OLBY, ROBERT CECIL.   "Francis Galton's Derivation of Mendelian Ratios in 1875."   _Heredity_ 20 (1965), 636-638.

2073. PARKES, ALAN S.   "The Galton-Korosi Correspondence."   _Journal of Biosocial Science_ 3 (1971), 461-472.

2074. PLATT, ROBERT.   "Darwin, Mendel, and Galton."   _Medical History_ 3 (1959), 87-99.

2075. REEVE, E. GAVIN.   "A Comment on Some of Sir Francis Galton's Observations and Inferences with Regard to Free-Will."   _Philosophy_ 46 (1971), 259-261.

2076. RYANS, D. G. "Francis Galton's Statistical Contributions." School and Society 48 (Sept. 3, 1938), 312-316.

2077. SPILLER, GUSTAV. "Francis Galton on Hereditary Genius." Sociological Review 24 (Jan.-Apr. 1932), 47-56, 155-164.

2078. SWINBURNE, R. G. "Galton's Law." Actes du XIe Congrès International d'Histoire des Sciences (Warsaw, 1965), 5 (1968), 340-343.

2079. SWINBURNE, R. G. "Galton's Law-Formulation and Development." Annals of Science 21 (1965), 15-31.

2080. TALBERT, ERNEST LYNN. "On Francis Galton's Contribution to the Psychology of Religion." Scientific Monthly 37 (1933), 53-54.

2081. WILKIE, J. S. "Galton's Contribution to the Theory of Evolution with Special Reference to His Use of Models and Metaphors." Annals of Science 11 (1955), 194-205.

4. Maine, Henry James Sumner

See introduction to section 1.

2082. BATZEL, VICTOR MERLYN. "Sir Henry James Sumner Maine: A Study in Naturalistic Law." Dissertation Abstracts 28 (1968), 2619A (University of Iowa, 1967).

2083. BOCK, KENNETH ELLIOTT. "Comparison of Histories: The Contribution of Henry Maine." Comparative Studies in Society and History 16 (Mar. 1974), 232-262.

2084. BOCK, KENNETH ELLIOTT. "The Moral Philosophy of Sir Henry Sumner Maine." Journal of the History of Ideas 37 (Jan.-Mar. 1976), 147-154.

2085.   COHN, B.   "From Indian Status to British Contract."
        Journal of Economic History 21 (1961), 613-628.

2086.   DERRETT, J. D. M.   "Sir Henry Maine and Law in
        India."   Juridical Review 4 (1959), 4-55.

2087.   FEAVER, GEORGE A.   From Status to Contract: A
        Biographical Study of Sir Henry Maine, 1822-88.
        London: Longmans, Green & Co., 1969.

2088.   FEAVER, GEORGE A.   "The Political Attitudes of Sir
        Henry Maine: Conscience of a 19th Century Con-
        servative."   Journal of Politics 27 (1965), 290-317.

2089.   GRAVESON, R. H.   "The Movement from Status to
        Contract."   Modern Law Review 4 (1940-41), 261-272.

2090.   HOEBEL, E. A.   "Fundamental Legal Concepts as
        Applied in the Study of Primitive Law."   Yale Law
        Journal 51 (1942), 951-966.

2091.   KIRK, R.   "Thought of Sir Henry Maine."   Review
        of Politics 15 (Jan. 1953), 86-96.

2092.   LANDMAN, J. H.   "Primitive Law, Evolution, and Sir
        Henry Maine."   Michigan Law Review 27 (1930),
        404-425.

2093.   OLDHAM, J. B.   Analysis of Maine's Ancient Law,
        with Notes.   Oxford: Blackwell, 1913.

2094.   ORENSTEIN, H.   "Ethnological Theories of Henry
        Sumner Maine."   American Anthropologist 70 (Apr.
        1968), 264-276.

2095.   PILLING, N.   "The Conservativism of Sir Henry
        Maine."   Political Studies 18 (Mar. 1970), 107-120.

2096.   POUND, ROSCOE.   "The Scope and Purpose of Socio-
        logical Jurisprudence."   Harvard Law Review 24
        (June 1911), 591-619; 25 (1911-12), 140-168, 489-
        516.

2097.   REDFIELD, R.   "Maine's 'Ancient Law' in the Light
        of Primitive Societies."   Western Political
        Quarterly 3 (1950), 574-589.

2098.   SMELLIE, K. B.   "Sir Henry Maine."   Economica 8
        (1928), 64-94.

2099.   SMITH, BRIAN C.   "Maine's Concept of Progress."
        Journal of the History of Ideas 24 (1963), 407-412.
2100.   THORNER, DANIEL.   "Sir Henry Maine (1822-1888)."
        In Herman Ausubel, J. Bartlet Brebner, and Erling
        M. Hunt (Eds.), Some Modern Historians of Britain:
        Essays in Honor of R. L. Schuyler. New York: The
        Dryden Press, 1951, 66-84.
2101.   TUPPER, CHARLES LEWIS.   "India and Sir Henry Maine."
        Journal of the Society of Arts 46 (1897-98), 390-
        405.
2102.   VINOGRADOFF, SIR PAUL.   "The Teaching of Maine."
        In Collected Papers.   2 vols.   Oxford: The
        Clarendon Press, 1928, Vol. 2, 173-189.

                    5.   Pearson, Karl

        See introduction to section 1; Science I.   2.
Method and Philosophy.

2103.   CAMP, BURTON H.   "Karl Pearson and Mathematical
        Statistics."   Journal of the American Statistical
        Association 28, No. 184 (Dec. 1933), 395-401.
2104.   DAVIS, H. T.   "E. S. Pearson. 'Karl Pearson: An
        Appreciation of Some Aspects of His Life and Work'."
        Isis 32 (1940), 158-164.   (Review)
2105.   DEWART, LESLIE S.   "The Development of Karl Pearson's
        Scientific Philosophy."   Ph.D. Dissertation,
        University of Toronto, 1954.
2106.   FILON, L. N. G.   "Karl Pearson as an Applied Mathe-
        matician."   Obituary Notices of Fellows of the
        Royal Society of London 2, No. 5 (Dec. 1936),
        104-110.
2107.   HALDANE, J. B. S.   "Karl Pearson, 1857-1957."
        Biometrika 44, Parts 3-4 (Dec. 1957), 303-313.
        (Also in Pearson, Egon Sharpe and M. G. Kendall,

Eds., _Studies in the History of Statistics and Probability: A Series of Papers_. London: Griffin, 1970, 427-437)

2108. HALDANE, J. B. S. et al. _Karl Pearson, 1857-1957: The Centenary Celebration at University College, London, 13 May 1957_. London: Biometrika Trustees, 1958.

2109. MAHALANOBIS, P. C. "Karl Pearson." _Sankhya_ 2, Part 4 (1936), 363-378.

2110. MAHALANOBIS, P. C. "A Note on the Statistical and Biometric Writings of Karl Pearson." _Sankhya_ 2, Part 4 (1936), 411-422.

2111. MONTAGU, M. F. ASHLEY. "Karl Pearson and the Historical Method in Ethnology." _Isis_ 34 (1943), 211-214.

2112. MORANT, G. M. "Karl Pearson." _Man_ 36, No. 118 (June 1936), 89-92.

2113. NALLETAMBY, M. "An Analysis of Karl Pearson's 'The Grammar of Science'." M.Sc. Thesis, University of London, 1955.

2114. NOCK, REV. ALBERT JAY. "New Science and Its Findings: Discoveries by Karl Pearson." _American Magazine_ 73 (Mar. 1912), 577-583.

2115. NORTON, B. J. "Biology and Philosophy: The Methodological Foundations of Biometry." _Journal of the History of Biology_ 8 (1975), 85-93.

2116. NORTON, B. J. "Metaphysics and Population Genetics: Karl Pearson and the Background to Fisher's Multifactorial Theory of Inheritance." _Annals of Science_ 32 (Nov. 1975), 537-553.

2117. PEARL, RAYMOND. "Karl Pearson." _Journal of the American Statistical Association_ 31, No. 196 (Dec. 1936), 653-664.

2118. PEARSON, EGON SHARPE. "Karl Pearson: An Appreciation of Some Aspects of His Life and Work." _Biometrika_ 28 (1936), 193-257; 29 (1937), 161-248.

(Book with same title, Cambridge: Cambridge University Press; New York: Macmillan, 1938; see review by H. T. Davis above)

2119.   PEARSON, EGON SHARPE. "Some Incidents in the Early History of Biometry and Statistics, 1890-94." Biometrika 52, Parts 1-2 (June 1965), 3-18. (Also in E. S. Pearson, and M. G. Kendall, Eds., Studies in the History of Statistics and Probability: A Series of Papers. London: Griffin, 1970, 323-338)

2120.   PEARSON, EGON SHARPE. "Some Reflexions on Continuity in the Development of Mathematical Statistics, 1885-1920." Biometrika 54, Parts 3-4 (Dec. 1967), 341-355. (Reprinted in E. S. Pearson and M. G. Kendall, Eds., Studies in the History of Statistics and Probability: A Series of Papers. London: Griffin, 1970, 339-353)

2121.   RIDDLE, CHAUNCEY CAZIER. "Karl Pearson's Philosophy of Science." Dissertation Abstracts 19 (1959), 3326 (Columbia University, 1958).

2122.   SEMMEL, BERNARD. Imperialism and Social Reform: English Social-Imperial Thought 1895-1914. Cambridge: Harvard University Press, 1960. (Includes Benjamin Kidd)

2123.   SEMMEL, BERNARD. "Karl Pearson: Socialist and Darwinist." British Journal of Sociology 9, No. 2 (June 1958), 111-125.

2124.   STOUFFER, SAMUEL A. "Karl Pearson - An Appreciation on the 100th Anniversary of His Birth." Journal of the American Statistical Association 53, No. 281 (Mar. 1958), 23-27.

2125.   THIELE, JOACHIM. "Karl Pearson, Ernst Mach, John B. Stallo: Briefe aus den Jahren 1897 bis 1904." Isis 60 (1969), 535-542.

2126.   WALKER, HELEN M. "The Contributions of Karl Pearson." Journal of the American Statistical Association 53, No. 281 (Mar. 1958), 11-22.

2127.   WILKS, S. S.   "Karl Pearson: Founder of the
        Science of Statistics."   Scientific Monthly 53, No.
        2 (Sept. 1941), 249-253.

2128.   YULE, G. UDNY.   "Karl Pearson."   Obituary Notices
        of Fellows of the Royal Society of London 2, No. 5
        (Dec. 1936), 73-104.

6.   Spencer, Herbert

See introduction to section 1; XIII.   Evolution and
Psychology.

2129.   ANDRESKI, STANISLAV. (Ed.)   Herbert Spencer: Struc-
        ture, Function and Evolution.   London: Joseph,
        1971.

2130.   BAKER, WILLIAM J.   "'A Problematical Thinker' to a
        'Sagacious Philosopher':   Some Unpublished George
        Henry Lewes-Herbert Spencer Correspondence."
        English Studies 56 (1975), 217-221.

2131.   BANNERMAN, LLOYD CHARLES FRANCIS.   "Spencer's
        Philosophical Interpretation of Biological Evolu-
        tion."   Ph.D. Dissertation, University of Toronto,
        1963.

2132.   BENNE, K. D.   "The Educational Outlook of Herbert
        Spencer."   Harvard Educational Review 10 (Oct.
        1940), 436-453.

2133.   BERNARD, L. L.   "Herbert Spencer: The Man and His
        Age."   South Atlantic Quarterly 21 (July 1922),
        241-251.

2134.   BERNARD, L. L.   "Herbert Spencer's Work in the
        Light of His Own Life."   Monist 31 (1921), 1-35.

2135.   BLIAKHER, LEONID I.   "Die Diskussion zwischen
        Spencer und Weismann über die Bedeutung der
        natürlichen Zuchtwahl und der direkten Anpassung
        für die Evolution."   Zeitschrift für Geschichte der

Naturwissenschaften, Technik und Medizin 10, No. 2
(1973), 50-58.

2136. BOUTROUX, EMILE. Religion According to Herbert
Spencer. Trans. A. S. Mories. Edinburgh: O.
Schulze and Co., 1907.

2137. BOWLER, PETER J. "Herbert Spencer and 'Evolution':
An Additional Note." Journal of the History of
Ideas 36 (1975), 367.

2138. BURLINGAME, R. "America's First Evangelist of
Science." Popular Science 150 (May 1947), 136-141.

2139. BURROW, JOHN W. "Herbert Spencer: The Philosopher
of Evolution." History Today 8 (Oct. 1958), 676-
683.

2140. CARBAUGH, DANIEL CARTER. "Biological Analogy in
the Theories of François Quesnay and Herbert
Spencer." Dissertation Abstracts International 30
(1969), 1320-21A (University of Missouri-Kansas
City, 1969).

2141. CARNEIRO, ROBERT L. "Herbert Spencer's 'The Study
of Sociology' and the Rise of Social Science in
America." Proceedings of the American Philosophical
Society 118 (1974), 540-554.

2142. CARNEIRO, ROBERT L. "Structure, Function, and
Equilibrium in the Evolutionism of Herbert Spencer."
Journal of Anthropological Research 29 (Summer
1973), 77-95.

2143. CAVENAGH, F. A. (Ed.) Herbert Spencer on Education. Cambridge: Cambridge University Press,
1932.

2144. COMPAYRE, GABRIEL. Herbert Spencer and Scientific
Education. Trans. Maria E. Findlay. New York: T.
Y. Crowell & Co., 1907.

2145. COOLEY, CHARLES HORTON. "Reflections upon the
Sociology of Herbert Spencer." American Journal of
Sociology 26 (Sept. 1920), 129-145.

2146. DENTON, GEORGE B.   "Early Psychological Theories of
       Herbert Spencer."   The American Journal of Psy-
       chology 32 (1921), 5-15.

2147. DIACONIDE, ELIAS.   Etude critique sur la sociologie
       de Herbert Spencer.   Paris: Libraire générale de
       droit et de jurisprudence, 1938.

2148. DUCASSE, PIERRE.   "La synthèse positiviste: Comte
       et Spencer."   Revue de Synthèse 67 (1950), 155-187.

2149. EISEN, SYDNEY.   "Frederic Harrison and Herbert
       Spencer: Embattled Unbelievers."   Victorian
       Studies 12 (Sept. 1968), 33-56.

2150. EISEN, SYDNEY.   "Herbert Spencer and the Spectre of
       Comte."   Journal of British Studies 7 (Nov. 1967),
       48-67.

2151. ELLIOT, HUGH.   Herbert Spencer.   London: Constable
       and Company Ltd., 1917.

2152. FREEMAN, DEREK.   "The Evolutionary Theories of
       Charles Darwin and Herbert Spencer."   Current
       Anthropology 15 (1974), 211-237.   (Comments by
       various scholars and Freeman's reply)

2153. GENZ, WILHELM.   "Der Agnostizismus Herbert Spencers
       mit Ruecksicht auf August Comte und Friedr.   Alb.
       Lange."   Ph.D. Dissertation, Universität Greifswald,
       1902.

2154. GREEN, PETER.   "Religious Controversy: I.   The
       Knowable and the Unknowable."   Twentieth Century
       151 (1952), 225-231.

2155. GREENE, JOHN C.   "Biology and Social Theory in the
       Nineteenth Century: Auguste Comte and Herbert
       Spencer."   In Marshall Clagett (Ed.), Critical
       Problems in the History of Science.   Madison,
       Wisconsin: University of Wisconsin Press, 1959,
       419-446.

2156. HARRIS, S. HUTCHINSON.   "Herbert Spencer's Socio-
       logy."   Times Literary Supplement No. 1762 (Nov. 9,
       1935), 722.

2157.   HARRISON, FREDERIC.   "Spencer's Life."   _Positivist_
        Review 16 (July 1908), 148-149.

2158.   HART, ALAN.   "The Synthetic Epistemology of Herbert
        Spencer."   _Dissertation Abstracts_ 26 (1966), 7366
        (University of Pennsylvania, 1965).

2159.   HASSEL, DAVID J.   "Herbert Spencer Centenary."
        America 103 (Aug. 27, 1960), 578.

2160.   HOOPER, CHARLES E.   "Religions of Comte and
        Spencer: A New Synthesis."   _Open Court_ 29 (Oct.
        1915), 620-628.

2161.   HOPPS, J. P.   "Spencer as Theist."   _Westminster_
        Review 168 (Sept. 1907), 314-318.

2162.   HOWERTH, I. W.   "Did Spencer Anticipate Darwin?"
        Science 43 (Mar. 31, 1916), 462-464.

2163.   HUDSON, WILLIAM HENRY.   _Herbert Spencer_.   London:
        A. Constable & Co., Ltd., 1908.

2164.   HUGHESDON, P. J.   "Spencer, Darwin and the Evolu-
        tion-hypothesis."   _Sociological Review_ 17 (Jan.
        1925), 31-44.

2165.   JAMES, WILLIAM.   "Herbert Spencer's Autobiography."
        In _Memories and Studies_. New York: Longmans,
        Green and Co., 1911, 107-142.

2166.   JEFFERSON, SIR GEOFFREY.   "Variations on a Neuro-
        logical Theme: Cortical Localization."   In
        Selected Papers. London: Pitman Medical Pub. Co.,
        1960, 35-44. (Includes Hughlings Jackson and
        Spencer; Spencer on localization and phrenology)

2167.   JONES, ROBERT ALUN.   "Comte and Spencer: A Priority
        Dispute in Social Science."   _Journal of the History_
        of Behavioral Sciences 6 (1970), 241-254.

2168.   JORDAN, E.   "Unknowable of Herbert Spencer."
        Philosophical Review 20 (May 1911), 291-309.

2169.   KATOPE, CHRISTOPHER G.   "'Sister Carrie' and
        Spencer's 'First Principles'."   _American Literature_
        41 (Mar. 1969), 64-75. (T. Dreiser's _Sister Carrie_)

2170. KROL, LEONILLA. "Spencer's Meaning of Structure."
       Organon (Warsaw), 3 (1966), 201-218.
2171. LAMAR, LILLIE B. "Herbert Spencer, Interpreter of
       Science." Ph.D. Dissertation, University of Texas
       at Austin, 1953.
2172. LEROY, JEAN F. "Naudin, Spencer et Darwin dans
       l'histoire des théories de l'hérédité." Actes du
       XIe Congrès International d'Histoire des Sciences
       (Warsaw, 1965), 5 (1968), 64-69.
2173. MEDAWAR, PETER BRIAN. "Herbert Spencer and the Law
       of General Evolution." In The Art of the Soluble.
       London: Methuen, 1967, 39-58.
2174. MEDAWAR, PETER BRIAN. "Onwards from Spencer:
       Evolution and Evolutionism." Encounter 21 (Sept.
       1963), 35-43.
2175. MELDOLA, RAPHAEL. Evolution, Darwinian and
       Spencerian: The Herbert Spencer Lecture Delivered
       at the Museum, 8 December 1910. Oxford: Clarendon
       Press, 1910.
2176. MUNRO, THOMAS. "Evolution and Progress in the
       Arts: A Reappraisal of Herbert Spencer's Theory."
       Journal of Aesthetics and Art Criticism 18 (1960),
       294-315.
2177. MUNRO, THOMAS. Evolution in the Arts and Other
       Theories of Culture History. Cleveland: The
       Cleveland Museum of Art, 1963.
2178. MURPHREE, IDUS LAVIGA. "Evolution: From Cosmic
       Progress to Human Reconstruction. The Concepts of
       Progress and Evolution in the Works of Spencer,
       Tylor, Lubbock, Morgan, and Veblen." Disserta-
       tion Abstracts 14 (1954), 155-156 (Columbia Uni-
       versity, 1953).
2179. MURRAY, REV. ROBERT HENRY. Studies in English
       Social and Political Thinkers of the Nineteenth
       Century. 2 vols. Cambridge: W. Heffer & Sons,
       1929. (See vol. 2)

2180. PEEL, J. D. Y. Herbert Spencer: The Evolution
of a Sociologist. New York: Basic Books, 1971.

2181. PEEL, J. D. Y. (Ed.) On Social Evolution: Selected
Writings. Chicago: University of Chicago Press,
1972.

2182. PEEL, J. D. Y. "Spencer and the Neo-evolution-
ists." Sociology 3 (May 1969), 173-192.

2183. PLATT, THOMAS WALTER. "Spencer and James on Mental
Categories: A Re-evaluation in Light of Modern
Biology." Dissertation Abstracts 27 (1967), 4302A
(University of Pennsylvania, 1966).

2184. PLOCHMANN, GEORGE KIMBALL. "Darwin or Spencer?"
Science 130 (Nov. 27, 1959), 1452-56.

2185. PRICE, ALAN. "Herbert Spencer and the Apotheosis
of Science." Educational Review 14 (1962), 87-97,
233-241.

2186. ROYCE, J. "Herbert Spencer and His Contribution to
the Concept of Evolution." International Quarterly
9 (June 1904), 335-365.

2187. RUMNEY, JUDAH. Herbert Spencer's Sociology.
London: Williams and Norgate, 1934.

2188. SARTON, GEORGE. "Herbert Spencer, 1820-1903."
Isis 3 (1921), 375-390.

2189. SCHOENWALD, RICHARD L. "Town Guano and 'Social
Statics'." Victorian Studies 11 (1968), 691-710.

2190. SEGUY, JEAN. "Herbert Spencer ou l'évolution des
formes religieuses." Archives de Sociologie des
Religions 14 (Jan.-June 1969), 29-35.

2191. SHELTON, H. S. "Spencer's Formula of Evolution."
Philosophical Review 19 (May 1910), 241-258.

2192. SHIPLEY, M. "Forty Years of a Scientific Friendship: Herbert Spencer and John Tyndall." Open
Court 34 (Apr. 1920), 252-255.

2193. SIMON, WALTER M. "Herbert Spencer and the 'Social
Organism'." Journal of the History of Ideas 21
(Apr. 1960), 294-299.

2194. STARK, WERNER. "Herbert Spencer's Three Sociol-
       ogies." American Sociological Review 26 (1961),
       515-521.

2195. STERN, BERNHARD JOSEPH. "Darwin on Spencer."
       Scientific Monthly 26 (1928), 180-181.

2196. THOMSON, JOHN ARTHUR. Herbert Spencer. London:
       J. M. Dent and Co., 1906.

2197. TILLET, ALFRED W. Herbert Spencer Betrayed.
       London: King, 1939.

2198. TROMPF, G. W. "Radical Conservatism in Herbert
       Spencer's Educational Thought." British Journal of
       Educational Studies 17 (Oct. 1969), 267-280.

2199. WALKER, N. T. "Sources of Herbert Spencer's Educa-
       tional Ideas." Journal of Educational Research 22
       (Nov. 1930), 299-308.

2200. YOUNG, ROBERT MAXWELL. "The Development of Herbert
       Spencer's Concept of Evolution." Actes du XIe Con-
       grès International d'Histoire des Sciences (Warsaw,
       1965), 2 (1968), 273-278.

2201. YOUNG, ROBERT MAXWELL. "The Functions of the
       Brain: Gall to Ferrier, (1808-1886)." Isis 59
       (1968), 251-268. (Includes A. Bain, J. H. Jackson)

IX. Evolution and Literature

General works on science and literature, and specialized studies on the impact of evolutionary thought on prose and poetry, are included in this category. There are some references to the work of H. G. Wells. See III. Ideas . . .; VII. Evolution, 1. General; categories X-XIII.

2202. ANON. "Darwin's Influence on Literature." World's Work 48 (Aug. 1924), 357-358.

2203. APPLEMAN, PHILIP. "Darwin and the Literary Critics." Dissertation Abstracts 15 (1955), 1618 (Northwestern University, 1955). (The effect of evolution on J. A. Symonds. W. Pater, L. Stephen)

2204. ARN, ROBERT M. "The Effect of Social Evolutionary Theory on Figurative Language in the Nineteenth-Century Novel." Ph.D. Dissertation, University of Cambridge, 1971.

2205. BARBER, OTTO. H. G. Wells' Verhältnis zum Darwinismus. Leipzig: B. Tauchnitz, 1934.

2206. BEACH, JOSEPH WARREN. The Concept of Nature in Nineteenth-Century English Poetry. New York: The Macmillan Co., 1936.

2207. BERGONZI, BERNARD. The Early H. G. Wells: A Study of the Scientific Romances. Toronto: University of Toronto Press, 1961.

2208. BUSH, DOUGLAS. Science and English Poetry: A Historical Sketch, 1590-1950. New York: Oxford University Press, 1950.

2209. COLWELL, MARY LOU MACKEY. "The Human Position: Hudson and the Darwinian Revolution." Dissertation Abstracts International 31 (1970), 2377A (University of Michigan, 1970).

2210. CRUM, RALPH BRINCKERHOFF. Scientific Thought in Poetry. New York: Columbia University Press, 1931.

2211.   GAINOR, MARY EVATT.   "Thunder on the Horizon:
        Hostility to Science in English Literature from
        1860 to 1900."   Ph.D. Dissertation, Harvard Uni-
        versity, 1974.

2212.   GIBBONS, TOM.   Rooms in the Darwin Hotel: Studies
        in English Literary Criticism and Ideas, 1880-1920.
        Nedlands, Western Australia: University of Western
        Australia Press, 1973.

2213.   HASTINGS, HARRY WORTHINGTON.   "The Scientific
        Spirit in the English Novel from 1850 to 1900."
        Ph.D. Dissertation, Harvard University, 1916.

2214.   HENKIN, LEO JUSTIN.   Darwinism in the English Novel
        1860-1910:   The Impact of Evolution on Victorian
        Fiction.   New York: Corporate Press, Inc., 1940.
        (New York: Russell & Russell, 1963).

2215.   HORNYANSKY, MICHAEL.   "Darwinism in Literature."
        In Herbert Hugh John Nesbitt (Ed.), Darwin in
        Retrospect.   Toronto: Ryerson Press, 1960, 67-86.

2216.   IRVINE, WILLIAM.   "The Influence of Darwin on
        Literature."   Proceedings of the American Philoso-
        phical Society 103 (1959), 616-628.

2217.   LEVINE, GEORGE AND WILLIAM ANTHONY MADDEN. (Eds.)
        The Art of Victorian Prose.   New York: Oxford
        University Press, 1968.   (Includes Darwin's Origin)

2218.   MCCRACKEN, ANDREW VANCE.   "The Theological Reac-
        tions of the Victorian Poets to the Natural Sciences
        and Evolutionism."   Ph.D. Dissertation, University
        of Chicago, 1932.

2219.   MÜLLER-SCHWEFE, GERHARD.   "Darwin and the Poets."
        In Gerhard Müller-Schwefe and Konrad Tuzinski
        (Eds.), Literatur-Kultur-Gesellschaft in England
        und Amerika.   Frankfurt: Diesterweg, 1966, 99-112.

2220.   PIZER, DONALD.   "Evolutionary Ideas in Late Nine-
        teenth-Century English and American Literary
        Criticism."   Journal of Aesthetics and Art Critic-
        ism 19 (1961), 305-310.

2221.   POTTER, GEORGE REUBEN.   "The Idea of Evolution in
        the English Poets from 1744 to 1832." Ph.D. Dis-
        sertation, Harvard University, 1922.
2222.   ROBINSON, EDWIN ARTHUR.   "The Influence of Science
        on George Meredith." Doctoral Dissertation, Ohio
        State University, 1936.
2223.   ROPPEN, GEORGE.   Evolution and Poetic Belief: A
        Study in Some Victorian and Modern Writers. Oslo:
        Oslo University Press, 1956. (Folcroft, Pa.:
        Folcroft Press, Inc., 1969)
2224.   SMIDT, K.   "Intellectual Quest of the Victorian
        Poets." English Studies 40 (Apr. 1959), 95-98.
2225.   STEVENSON, ARTHUR L.   "The Reflection of the Evolu-
        tionary Theory in English Poetry." Ph.D. Disserta-
        tion, University of California at Berkeley, 1925.
2226.   STEVENSON, LIONEL.   Darwin among the Poets.
        Chicago: The University of Chicago Press, 1932.
        (New York: Russell & Russell, 1963)
2227.   STEVENSON, LIONEL.   "Darwin and the Novel."
        Nineteenth-Century Fiction 15 (1960), 29-30.
2228.   SUSSMAN, HERBERT LEWIS.   Victorians and the
        Machine: The Literary Response to Technology.
        Cambridge, Mass.: Harvard University Press, 1968.
        (Includes S. Butler, H. G. Wells)
2229.   WARD, W. A.   "The Idea in Nature: A Study of the
        Thought of Walter Pater, with Reference to Hegel
        and the Theory of Evolution." Ph.D. Dissertation,
        St. John's College, University of Cambridge, 1964.

X. Evolution and Ethics

        This category includes studies of the impact of
scientific thought, particularly the theory of evolution,
on ethics and morals.   See III. Ideas . . . ; VII. Evolution, 1. General; categories VIII, IX, XI-XIII.

2230. ANNAN, NOEL GILROY. "Religious Controversy: II.
       Religion and Morality." The Twentieth Century 151
       (1952), 233-238.
2231. ANON. "Darwin Seen As Leading Revised Science of
       Ethics." Science News Letter 35 (Apr. 15, 1939),
       232.
2232. BALMFORTH, R. "Influence of the Darwinian Theory
       on Ethics." International Journal of Ethics 21
       (July 1911), 448-465.
2233. EDEL, ABRAHAM. Ethical Judgement: The Use of
       Science in Ethics. New York: The Free Press of
       Glencoe, 1955.
2234. FLEW, ANTONY G. N. Evolutionary Ethics. London:
       Macmillan, 1967.
2235. GANTZ, KENNETH F. "The Beginnings of Darwinian
       Ethics." University of Texas Publication No. 3926,
       July 8, 1939. Studies in English, 1939. Austin:
       University of Texas, 1939, 180-209.
2236. GANTZ, KENNETH F. "The Beginnings of Darwinian
       Ethics, 1859-71." Ph.D. Dissertation, University
       of Chicago, 1938.
2237. GLASS, BENTLEY. Science and Ethical Values.
       Chapel Hill: University of North Carolina Press,
       1965.
2238. HOBHOUSE, LEONARD TRELAWNEY. Morals in Evolution:
       A Study in Comparative Ethics. London: Chapman &
       Hall, Ltd., 1906.
2239. HOLMES, SAMUEL JACKSON. "Darwinian Ethics and Its
       Practical Applications." Science 90 (Aug. 11,
       1939), 117-123.

2240.   HUXLEY, JULIAN SORELL AND T. H. HUXLEY.   Touch-
        stones for Ethics, 1893-1943. New York: Harper,
        1947.

2241.   HYMAN, VIRGINIA R.   "The Illuminations of Time: A
        Study of the Influence of Ethical Evolution on the
        Novels of Thomas Hardy."   Dissertation Abstracts
        International 30 (1970), 4947A (Columbia Univer-
        sity, 1969).

2242.   IRVING, JOHN A.   "Evolution and Ethics."   Queen's
        Quarterly 55 (Winter 1948-49), 450-463.

2243.   KANNWISCHER, ARTHUR.   "Psychology and Ethics in
        John Stuart Mill's 'Logic'."   University of Pitts-
        burgh Abstracts of Dissertations 49 (1953), 25-30.

2244.   KAUTSKY, KARL.   Ethics and the Materialist Concep-
        tion of History.   Trans. John B. Askew.   Chicago:
        Charles H. Kerr and Co., 1909.   (Includes Darwinism)

2245.   KEITH, SIR ARTHUR.   Essays on Human Evolution.
        London: Watts and Co., 1946.

2246.   LEWIS, J.   Man and Evolution.   London: Lawrence
        and Wishart, 1962.

2247.   MCCLINTOCK, THOMAS.   "The Definition of Ethical
        Relativism."   Personalist 50 (Fall 1969), 435-447.
        (E. A. Westermarck)

2248.   PENN, STUART LEE.   "The Ethical Relativism of
        Edward Westermarck."   Ph.D. Dissertation, Yale Uni-
        versity, 1957.

2249.   PITCHER, ALVIN.   "Darwinism and Christian Ethics."
        Journal of Religion 40 (1960), 256-266.

2250.   QUILLIAN, W. F., JR.   The Moral Theory of Evolu-
        tionary Naturalism.   New Haven: Yale University
        Press, 1945.   (Includes Darwin, W. K. Clifford, L.
        Stephen)

2251.   QUINTON, ANTHONY M.   "Ethics and the Theory of
        Evolution."   In I. T. Ramsey (Ed.), Biology and
        Personality: Frontier Problems in Science, Philos-
        ophy and Religion.   New York: Barnes and Noble,
        Inc., 1965, 107-131.

2252.   RITTER, WILLIAM EMERSON.   Charles Darwin and the
        Golden Rule. New York: Storm Publishers, 1954.

2253.   RITTER, WILLIAM EMERSON.   "Darwin and the Golden
        Rule." Christian Register (Oct. 7, 1937), 580.

2254.   SORLEY, WILLIAM RITCHIE.   Moral Values and the
        Idea of God. Cambridge: Cambridge University
        Press, 1918.

2255.   TOULMIN, STEPHEN EDELSTON.   "World Stuff and Non-
        sense." Cambridge Journal No. 1 (May 1948), 465-473.

2256.   TUFTS, JAMES H.   "Darwin and Evolutionary Ethics."
        Psychological Review 16 (1909), 195-206.

2257.   WADDINGTON, C. H. (Ed.)   Science and Ethics.
        London: Allen & Unwin, 1942.

2258.   WAND, BERNARD.   "Evolution and the Basis of Moral
        Principles." In Herbert Hugh John Nesbitt (Ed.),
        Darwin in Retrospect. Toronto: Ryerson Press,
        1960, 35-47.

2259.   WESTERMARCK, EDWARD.   The Origin and Development
        of Moral Ideas. London: Macmillan & Co., 1906.

XI. Evolution and History

    This category looks at historical thinking, partic-
ularly as it was affected by the theory of evolution and
the idea of progress. See III. Ideas . . .; VII.
Evolution, 1. General; VIII. Evolution and Social and
Political Thought (all sections); XII. Evolution and
Anthropology; categories IX, X, XIII.

2260. BOCK, KENNETH ELLIOTT. The Acceptance of Histories:
        Toward a Perspective for Social Science. Berkeley:
        University of California Press, 1956. (A. Comte,
        J. S. Mill, H. Spencer, E. B. Tylor, J. G. Frazer,
        C. Kingsley, J. A. Froude)
2261. BURY, JOHN BAGNELL. The Idea of Progress: An
        Inquiry into Its Origin and Growth. London:
        Macmillan and Co., 1920. (Reprint: New York:
        Dover Publications, Inc., 1955)
2262. BUTLER, GIBBON FRANCIS. "John Henry Newman's Use
        of History in His Anglican Career, 1825-1845."
        Microfilm Abstracts 10, No. 4 (1950), 207-208
        (University of Illinois at Urbana-Champaign, 1950).
2263. CARPENTER, E. S. "The Role of Archeology in the
        19th Century Controversy between Developmentalism
        and Degeneration." Pennsylvania Archeologist 20
        (1950), 5-18.
2264. GINSBERG, MORRIS. The Idea of Progress: A Revalua-
        tion. Boston: Beacon Press, 1953.
2265. HABER, FRANCIS C. "The Darwinian Revolution in the
        Concept of Time." Studium Generale: Zeitschrift
        für die Einheit der Wissenschaften im Zusammenhang
        ihrer Begriffsbildungen und Forschungsmethoden 24
        (1971), 289-307.
2266. HADLEY, ARTHUR TWINING. "The Influence of Charles
        Darwin upon Historical and Political Thought."
        The Psychological Review 16 (1909), 143-151.

2267.   LORETAN, JOSEPH O. "The Appearance of Evolutionary
        Theory in Historical Writing with Particular
        Emphasis on Darwinism." Ph.D. Dissertation, Ford-
        ham University, 1930.

2268.   MARCUS, JOHN T. Heaven, Hell and History: A
        Survey of Man's Faith in History from Antiquity
        to the Present. New York: Macmillan, 1967.
        (Includes Darwinism)

2269.   SKLAIR, LESLIE. The Sociology of Progress. Lon-
        don: Routledge and Kegan Paul, Ltd., 1970.

2270.   TSANOFF, RADOSLAV A. "Evolution, Teleology and
        History." Rice Institute Pamphlet 46, No. 1 (Apr.
        1959), 32-52. (Darwin)

2271.   WAGAR, W. WARREN. Good Tidings: The Belief in
        Progress from Darwin to Marcuse. Bloomington:
        Indiana University Press, 1972.

2272.   WAGAR, W. WARREN. "Modern Views of the Origin of
        the Idea of Progress." Journal of the History of
        Ideas 28 (1967), 55-70.

XII. Evolution and Anthropology

Works on the development of anthropological theory
and anthropology as a discipline are included here. A
number of these deal with the influence of Darwin and with
the contributions of individual anthropologists, including
John Lubbock, Edward Burnett Tylor, and James George
Frazer. See III. Ideas . . . ; VII. Evolution, 1.
General; XI. Evolution and History; categories VIII-X,
XIII.

2273. ACKERKNECHT, ERWIN HEINZ. "On the Comparative
       Method in Anthropology." In Robert F. Spencer
       (Ed.), Method and Perspective in Anthropology.
       Minneapolis: The University of Minnesota Press,
       1954, 117-125.
2274. ACKERMAN, ROBERT. "Frazer on Myth and Ritual."
       Journal of the History of Ideas 36 (1975), 115-134.
2275. ANON. "Sir James Frazer O.M." Bulletin of the
       John Rylands Library 26, No. 1 (Oct.-Nov. 1941),
       16-18.
2276. BEIDELMAN, THOMAS O. W. Robertson Smith and the
       Sociological Study of Religion. Chicago: University
       of Chicago Press, 1974.
2277. BESTERMAN, THEODORE. A Bibliography of Sir James
       George Frazer. London: Macmillan and Co., Limited,
       1934.
2278. BREEN, T. H. "The Conflict in The Golden Bough:
       Frazer's Two Images of Man." South Atlantic
       Quarterly 66 (1967), 179-194.
2279. BURROW, JOHN W. "Evolution and Anthropology in the
       1860's: The Anthropological Society of London,
       1863-71." Victorian Studies 7 (1963-64), 137-154.
2280. BYNUM, WILLIAM FREDERICK. "Time's Noblest Off-
       spring: The Problem of Man in the British Natural
       Historical Sciences, 1800-1863." Ph.D. Thesis,
       University of Cambridge, 1974.

2281.   CHAUDHURI, NIRAD C.   Scholar Extraordinary: The
        Life of Professor the Rt. Hon. Friedrich Max Müller,
        PC. London: Chatto, 1974.
2282.   DOWNIE, ROBERT ANGUS. Frazer and "The Golden
        Bough." London: Gollancz, 1970.
2283.   DOWNIE, ROBERT ANGUS. James George Frazer: The
        Portrait of a Scholar. London: Watts & Co., 1940.
2284.   EVANS-PRITCHARD, EDWARD EVAN. "Religion and the
        Anthropologists." In Social Anthropology and Other
        Essays. New York: The Free Press of Glencoe,
        1962, 155-171.
2285.   FAVERTY, FREDERIC E.   Matthew Arnold the Ethnol-
        ogist. Evanston: Northwestern University Press,
        1951.
2286.   GOLDENWEISER, ALEXANDER ALEXANDROVITCH. History,
        Psychology and Culture. London: K. Paul, Trench,
        Trubner, 1933. (Includes J. G. Frazer)
2287.   GOLDMAN, IRVING. "Evolution and Anthropology."
        Victorian Studies 3 (1959-60), 55-75.
2288.   GRANT DUFF, URSULA. (Ed.)   The Life-Work of Lord
        Avebury (Sir John Lubbock) 1834-1913. London:
        Watts and Co., 1924.
2289.   GROSS, JOHN J. "After Frazer: The Ritualistic
        Approach to Myth." Western Humanities Review 5
        (1951), 379-391.
2290.   HAECKEL, E. "Charles Darwin as an Anthropologist."
        In Albert Charles Seward (Ed.), Darwin and Modern
        Science. Cambridge: University Press, 1909,
        137-151.
2291.   HARRIS, MARVIN. The Rise of Anthropological Theory.
        New York: Thomas Y. Crowell Co., 1968. (Includes
        E. B. Tylor, J. S. Mill, H. Spencer, A. Comte, J.
        Lubbock)
2292.   HODGEN, MARGARET T. "Anthropology in the BAAS: Its
        Inception." Scientia 108 (1973), 803-811.

2293.   HODGEN, MARGARET T.   "The Doctrine of Survivals:
        The History of an Idea."   American Anthropologist
        33, No. 3 (July-Sept. 1931), 307-324.

2294.   HUTCHINSON, HORACE G.   Life of Sir John Lubbock,
        Lord Avebury.   2 vols.   London: Macmillan, 1914.

2295.   HUXLEY, FRANCIS.   "Frazer within the Bloody Wood."
        New Statesman and Nation 59 (1960), 561-562.
        (Frazer's The Golden Bough)

2296.   HYMAN, STANLEY EDGAR.   "After the Great Metaphors."
        American Scholar 31 (Spring 1962), 236-258.   (In-
        cludes Darwin, K. Marx, J. G. Frazer, S. Freud)

2297.   HYMAN, STANLEY EDGAR.   The Tangled Bank: Darwin,
        Marx, Frazer and Freud as Imaginative Writers.   New
        York: Atheneum, 1962.

2298.   JARVIE, I. C.   The Revolution in Anthropology.
        London: Routledge and K. Paul, 1964.

2299.   KARDINER, ABRAM AND EDWARD PREBLE.   They Studied
        Man.   Cleveland: World Publishing Co., 1961.
        (Includes Darwin, H. Spencer, E. B. Tylor, J. G.
        Frazer)

2300.   LANG, ANDREW.   "Lord Avebury on Marriage, Totemism
        and Religion."   Folklore 22 (1911), 402-425.

2301.   LEACH, EDMUND RONALD, I.C. JARVIE et al.   "Frazer
        and Malinowski: A CA Discussion."   Current Anthro-
        pology 7 (1966), 560-576.   (Reprint, with addi-
        tional material, of an article in Encounter 25,
        1965, 24-36; 26, 1966, 53-56, 92-93)

2302.   LEACH, EDMUND RONALD.   "Golden Bough or Gilded
        Twig?"   Daedalus 90 (1959), 371-387.

2303.   LOWIE, ROBERT H.   "Edward B. Tylor."   American
        Anthropologist 19, No. 2 (Apr.-June, 1917), 262-
        268.

2304.   MALINOWSKI, BRONISLAW.   "Sir James George Frazer:
        A Biographical Appreciation."   In A Scientific
        Theory of Culture, and Other Essays.   Chapel Hill,
        North Carolina: University of North Carolina
        Press, 1944, 177-221.

2305.   MARETT, ROBERT RANULPH.   James George Frazer 1854-
        1941. London: Oxford University Press, 1941.

2306.   MARETT, ROBERT RANULPH. Tylor. New York: Wiley,
        1936.

2307.   MARRECO, BARBARA WHITCHURCH FREIRE.   "A Biblio-
        graphy of Edward Burnett Tylor from 1861 to 1907."
        In N. W. Thomas (Ed.), Anthropological Essays
        Presented to Edward Burnett Tylor in Honour of His
        75th Birthday. Oxford: Clarendon Press, 1907,
        375-409.

2308.   MEGGERS, BETTY J. (Ed.) Evolution and Anthropology:
        A Centennial Appraisal. Washington, D.C.:   The
        Anthropological Society of Washington, 1959.

2309.   MURPHREE, IDUS LAVIGA.   "Evolution: From Cosmic
        Progress to Human Reconstruction. The Concepts of
        Progress and Evolution in the Works of Spencer,
        Tylor, Lubbock, Morgan, and Veblen." Disserta-
        tion Abstracts 14 (1954), 155-156 (Columbia Uni-
        versity, 1953).

2310.   MURPHREE, IDUS LAVIGA.   "The Evolutionary Anthro-
        pologists: The Progress of Mankind. The Concepts
        of Progress and Culture in the Thought of John
        Lubbock, Edward B. Tylor, and Lewis H. Morgan."
        Proceedings of the American Philosophical Society
        105 (1961), 265-300.

2311.   MURRAY, GILBERT, VICTOR WHITE AND ALEXANDER MACBEATH.
        "The Author of the 'Golden Bough'." Listener 51
        (1954), 13-14, 137-139, 217-218.

2312.   MYRES, SIR J. L. "A Hundred Years of Anthropology
        in Britain." Nature 152, No. 3861 (Oct. 30, 1943),
        493-495.

2313.   ODOM, HERBERT H. "Generalizations on Race in
        Nineteenth-Century Physical Anthropology." Isis
        58 (1967), 5-18.

2314.   OPLER, M. E.   "Cause, Process, and Dynamics in the
        Evolutionism of E. B. Tylor." Southwestern Journal
        of Anthropology 20 (Summer 1964), 123-144.

2315.   OPLER, M. E.   "Tylor's Application of Evolutionary
        Theory to Public Issues of His Day." Anthropo-
        logical Quarterly 41 (Jan. 1968), 1-8.

2316.   PEAR, T. M.   "Some Early Relations between English
        Ethnologists and Psychologists." The Journal of
        the Royal Anthropological Institute of Great
        Britain and Ireland 90 (1960), 227-237.

2317.   PECKHAM, MORSE.   "The Romantic Birth of Anthro-
        pology." In Victorian Revolutionaries: Specula-
        tions on Some Heroes of a Culture Crisis. New
        York: George Braziller, 1970, 175-234.

2318.   PENNIMAN, THOMAS KENNETH. A Hundred Years of
        Anthropology. London: Duckworth, 1935. (Third
        edition, revised. London: Duckworth, 1965)

2319.   PUMPHREY, R. J.   "The Forgotten Man: Sir John
        Lubbock." Notes and Records of the Royal Society
        of London 13 (1958), 49-58.

2320.   RIDGEWAY, SIR WILLIAM.   "The Methods of Mannhardt
        and Sir J. G. Frazer As Illustrated by the Writings
        of the Mistress of Girton (Miss Phillpotts, O.B.E.),
        Miss Jessie Weston, and Dr. B. Malinowski."
        Proceedings of the Cambridge Philological Society
        Nos. 124-126 (1924), 6-19.

2321.   SMITH, JONATHAN ZITTELL.   "The Glory, Jest and
        Riddle: James George Frazer and the 'Golden Bough'."
        Dissertation Abstracts International 35 (1975),
        5519A (Yale University, 1969).

2322.   SMITH, JONATHAN ZITTELL.   "When the Bough Breaks."
        History of Religions 12 (1973), 342-371. (On J. G.
        Frazer's Golden Bough)

2323.   SOMKIN, FRED.   "The Contributions of Sir John
        Lubbock, F.R.S., to the 'Origin of Species': Some
        Annotations to Darwin." Notes and Records of the
        Royal Society of London 17 (1962), 183-191.

2324.   STOCKING, GEORGE W., JR.   "'Cultural Darwinism' and
        'Philosophical Idealism' in E. B. Tylor: A Special
        Plea for Historicism in the History of Anthro-
        pology." Southwestern Journal of Anthropology 21
        (Summer 1965), 130-147.
2325.   STOCKING, GEORGE W., JR.   "Matthew Arnold, E. B.
        Tylor, and the Uses of Invention." American Anthro-
        pologist 65 (1963), 783-799.
2326.   STOCKING, GEORGE W., JR.   Race, Culture and Evolu-
        tion: Essays in the History of Anthropology. New
        York: Free Press, 1968.
2327.   STOCKING, GEORGE W., JR.   "What's in a Name? The
        Origins of the Royal Anthropological Institute
        (1837-71)." Man: Journal of the Royal Anthropo-
        logical Institute 6 (1971), 369-390.
2328.   VICKERY, JOHN B.   "'The Golden Bough': Impact and
        Archetype." Virginia Quarterly Review 39 (1963),
        37-57.
2329.   VICKERY, JOHN B.   The Literary Impact of "The Golden
        Bough." Princeton: Princeton University Press,
        1973.
2330.   WEBER, GAY.   "Science and Society in Nineteenth
        Century Anthropology." History of Science 12
        (1974), 260-283.
2331.   WEISINGER, H.   "The Branch That Grew Full Straight."
        Daedalus 90 (1959), 388-399. (J. G. Frazer)

XIII. Evolution and Psychology

This category ranges beyond the impact of Darwin's
theory of evolution. It lists general works on the
development of psychology in the nineteenth century and
a number of studies in specific areas including phrenology,
cerebral localization, physiology, and neurology. There
are entries on Franz Joseph Gall, George Combe, Herbert
Spencer, John Stuart Mill, George John Romanes, Francis
Galton, Alexander Bain, James Ward, and John Hughlings
Jackson, among others. See III. Ideas . . . ; VII.
Evolution, 1. General; categories VIII-XII.

2332. ACKERKNECHT, ERWIN HEINZ AND HENRI V. VALLOIS.
       Franz Joseph Gall: Inventor of Phrenology, and His
       Collection. Trans. Claire St. Léon. Madison:
       University of Wisconsin Medical School, 1956.
2333. ANGEL, RONALD W. "Jackson, Freud, Sherrington on
       the Reaction of Brain and Mind." American Journal
       of Psychiatry 118 (1961), 193-197.
2334. ANGELL, JAMES R. "The Influence of Darwin on
       Psychology." Psychological Review 16 (1909),
       152-169.
2335. ANON. "1821: The Phrenologist." The Nation and
       the Athenaeum (London) 29 (June 18, 1921), 445.
2336. BALDWIN, JAMES M. "Sketch of the History of Psychol-
       ogy." Psychological Review 12 (1905), 144-165.
2337. BARTLETT, FREDERICK CHARLES. "James Ward, 1843-
       1925." American Journal of Psychology 36 (1925),
       449-453.
2338. BENTLEY, M. "The Psychological Antecedents of
       Phrenology." Psychological Monographs 21, No. 4,
       Whole no. 92 (1916), 102-115.
2339. BLONDEL, CHARLES AIME ALFRED. La Psycho-Physiolo-
       gie de Gall. Paris: F. Alcan, 1914.

2340.   BORING, EDWIN G.   A History of Experimental
        Psychology. New York: Century, 1929.

2341.   BORING, EDWIN G.   Sensation and Perception in
        the History of Experimental Psychology. New York,
        London: D. Appleton-Century, Company, Incorporated,
        1942.

2342.   BRAZIER, MARY A. B.   "Rise of Neurophysiology in
        the 19th Century." Journal of Neurophysiology 20
        (1957), 212-226.

2343.   BREMNER, JEAN P.   "George Combe (1788-1858): The
        Pioneer of Physiology Teaching in British Schools."
        School Science Review 38 (1956), 48-52.

2344.   BRETT, GEORGE SIDNEY.   A History of Psychology.   3
        vols. London: G. Allen & Company, Ltd., 1912-21.
        (Republished as Brett's History of Psychology.
        Edited by R. S. Peters, New York: Macmillan & Co.,
        1953)

2345.   CANTOR, G. N.   "The Edinburgh Phrenology Debate:
        1803-1828." Annals of Science 32 (1975), 195-218.

2346.   CARDNO, J. A.   "Bain and Physiological Psychology."
        Australian Journal of Psychology 7 (1955), 108-119.

2347.   CARDNO, J. A.   "Bain as a Social Psychologist."
        Australian Journal of Psychology 8 (1956), 66-76.

2348.   CARMICHAEL, LEONARD.   "Sir Charles Bell: A Contri-
        bution to the History of Physiological Psychology."
        Psychology Review 33 (1926), 188-217.

2349.   COLLINS, PHILIP.   "When Morals Lay in Lumps."
        Listener 90 (Aug. 16, 1973), 213-215. (Includes
        phrenology and J. S. Mill)

2350.   CROSS, ROBERT C.   "Alexander Bain." Aberdeen
        University Review 44 (Spring 1971), 1-9.

2351.   DALLENBACH, KARL M.   "The History and Derivation of
        the Word 'Function' as a Systematic Term in Psy-
        chology." American Journal of Psychology 26 (1915),
        473-484.

2352.   DAVIDSON, WILLIAM L.   "Prof. Bain's Philosophy."
        Mind 13 (1904), 161-179.
2353.   DAVIDSON, WILLIAM L.   "Professor Bain."   Mind 13
        (1904), 151-155.
2354.   DENNIS, WAYNE. (Ed.)   Readings in the History of
        Psychology.   New York: Appleton-Century-Crofts,
        1948.
2355.   DENTON, GEORGE B.   "Early Psychological Theories of
        Herbert Spencer."   American Journal of Psychology
        32 (1921), 5-15.
2356.   DICKINSON, ZENAS CLARK.   "Utilitarian Psychology:
        The Two Mills and Bain."   In Economic Motives: A
        Study in the Psychological Foundation of Economic
        Theory.   Cambridge: Harvard University Press,
        1922, 67-80.
2357.   ENGELHARDT, H. TRISTRAM, JR.   "John Hughlings
        Jackson and the Mind-Body Relation."   Bulletin of
        the History of Medicine 49 (1975), 137-151.
2358.   FELTES, N. N.   "Phrenology from Lewes to George
        Eliot."   Studies in the Literary Imagination 1, No.
        1 (1968), 13-22.
2359.   FISCH, MAX H.   "Alexander Bain and the Genealogy of
        Pragmatism."   Journal of the History of Ideas 15
        (1954), 413-444.
2360.   FISHMAN, STEPHEN MICHAEL.   "James and Lewes on
        Unconscious Judgement."   Journal of the History of
        Behavioral Sciences 4 (1968), 335-348.
2361.   FLUGEL, J. C.   A Hundred Years of Psychology,
        1833-1933.   London: Duckworth, 1933.
2362.   FRENCH, RICHARD D.   "Some Concepts of Nerve Struc-
        ture and Function in Great Britain, 1875-1885:
        Background to Sir Charles Sherrington and the
        Synapse Concept."   Medical History 14 (1970),
        154-165.   (Includes G. J. Romanes)
2363.   FULLERTON, GEORGE STUART.   "The Influence of Darwin
        on the Mental and Moral Sciences."   Proceedings

of the American Philosophical Society 48 (1909), xxv-xxxvii.

2364. GHISELIN, MICHAEL T. "Darwin and Evolutionary Psychology." Science 179 (1973), 964-968.

2365. GIUSTINO, DAVID A. DE. Conquest of Mind: Phrenology and Victorian Social Thought. London: Croom Helm; Totowa, N. J.: Rowman and Littlefield, 1975.

2366. GIUSTINO, DAVID A. DE. "Phrenology in Britain, 1815-1855: A Study of George Combe and His Circle." Dissertation Abstracts International 30 (1970), 5375A (University of Wisconsin, 1969).

2367. GIUSTINO, DAVID A. DE. "Reforming the Commonwealth of Thieves: British Phrenologists and Australia." Victorian Studies 15 (June 1972), 439-461.

2368. GRANT, A. CAMERON. "Combe on Phrenology and Free Will: A Note on XIXth Century Secularism." Journal of the History of Ideas 26 (1965), 141-147.

2369. GRANT, A. CAMERON. "George Combe and His Circle, with Particular Reference to His Relations with the United States of America." Ph.D. Dissertation, University of Edinburgh, 1961.

2370. GRANT, A. CAMERON. "New Light on an Old View." Journal of the History of Ideas 29 (Apr.-June 1968), 293-301. (G. Combe, Robert Owen)

2371. GRAY, PHILIP HOWARD. "The Morgan-Romanes Controversy: A Contradiction in the History of Comparative Psychology." Proceedings of the Montana Academy of Sciences 23 (1963), 225-230.

2372. GRAY, PHILIP HOWARD. "Prerequisite to an Analysis of Behaviorism: The Conscious Automaton Theory from Spalding to William James." Journal of the History of Behavioral Sciences 4 (1968), 365-376. (Includes G. J. Romanes)

2373. GRAY, PHILIP HOWARD. "Spalding and His Influence on Research in Developmental Behavior." Journal of

the History of Behavioral Sciences 3 (1967), 168-
179.

2374.   GREENBLATT, SAMUEL H.   "Hughlings Jackson's First
Encounter with the Work of Paul Broca: The Physiological and Philosophical Background." Bulletin of
the History of Medicine 44 (1970), 555-570.

2375.   GREENBLATT, SAMUEL H.   "The Major Influences on the
Early Life and Work of John Hughlings Jackson."
Bulletin of the History of Medicine 39 (1965),
346-376.

2376.   GREENWAY, A. P.   "The Incorporation of Action into
Associationism: The Psychology of Alexander Bain."
Journal of the History of Behavioral Sciences 9
(1973), 42-52.

2377.   GRINDER, ROBERT E.   A History of Genetic Psychology: The First Science of Human Development.   New
York, London, Sydney: John Wiley and Sons, Inc.,
1967.   (Includes J.-B. de Lamarck, Darwin, H.
Spencer, T. Huxley, H. Drummond, G. Romanes)

2378.   HALDANE, J. B. S.   "Introducing Douglas Spalding."
British Journal of Animal Behaviour 2 (1954), 1.

2379.   HAMLYN, D. W.   "Bradley, Ward, and Stout."   In
Benjamin B. Wolman (Ed.), Historical Roots of Contemporary Psychology.   New York: Harper and Row,
1968, 298-320.

2380.   HARVEY, NIGEL.   "Phrenology."   Listener 90 (Aug.
30, 1973), 283.   (Letter to editor on J. S. Mill
and phrenology)

2381.   HEARNSHAW, LESLIE SPENCER.   A Short History of
British Psychology, 1840-1940.   London: Methuen &
Co., 1964.

2382.   HERRNSTEIN, RICHARD J. AND EDWIN G. BORING.   (Eds.)
A Source Book in the History of Psychology.   Cambridge: Harvard University Press, 1965.

2383.   HICKS, GEORGE. DAWES.   "Professor Ward's
Psychology."   Mind 30 (1921), 1-24.

2384.   HITZIG, EDUARD. "Hughlings Jackson and the Cor-
        tical Motor Centres in the Light of Physiological
        Research." Brain 23 (1900), 545-581.

2385.   HOLLANDER, BERNARD. "In Commemoration of Francis
        Joseph Gall (1758-1828)." Ethnological Journal 13
        (1928), 51-64.

2386.   HOWARD, D. T. "The Influence of Evolutionary
        Doctrine on Psychology." Psychology Review 34
        (1927), 305-312.

2387.   HUARD, PIERRE. "Quelques aspects de Paul Broca
        (1824-1880)." Clio Medica 1 (1966), 289-301.

2388.   JEFFERSON, SIR GEOFFREY. "The Contemporary Re-
        action to Phrenology." In Selected Papers. Lon-
        don: Pitman Medical Publishing Co. Ltd., 1960,
        94-112. (Includes F. J. Gall, J. Spurzheim, G.
        Combe, the reception of phrenology in Great
        Britain)

2389.   JOHNSON, M. L. "George Combe and George Eliot."
        Westminster Review 166 (Nov. 1906), 557-568.

2390.   KNOTT, J. "Franz Joseph Gall and Phrenology."
        Westminster Review 166 (Aug. 1906), 150-163.

2391.   LAIRD, J. "James Ward's Account of the Ego."
        Monist 36 (1926), 90-110.

2392.   LANTERI-LAURA, GEORGES. Histoire de la phrénolo-
        gie: l'homme et son cerveau selon F. J. Gall.
        Paris: Presses Universitaires de France, 1970.

2393.   LASSEK, ARTHUR M. The Unique Legacy of Doctor
        Hughlings Jackson. Springfield, Illinois: Thomas,
        1970.

2394.   LEIGH, DENIS. The Historical Development of
        British Psychiatry. New York: Pergamon Press,
        1961.

2395.   LEROUX, EMMANUEL. "James Ward's Doctrine of Ex-
        perience." Monist 36 (1926), 70-89.

2396.   LESKY, ERNA. "Structure and Function in Gall."
        Bulletin of the History of Medicine 44 (1970),
        297-314.

2397.   LEVIN, MAX.   "The Mind-Brain Problem and Hughlings
        Jackson's Doctrine of Concomitance."   American
        Journal of Psychiatry 116 (1960), 718-722.
2398.   LEWIS, AUBREY.   "Henry Maudsley, His Work and
        Influence."   Journal of Mental Science 97 (1951),
        259-277.
2399.   LOPEZ PIÑERO, JOSE MARIA.   "Condicionamientos
        históricos de la evolución de la obra de John Hugh-
        lings Jackson."   Episteme 6 (1972), 266-293.
2400.   LOPEZ PIÑERO, JOSE MARIA.   John Hughlings Jackson,
        1835-1911: evolucionismo y neurología.   Madrid:
        Editorial Moneda y Crédito, 1973.
2401.   LYON, JUDSON S.   "Romantic Psychology and the Inner
        Senses: Coleridge."   PMLA 81 (1966), 246-260.
2402.   MCLAREN, ANGUS.   "Phrenology: Medium and Message."
        Journal of Modern History 46 (1974), 86-97.
        (Phrenology and the acceptance by the lower classes
        of radicalism and free thought)
2403.   MAGOUN, H. W.   "Darwin and Concepts of Brain Func-
        tion."   In A. Fessard, R. W. Gerard, J. Konorski
        (Eds.), The Council for International Organizations
        of Medical Sciences, Brain Mechanisms and Learning.
        Springfield, Illinois: Charles C. Thomas, 1961,
        1-20.
2404.   MISCHEL, THEODORE.   "'Emotion' and 'Motivation' in
        the Development of English Psychology: D. Hartley,
        James Mill, A. Bain."   Journal of the History of
        Behavioral Sciences 2 (1966), 123-144.
2405.   MURPHY, GARDNER.   Historical Introduction to Modern
        Psychology.   London: Kegan, Paul & Co., 1929.
2406.   PARSSINEN, T. M.   "Popular Science and Society:
        The Phrenology Movement in Early Victorian Britain."
        Journal of Social History 8 (Fall 1974), 1-20.
2407.   PEEL, E. A.   "The Permanent Contribution of Francis
        Galton to Psychology."   British Journal of Educa-
        tional Psychology 24 (Feb. 1954), 9-16.

2408. POYNTER, FREDERICK NOEL LAWRENCE. (Ed.) The History and Philosophy of Knowledge of the Brain and Its Functions: An Anglo-American Symposium, London, July 15th-17th, 1957. Oxford: Blackwell, 1958. (Includes F. J. Gall, phrenology, H. Jackson)

2409. RIESE, WALTHER AND E. C. HOFF. "A History of the Doctrine of Cerebral Localization." Journal of the History of Medicine and Allied Sciences 5 (1950), 51-71; 6 (1951), 439-470.

2410. RIESE, WALTHER. "The Sources of Jacksonian Neurology." Journal of Nervous and Mental Disease 124 (1956), 125-134.

2411. SENSEMAN, WILFRED M. "Charlotte Brontë's Use of Physiognomy and Phrenology." Papers of Michigan Academy of Science, Arts, and Letters 38 (1953), 475-486.

2412. SHAPIN, STEVEN. "Phrenological Knowledge and the Social Structure of Early 19th-Century Edinburgh." Annals of Science 32 (1975), 219-243.

2413. SHEARER, NED A. "Alexander Bain and the Classification of Knowledge." Journal of the History of Behavioral Sciences 10 (1974), 56-73.

2414. SHERRINGTON, CHARLES S. "Sir David Ferrier, 1843-1928." Proceedings of the Royal Society 103B (1928), viii-xvi.

2415. SOKAL, MICHAEL M. "Psychology at Victorian Cambridge: The Unofficial Laboratory of 1887-1888." Proceedings of the American Philosophical Society 116 (1972), 145-147. (James Ward, James McKeen Cattell)

2416. SPOERL, HOWARD D. "Faculties versus Traits: Gall's Solution." Character and Personality 4 (1935-36), 216-231.

2417. STENGEL, E. A. "Hughlings Jackson's Influence in Psychiatry." British Journal of Psychiatry 109 (1963), 348-355.

2418.   STOUT, G. F.   "Ward As a Psychologist."   Monist 36
        (1926), 20-55.

2419.   TEMKIN, OWSEI.   "Gall and the Phrenological Move-
        ment."   Bulletin of the History of Medicine 21
        (1947), 275-321.

2420.   TEMKIN, OWSEI.   "Remarks on the Neurology of Gall
        and Spurzheim."   In Edgar Ashworth Underwood (Ed.),
        Science, Medicine and History: Essays on the
        Evolution of Scientific Thought and Medical Prac-
        tice Written in Honor of Charles Singer. 2 vols.
        London, New York: Oxford University Press, 1953,
        Vol. 2, 282-289.

2421.   TOLPIN, MARTHA.   "The Darwinian Influence on
        Psychological Definitions of Feminity in England,
        1871-1914." Ph.D. Dissertation, Harvard University,
        1972.

2422.   WALKER, A. E.   "The Development of the Concept of
        Cerebral Localisation in the Nineteenth Century."
        Bulletin of the History of Medicine 31 (1957),
        99-121.

2423.   WALSH, ANTHONY A.   "George Combe: A Portrait of a
        Heretofore Generally Unknown Behaviorist." Journal
        of the History of Behavioral Sciences 7 (1971),
        269-278.

2424.   WALSHE, F. M. R.   "Contributions of John Hughlings
        Jackson to Neurology." Archives of Neurology 5
        (1961), 119-131.

2425.   WARREN, HOWARD CROSBY.   A History of the Associa-
        tion Psychology from Hartley to Lewes. London:
        Constable & Co., 1921.

2426.   WILLIAMS, D. G.   "Alexander Bain as an Educational
        Psychologist." Aberdeen University Review 45
        (1973-74), 380-389.

2427.   WROBEL, ARTHUR.   "Orthodoxy and Respectability in
        Nineteenth-Century Phrenology." Journal of Popular
        Culture 9 (1975), 38-50.

2428.   YOUNG, GEORGE MALCOLM.   "Victorian Psychology."
        Times Literary Supplement (Jan. 25, 1936), 75.
2429.   YOUNG, ROBERT MAXWELL.   "Animal Soul."   In Paul
        Edwards (Ed.), The Encyclopedia of Philosophy.   8
        vols. New York: Collier-Macmillan, 1967, Vol. 1,
        122-127.
2430.   YOUNG, ROBERT MAXWELL.   "The Functions of the
        Brain: Gall to Ferrier (1808-1886)."   Isis 59
        (1968), 251-268.
2431.   YOUNG, ROBERT MAXWELL.   Mind, Brain and Adapta-
        tion in the Nineteenth Century: Cerebral Localiza-
        tion and Its Biological Context from Gall to Ferrier.
        Oxford: Clarendon Press, 1970.
2432.   YOUNG, ROBERT MAXWELL.   "The Role of Psychology in
        the Nineteenth-Century Evolutionary Debate."   In
        Mary Henle, Julian Jaynes and John J. Sullivan
        (Eds.), Historical Conceptions of Psychology.   New
        York: Springer Pub. Co., 1973, 180-204.
2433.   YOUNG, ROBERT MAXWELL.   "Scholarship and the History
        of the Behavioural Sciences."   History of Science 5
        (1966), 1-51.
2434.   ZANGWILL, O. L.   "The Cerebral Localization of
        Psychological Functions."   Advancement of Science
        20 (1963-64), 335-344.

PART C. RELIGION - IDEAS AND INSTITUTIONS

XIV. Church of England

1. General

Listed in this section are general works on the
Church of England and studies of individual Anglican
clergymen. Specific subjects covered include Anglo-
Catholicism, the High Church, the episcopate, educational
institutions for clergymen, internal church controversies,
relations between Church and State, religious communities,
the social role of the Church, and the relationship between the Anglican Church and other churches. There are
also a few entries on missionary activities and on the
Salvation Army. See II. Religion; III. Ideas. . . ;
IV. Education; XVII. Christian Socialism; other
categories and sections on religion.

2435. ABEL, MRS. EMILY K. "Canon Barnett and the First
        Thirty Years of Toynbee Hall." Ph.D. Dissertation,
        University of London, 1969.
2436. ADDISON, WILLIAM GEORGE. "Church, State, and
        Mr. Gladstone." Theology 39 (1939), 362-370,
        439-446.
2437. ADDISON, WILLIAM GEORGE. The English Country
        Parson. London: J. M. Dent, 1947.
2438. ADDISON, WILLIAM GEORGE. J. R. Green. London:
        Society for Promoting Christian Knowledge, 1946.
2439. ADDLESHAW, GEORGE WILLIAM OUTRAM AND FREDERICK
        ETCHELLS. The Architectural Setting of Anglican
        Worship: An Inquiry into the Arrangements for
        Public Worship in the Church of England from the
        Reformation to the Present Day. London: Faber and
        Faber, 1948.
2440. ADLARD, JOHN. "The Failure of Francis Kilvert."
        Michigan Quarterly Review 13 (1974), 130-137.

2441. ADY, CECILIA M. "The Post-Reformation Episcopate in England: (ii) From the Restoration to the Present Day." In Kenneth Escott Kirk (Ed.), The Apostolic Ministry: Essays on the History and Doctrine of Episcopacy. London: Hodder and Stoughton Limited, 1946, 433-460.

2442. ALDRICH, R. E. "H. H. Milman and Popular Education, 1846." British Journal of Educational Studies 21 (June 1973), 172-179.

2443. ALLCHIN, ARTHUR MACDONALD. "The Revival of the 'Religious Life' in the Church of England during the Nineteenth Century." B. Litt. Dissertation, University of Oxford, 1956.

2444. ALLCHIN, ARTHUR MACDONALD. The Silent Rebellion: Anglican Religious Communities, 1845-1900. London: SCM Press, 1958.

2445. ALTHOLZ, JOSEF LEWIS. "Gladstone and the Vatican Decrees." The Historian 25 (May 1963), 312-324.

2446. ANDERSON, OLIVE. "Gladstone's Abolition of Compulsory Church Rates: A Minor Political Myth and Its Historiographical Career." Journal of Ecclesiastical History 25 (Apr. 1974), 185-198.

2447. ANDREWS, J. H. B. "Essay in Historical Revision." Theology 77 (Jan. 1974), 27-37. (Church and clergy in the early 19th century)

2448. ANON. "The Episcopate of Bishop Creighton." Church Quarterly Review 52 (1901), 84-100.

2449. ANON. "George Ridding, First Bishop of Southwell." Church Quarterly Review 60 (1905), 241-285.

2450. ANON. "Mandell Creighton." Quarterly Review 193 (1901), 584-622.

2451. ANSON, PETER FREDERICK. The Call of the Cloister: Religious Communities and Kindred Bodies in the Anglican Communion. New York: Macmillan, 1953.

2452. ANSON, PETER FREDERICK. "The Foundation of Our Community." Pax 29 (1939), 25-35, 97-109. (Anglican Benedictines of Caldey, 1892-1913)

2453.   ARMSTRONG, ANTHONY.   The Church of England, the
        Methodists and Society, 1700-1850.   New York:
        Rowman and Littlefield, 1973.
2454.   ARNOLD, RALPH.   The Whiston Matter: The Reverend
        Robert Whiston versus the Dean and Chapter of
        Rochester.   London: R. Hart-Davis, 1961.
2455.   ATLAY, JAMES BERESFORD.   The Life of the Right
        Reverend Ernest Roland Wilberforce: First Bishop
        of Newcastle-on-Tyne and Afterward Bishop of
        Chichester.   London: Smith, Elder and Co., 1912.
2456.   ATTWATER, DONALD.   Father Ignatius of Llanthony:
        A Victorian.   London: Cassell and Company, Ltd.,
        1931.   (Joseph Lyne)
2457.   BAHLMAN, DUDLEY W. R.   "The Queen, Mr. Gladstone,
        and Church Patronage."   Victorian Studies 3 (1959-
        60), 349-380.
2458.   BALLEINE, GEORGE REGINALD.   The Layman's History
        of the Church of England.   London: Longmans, 1923.
2459.   BARING-GOULD, SABINE.   The Church Revival:
        Thoughts Thereon and Reminiscences.   London:
        Methuen, 1914.
2460.   BARNETT, HENRIETTA OCTAVIA.   Canon Barnett: His
        Life, Work, Friends.   2 vols.   London: J. Murray,
        1918.
2461.   BARRATT, D. M.   "Correspondence of Thomas Burgess,
        Bishop of Salisbury (1756-1837)."   Bodleian Library
        Record 3 (Dec. 1951), 274-278.
2462.   BATTISCOMBE, GEORGINA.   "Gerald Wellesley: A
        Victorian Dean."   History Today 19 (1969), 159-166.
2463.   BECKETT, J. C.   "Select Documents, XXII: Glad-
        stone, Queen Victoria, and the Disestablishment of
        the Irish Church, 1868-9."   Irish Historical
        Studies 13 (1962), 38-47.
2464.   BELL, GEORGE KENNEDY ALLEN.   A Brief Sketch of
        the Church of England.   London: S.C.M., 1929.

2465. BELL, GEORGE KENNEDY ALLEN. Randall Davidson,
       Archbishop of Canterbury. 2 vols. London: Oxford
       University Press, H. Milford, 1935.

2466. BELL, P. M. H. Disestablishment in Ireland and
       Wales. London, S.P.C.K., 1969.

2467. BENSON, EDWARD FREDERIC. As We Were: A Victorian
       Peep-Show. London, New York: Longmans, Green and
       Co., 1930. (Edward White Benson)

2468. BENSON, EDWARD FREDERIC. Our Family Affairs,
       1867-96. London, New York: Cassell and Company,
       Ltd., 1920. (Edward White Benson)

2469. BENTLEY, JAMES. "The Bishops, 1860-1960: An Elite
       in Decline." A Sociological Yearbook of Religion
       in Britain (1972), 161-183.

2470. BERTOUCH, BEATRICE DE. The Life of Father Ignatius,
       O.S.B.: The Monk of Llanthony. London: Methuen
       and Company, 1904.

2471. BEST, GEOFFREY FRANCIS ANDREW. "Church and State
       in English Politics, 1800-33." Ph.D. Dissertation,
       University of Cambridge, 1955.

2472. BEST, GEOFFREY FRANCIS ANDREW. "The Constitutional
       Revolution, 1828-32, and Its Consequences for the
       Established Church." Theology 62 (June 1959),
       226-234.

2473. BEST, GEOFFREY FRANCIS ANDREW. "The Protestant
       Constitution and Its Supporters, 1800-1829."
       Transactions of the Royal Historical Society (5th
       Series) 8 (1959), 105-127.

2474. BEST, GEOFFREY FRANCIS ANDREW. Temporal Pillars:
       Queen Anne's Bounty, the Ecclesiastical Commis-
       sioners, and the Church of England. Cambridge,
       England: University Press, 1964.

2475. BILL, EDWARD GEOFFREY WATSON. (Ed.) Anglican
       Initiatives in Christian Unity. London: S.P.C.K.,
       1967. (Includes the Church of England's relations
       with the Roman Catholic Church, Free Churches,

Orthodox Churches, Old Catholic Churches, and
Lutheran and Reformed Churches)

2476. BINNALL, PETER B. G. "The Reverend Henry Walter,
Rector of Haselbury Bryan, 1821-1859." Somerset
and Dorset Notes and Queries 24, Pt. 216 (Mar.
1943), 10-14.

2477. BLAND, J. The Development of the Church of England. London: Elliot Stock, 1919.

2478. BODENHEIMER, F. S. "Canon Henry Baker Tristram of
Durham (1822-1906)." Durham University Journal 49
(1957), 95-97.

2479. BOULTER, BENJAMIN CONSITT. The Anglican Reformers.
London: P. Allan, 1933. (Lesser figures among
High Churchmen)

2480. BOWEN, DESMOND. "Anglo-Catholicism in Victorian
England." Canadian Journal of Theology 12 (Jan.
1966), 35-49.

2481. BOWEN, DESMOND. The Idea of the Victorian Church:
A Study of the Church of England 1833-1889. Mon-
treal: McGill University Press, 1968.

2482. BRACKWELL, C. "The Church of England and Social
Reform, 1830-1850." M.A. Dissertation, University
of Birmingham, 1949.

2483. BRADSHAW, PAUL F. The Anglican Ordinal: Its
History and Development from the Restoration to the
Present. London: S.P.C.K., 1971.

2484. BRANDRETH, HENRY REYNAUD TURNER. "Episcopi
Vagantes" and the Anglican Church. London:
S.P.C.K., 1947. (The Order of Corporate Reunion in
the 19th and 20th centuries)

2485. BREMOND, HENRI. "L'Evolution du clergé anglican:
William Charles Lake (1817-1897)." Etudes publiées
par des PP. de la Compagnie de Jésus 93 (1902),
793-817.

2486. BRIGGS, JOHN HENRY YORK. "Church, Clergy and
Society in Victorian Britain." Baptist Quarterly
23 (Jan. 1970), 223-233.

2487. BRILIOTH, YNGVE TORGNY. "La Renaissance euchar-
istique anglicane." Oecumenica 1, No. 2 (juin
1934), 119-133.

2488. BRITTEN, J. "Anglicanism Sixty Years Ago."
Dublin Review 146 (Apr. 1910), 345-370.

2489. BROCK, WILLIAM H. "The Fortieth Article of Reli-
gion and the F.R.S. Who Fairly Represents Science.
The Declaration of Students of the Natural and
Physical Sciences, 1865." Clio (University of
Leicester History Society), No. 6 (1974), 15-21.
(Later version by Brock and R. M. Macleod, "The
Scientists' Declaration . . . ." The British Journal
for the History of Science 9, No. 31, Mar. 1976,
39-66)

2490. BROMLEY, JOHN. The Man of Ten Talents: A Portrait
of Richard Chenevix Trench (1807-1886), Philologist, Poet, Theologian, Archbishop. London:
S.P.C.K., 1959.

2491. BROSE, OLIVE J. Church and Parliament: The Re-
shaping of the Church of England, 1828-1860.
Stanford, California: Stanford University Press,
1959.

2492. BROSE, OLIVE J. "The Irish Precedent for English
Church Reform: The Church Temporalities Act of
1833." Journal of Ecclesiastical History 7 (1956),
204-225.

2493. BROSE, OLIVE J. "The Survival of the Church of
England as by Law Established - 1828-1860." Dis-
sertation Abstracts 16 (1956), 1669 (Columbia
University, 1956).

2494. BROWN, CHARLES KENNETH FRANCIS. A History of
the English Clergy, 1800-1900. London: Faith
Press, 1953.

2495. BROWN, H. M. "High Church Tradition in Cornwall."
Church Quarterly Review 150 (1950), 69-80.

2496.   BRUNDAGE, A.   "John Richard Green and the Church:
        The Making of a Social Historian." Historian 35
        (Nov. 1972), 32-42.

2497.   BULLOCK, FREDERICK WILLIAM BAGSHAWE.   The History
        of Ridley Hall, Cambridge. 2 vols. Cambridge,
        England: Printed for the Council of Ridley Hall at
        the University Press, 1941-53. (Educational insti-
        tution for prospective clergymen)

2498.   BULLOCK, FREDERICK WILLIAM BAGSHAWE.   A History
        of Training for the Ministry of the Church of
        England in England and Wales from 1800 to 1874.
        St. Leonards-on-Sea: Budd and Gillatt, 1955.

2499.   BURDETT, OSBERT.   The Rev. Smith, Sydney. London:
        Chapman and Hall, Ltd., 1934.

2500.   BURGESS, H. J.   Enterprise in Education: The
        Story of the Work of the Established Church in the
        Education of the People Prior to 1870. London:
        National Society, S.P.C.K., 1958.

2501.   BURLEIGH, JOHN HENDERSON SEAFORTH.   "Henry Francis
        Lyte, 1793-1847." Evangelical Quarterly 20, No. 1
        (Jan. 1948), 16-21.

2502.   CALDER-MARSHALL, ARTHUR.   The Enthusiast. London:
        Faber Press, 1962.   (The Rev. Joseph Lyne, or
        Father Ignatius of Llanthony)

2503.   CAMERON, ALLAN THOMAS.   The Religious Communities
        of the Church of England. London: Faith Press,
        1918.

2504.   CARPENTER, EDWARD FREDERICK.   Cantuar: The
        Archbishops in Their Office. London: Cassell,
        1971. (Includes W. Howley, J. B. Sumner, A. C.
        Tait, E. W. Benson, F. Temple, R. Davidson;
        Cantuar refers to Canterbury)

2505.   CARPENTER, SPENCER CECIL.   Church and People,
        1789-1889: A History of the Church of England from
        William Wilberforce to "Lux Mundi." London:
        Society for the Promotion of Christian Knowledge;
        New York: The Macmillan Co., 1933.

2506. CARPENTER, SPENCER CECIL. Winnington-Ingram:
       The Biography of Arthur Foley Winnington-Ingram,
       Bishop of London, 1901-1939. London: Hodder and
       Stoughton, 1949.

2507. CASTLE, JOHN. Cambridge Churchmen: An Account
       of the Anglo-Catholic Tradition at Cambridge. Cam-
       bridge: A. R. Mowbray, 1951.

2508. CHADWICK, OWEN. Edward King: Bishop of Lincoln,
       1885-1910. Lincoln: Friends of Lincoln Cathedral,
       1968.

2509. CHADWICK, OWEN. The Founding of Cuddesdon. Ox-
       ford: Printed by C. Batey at the University Press,
       1954. (Educational institution for prospective
       clergymen)

2510. CHADWICK, OWEN. The Victorian Church. 2 parts.
       Volumes 7 and 8 of J. C. Dickinson (Ed.), An Ec-
       clesiastical History of England. London: Adam &
       Charles Black; New York: Oxford University Press,
       1966-1970. (Second ed., part 1, Adam & Charles
       Black, 1970; Third ed., Black, 1971)

2511. CHADWICK, OWEN. Victorian Miniature. London:
       Hodder and Stoughton, 1960. (An account of
       an obscure East Anglian parson and his squire)

2512. CLARKE, CHARLES PHILLIP STEWART. Short History
       of the Church of England: From the Earliest Times
       to the Present Day. London: Longmans, 1929.

2513. CLAYTON, JOSEPH. The Bishops as Legislators:
       A Record of Votes and Speeches Delivered by the
       Bishops of the Established Church in the House of
       Lords during the Nineteenth Century. London: A.
       C. Fifield, 1906.

2514. CLAYTON, JOSEPH. Father Dolling: A Memoir.
       London: W. Gardner, Darton and Co., 1902.

2515. CLAYTON, JOSEPH. Father Stanton of St. Alban's,
       Holborn: A Memoir. London: Wells, Gardner,
       Darton, 1913.

2516.   COCKSHUT, ANTHONY O. J.   Anglican Attitudes: A
        Study of Victorian Religious Controversies.   Lon-
        don: Collins, 1959.

2517.   COLEMAN, ARTHUR MAINWARING et al.   "Church
        Services."   Notes and Queries 191 (1946), 214;
        192 (1947), 40-41, 86-87, 239, 350.   (Changes in
        Church of England services since the Oxford Move-
        ment)

2518.   COLEMAN, B. I.   "Anglican Church Extension and
        Related Movements, c. 1800-60, with Special
        Reference to London."   Ph.D. Dissertation, Uni-
        versity of Cambridge, 1968.

2519.   COLLINS, PHILIP.   "The Rev. John Chippendale
        Montesquieu Bellew."   Listener 86 (1971), 716-718.

2520.   COLSON, PERCY.   Life of the Bishop of London:
        An Authorised Biography.   London: Jarrolds, 1935.
        (A. F. Winnington-Ingram)

2521.   COOK, MICHAEL. (Ed.)   The Diocese of Exeter in
        1821:   Bishop Carey's Replies to Queries before
        Visitation.   2 vols.   Torquay: The Society,
        1958-60.   (The Devon and Cornwall Record Society)

2522.   COOLEN, GEORGES.   Histoire de l'Eglise d'Angle-
        terre.   Paris: Bloud & Gay, 1932.

2523.   COOLIDGE, C. W.   "The Finances of the Church of
        England, 1830-1880."   Ph.D. Dissertation, Uni-
        versity of Dublin, 1958.

2524.   COOMBS, JOYCE.   Judgement on Hatcham: The History
        of a Religious Struggle, 1877-86.   London: Faith
        Press, 1969.   (Arthur Tooth of St. James, Hatcham,
        who was imprisoned for ritualism)

2525.   COOMBS, JOYCE.   "William Henry Whitworth, Victorian
        Minister, 1834-1885."   Church Quarterly Review 168
        (1967), 190-203.

2526.   COVERT, JAMES THAYNE, II.   "Mandell Creighton and
        English Education."   Dissertation Abstracts 29
        (1968), 207-208A (University of Oregon, 1967).

2527.   CRATCHLEY, W. J.   "Edward Copleston, Bishop of
        Llandaff."   B. Litt. Dissertation, University of
        Oxford, 1938.

2528.   CRATCHLEY, W. J.   "The Trials of R. D. Hampden."
        Theology 35 (1937), 211-226.

2529.   CROPPER, MARGARET BEATRICE.   Shining Lights: Six
        Anglican Saints of the Nineteenth Century.   London:
        Darton, Longman, and Todd, 1963.   (Lord Shaftes-
        bury, John C. Patteson, C. Rossetti, Edward King,
        Mother Cecile, Mary Brown)

2530.   CROSS, FRANK LESLIE.   Darwell Stone, Churchman
        and Counsellor.   London: Dacre Press, 1943.

2531.   CROSS, FRANK LESLIE.   Preaching in the Anglo-
        Catholic Revival.   London: Society for Promoting
        Christian Knowledge, 1933.

2532.   CUMING, G. J.   A History of the Anglican Liturgy.
        London: Macmillan, 1969.

2533.   CUNNINGHAM, J. M.   "Bishop Longley's Visitation
        Returns, 1836-56."   Ripon Diocesan Gazette 17
        (1946), 28-29, 61-62, 69-70.

2534.   CURR, H. S.   "Gladstone and the Bible."   Churchman
        55, No. 4 (Oct.-Dec. 1941), 243-252.

2535.   CURTIS, WILLIAM REDMOND.   The Lambeth Conferences:
        The Solution for Pan-Anglican Organization.   New
        York: Columbia University Press; London: P. S.
        King and Son, Ltd., 1942.

2536.   DAHMEN, GUNNAR.   "William Gladstone och ritualismen
        i England."   Svensk Teologisk Kvartalskrift 28
        (1952), 106-114.

2537.   DARK, SIDNEY.   Archbishop Davidson and the English
        Church.   London: P. Allan and Co., Ltd., 1929.

2538.   DARK, SIDNEY.   "Emancipation and the Catholic
        Movement in the Church of England."   Dublin Review
        184 (Apr. 1929), 287-294.

2539.   DARK, SIDNEY.   Lord Halifax: A Tribute.   London
        and Oxford: A. R. Mowbray and Co., Ltd.; Mil-

waukee: Morehouse Publishing Co., 1934. (Anglo-
Catholicism)

2540. DARK, SIDNEY. Wilson Carlile: The Laughing
Cavalier of Christ. London: J. Clarke, 1944.

2541. DAVIDSON, RANDALL THOMAS. "Henry Wace." Empire
Review 39 (1924), 135-140.

2542. DAVIDSON, RANDALL THOMAS. (Ed.) The Six Lambeth
Conferences, 1867-1920. London: Society for
Promoting Christian Knowledge, 1929.

2543. DAVIES, GEORGE COLLISS BOARDMAN. Henry Phillpotts: Bishop of Exeter, 1778-1869. London:
Society for the Promotion of Christian Knowledge,
1954.

2544. DAVIES, HORTON. "Dean Inge: The Outspoken Oracle."
Religion in Life 31 (Spring 1962), 244-253.

2545. DEANE, ANTHONY CHARLES. (Ed.) Pillars of the
English Church: Biographical Studies of Eminent
Churchmen. London and Oxford: A. R. Mowbray and
Co., Ltd.; Milwaukee: Morehouse Publishing Co.,
1934. (Includes R. W. Church, S. Wilberforce, F.
Temple, C. Gore, T. Arnold, F. D. Maurice, C.
Kingsley, H. S. Holland, C. Simeon, R. Dolling, W.
Hook)

2546. DEARMER, PERCY. Everyman's History of the English
Church. London, Oxford: A. R. Mowbray and Co.,
1909.

2547. DELL, ROBERT S. "Social and Economic Theories and
Pastoral Concerns of a Victorian Archbishop."
Journal of Ecclesiastical History 16 (1965), 196-
208. (J. B. Sumner)

2548. DENISON, K. M. "The Origins and Early Growth of
Anglican Sisterhoods in the Nineteenth Century."
Ph.D. Dissertation, University of Cambridge, 1971.

2549. DIBDIN, SIR LEWIS TONNA AND STANFORD EDWIN DOWNING.
The Ecclesiastical Commission: A Sketch of Its
History and Work. London: Macmillan and Co.,
1919.

2550. DICKINSON, B. H. C. Sabine Baring-Gould: Squarson, Writer and Folklorist, 1834-1924. Newton Abbot: David and Charles, 1970.

2551. DONALDSON, AUGUSTUS BLAIR. The Bishopric of Truro: The First Twenty-Five Years, 1877-1902. London: Rivingtons, 1902. (E. W. Benson, G. H. Wilkinson, John Gott)

2552. DOWDELL, VICTOR LYLE. Aristotle and Anglican Religious Thought. Ithaca, N.Y.: Cornell University Press, 1942.

2553. DRUMMOND, ANDREW LANDALE. "Father Ignatius, 1837-1908." Church Quarterly Review 151 (1950), 63-86. (Joseph Lyne)

2554. DUNKLEY, E. H. "Robert Scott, Dean of Rochester, 1870-1887." Friends of Rochester Cathedral Annual Report 4 (Feb. 1939), 23-28.

2555. DUNSTAN, G. R. et al. Aspects de l'anglicanisme: Colloque de Strasbourg, 14-16 juin 1972. Paris: Presses Universitaires de France, 1974. (Christian Socialists T. Hancock and S. Headlam; William Tuckwell; W. E. Gladstone and H. H. Henson on the idea of a national church)

2556. DUTHIE, DAVID WALLACE. (Ed.) A Bishop in the Rough. London: Smith, Elder and Co., 1909. (John Sheepshanks)

2557. EDWARDS, ALFRED GEORGE. Landmarks in the History of the Welsh Church. London: J. Murray, 1912.

2558. EDWARDS, DAVID L. Leaders of the Church of England, 1828-1944. London: Oxford University Press, 1971. (Includes T. Arnold, M. Arnold, J. H. Newman, J. Keble, S. Wilberforce, A. C. Tait, Lord Shaftesbury, F. D. Maurice, W. E. Gladstone, E. W. Benson, J. B. Lightfoot, B. F. Westcott, M. Creighton, R. Davidson, C. Gore, H. H. Henson, F. Temple, W. Temple)

2559. ELLIOTT-BINNS, LEONARD ELLIOTT. The Story of England's Church. London: S.P.C.K., 1945.

2560.   ELLSWORTH, LIDA E.   "Charles Lowder and the
        Ritualist Movement."   Ph.D. Dissertation, Uni-
        versity of Cambridge, 1975.

2561.   ELTON, GODFREY.   Edward King and Our Times.   Lon-
        don: G. Bles, 1958.

2562.   EMBRY, JAMES.   The Catholic Movement and the
        Society of the Holy Cross.   London: The Faith
        Press, Ltd.; Milwaukee: Morehouse Publishing Co.,
        1931.   (An Anglican religious community)

2563.   EVANS, JOHN H.   Churchman Militant: George
        Augustus Selwyn, Bishop of New Zealand and Lich-
        field.   London: Allen and Unwin, 1964.

2564.   FAGAN, E. F.   "The Religious Life of Mr. Glad-
        stone."   Church Quarterly Review 155 (1954), 16-21.

2565.   FAIR, JOHN D.   "The Irish Disestablishment Con-
        ference of 1869."   Journal of Ecclesiastical His-
        tory 26 (Oct. 1975), 379-394.

2566.   FALLOWS, W. G.   Mandell Creighton and the English
        Church.   London: Oxford University Press, 1964.

2567.   FARRAR, REGINALD.   The Life of Frederic William
        Farrar, Sometime Dean of Canterbury.   London: James
        Nisbet and Co., 1905.

2568.   FINLAYSON, ARTHUR ROBERT MORRISON.   Life of Canon
        Fleming: Vicar of St. Michael's, Chester Square,
        Canon of York, Chaplain in Ordinary to the King.
        London: James Nisbet and Co., 1909.

2569.   FITZGERALD, MAURICE HENRY.   A Memoir of Herbert
        Edward Ryle, K.C.V.O., D.D., Sometime Bishop of
        Winchester and Dean of Westminster.   London:
        Macmillan and Co., Ltd., 1928.

2570.   FLINDALL, R. P.   "Anglican and Roman Attitudes:
        1825-1875."   Church Quarterly Review 169 (Apr.-June
        1968), 206-215.

2571.   FLINDALL, R. P. (Ed.)   The Church of England,
        1815-1848:   A Documentary History.   London:
        S.P.C.K., 1972.

2572.   FLINDALL, R. P.   "The Parish Priest in Victorian
        England."   Church Quarterly Review 168 (July-Sept.
        1967), 296-306.   (Summary of eight meetings of the
        Flegg Deanery Clerical Society in the later 1850's)

2573.   FLOYER, JOHN KESTELL.   Studies in the History
        of English Church Endowments.   London: Macmillan
        and Co., Limited, 1917.

2574.   FOX, ADAM.   Dean Inge.   London: J. Murray, 1960.

2575.   GARBETT, CYRIL FORSTER.   Church and State in Eng-
        land.   London: Hodder and Stoughton, 1950.

2576.   GASH, NORMAN.   "Church and Dissent."   In Reaction
        and Reconstruction in English Politics 1832-1852.
        Oxford: At the Clarendon Press, 1965, 60-118.

2577.   GIROUARD, MARK.   "A Speculating Clergyman:
        Dr. Walker in Cornwall and London."   Country Life
        158 (1975), 842-845.   (Dr. Samuel Walker, London
        property speculator and Rector of St. Columb Major,
        Cornwall)

2578.   GLOYN, CYRIL KENNARD.   The Church in the Social
        Order: A Study of Anglican Social Theory from
        Coleridge to Maurice.   Forest Grove, Oregon:
        Pacific University, 1942.

2579.   GODFREY, JOHN.   "Victorian Rector on His Rounds."
        Theology 73 (Dec. 1970), 551-556.   (Robert Burr
        Bourne)

2580.   GOSSE, SIR EDMUND WILLIAM.   "Mandell Creighton
        1843-1901."   In Portraits and Sketches.   London:
        W. Heinemann, 1912, 163-196.

2581.   GOWING, ELLIS NORMAN.   John Edwin Watts-Ditch-
        field, First Bishop of Chelmsford.   London: Hodder
        and Stoughton, 1926.

2582.   GRAHAM, EDWARD.   The Harrow Life of Henry Montagu
        Butler.   London, New York: Longmans, Green and Co.,
        1920.

2583.   GRAY, ARTHUR ROMEYN.   "Archibald Campbell Tait."
        Sewanee Review 15 (1907), 385-408.

2584.   GRAY, ARTHUR ROMEYN.   "Mandell Creighton, Pastor,
        Scholar and Man."   Sewanee Review 15 (1907), 227-
        243.

2585.   GREAVES, R. W.   "Jerusalem Bishopric, 1841."
        English Historical Review 64 (July 1949), 328-352.

2586.   GRIERSON, JANET.   Isabella Gillmore: Sister to
        William Morris.   London: S.P.C.K., 1962.

2587.   GRIEVE, ALASTAIR, AND LINDSAY ERRINGTON.   "The
        Pre-Raphaelite Brotherhood and the Anglican High
        Church."   Burlington Magazine 3 (1969), 294-295,
        521-522.

2588.   GRIFFINHOOFE, CHARLES GEORGE.   "Benjamin Webb and
        St. Andrew's Wells Street."   Church Quarterly
        Review 79 (1915), 36-57.

2589.   HARRIS, RANSOM BAINE.   "The Christian Neoplatonism
        of William Ralph Inge."   Dissertation Abstracts
        International 32 (1971), 3411A (Temple University,
        1971).

2590.   HART, ARTHUR TINDAL.   The Country Priest in English
        History.   London: Phoenix House, 1959.

2591.   HART, ARTHUR TINDAL.   The Curate's Lot: The Story
        of the Unbeneficed English Clergy.   London: J.
        Baker, 1970.

2592.   HART, ARTHUR TINDAL AND EDWARD FREDERICK CARPENTER.
        The Nineteenth Century Country Parson (circa 1832-
        1900).   Shrewsbury: Wilding, 1954.

2593.   HEAD, FREDERICK WALDEGRAVE.   Six Great Anglicans:
        A Study of the History of the Church of England in
        the Nineteenth Century.   London: Student Christian
        Movement, 1929.   (C. Simeon, J. Keble, W. Hook, F.
        W. Robertson, C. Kingsley, S. Barnett)

2594.   HEALY, V.   "Liberalism and Church and State in
        England."   Historical Bulletin 34 (Nov. 1955),
        3-10.

2595.   HEENEY, BRIAN.   "A Theory of Pastoral Ministry in
        the Mid-Victorian Church of England."   Historical

Magazine of the Protestant Episcopal Church 43
(1974), 215-230.

2596. HEGARTY, W. J. "Gladstone's Attitude to Catholicism." Irish Ecclesiastical Record 86 (1956),
26-42.

2597. HELM, ROBERT M. The Gloomy Dean: The Thought
of William Ralph Inge. Winston-Salem: Blair,
1962.

2598. HENDERSON, L. O. "The Church of England in Its
Relations with Parliament and Crown, 1825-45."
Ph.D. Dissertation, University of London, 1961.

2599. HENSON, HERBERT HENSLEY. Bishoprick Papers.
London, New York: G. Cumberlege, Oxford University
Press, 1946. (Includes papers on the word
"Protestant" 1529-1929, the Thirty-Nine Articles,
J. B. Lightfoot, the Oxford Movement, religion and
education since 1870)

2600. HENSON, HERBERT HENSLEY. The Church of England.
Cambridge: Cambridge University Press, 1939.

2601. HERBERT, CHARLES. Twenty-Five Years as Bishop
of London. London: W. Gardner, Darton and Co.,
Ltd., 1926. (Arthur Winnington-Ingram)

2602. HERKLOTS, HUGH GERARD GIBSON. The Church of England and the American Episcopal Church: From the
First Voyages of Discovery to the First Lambeth
Conference. London: Mowbray; New York: More-
house-Barlow, 1966.

2603. HERKLOTS, HUGH GERARD GIBSON. Frontiers of the
Church: The Making of the Anglican Communion.
London: E. Benn, 1960.

2604. HEUSS-BURCKHARDT, URSULA. Gladstone und das
Problem der Staatskirche. Zürich: Europa Verlag,
1957.

2605. HILL, SIR FRANCIS. "Squire and Parson in Early
Victorian Lincolnshire." History 58 (1973),
337-349.

2606.   HINCHLIFF, PETER BINGHAM.   The One-sided Reci-
        procity: A Study in the Modification of the Es-
        tablishment. London: Darton, Longman and Todd,
        1966, 138-179.

2607.   HOLE, CHARLES. A Manual of English Church History.
        London: Longmans, 1910.

2608.   HOLLAND, HENRY SCOTT.   "The Life of Edward White
        Benson, Archbishop of Canterbury." Journal of
        Theological Studies 2 (1901), 26-48.

2609.   HOPKINSON, D. M.   "Parson Hawker of Morwenstow."
        History Today 18 (1968), 38-44.

2610.   HOW, FREDERICK DOUGLAS.   Archbishop Maclagan.
        London: W. Gardner, Darton, 1911.

2611.   HOW, FREDERICK DOUGLAS. A Memoir of Bishop Sir
        Lovelace Tomlinson Stamer. London: Hutchinson and
        Co., 1910.

2612.   HOWARD, WILLIAM CECIL J. P. J. P., 8TH EARL OF
        WICKLOW.   "The Monastic Revival in the Anglican
        Communion." Studies 42 (1953), 420-432.

2613.   HOWARD-FLANDERS, W.   The Church of England and
        Her Reformations. London: Heath, Cranton, 1932.

2614.   HUGHES, EDWARD. "The Bishops and Reform, 1831-3:
        Some Fresh Correspondence." English Historical
        Review 56 (July 1941), 459-490.

2615.   HUTTON, WILLIAM HOLDEN. William Stubbs, Bishop
        of Oxford, 1825-1901. London: A. Constable and
        Co., Ltd., 1906.

2616.   INGRAM, KENNETH. "A Century of Anglo-Catholicism."
        Bookman (London) 84 (June 1933), 140-141.

2617.   JACKSON, J. R. "Dean Farrar: A Study in Nineteenth-
        Century Anglicanism." Ph.D. Dissertation, University
        of Edinburgh, 1957.

2618.   JAEGER, HENRY-EVRARD. Anglikanismus. Mainz:
        Matthias-Grünewald, 1972.

2619.   JASPER, RONALD CLAUD DUDLEY. Arthur Cayley Head-
        lam: Life and Letters of a Bishop. London: Faith
        Press, 1960.

2620. JASPER, RONALD CLAUD DUDLEY. Prayer Book Revision
       in England 1800-1900. London: S.P.C.K., 1954.

2621. JELF, KATHARINE FRANCES. George Edward Jelf:
       A Memoir by His Wife. London: Skeffington and
       Son, 1909.

2622. JENKINS, CLAUDE. "William Howley, Archbishop of
       Canterbury, 1828-1848." Canterbury Cathedral
       Chronicle 41 (Oct. 1945) 10-13.

2623. JENNINGS, DEREK ANDREW. The Revival of the Con-
       vocation of York, 1837-1861. York: St. Anthony's
       Press, 1975.

2624. JESSOPP, A. A Short History of the Church of
       England. London: S.P.C.K., 1922.

2625. JOANNA, SISTER. "The Deaconess Community of
       St. Andrew." Journal of Ecclesiastical History 12
       (1961), 215-230. (An Anglican sisterhood)

2626. JONES, DAVID AMBROSE. A History of the Church
       in Wales. Carmarthen: W. Spurrell and Son, 1926.

2627. JONES, F. W. "Social Concern in the Church of
       England, As Revealed in Its Pronouncements on
       Social and Economic Matters, Especially during the
       Years 1880-1940." Ph.D. Dissertation, University
       of London, 1968.

2628. JONES, FRANCIS. "A Victorian Bishop of Llandaff."
       National Library of Wales Journal 19 (1975), 14-56.
       (Richard Lewis of Henllan)

2629. JOSAITIS, NORMAN F. Edwin Hatch and Early Church
       Order. Gembloux, Belgique: J. Duculot, 1971.

2630. K., M. B., H. G. AND J. F. B.-B. Henry Barclay
       Swete, D.D., F.B.A., Sometime Regius Professor of
       Divinity, Cambridge: A Remembrance. London:
       Macmillan, 1918.

2631. KEMP, ERIC WALDRAM. An Introduction to Canon
       Law in the Church of England. London: Hodder and
       Stoughton, 1957. (Largely pre-nineteenth century)

2632. KENDALL, J. F. A Short History of the Church
       of England. London: Black, 1910.

2633.   KENYON, JOHN PETER BLYTHE. "High Churchmen and
        Politics, 1845-1865." Dissertation Abstracts 29
        (1969), 2643-44A (University of Toronto, 1967).

2634.   KERR, ELEANOR. Hunting Parson: The Life and
        Times of the Rev. John Russell. London: Herbert
        Jenkins, 1963.

2635.   KITSON CLARK, GEORGE SIDNEY ROBERTS. Churchmen and
        the Condition of England 1832-1885. London:
        Methuen, 1973.

2636.   KNAPLUND, P. "William Ewart Gladstone, the
        Christian Statesman." Church Quarterly Review 162
        (1961), 467-475.

2637.   KNOX, WILFRED LAWRENCE. The Catholic Movement
        in the Church of England. London: Allan, 1923.
        (Anglo-Catholicism)

2638.   KNOX, WILFRED LAWRENCE AND ALEC ROPER VIDLER.
        The Development of Modern Catholicism. London:
        Allan, 1933. (Anglo-Catholicism from the middle of
        the nineteenth century to the present)

2639.   KOSZUL, A. "Etudes religieuses: le cas de Glad-
        stone." La Quinzaine 12 (1905), 437-454.

2640.   L., H. R. "Richard Cobbold of Wortham." East
        Anglian Miscellany (July-Dec. 1942), 28-34.

2641.   LAKE, KATHARINE. (Ed.) Memorials of William
        Charles Lake, Dean of Durham, 1869-1894. London:
        E. Arnold, 1901.

2642.   LAMBERT, RICHARD STANTON. The Cobbett of the
        West: A Study of Thomas Latimer and the Struggle
        between the Pulpit and Press at Exeter. London:
        Nicholson and Watson, Limited, 1939.

2643.   LARKIN, EMMET. "Church and State in Ireland in the
        Nineteenth Century." Church History 31 (Sept.
        1962), 294-306.

2644.   LATHBURY, DANIEL C. (Ed.) Correspondence on Church
        and Religion, by William Ewart Gladstone. 2 vols.
        London: J. Murray, 1910.

2645. LEFANU, WILLIAM RICHARD. Queen Anne's Bounty: A Short Account of Its History and Work. London: Macmillan, 1921.

2646. LEGER, A. "Un Homme d'Etat chrétien, William-Ewart Gladstone." Le Correspondant 215 (1904), 611-641.

2647. LEGG, JOHN WICKHAM. English Church Life from the Restoration to the Tractarian Movement. London: Longmans, Green and Co., 1914.

2648. LERRY, GEORGE GEOFFREY. Alfred George Edwards, Archbishop of Wales. Oswestry: Woodall, Minshall, Thomas and Co., 1940.

2649. LEWIS, C. J. "The Disintegration of the Tory-Anglican Alliance in the Struggle for Catholic Emancipation." Church History 29 (1960), 25-43.

2650. LIVINGSTON, JAMES C. The Ethics of Belief: An Essay on the Victorian Religious Conscience. Tallahassee, Fla.: American Academy of Religion, 1974.

2651. LOCKHART, JOHN GILBERT. Charles Lindley, Viscount Halifax. 2 vols. London: Geoffrey Bles, 1935-36. (Anglo-Catholicism)

2652. LOCKHART, JOHN GILBERT. Cosmo Gordon Lang. London: Hodder and Stoughton, 1949.

2653. LONGDEN, HENRY ISHAM. Northampton and Rutland Clergy from 1500. 16 vols. Northampton: Archer and Goodman, 1938-52.

2654. LONGFORD, ELIZABETH. "Queen Victoria's Religious Life." Wiseman Review No. 492 (1962), 107-126.

2655. LUNDEEN, THOMAS BAILEY. "The Bench of Bishops: A Study of the Secular Activities of the Bishops of the Church of England and of Ireland, 1801-1871." Dissertation Abstracts 24 (1963), 2447-48 (State University of Iowa, 1963).

2656. MCCARTHY, EDWARD J., S.J. "The Theology of the Eucharist as Sacrifice According to Darwell Stone." Dissertation Abstracts International 32 (1971), 2183A (Catholic University of America, 1971).

2657.   MCCARTHY, MICHAEL JOHN FITZGERALD.   Church and
        State in England and Wales 1829-1906.   Dublin:
        Hodges, Figgis and Co.; London: Simpkin, Marshall,
        Hamilton, Kent and Co., 1906.

2658.   MCCLATCHEY, DIANA.   Oxfordshire Clergy 1777-1869:
        A Study of the Established Church and of the Role
        of Its Clergy in Local Society.   Oxford: Clarendon
        Press, 1960.

2659.   MCDOWELL, ROBERT BRENDAN.   "The Anglican Episco-
        pate, 1780-1945."   Theology 50, No. 324 (June
        1947), 202-209.   (Changes in social composition)

2660.   MCGEE, EARL WILLIAM.   "The Anglican Church and
        Social Reform, 1830-1850."   Dissertation Abstracts
        20 (1960), 3710-11 (University of Kentucky, 1952).

2661.   MCGRATH, ALBERTUS M., SISTER.   "The History of the
        Anglo-Catholic Movement, 1850-1875."   Ph.D. Dis-
        sertation, University of Wisconsin, 1947.

2662.   MACKAY, HENRY FALCONAR BARCLAY.   Saints and Leaders.
        London: P. Allan and Co., Ltd., 1928.   (Includes
        Charles Lowder, Robert Dolling, Edward King, Arthur
        Stanton, Richard Benson, and Frank Weston)

2663.   MACKERNESS, ERIC DAVID.   The Heeded Voice: Studies
        in the Literary Status of the Anglican Sermon,
        1830-1900.   London: Heffer, 1959.

2664.   MCNEILL, JOHN T.   "Anglicanism on the Eve of the
        Oxford Movement."   Church History 3 (1934), 95-114.

2665.   MALDEN, RICHARD HENRY.   "The Church in the Nine-
        teenth Century."   In Essays Mainly on the Nine-
        teenth Century Presented to Sir Humphrey Milford.
        London: Oxford University Press, 1948, 97-112.

2666.   MALDEN, RICHARD HENRY.   The English Church and
        Nation.   London: S.P.C.K., 1952.   (A survey from
        the Roman occupation to the twentieth century)

2667.   MARSH, PETER T.   "Anglican Attitudes to Church,
        State and Society during Mr. Gladstone's First
        Ministry, 1868-74."   Ph.D. Dissertation, University
        of Cambridge, 1962.

2668. MARSH, PETER T. "The Primate and the Prime Minister: Archbishop Tait, Gladstone, and the National Church." Victorian Studies 9 (1965), 113-140.

2669. MARSH, PETER T. The Victorian Church in Decline: Archbishop Tait and the Church of England, 1866-1882. London: Routledge and Kegan Paul, 1969.

2670. MASON, ARTHUR JAMES. Life of William Edward Collins, Bishop of Gibraltar. London, New York: Longmans, Green and Co., 1912.

2671. MASON, ARTHUR JAMES. Memoir of George Howard Wilkinson, Bishop of St. Andrews, Dunkeld and Dunblane, and Primus of the Scottish Church, formerly Bishop of Truro. 2 vols. London, New York: Longmans, Green and Co., 1909.

2672. MASSINGHAM, BETTY. Turn on the Fountains: The Life of Dean Hole. London: Gollancz, 1974.

2673. MATHESON, PERCY EWING. The Life of Hastings Rashdall. London: Oxford University Press, H. Milford, 1928.

2674. MATHIESON, WILLIAM LAW. English Church Reform, 1815-1840. London, New York: Longmans, Green and Co., 1923.

2675. MAYOR, STEPHEN H. "The Anglo-Catholic Understanding of the Ministry: Some Protestant Comments." Church Quarterly 2 (Oct. 1969), 152-159.

2676. MAYOR, STEPHEN H. "Discussion of the Ministry in Late Nineteenth-Century Anglicanism." Church Quarterly 2 (July 1969), 54-62. (J. B. Lightfoot)

2677. MEACHAM, STANDISH. "The Church in the Victorian City." Victorian Studies 11 (1968), 359-378.

2678. MEISEL, D. M. "The Contribution of John Neville Figgis (1866-1919) to the Religious Thought of His Period." Ph.D. Dissertation, University of Edinburgh, 1954.

2679. MERMAGEN, ROBERT P. H. "The Established Church in England and Ireland: Principles of Church Reform."

Journal of British Studies 3, No. 2 (May 1964),
143-147. (The 1830's)

2680. MICKLEWRIGHT, FREDERICK HENRY AMPHLETT. "Clerical
Delinquency in the Early Nineteenth Century: The
Case of Dr. Free." _Notes and Queries_ 18 (May
1971), 175-183.

2681. MIDDAUGH, JOHN T. "The Reform of the Church of
England, 1830-1841." Ph.D. Dissertation, Case
Western Reserve University, 1950.

2682. MILMAN, ARTHUR. _Henry Hart Milman, D.D., Dean
of St. Paul's: A Biographical Sketch_. London: J.
Murray, 1900.

2683. MOBERLY, CHARLOTTE ANNE ELIZABETH. _Dulce Domum:
George Moberly_. London: J. Murray, 1911.

2684. MOLE, DAVID E. H. "Challenge to the Church." In
H. J. Dyos and Michael Wolff (Eds.), _The Victorian
City_. London and Boston: Routledge and Kegan
Paul, 1973, Vol. 2, 815-836. (The Industrial
Revolution and expanding cities)

2685. MOLE, DAVID E. H. "The Church of England and
Society in Birmingham, c. 1830-66." Ph.D. Dissertation, University of Cambridge, 1961.

2686. MOLESWORTH, SIR GUILFORD LINDSEY. _Life of John
Edward Nassau Molesworth, D.D., an Eminent Divine
of the Nineteenth Century, by His Youngest Son_.
London: Longmans, Green, 1915.

2687. MONTGOMERY, HENRY HUTCHINSON. "Randall Thomas
Davidson." _Church Quarterly Review_ 111 (1931),
1-14.

2688. MOORE, MARY. _Winfrid Burrows, 1858-1929: Bishop
of Truro, 1912-1919, Bishop of Chichester, 1919-1929_.
London: Society for Promoting Christian Knowledge,
1932.

2689. MOORMAN, JOHN RICHARD HUMPIDGE. _A History of
the Church in England_. London: A. and C. Black,
1953.

2690. MORGAN, D. H. J. "Social and Educational Back-
       grounds of English Diocesan Bishops, 1860-1960."
       M.A. Dissertation, University of Hull, 1963.

2691. MORRISH, P. S. "County and Urban Dioceses:
       Nineteenth-Century Discussions on Ecclesiastical
       Geography." Journal of Ecclesiastical History 26
       (1975), 279-300.

2692. MORRISH, P. S. "New Anglican Bishoprics, 1836-1919,
       Their Creation and Endowment in England, with
       Special Reference to Truro and Birmingham." M.A.
       Dissertation, University of London, 1965.

2693. MORSE-BOYCOTT, REV. DESMOND LIONEL. They Shine
       like Stars. London, New York: Skeffington, 1947.
       (The Anglo-Catholic movement since 1845)

2694. MOSS, CLAUDE BEAUFORT. The Orthodox Revival:
       1833-1933. Oxford and London: A. R. Mowbray and
       Co., 1933.

2695. MUNDY, P. D. AND FREDERICK HENRY AMPHLETT MICKLE-
       WRIGHT. "William Herbert, Dean of Manchester,
       1778-1847." Notes and Queries 192 (1947), 257-258,
       370; 193 (1948), 85-86.

2696. MURPHY, OSWALD J. "Ignatius of Llanthony." Clergy
       Review 29, No. 1 (Jan. 1948), 27-33. (Joseph Lyne)

2697. NEWSOME, DAVID H. Godliness and Good Learning:
       Four Studies on a Victorian Ideal. London: Murray,
       1961. (Includes Bishop Westcott, A. P. Stanley, J.
       P. Lee, M. W. Benson, C. Kingsley, T. Hughes)

2698. NIAS, JOHN CHARLES SOMERSET. Flame from an Oxford
       Cloister: The Life and Writings of Philip Napier
       Waggett, 1862-1939, Scientist, Religious, Theo-
       logian, Missionary, Philosopher, Diplomat, Author,
       Orator, Poet. London: Faith Press, 1961.

2699. NIAS, JOHN CHARLES SOMERSET. "Gorham and Baptismal
       Regeneration." Theology 53 (1950), 3-11.

2700. NIAS, JOHN CHARLES SOMERSET. Gorham and the Bishop
       of Exeter. London: S.P.C.K., 1951.

2701. NICHOLLS, DAVID GWYN. "Authority in Church and
       State: Aspects of the Thought of J. N. Figgis and
       His Contemporaries." Ph.D. Dissertation, Uni-
       versity of Cambridge, 1962.

2702. NICHOLLS, DAVID GWYN. Church and State in Britain
       since 1820. London: Routledge and Kegan Paul; New
       York: Humanities Press, 1967.

2703. NORTHCOTT, CECIL. "Countryman Cleric: James
       Morton of Holbeach." Contemporary Review 205
       (1964), 599-601.

2704. NORWOOD, PERCY VARNEY. "A Victorian Primate."
       Church History 14 (Mar. 1945), 3-16. (A. C. Tait)

2705. OLLARD, SIDNEY LESLIE AND FRANK LESLIE CROSS.
       The Anglo-Catholic Revival in Outline. London:
       Society for the Promotion of Christian Knowledge,
       1933.

2706. OLSEN, GERALD W. "Pub and Parish: The Beginnings
       of Temperance Reform in the Church of England,
       1835-1875." Dissertation Abstracts International
       33 (1972), 1122A (University of Western Ontario,
       1972).

2707. OSBORNE, CHARLES EDWARD. The Life of Father Doll-
       ing. London: E. Arnold, 1903.

2708. OVERTON, JOHN HENRY. The Anglican Revival. Lon-
       don: Blackie & Son, Ltd., 1897.

2709. PAGET, ELMA KATIE. Henry Luke Paget: Portrait
       and Frame. London: Longmans, Green and Co., 1939.

2710. PARKER, W. M. "Dean Milman and 'The Quarterly
       Review'." Quarterly Review 293 (1955), 30-43.

2711. PATTERSON, MELVILLE WATSON. A History of the
       Church of England. London, New York: Longmans,
       Green and Co., 1909.

2712. PATTERSON, ROBERT LEYBURNE, JR. "The Crisis of the
       Unreformed Church of England, 1828-1833." Ph.D.
       Dissertation, Yale University, 1960.

2713.   PAUL, HERBERT WOODFIELD. "The Late Bishop of
        London: A Personal Impression." Nineteenth Cen-
        tury 50 (1901), 103-113. (M. Creighton)

2714.   PHILIP, ADAM. The Ancestry of Randall Thomas
        Davidson, D.D. London: E. Stock, 1903.

2715.   PHILLIPS, CHARLES STANLEY et al. Walter Howard
        Frere, Bishop of Truro. London: Faber and Faber,
        1947.

2716.   PINNINGTON, JOHN E. "Bishop Blomfield and
        St Barnabas', Pimlico: The Limits of Ecclesias-
        tical Authority." Church Quarterly Review 168
        (July-Sept. 1967), 289-296.

2717.   PINNINGTON, JOHN E. "Bishop Phillpotts and the
        Rubrics." Church Quarterly Review 169 (Apr.-June
        1968), 167-178.

2718.   PINNINGTON, JOHN E. "The Church of Ireland's
        Apologetic Position in the Years before Disestab-
        lishment." Irish Ecclesiastical Record 108 (1967),
        303-325.

2719.   PINNINGTON, JOHN E. "Living with Catholic Emanci-
        pation: Some Anglican Reactions." Dublin Review
        241 (Summer 1967), 154-161.

2720.   PINNINGTON, JOHN E. "The Vocation of Anglicanism."
        Dublin Review 242 (Spring 1968), 38-52.

2721.   PLOMER, WILLIAM CHARLES FRANKLYN. "Francis Kilvert
        and His Diary." Essays by Divers Hands: Transac-
        tions of the Royal Society of Literature 38 (1975),
        78-92.

2722.   PLOMER, WILLIAM CHARLES FRANKLYN. (Ed.) Kilvert's
        Diary: Selections from the Diary of the Rev. Francis
        Kilvert. 3 vols. London: Cape, 1938-40.

2723.   POLLET, J.-V.-M. "L'Anglicanisme libéral et le
        mouvement oecuménique." Revue de l'Université
        d'Ottawa 17, No. 2 (avril-juin 1947), 219-238.
        (From the middle of the 19th century)

2724.   PORT, MICHAEL HARRY.   Six Hundred New Churches:
        A Study of the Church Building Commission, 1818-
        1856, and Its Church Building Activities.   London:
        S.P.C.K., 1961.

2725.   PRECLIN, EDMOND.   "Les rapports de l'Eglise et de
        l'Etat en Angleterre depuis 1830."   Etudes Histor-
        iques New Series No. 2 (1948), 15-34.

2726.   PRESTON, RONALD.   "A Century of Anglican Social
        Thought."   Modern Churchman 32, Nos. 10-12 (Jan.-
        Mar. 1943), 337-347.

2727.   PROCTER, FRANCIS.   A New History of the Book of
        Common Prayer: With a Rationale of Its Offices,
        on the Basis of the Former Work by Francis
        Procter . . . .   Revised and rewritten by Walter
        Howard Frere.   London, New York: Macmillan, 1901.

2728.   PURCELL, WILLIAM ERNEST.   Onward Christian Soldier:
        A Life of Sabine Baring-Gould, Parson, Squire,
        Novelist, Antiquary, 1834-1924.   London, New York:
        Longmans, Green, 1957.

2729.   R., J. D.   "Dean Farrar as Headmaster."   Cornhill
        Magazine 14 (1903), 597-608.

2730.   RABY, F. J. E.   "John Neville Figgis, Prophet,
        1866-1919."   Theology 40, No. 239 (May 1940),
        325-332.

2731.   RAMSEY, ARTHUR MICHAEL.   From Gore to Temple:
        The Development of Anglican Theology Between "Lux
        Mundi" and the Second World War.   London: Longmans
        & Green, 1960.

2732.   RAMSEY, ARTHUR MICHAEL.   The Gospel and the Catholic
        Church.   London, New York: Longmans, Green and
        Co., 1936.

2733.   RANDOLPH, BERKELEY WILLIAM AND JAMES WESTON TOWNROE.
        The Mind and Work of Bishop King.   London, Oxford:
        A. R. Mowbray and Co., Ltd.; Milwaukee: The Young
        Churchman Co., 1918.

2734.   RECKITT, MAURICE BENINGTON.   Church and Society
        in England from 1800.   London: Allen and Unwin,
        1940.

2735.   RECKITT, MAURICE BENINGTON.   Maurice to Temple:
        A Century of the Social Movement in the Church of
        England.   London: Faber and Faber, 1947.   (In-
        cludes C. Kingsley, Christian Socialism)

2736.   REES, G.   "Connop Thirlwall, Liberal Anglican."
        Journal of the Historical Society of the Church in
        Wales 14 (1964), 66-76.

2737.   REFFOLD, A. E.   The Audacity to Live: A Résumé
        of the Life and Work of Wilson Carlile.   London,
        Edinburgh: Marshall, Morgan and Scott, 1938.

2738.   REFFOLD, A. E.   Wilson Carlile, 1847-1942:   Priest,
        Prophet, Evangelist.   London: Church Book Room
        Press, 1947.

2739.   REXROTH, KENNETH.   "Evolution of Anglo-Catholicism."
        Continuum 7 (Summer 1969), 345-360.

2740.   REYNOLDS, MICHAEL.   Martyr of Ritualism: Father
        Mackonochie of St. Alban's Holborn.   London: Faber
        and Faber, 1965.

2741.   RICHARDS, GWYNFRYN.   "The Rural Deanery of
        Arllechwedd: An Old Chapter Minute Book."   Na-
        tional Library of Wales Journal 18 (Winter 1973),
        149-180.   (Patronage, tithes, Sunday schools,
        Church property and Church reform from 1873-1910)

2742.   RICHARDS, NOEL JUDD.   "Disestablishment of the
        Anglican Church in England in the Late Nineteenth
        Century: Reasons for Failure."   Journal of Church
        and State 12 (Spring 1970), 193-211.

2743.   RICKARDS, EDITH C.   Bishop Moorhouse of Melbourne
        and Manchester.   London: J. Murray, 1920.

2744.   RIDDING, LADY LAURA ELIZABETH.   George Ridding:
        Schoolmaster and Bishop.   London: E. Arnold, 1908.

2745.   ROBERTS, M. J. D.   "The Role of the Laity in the
        Church of England, c. 1850-1885."   Ph.D. Disserta-
        tion, University of Oxford, 1974.

2746.   ROSE, ELLIOT. "The Stone Table in the Round
        Church." Victorian Studies 10 (1966), 119-144.
        (Cambridge Camden Society)

2747.   ROWAN, EDGAR. Wilson Carlile and the Church Army.
        London: Hodder and Stoughton, 1905.

2748.   RUSSELL, GEORGE WILLIAM ERSKINE. Arthur Stanton:
        A Memoir. London, New York: Longmans, Green and
        Co., 1917.

2749.   RUSSELL, GEORGE WILLIAM ERSKINE. Edward King,
        Sixtieth Bishop of Lincoln: A Memoir. London:
        Smith, Elder and Co., 1912.

2750.   RUSSELL, GEORGE WILLIAM ERSKINE. Sydney Smith.
        London: Macmillan and Co., Limited, 1905.

2751.   SALOMON, R. G. "Mother Church - Daughter Church -
        Sister Church: The Relations of the Protestant
        Episcopal Church and the Church of England in the
        19th Century." Historical Magazine of the Protes-
        tant Episcopal Church 21 (1952), 417-446.

2752.   SANDERS, J. N. "Zwist in der englischen Hochkirche."
        Hamburger akademische Rundschau 2, Heft 9-10 (März-
        April 1948), 485-488. (Dissension in the High
        Church since the publication of Essays and Reviews)

2753.   SAVIDGE, ALAN. The Parsonage in England: Its
        History and Architecture. London: S.P.C.K., 1964.

2754.   SCHAEFER, PAUL. The Catholic Regeneration of
        the Church of England. Trans. Ethel Scheffauer.
        London: Williams and Norgate, 1935.

2755.   SCOTT, JUDITH G. "Ecclesiological Influences in
        England, 1846-1963." Scottish Ecclesiological
        Society Transactions 15 (1965), 29-32.

2756.   SCOTT, SIDNEY HERBERT. Modernism in Anglo-
        Catholicism. London: Talbot and Co., 1933.

2757.   SCULLY, FRANCIS MICHAEL. L'Evolution de l'opinion
        anglaise sur les rapports entre l'église établie
        d'Angleterre et l'état à l'époque des réformes,
        1829-1839. Paris: Les Presses Modernes, 1938.

2758.   SCULLY, FRANCIS MICHAEL.   "Relations between Church
        and State in England between 1829 and 1839."
        B. Litt. Dissertation, University of Oxford, 1935.

2759.   SHEARMAN, HUGH.   How the Church of Ireland Was Dis-
        established.   Belfast: Church of Ireland Dises-
        tablishment Centenary Committee, 1970.

2760.   SHIMAN, LILIAN L.   "The Church of England Temper-
        ance Society in the Nineteenth Century."   Historical
        Magazine of the Protestant Episcopal Church 41
        (1972), 179-195.

2761.   SICHEL, EDITH.   "Canon Ainger."   Quarterly Review
        202 (1905), 169-196.

2762.   SICHEL, EDITH.   "Canon Ainger: A Personal Impres-
        sion."   Monthly Review 14 (1904), 64-74.

2763.   SILVESTER, JAMES. (Ed.)   A Champion of the Faith:
        A Memoir of the Rev. Chas. Henry Hamilton Wright,
        D.D., Ph.D., with Extracts from His Writings and
        Journals.   London: C. J. Thynne, 1917.

2764.   SIMON, ALAN.   "Church Disestablishment as a Factor
        in the General Election of 1885."   Historical
        Journal 18 (1975), 791-820.

2765.   SIMON, W. G.   "The Bishops and the Anglican Es-
        tablishment, 1660-1865."   Ph.D. Dissertation,
        University of Wisconsin, 1955.

2766.   SIMPSON, WILLIAM JOHN SPARROW.   The Contribution
        of Cambridge to the Anglo-Catholic Revival.   Lon-
        don: Society for the Promotion of Christian
        Knowledge, 1933.

2767.   SIMPSON, WILLIAM JOHN SPARROW.   The History of
        the Anglo-Catholic Revival from 1845.   London: G.
        Allen and Unwin, Ltd., 1932.

2768.   SISTERS OF THE CHURCH.   A Valiant Victorian: The
        Life and Times of Mother Emily Ayckbowm, 1836-1900,
        of the Community of the Sisters of the Church.
        London: A. R. Mowbray, 1964.

2769.   SKIPTON, HORACE PITT KENNEDY.   "Community Life in
        the Church of England since the Reformation."
        Church Quarterly Review 86 (1918), 77-98.

2770.   SMITH, ALAN W.   The Established Church and Popular
        Religion 1750-1850.   London: Longman, 1971.

2771.   SMITH, H. K.   "The Relations Between the Church of
        England and the State from 1838 to 1870."   M.A.
        Dissertation, University of Sheffield, 1946.

2772.   SMYTH, CHARLES HUGH EGERTON.   The Art of Preach-
        ing: A Practical Survey of Preaching in the Church
        of England, 747-1939.   London: Society for Promot-
        ing Christian Knowledge; New York: Macmillan
        Company, 1940.

2773.   SMYTH, CHARLES HUGH EGERTON.   The Church and the
        Nation: Six Studies in the Anglican Tradition.
        London: Hodder and Stoughton, 1962.   (Evangelicals,
        Tractarians)

2774.   SMYTH, CHARLES HUGH EGERTON.   Dean Milman, 1791-1868.
        London: S.P.C.K., 1949.

2775.   SOCKMAN, RALPH WASHINGTON.   The Revival of Conven-
        tual Life in the Church of England in the Nine-
        teenth Century.   New York: W. D. Gray, 1917.

2776.   SOLOWAY, RICHARD ALLEN.   "Episcopal Perspectives
        and Religious Revivalism in England 1784-1851."
        Historical Magazine of the Protestant Episcopal
        Church 40 (1971), 27-61.

2777.   SOLOWAY, RICHARD ALLEN.   Prelates and People:
        Ecclesiastical Social Thought in England 1783-1852.
        London: Routledge and K. Paul; Toronto: Uni-
        versity of Toronto Press, 1969.

2778.   STEPHEN, M. D.   "Gladstone and the Anglican Church
        in England, 1868-1874."   M. Litt. Dissertation,
        University of Cambridge, 1955.

2779.   STEPHEN, M. D.   "Gladstone and the Composition of
        the Final Court in Ecclesiastical Cases, 1850-1873."
        Historical Journal 9 (1966), 191-200.

2780. STEPHEN, M. D. "Gladstone's Ecclesiastical Patronage,
       1868-1874." Historical Studies (Australia & New
       Zealand) 11, No. 42 (1964), 145-162.

2781. STEPHEN, M. D. "Liberty, Church and State:
       Gladstone's Relations with Manning and Acton,
       1832-70." Journal of Religious History 1 (Dec.
       1961), 217-232.

2782. STEPHENSON, ALAN MALCOLM GEORGE. "Archbishop
       Vernon Harcourt, 1807-1847." Studies in Church
       History 4 (1967), 143-154.

2783. STEPHENSON, ALAN MALCOLM GEORGE. The First
       Lambeth Conference 1867. London: Published for
       the Church Historical Society, by S.P.C.K., 1967.

2784. STEPHENSON, ALAN MALCOLM GEORGE. "The Formation of
       the See of Ripon and the Episcopate of Its First
       Bishop, Charles Thomas Longley." B. Litt. Dis-
       sertation, University of Oxford, 1960.

2785. STEPHENSON, ALAN MALCOLM GEORGE. "Nineteenth-
       Century Plea for More Bishops: Thomas Sims on
       Episcopacy and Synods." Modern Churchman 4 (Apr.
       1961), 178-181.

2786. STEWART, HERBERT LESLIE. A Century of Anglo-
       Catholicism. London and Toronto: J. M. Dent and
       Sons Ltd., 1929.

2787. STEWART, HERBERT LESLIE. "James Anthony Froude and
       Anglo-Catholicism." American Journal of Theology
       22 (1918), 253-273.

2788. STOCK, EUGENE. The English Church in the Nine-
       teenth Century. London, New York: Longmans, Green
       and Co., 1910.

2789. STORR, VERNON FAITHFUL. Freedom and Tradition:
       A Study of Liberal Evangelicalism. London: Nisbet
       and Co., Ltd., 1940. (Includes the Oxford Move-
       ment, biblical criticism)

2790. STOWELL, HILDA MARY. George Isaac Huntingford:
       Warden of Winchester College, 1789-1832, Bishop of

Gloucester, 1802, Bishop of Hereford, 1815. Southampton: The Author, the Mound, West End, 1970.

2791. STRANKS, CHARLES JAMES. Dean Hook. London: A. R.
Mowbray, 1954.

2792. SUMNER, MARY ELIZABETH. Memoir of George Henry
Sumner, D.D., Bishop of Guildford. Winchester:
Warren and Son, Ltd., 1910.

2793. SWANSTON, HAMISH F. G. "Archbishop Tait: The Law
and the Order." Theology 78 (Nov. 1975), 592-600.

2794. SYKES, NORMAN. Church and State in England since
the Reformation. London: Benn, 1929.

2795. TAVARD, GEORGES HENRI. The Quest for Catholicity:
A Study in Anglicanism. London: Burns and Oates,
1963. (Anglo-Catholic theology from the Reformation)

2796. TAYLOR, BRIAN. "Bishop Hamilton (1808-69)."
Church Quarterly Review 155, No. 316 (Oct.-Dec.
1954), 235-248.

2797. THIRLWALL, JOHN CONNOP. Connop Thirlwall, Historian and Theologian. London: Society for
Promoting Christian Knowledge, 1936.

2798. THIRLWALL, JOHN CONNOP. "Connop Thirlwall, Historian and Theologian." Yale University Library
Gazette 41 (1967), 183-192.

2799. THOMAS, WILLIAM BRYN. Church of England Finance,
1836-1936: Uncensored Facts Concerning Church
Finance. A Study in the Economic Potential of
the Church of England. London: Church Book Room
Press, 1947.

2800. THOMPSON, KENNETH A. Bureaucracy and Church Reform: The Organizational Response of the Church
of England to Social Change, 1800-1965. London:
Oxford University Press, 1970.

2801. THUREAU-DANGIN, PAUL. The English Catholic Revival in the Nineteenth Century. Revised and
re-edited from a translation by Wilfred Wilber-

force. 2 vols. London: Simpson, Marshall and Co., 1914.

2802. THUREAU-DANGIN, PAUL. "Le Mouvement ritualiste dans l'église anglicane: I. L'Origine et les premières luttes du ritualisme; II. La Persécution; III. Suite de la persécution; IV. La Faillite de la persécution." Revue des Deux Mondes, 5th Period, 26 (1905), 834-859; 27 (1905), 116-151, 295-330, 567-606.

2803. TUCKER, MAURICE GRAHAME. John Neville Figgis: A Study. London: S.P.C.K., 1950.

2804. TUCKWELL, WILLIAM. Pre-Tractarian Oxford: A Reminiscence of the Oriel "Noetics." London: Smith, Elder and Co., 1909. (Includes J. Eveleigh, E. Copleston, R. Whately, T. Arnold, R. Hampden, Edward Hawkins, Baden Powell, Blanco White)

2805. TURNER, ARTHUR. "Hymn-writer for the Multitude: Henry Francis Lyte, 1793-1847." Chambers Journal, 8th Series, 16 (Nov. 1947), 681-683.

2806. TURNER, J. M. "J. N. Figgis: Anglican Prophet." Theology 18 (1975), 538-544.

2807. TURNER, ROBERT T. "Tithe Reform in the English Church, 1830-1836." Historical Magazine of the Protestant Episcopal Church 32 (1954), 143-166.

2808. VIDLER, ALEXANDER ROPER. The Church in an Age of Revolution: 1789 to the Present Day. Grand Rapids: W. B. Eerdman, 1961.

2809. VIDLER, ALEXANDER ROPER. The Orb and the Cross: A Normative Study in the Relations of Church and State. London: S.P.C.K., 1945. (Gladstone's The Church in Its Relations with the State)

2810. VOLL, DIETER. Catholic Evangelicalism: The Acceptance of Evangelical Traditions by the Oxford Movement during the Second Half of the Nineteenth Century. Trans. Veronica Ruffer. London: Faith Press, 1963.

2811. WAGNER, DONALD OWEN. The Church of England and
       Social Reform since 1854. New York: Columbia Uni-
       versity Press, 1930.
2812. WAKE, JOAN. A Northamptonshire Rector: The Life
       of Henry Isham Longden, Scholar, Sportsman, Priest,
       1859-1942. Northampton: Archer and Goodman, 1943.
2813. WAKEMAN, H. O. An Introduction to the History
       of the Church of England from the Earliest Times
       to the Present Day. London: Rivington, 1914.
2814. WALKER, DAVID. "The Welsh Church and Disestab-
       lishment." Modern Churchman 14 (Jan. 1971),
       139-154.
2815. WALL, REV. JAMES. "Converting the Pope." Cornhill
       Magazine 151 (1935), 90-99. (Attempt of Canon
       George Townsend to convert Pius IX to Anglicanism
       in 1850)
2816. WAND, JOHN WILLIAM CHARLES. The Second Reform.
       London: Faith Press; New York: Morehouse-Gorham
       Co., 1953. (Reform of the Church of England in the
       first half of the nineteenth century compared with
       the first reform period, or the Reformation)
2817. WARD, WILLIAM REGINALD. "The Cost of Establish-
       ment: Some Reflections on Church Building in
       Manchester." Studies in Church History 3 (1966),
       277-289.
2818. WARD, WILLIAM REGINALD. "The French Revolution and
       the English Churches: A Case Study in the Impact
       of Revolution upon the Church." Miscellanea His-
       toriae Ecclesiasticae 4 (1972), 55-84.
2819. WARD, WILLIAM REGINALD. "The Tithe Question in
       England in the Early Nineteenth Century." Journal
       of Ecclesiastical History 16 (1965), 67-81.
2820. WARD, WILLIAM REGINALD. Victorian Oxford. London:
       F. Cass, 1965.
2821. WARRE-CORNISH, FRANCIS. The English Church in
       the Nineteenth-Century. 2 vols. London: Mac-
       millan & Co., 1910.

2822.   WATSON, EDWARD WILLIAM. The Church of England.
        London: Williams and Norgate; New York: H. Holt
        and Co., 1914. (See 2843)

2823.   WATSON, EDWARD WILLIAM. Life of Bishop John Words-
        worth. London, New York: Longmans, Green and Co.,
        1915.

2824.   WATSON, W. L. R. "Archbishop Richard Whately,
        1787-1863." Theology 66 (1963), 405-410.

2825.   WATT, MARGARET HEWITT. The History of the Parson's
        Wife. London: Faber and Faber Limited, 1943.
        (Wives of clergymen)

2826.   WEBB, CLEMENT CHARLES JULIAN. "Benjamin Webb."
        Church Quarterly Review 75 (1913), 329-348.

2827.   WEBSTER, ALAN B. Joshua Watson: The Story of
        a Layman, 1771-1855. London: Society for the
        Promotion of Christian Knowledge, 1954. (Non-
        Tractarian High Church)

2828.   WELCH, P. J. "Anglican Churchmen and the Estab-
        lishment of the Jerusalem Bishopric." Journal of
        Ecclesiastical History 8 (Oct. 1957), 193-204.

2829.   WELCH, P. J. "Bishop Blomfield." Ph.D. Disserta-
        tion, University of London, 1952.

2830.   WELCH, P. J. "Bishop Blomfield and Church Exten-
        sion in London." Journal of Ecclesiastical History
        4 (1953), 203-215.

2831.   WELCH, P. J. "Bishop Blomfield and the Development
        of Tractarianism in London." Church Quarterly
        Review 155, No. 317 (1954), 332-344.

2832.   WELCH, P. J. "Blomfield and Peel: A Study in
        Cooperation between Church and State, 1841-46."
        Journal of Ecclesiastical History 12 (1961), 71-84.

2833.   WELCH, P. J. "Contemporary Views on the Proposals
        for the Alienation of Capitular Property in England
        (1832-1840)." Journal of Ecclesiastical History 5
        (1954), 184-195.

2834. WELCH, P. J. "The Difficulties of Church Extension
      in Victorian London." Church Quarterly Review 166
      (July-Sept. 1965), 302-315. (Problems of the
      Church in an urban setting)

2835. WELCH, P. J. "The Revival of an Active Convocation
      of Canterbury, 1852-1855." Journal of Ecclesias-
      tical History 10 (1959), 188-197.

2836. WELCH, P. J. "The Significance of Bishop C. J.
      Blomfield." Modern Churchman 45, No. 4 (1955),
      336-344.

2837. WELCH, P. J. "The Two Episcopal Resignations of
      1856: Blomfield and Maltby." Church Quarterly
      Review 165 (Jan.-Mar. 1964), 17-27.

2838. WELSBY, PAUL A. "Church and People in Victorian
      Ipswich." Church Quarterly Review 164 (Apr.-June
      1963), 207-217.

2839. WHITE, JAMES FLOYD. The Cambridge Movement: The
      Ecclesiologists and the Gothic Revival. Cambridge,
      England: Cambridge University Press, 1962. (Cam-
      bridge Camden Society)

2840. WICKHAM, A. K. "An Anglican Diehard." Cornhill
      Magazine 65 (1928), 609-624. (George Denison)

2841. WICKHAM, EDWARD RALPH. Church and People in an
      Industrial City. London: Lutterworth Press, 1957.
      (Sheffield)

2842. WILLIAMS, ALWYN T. P. The Anglican Tradition in
      the Life of England. London: S.C.M. Press, 1947.

2843. WILLIAMS, ALWYN T. P. "Epilogue," in E. W. Watson,
      The Church of England. 2nd ed., London: University
      Press, 1944.

2844. WILLIAMS, RONALD RALPH. "A Neglected Victorian
      Divine: Vaughan of Llandaff." Church Quarterly
      Review 154, No. 310 (Jan.-Mar. 1953), 72-85.

2845. WILLIAMS, THOMAS JAY. "The Beginnings of Anglican
      Sisterhoods: Pt. 1, The Revival of Sisterhoods in
      England." Historical Magazine of the Protestant
      Episcopal Church 16, No. 4 (Dec. 1947), 350-361.

2846. WILLIAMS, THOMAS JAY AND ALLAN WALTER CAMPBELL.
       The Park Village Sisterhood. London: S.P.C.K.,
       1965. (The revival of Anglican religious com-
       munities)

2847. WILLIAMS, THOMAS JAY. Priscilla Lydia Sellon:
       The Restorer after Three Centuries of the Religious
       Life in the English Church. London: S.P.C.K.,
       1950.

2848. WILLIAMSON, MARY PAULA. "Anglicanism Is Not
       Catholicism." Catholic World 158 (1944), 474-481.

2849. WITHYCOMBE, R. S. M. "The Development of Con-
       stitutional Autonomy in the Established Church in
       Later Victorian England." Ph.D. Dissertation, Uni-
       versity of Cambridge, 1970.

2850. WOOD, A. C. "A Nottingham Archdeacon's Letter Book
       of 1832." Transactions of the Thoroton Society 57
       (1954), 43-47. (George Wilkins)

2851. WOODGATE, MILDRED VIOLET. Father Benson: Founder
       of the Cowley Fathers. London: Bles, 1953.

2852. WOODRUFF, CLINTON ROGERS. "The Part of Dr. Routh
       in Dr. Seabury's Consecration." Historical Maga-
       zine of the Protestant Episcopal Church 9, No. 3
       (Sept. 1940), 231-246. (Martin Routh and Bishop
       Samuel Seabury)

2853. WORDEN, MARGARET A. "Conflicts Over Religious
       Inquiry among the Anglican Clergy in the Eighteen-
       Sixties." Ph.D. Dissertation, University of
       Oxford, 1968.

2854. WRIGHT, CHARLES JAMES. "One of Many: Bishop Jeune
       of Peterborough, Anglo-French Oxford Reformer."
       London Quarterly and Holborn Review 35 (Jan. 1966),
       57-63.

2855. YOUELL, G. "The Anglican Establishment, 1830-1930."
       M.A. Dissertation, University of Keele, 1969.

2.   Evangelicalism

This section includes studies of the Evangelical party in the Church of England, the Clapham Sect, and the ideas and activities of individual Evangelicals, e.g., William Wilberforce, Charles Simeon, and Hannah More. There are also a number of entries on the Evangelical revival before 1800. See II. Religion; other sections in this category (XIV).

2856.   AGLIONBY, FRANCIS KEYES.   The Life of Edward Henry Bickersteth: Bishop and Poet. London, New York: Longmans, Green, and Co., 1907.

2857.   BALLEINE, GEORGE REGINALD. A History of the Evangelical Party in the Church of England. London, New York: Longmans, Green, and Co., 1908.

2858.   BARLOW, MARGARET. (Ed.) The Life of William Hagger Barlow, D.D., Late Dean of Peterborough: With an Introduction by the Bishop of Liverpool and Chapters by the Bishop of Durham, the Dean of Canterbury and Others. London: G. Allen and Sons, 1910.

2859.   BENTLEY, ANNE. "The Transformation of the Evangelical Party in the Church of England in the Later Nineteenth Century." Ph.D. Dissertation, University of Durham, 1971.

2860.   BERNSTEIN, J. A. "Beauty and the Law: Shaftesbury's Relation to Christianity and the Enlightenment." Ph.D. Dissertation, Harvard University, 1970.

2861.   BERWICK, GEOFFREY. "Close of Cheltenham: Parish Pope." Theology 39 (July-Dec. 1939), 193-201, 276-285. (Francis Close)

2862.   BEST, GEOFFREY FRANCIS ANDREW. "The Evangelicals and the Established Church in the Early 19th Century." Journal of Theological Studies 10 (Apr. 1959), 63-78.

2863. BEST, GEOFFREY FRANCIS ANDREW. Shaftesbury.
      London: B. T. Batsford, 1964.
2864. BOAS, GUY. "Two Thoughts on William Wilberforce."
      English 12 (1959), 210.
2865. BREADY, JOHN WESLEY. "The Influence of Christianity
      on Social Progress, As Illustrated by the Career of
      Lord Shaftesbury." Ph.D. Dissertation, University
      of London, 1927.
2866. BROWN, FORD KEELER. "Fathers of the Victorians."
      Virginia Quarterly Review 12 (July 1936), 416-429.
2867. BROWN, FORD KEELER. Fathers of the Victorians:
      The Age of Wilberforce. Cambridge: Cambridge
      University Press, 1961. (See 2926)
2868. BROWN, IAN W.   "The Anglican Evangelicals in
      British Politics, 1780-1833." Dissertation
      Abstracts 26 (1966), 7276 (Lehigh University,
      1965).
2869. BROWN-SERMAN, STANLEY. "The Evangelicals and the
      Bible." Historical Magazine of the Protestant
      Episcopal Church 12, No. 2 (June 1943), 157-179.
      (Thomas Scott and C. Simeon)
2870. CASSON, J. S. "John Charles Ryle and the Evangelical
      Party in the Nineteenth Century." M.Phil. Disserta-
      tion, University of Nottingham, 1969.
2871. CLARK, MARMADUKE GUTHRIE BARTLETT. John Charles
      Ryle, 1816-1900: First Bishop of Liverpool.
      London: Church Book Room, 1947.
2872. CLARKE, WILLIAM KEMP LOWTHER. Eighteenth Century
      Piety. New York: Macmillan Co., 1944.
2873. CLEGG, HERBERT. "Evangelicals and Tractarians."
      Historical Magazine of the Protestant Episcopal
      Church 35 (1966), 111-153, 237-294; 36 (1967),
      127-178.
2874. COUPLAND, REGINALD. "The Memory of Wilberforce."
      Hibbert Journal 32 (Oct. 1933), 94-103. (W.
      Wilberforce)

2875.   COUPLAND, REGINALD.   Wilberforce: A Narrative.
        Oxford: Clarendon Press, 1923. (W. Wilberforce)

2876.   DAVIES, GEORGE COLLISS BOARDMAN. "Early Evangelicals."
        Church Quarterly Review 155, No. 315 (1954), 121-130.

2877.   DAVIES, GEORGE COLLISS BOARDMAN. The First
        Evangelical Bishop: Some Aspects of the Life of
        Henry Ryder. London: Tyndale Press, 1958.

2878.   DOWNER, ARTHUR CLEVELAND. A Century of Evangelical
        Religion in Oxford. London: Church Book Room,
        1938.

2879.   ELLIOTT-BINNS, LEONARD ELLIOTT. The Early
        Evangelicals: A Religious and Social Study.
        London: Lutterworth Press, 1953.

2880.   ELLIOTT-BINNS, LEONARD ELLIOTT. The Evangelical
        Movement in the English Church. London: Methuen
        and Co., Ltd., 1928.

2881.   FIGGIS, JOHN BENJAMIN. Keswick from Within.
        London: Marshall Bros., 1914.

2882.   FORSTER, EDWARD MORGAN. Marianne Thornton, 1797-
        1887. London: Arnold, 1956.

2883.   FOSKETT, R. "John Kaye and the Diocese of Lincoln."
        Ph.D. Dissertation, University of Nottingham, 1957.

2884.   FRANKE, W. "Christlicher Konservativismus und
        Aufklärungskritik bei William Wilberforce." Anglia
        88 (1970), 488-502.

2885.   FURNEAUX, ROBIN. William Wilberforce. London:
        Hamish Hamilton, 1974.

2886.   GARRATT, EVELYN R. Life and Personal Recollections
        of Samuel Garratt. Pt. I. A Memoir by His Daughter,
        E. R. Garratt. Pt. II. Personal Recollections by
        Himself. London: J. Nisbet and Co., Limited, 1908.

2887.   GILL, JOHN CLIFFORD. Parson Bull of Byerley.
        London: S.P.C.K., 1963. (George Bull)

2888.   HARDMAN, B. E. "The Evangelical Party in the
        Church of England, 1855-65." Ph.D. Dissertation,
        University of Cambridge, 1964.

2889.   HARFORD, CHARLES FORBES. (Ed.) The Keswick
        Convention: Its Message, Its Method and Its Men.
        London: Marshall Bros., 1907.
2890.   HARFORD, JOHN BATTERSBY AND FREDERICK CHARLES
        MACDONALD. Handley Carr Glyn Moule: Bishop of
        Durham. London: Hodder and Stoughton Limited,
        1922.
2891.   HARMAN, A. C. "Dean Vaughan and His Men." Theology
        35, No. 205 (1937), 103-111. (Charles Vaughan)
2892.   HEASMAN, KATHLEEN J. Evangelicals in Action: An
        Appraisal of Their Social Work in the Victorian Era.
        London: G. Bles, 1962.
2893.   HENNELL, MICHAEL AND ARTHUR POLLARD. (Eds.) Charles
        Simeon, 1759-1836: Essays Written in Commemora-
        tion of His Bi-centenary. London: S.P.C.K., 1959.
2894.   HENNELL, MICHAEL. John Venn and the Clapham Sect.
        London: Lutterworth Press, 1958.
2895.   HENNELL, MICHAEL. William Wilberforce, 1759-1833.
        London: Lutterworth Press, 1947.
2896.   HENSON, HERBERT HENSLEY. Sibbes and Simeon: An
        Essay on Patronage Trusts. London: Hodder and
        Stoughton, 1932. (Richard Sibbes, 1577-1635,
        Puritan divine)
2897.   HESELTINE, GEORGE COULEHAN. "William Wilberforce."
        English Review 57 (July 1933), 59-65.
2898.   HOPKINS, M. A. Hannah More and Her Circle. New
        York: Longmans and Green, 1947.
2899.   HOWSE, ERNEST MARSHALL. Saints in Politics: The
        "Clapham Sect" and the Growth of Freedom. Toronto:
        University of Toronto Press, 1952.
2900.   INSKIP, JAMES T. The Evangelical Influence in
        English Life. New York and London: Macmillan,
        1933. (Survey of Evangelicalism before and after
        John Wesley)
2901.   JAMES, M. G. "The Clapham Sect: Its History and
        Influence." Ph.D. Dissertation, University of
        Oxford, 1950.

2902.   JERKINS, WILFRED J.   William Wilberforce: A
        Champion of Freedom.   London: Epworth Press, 1932.

2903.   JOHNSON, J. ANGLIN.   "William Wilberforce, 1759-
        1833."   Contemporary Review 196 (Aug.-Sept. 1959),
        105-108.

2904.   JONES, HERBERT GRESFORD.   Francis James Chavasse,
        1846-1928:   Bishop of Liverpool.   London: Church
        Book Room Press, 1948.

2905.   JONES, JOHN WESTBURY.   "Figgis of Brighton":   A
        Memoir of a Modern Saint.   London: Marshall Bros.,
        1917.

2906.   JONES, M. G.   Hannah More, 1745-1833.   Cambridge:
        Cambridge University Press, 1953.

2907.   KENT, JOHN HENRY SOMERSET.   "William Wilberforce."
        London Quarterly and Holborn Review (Jan. 1967),
        64-68.

2908.   KIERNAN, V.   "Evangelicalism and the French Revolu-
        tion."   Past and Present 1 (1952), 44-56.

2909.   KIRK-SMITH, HAROLD.   William Thomson, Archbishop of
        York: His Life and Times, 1819-90.   London:
        Published for the Church Historical Society by
        S.P.C.K., 1958.

2910.   LANCELOT, JOHN BENNETT.   Francis James Chavasse:
        Bishop of Liverpool.   Oxford: B. Blackwell, 1929.

2911.   LARTER, L.   "William Wilberforce; 1759-1833."
        Church Quarterly Review 160 (Oct.-Dec. 1959),
        483-490.

2912.   LAW, ALICE.   "The Achievement of Wilberforce."
        Fortnightly Review 139 (June 1933), 749-758.
        (W. Wilberforce)

2913.   LEEDS, HERBERT.   Life of Dean Lefroy.   Norwich: H.
        J. Vince; London: Jarrold and Sons, 1909.
        (William Lefroy)

2914.   LOANE, MARCUS L.   Cambridge and the Evangelical
        Succession.   London: Lutterworth Press, 1952.
        (Includes W. Grimshaw, J. Berridge, H. Venn, C.
        Simeon, Henry Martyn)

2915. LOANE, MARCUS L. Handley Carr Glyn Moule, 1841-
       1920. London: Church Book Room Press, 1947.

2916. LOANE, MARCUS L. John Charles Ryle, 1816-1900: A
       Short Biography. London: J. Clarke, 1953.

2917. LOANE, MARCUS L. Makers of Our Heritage: A Study
       of Four Evangelical Leaders. London: Hodder and
       Stoughton, 1967. (J. C. Ryle, H. C. G. Moule, E. A.
       Knox, H. W. K. Mowll)

2918. LOANE, MARCUS L. Oxford and the Evangelical
       Succession. London: Lutterworth Press, 1950.
       (Includes G. Whitefield, J. Newton, T. Scott, R.
       Cecil, D. Wilson)

2919. MARTIN, WARREN BRYAN. "The 'Saints' of Clapham:
       Their Motivation and Their Work." Dissertation
       Abstracts 14 (1954), 1104-05 (Boston University
       School of Theology, 1954).

2920. MASTERS, DONALD C. "George Eliot and the
       Evangelicals." Dalhousie Review 41 (1961-62),
       505-512.

2921. MEACHAM, STANDISH. "The Evangelical Inheritance."
       Journal of British Studies 3 (Nov. 1963), 88-104.

2922. MEACHAM, STANDISH. Henry Thornton of Clapham,
       1760-1815. Cambridge: Harvard University Press,
       1964.

2923. MEAKIN, ANNETTE M. B. Hannah More: A Biographical
       Study. London: Smith, Elder, and Co., 1911.

2924. MOLE, DAVID E. H. "John Cale Miller: A Victorian
       Rector of Birmingham." Journal of Ecclesiastical
       History 17 (1966), 95-103.

2925. MOULE, HANDLEY CARR GLYN. The Evangelical School
       in the Church of England: Its Men and Its Works in
       the Nineteenth Century. London: J. Nisbet and
       Co., Limited, 1901.

2926. NEWSOME, DAVID H. "Father and Sons." Historical
       Journal 6 (1963), 295-310. (Review of Ford K.
       Brown, Fathers of the Victorians, 2867)

2927.   NICHOLLS, E.   "Wilberforce Centenary."   Bookman
        (London) 84 (Aug. 1933), 256.   (W. Wilberforce)

2928.   O'RORKE, LUCY ELIZABETH.   The Life and Friendships
        of Catherine Marsh.   London, New York: Longmans,
        Green and Co., 1917.

2929.   PEASTON, ALEXANDER ELLIOTT.   The Prayer Book
        Revisions of the Victorian Evangelicals.   Dublin:
        Association for Promoting Christian Knowledge,
        1963.

2930.   PICKERING, SAMUEL F., JR.   "Literature and Theology:
        The 'Christian Observer' and the Novel, 1802-1822."
        Historical Magazine of the Protestant Episcopal
        Church 43 (Mar. 1974), 29-43.

2931.   PIERSON, ARTHUR TAPPAN.   Forward Movements of the
        Last Half Century.   New York and London: Funk and
        Wagnalls Company, 1900.

2932.   POLLARD, ARTHUR.   "Evangelical Parish Clergy,
        1820-1840."   Church Quarterly Review 159 (July-Sept.
        1958), 387-395.

2933.   POLLOCK, JOHN CHARLES.   The Keswick Story: The
        Authorized History of the Keswick Convention.
        London: Hodder and Stoughton, 1964.

2934.   REYNOLDS, JOHN STEWART.   Canon Christopher of
        St. Aldgate's Oxford.   Abingdon: Abbey Press,
        1967.   (Alfred M. W. Christopher)

2935.   REYNOLDS, JOHN STEWART.   The Evangelicals at Oxford
        1735-1871:   A Record of an Unchronicled Movement.
        Oxford: Blackwell, 1953.

2936.   RICHARDSON, ANNA S.   Forty Years' Ministry in East
        London: Memoir of the Rev. Thomas Richardson.
        London: Hodder and Stoughton, 1903.

2937.   RICHARDSON, WILLIAM.   "The Sentimental Journey of
        Hannah More: Propagandist and Shaper of Victorian
        Attitudes."   Revolutionary World 11-13 (1975),
        228-239.

2938.   RUSSELL, GEORGE WILLIAM ERSKINE. A Short History
        of the Evangelical Movement. London: A. R. Mowbray
        and Co., Ltd., 1915.

2939.   SHORT, KENNETH RICHARD M. "Baptist Wriothesley
        Noel: Anglican-Evangelical-Baptist." Baptist
        Quarterly 20 (Apr. 1963), 51-61.

2940.   SILVESTER, JAMES. William Wilberforce: Christian
        Liberator. London: Mitre Press, 1934.

2941.   SKINNER, B. E. Henry Francis Lyte: Brixham's Poet
        and Priest. Exeter: Exeter University Press,
        1974.

2942.   SMELLIE, ALEXANDER. Evan Henry Hopkins. London:
        Marshall Bros., 1920.

2943.   SMITH, H. K. "Some Aspects of the Career of William
        Thomson as Archbishop of York (1863-1890)." Ph.D.
        Dissertation, University of Sheffield, 1953.

2944.   SMYTH, CHARLES HUGH EGERTON. "The Evangelical
        Movement in Perspective." Cambridge Historical
        Journal 7 (1943), 160-174.

2945.   SMYTH, CHARLES HUGH EGERTON. Simeon and Church
        Order: A Study of the Origins of the Evangelical
        Revival in Cambridge in the Eighteenth Century.
        Cambridge: Cambridge University Press, 1940.

2946.   SPRING, DAVID. "The Clapham Sect: Some Social and
        Political Aspects." Victorian Studies 5 (Sept.
        1961), 35-48.

2947.   STEVENSON, HERBERT FREDERICK. (Ed.) Keswick's
        Authentic Voice: Sixty-five Dynamic Addresses
        Delivered at the Keswick Convention, 1875-1947.
        London, Edinburgh: Marshall, Morgan and Scott,
        1959.

2948.   STEVENSON, HERBERT FREDERICK. (Ed.) Keswick's
        Triumphant Voice: Forty-eight Outstanding Addresses
        Delivered at the Keswick Convention, 1882-1962.
        Grand Rapids: Zondervan Pub. House, 1963.

2949. STEVENSON, HERBERT FREDERICK. (Ed.) The Ministry
       of Keswick: A Selection from the Bible Readings
       Delivered at the Keswick Convention. Grand Rapids:
       Zondervan Pub. House, 1963.
2950. SWIFT, DAVID E. "Charles Simeon and J. J. Gurney:
       A Chapter in Anglican-Quaker Relations." Church
       History 29 (1960), 167-186.
2951. TAIT, ARTHUR JAMES. Charles Simeon and His Trust.
       London: S.P.C.K., 1936.
2952. TELFORD, JOHN. A Sect That Moved the World: Three
       Generations of Clapham Saints and Philanthropists.
       London: Charles H. Kelly, 1907. (Includes H.
       Venn, H. Thornton, W. Wilberforce and the Victorian
       inheritors of the tradition)
2953. TOON, PETER. "J. C. Ryle and Comprehensiveness."
       Churchman 89 (Oct.-Dec. 1975), 276-283.
2954. WALSH, J. D. "Joseph Milner's Evangelical Church
       History." Journal of Ecclesiastical History 10
       (1959), 175-187.
2955. WALSH, J. D. "The Magdalene Evangelicals."
       Church Quarterly Review 159 (1958), 499-511.
2956. WARNER, OLIVER. William Wilberforce and His Times.
       London: Batsford, 1962.
2957. WARREN, MAX ALEXANDER CUNNINGHAM. (Ed.) Charles
       Simeon, 1759-1836. London: Church Book Room
       Press, 1949.
2958. WARREN, MAX ALEXANDER CUNNINGHAM. (Ed.) To Apply
       the Gospel: Selections from the Writings of Henry
       Venn. Grand Rapids, Mich.: Eerdman's Press, 1971.
2959. WEBSTER, DOUGLAS. "Charles Simeon and the Liturgy."
       Theology 54, (Aug. 1951), 296-301.
2960. WILLS, WILTON D. "The Established Church in the
       Diocese of Llandaff, 1850-70: A Study of the
       Evangelical Movement in the South Wales Coalfield."
       Welsh History Review 4 (June 1969), 235-267.

2961.   WINDLE, SIR BERTRAM COGHILL ALAN.   "Not Peace, but
         a Sword."   Catholic World 122 (Feb. 1926), 630-633.
         (The Wilberforce family)

2962.   WOOD, ARTHUR SKEVINGTON.   Thomas Haweis, 1734-1820.
         London: Published for the Church Historical Society
         by S.P.C.K., 1957.

2963.   YOUNG, HOWARD V., JR.   "The Evangelical Clergy in
         the Church of England, 1790-1850."   Dissertation
         Abstracts 19 (1959), 1734 (Brown University, 1958).

2964.   ZABRISKIE, ALEXANDER CLINTON. (Ed.)   Anglican
         Evangelicalism.   Philadelphia: The Church Historical
         Society, 1943.

2965.   ZABRISKIE, ALEXANDER CLINTON.   "The Rise and Main
         Characteristics of the Anglican Evangelical Movement
         in England and America."   Historical Magazine of the
         Protestant Episcopal Church 12 (1943), 81-115.

3.   Oxford Movement

        Studies of the growth and nature of the Oxford
Movement and its impact both within and beyond the Church
of England are listed under this heading.   While works on
John Henry Newman and Edward Bouverie Pusey appear in
category XIX, those on other prominent figures in the
Movement are assembled here, e.g., Richard William Church,
Frederick William Faber, Hurrell Froude, William George
Ward, John Keble, and Henry Parry Liddon.   Included also
are writings on fiction and poetry related to Trac-
tarianism (especially the work of Charlotte Yonge), on
Tractarian converts to Catholicism (except for Henry
Edward Manning, who is dealt with more fully under
Catholicism), and on newcomers to the Movement after 1845,
the year of J. H. Newman's conversion to Roman
Catholicism.   See II.   Religion; other sections of this

category (XIV), particularly 1. General; XVI.   Catholicism;
XIX.   Varieties of Belief. . ., especially 17. Newman,
John Henry, and 18. Pusey, Edward Bouverie.

2966.   ABERCROMBIE, NIGEL J.   "Some Directions of the
        Oxford Movement."   Dublin Review 193 (1933), 74-84.
2967.   ACLAND, JOHN EDWARD.   A Layman's Life in the Days
        of the Tractarian Movement: In Memoriam, Arthur
        Troyte.   London: J. Parker and Co., 1904.
2968.   ADAMS, WILLIAM SETH.   "William Palmer of Worcester,
        1803-1885:   The Only Really Learned Man among
        Them."   Dissertation Abstracts International 34
        (1974), 5296A (Princeton University, 1973).
2969.   ADCOCK, A. C.   "The Divinity of Christ, 1866-1966."
        Hibbert Journal 65 (1967), 134-138.   (Compares H.
        P. Liddon's 1866 Bampton Lectures with those of
        David E. Jenkins in 1966)
2970.   ADDINGTON, RALEIGH. (Ed.)   Faber: Selected Letters
        by Frederick William Faber 1833-1863.   London: D.
        Brown and Sons, 1974.
2971.   ADDLESHAW, S.   "The High Church Movement in
        Victorian Fiction: Charlotte M. Yonge."   Church
        Quarterly Review 120, No. 239 (Apr. 1935), 54-73.
2972.   ALLCHIN, ARTHUR MACDONALD.   "The Oxford Movement."
        One in Christ 1 (1965), 43-52.
2973.   ALLEN, LOUIS.   "Gladstone et Montalembert:
        correspondance inédite."   Revue de Littérature
        Comparée 30 (1956), 28-52.
2974.   ALLEN, LOUIS.   [Letter to the Editor on the Oxford
        Movement and the Vatican.]   Victorian Studies 3
        (1959-60), 458-459.   (See article by Charles T.
        Dougherty, below)
2975.   ALLIES, MARY HELEN AGNES.   Thomas William Allies,
        1813-1903.   London: Burns and Oates, 1907.

2976. ANGLO-CATHOLIC CONGRESS. Report of the Oxford Movement Centenary Congress, July 1933. London: Catholic Lit. Assoc., 1933.

2977. ANON. "Canon Carter of Clewer." Church Quarterly Review 53 (1902), 416-432. (Thomas Thelluson Carter)

2978. ANON. "The Dean of the Tractarians." Times Literary Supplement No. 1641 (July 13, 1933), 469-470. (R. W. Church; see 3029)

2979. ANON. "John Keble." Dublin Review 143 (Oct. 1908), 376-383.

2980. ANON. "The Origin and Historical Basis of the Oxford Movement." Quarterly Review 205 (1906), 196-214.

2981. ANON. "The Oxford Movement, 1833-1933." Spectator 151 (July 14, 1933), 37-38.

2982. ANON. "Richard Waldo Sibthorp, 1792-1879." Lincoln Diocesan Magazine 64, No. 20 (Aug. 1948), 210-212, 222.

2983. ARMSTRONG, HERBERT BENJAMIN JOHN. (Ed.) Armstrong's Norfolk Diary: Further Passages from the Diary of the Reverend Benjamin John Armstrong, Vicar of East Dereham, 1850-88. London: Hodder and Stoughton, 1963.

2984. ARMSTRONG, HERBERT BENJAMIN JOHN. (Ed.) A Norfolk Diary: Passages from the Diary of the Rev. Benjamin John Armstrong M.A. (Cantab.), Vicar of East Dereham, 1850-88. London: G. G. Harrap, 1949.

2985. BAAR, WILLIAM H. "John Mason Neale and His Contribution to the Catholic Revival of the Church of England." Ph.D. Dissertation, Yale University, 1953.

2986. BAAR, WILLIAM, H. "John Mason Neale (1818-1886)." Historical Magazine of the Protestant Episcopal Church 28 (Sept. 1959), 222-256.

2987.   BAILEY, SARAH. "Charlotte Mary Yonge." Cornhill
        Magazine 150 (1934), 188.

2988.   BAKER, JOSEPH ELLIS. The Novel and the Oxford
        Movement. Princeton: Princeton University Press,
        1932.

2989.   BAKER, WILLIAM J. "Hurrell Froude and the Reformers."
        Journal of Ecclesiastical History 21 (July 1970),
        243-259.

2990.   BARMANN, LAWRENCE F. "The Liturgical Dimension of
        the Oxford Tracts, 1833-1841." Journal of British
        Studies 7, No. 2 (1968), 92-113.

2991.   BARNES, WILLIAM EMERY. After the Celebration
        of the Oxford Movement: Some Considerations.
        Cambridge: Bowes & Bowes, 1933.

2992.   BARRY, WILLIAM FRANCIS. "Centenary of William
        George Ward." Dublin Review 151 (July 1912), 1-24.

2993.   BATTISCOMBE, GEORGINA AND MARGHANITA LASKI. (Eds.)
        A Chaplet for Charlotte Yonge: Papers by Georgina
        Battiscombe and Others. London: Cresset Press,
        1965.

2994.   BATTISCOMBE, GEORGINA. Charlotte Mary Yonge:
        The Story of an Uneventful Life. London: Constable
        and Co., 1943.

2995.   BATTISCOMBE, GEORGINA. John Keble: A Study in
        Limitations. London: Constable, 1963.

2996.   BEACH, ARTHUR G. "The Attitude of Writers of the
        English Romantic Movement to the Medieval Church
        and the Catholic Church and Their Influence on the
        Oxford Movement." Ph.D. Dissertation, University
        of Michigan, 1913.

2997.   BEEK, WILLEM JOSEPH ANTOINE MARIE. John Keble's
        Literary and Religious Contribution to the Oxford
        Movement. Nijmegen: 1959.

2998.   BELLOC, ELIZABETH. "Frederick Faber, 1814-63."
        Studies 36, No. 142 (June 1947), 163-174.

2999.   BENAS, BERTRAM B.   "The Oxford Movement from an
        Anglo-Jewish Standpoint."   Jewish Review No. 8
        (Mar.-June 1934), 46-58.

3000.   BENNETT, FREDERICK.   The Story of W. J. E. Bennett,
        Founder of S. Barnabas', Pimlico, and Vicar of
        Froome-Selwood, and of His Part in the Oxford
        Church Movement of the Nineteenth Century.   London,
        New York: Longmans, Green, and Co., 1909.

3001.   BLAIR, SIR DAVID OSWALD HUNTER.   "Memories of Old
        Oxford."   Dublin Review 195 (Oct. 1934), 227-239.
        (The Oxford Movement in the second half of the
        nineteenth century)

3002.   BOLTON, J. R. G.   "Polemics without Literature."
        Bookman (London), 84 (1933), 142-143.

3003.   BOULTER, BENJAMIN CONSITT.   The Anglican Reformers.
        London: P. Allan, 1933.   (Essays on lesser figures
        among Tractarians)

3004.   BOWEN, T. H.   "A Welsh Tractarian."   Welsh Review
        6, No. 2 (Summer 1947), 101-105.   (Isaac Williams)

3005.   BRANDRETH, HENRY RENAUD TURNER.   Dr. Lee of Lambeth:
        A Chapter in Parenthesis in the History of the
        Oxford Movement.   London: S.P.C.K., 1951.

3006.   BRANDRETH, HENRY RENAUD TURNER.   The Oecumenical
        Ideals of the Oxford Movement.   London: S.P.C.K.,
        1947.

3007.   BRASH, WILLIAM BARDSLEY.   "The Oxford Movement."
        London Quarterly and Holborn Review 158 (Apr.
        1933), 145-156.

3008.   BREGY, KATHERINE MARIE CORNELIA.   "Some Fruits of
        the Oxford Movement."   Catholic World 137 (Sept.
        1933), 684-688.

3009.   BRENDON, PIERS.   Hurrell Froude and the Oxford
        Movement.   London: Elek, 1974.

3010.   BRENDON, PIERS.   "Newman, Keble and Froude's
        'Remains'."   English Historical Review 87 (Oct.
        1972), 697-716.

3011.   BRILIOTH, YNGVE TORGNY. The Anglican Revival:
        Studies in the Oxford Movement. New York: Longmans,
        Green & Co., 1925.

3012.   BRILIOTH, YNGVE TORGNY. Three Lectures on Evangeli-
        calism and the Oxford Movement. London: Oxford
        University Press, 1934.

3013.   BRISCOE, JOHN FETHERSTONHAUGH AND HENRY FALCONAR
        BARCLAY MACKAY. A Tractarian at Work: A Memoir
        of Dean Randall. London and Oxford: Mowbray and
        Co., Ltd., 1932.

3014.   BROMLEY, J. F. "John Keble, 1792-1866." Central
        Literary Magazine 40 (Dec. 1966), 34-42.

3015.   BROWN, W. J. "The Deanery, Hadleigh, Suffolk."
        Church Quarterly Review 144 (July 1947), 205-208.
        (History of the Deanery of Hadleigh and its links
        with H. J. Rose and the Oxford Movement)

3016.   BRUNTON, B. R. "The Life of Richard Hurrell Froude
        to 1833." M.A. Dissertation, University of Exeter,
        1971.

3017.   BURCH, VACHER. "Newman and the Vision Keble Saw."
        Theology 27 (Sept. 1933), 130-139.

3018.   BURDETT, OSBERT. "What of the Oxford Movement?"
        Nineteenth Century and After 114 (Aug. 1933),
        215-225.

3019.   BURROWS, MARGARET FLORENCE. Robert Stephen Hawker:
        A Study of His Thought and Poetry. Oxford:
        B. Blackwell, 1926.

3020.   BURY, SHIRLEY. "Pugin and the Tractarians."
        Connoisseur 179 (1972), 15-20.

3021.   BUTLER, P. "Irvingism as an Analogue of the Oxford
        Movement." Church History 6 (1937), 101-112.

3022.   CAMPBELL, MARY FRANCES T., SISTER. "The Oxford
        Movement: Success or Failure?" Ph.D. Dissertation,
        Fordham University, 1935.

3023.   CANADY, CHARLES ELI. "A Comparative Study of the
        Piety of George Herbert and John Keble." Disserta-

tion Abstracts 24 (1963), 1262-63 (Temple University, 1963).

3024. CAPONIGRI, A. R. "The Oxford Movement and New Humanism." Modern Schoolman 13 (Nov. 1935), 12-14.

3025. CARTER, JANE FRANCES MARY. Life and Work of the Rev. T. T. Carter. London, New York: Longmans, Green, 1911.

3026. CASSIDY, JAMES FRANCIS. Life of Father Faber, Priest of the Oratory of St. Philip Neri. London: Sands, 1946.

3027. CAUL, D. F. "The Oxford Movement." At-one-ment 1 (1959), 26-36.

3028. CECIL, ALGERNON. "Dean Church." Monthly Review 20 (1905), 139-149.

3029. CECIL, ALGERNON. "The Dean of the Tractarians." Times Literary Supplement (July 20, 1933), 496. (R. W. Church)

3030. CECIL, LORD HUGH. "The Oxford Movement: Our Debt to the Tractarians." Spectator 105 (May 5 and 12, 1933), 634-635, 684.

3031. CHADWICK, OWEN. "The Limitations of Keble." Theology 67 (1964), 46-52.

3032. CHADWICK, OWEN. (Ed.) The Mind of the Oxford Movement. London: A. and C. Black, 1960.

3033. CHAPMAN, RAYMOND. Faith and Revolt: Studies in the Literary Influence of the Oxford Movement. London: Weidenfeld and Nicolson, 1970.

3034. CHAPMAN, RONALD. Father Faber. Westminster, MD.: Newman Press, 1961.

3035. CHURCH, R. W. The Oxford Movement, 1833-1845. London: Macmillan, 1891.

3036. CIOFFI, PAUL LAWRENCE. "The Mediatorial Principle of the Incarnation Analogously Extended to Church and Sacraments in the Writings of Robert Isaac Wilberforce (1802-1857)." Dissertation Abstracts 34

(1973), 2013-14A (Catholic University of America, 1973).

3037.   CLARKE, CHARLES PHILLIP STEWART.   The Oxford Movement and After.   London and Oxford: Mowbray & Co., Ltd., 1932.

3038.   CLEGG, HERBERT.   "Evangelicals and Tractarians." Historical Magazine of the Protestant Episcopal Church 35 (1966), 111-153, 237-294; 36 (1967), 127-178.

3039.   CLEGG, HERBERT.   "Froude's 'Remains'." Church Quarterly Review 167 (1966), 166-179.   (R. H. Froude)

3040.   COBHAM, J. O.   "E. C. Hoskyns: Sunderland Curate." Church Quarterly Review 158 (July-Sept. 1957), 280-295.

3041.   COLEMAN, ARTHUR MAINWARING.   "'Puseyite' as a General Term of Abuse."   Notes and Queries 166 (Apr. 7, 1934), 241-242; 166 (June 16, 1934), 428-429.

3042.   COLERIDGE, CHRISTABEL.   Charlotte Mary Yonge. London: Macmillan and Co., Limited, 1903.

3043.   CORNISH, J. G.   "Keble and the Cornishes of Calcombe." Devon and Cornwall Notes and Queries 18, Pt. 4 (Oct. 1934), 337-349.

3044.   CRANNY, T.   "A Study in Contrasts: Newman and Faber."   American Ecclesiastical Review 129 (1953), 300-313.

3045.   CROSS, FRANK LESLIE.   The Oxford Movement and the Seventeenth Century.   London: Society for Promoting Christian Knowledge, 1933.

3046.   CROSS, FRANK LESLIE.   The Tractarians and Roman Catholicism.   London: Society for Promoting Christian Knowledge, 1933.

3047.   CUMMINS, E. A.   "Catholics in a Protestant Church." American Mercury 29 (June 1933), 157-164.

3048.   DANIEL-ROPS, HENRY.   "Le Mouvement d'Oxford." Revue des Deux Mondes (Mar. 1, 1960), 3-17.

3049. DANIELS, WILLIAM EDWARD. John Keble, 1792-1866:
With Some Reference to the Oxford Movement. London:
Church Book Room Press, 1948.

3050. DARK, SIDNEY. "The Oxford Centenary." Saturday
Review 156 (July 8, 1933), 35-37.

3051. DAVIES, D. S. "Tractarianism: With Especial
Reference to Its Social Implications and Achievements." M.A. Dissertation, University of Leeds,
1950.

3052. DAWSON, CHRISTOPHER HENRY. "The Main Issue of the
Oxford Movement." Catholic World 138 (Feb. 1934),
612-613. (Excerpt from The Spirit of the Oxford
Movement)

3053. DAWSON, CHRISTOPHER HENRY. The Spirit of the Oxford
Movement. London: Sheed & Ward, 1933.

3054. DEARING, TREVOR. Wesleyan and Tractarian Worship:
An Ecumenical Study. London: Epworth Press;
S.P.C.K., 1966.

3055. DILWORTH-HARRISON, TALBOT. Every Man's Story
of the Oxford Movement. London and Oxford:
Mowbray, 1932.

3056. DIMOND, SYDNEY GEORGE. "The Oxford Movement and
Anglo-Catholicism, II." London Quarterly and Holborn
Review 159 (Jan. 1934), 54-64.

3057. DIMOND, SYDNEY GEORGE. "The Philosophy and Theology
of the Oxford Movement and Anglo-Catholicism."
London Quarterly and Holborn Review 158 (Oct.
1933), 433-446.

3058. DONALD, GERTRUDE. Men Who Left the Movement:
John Henry Newman, Thomas W. Allies, Henry Edward
Manning, Basil William Maturin. London: Burns,
Oates, and Washbourne Ltd., 1933.

3059. DONALDSON, AUGUSTUS BLAIR. Five Great Oxford
Leaders: Keble, Newman, Pusey, Liddon and Church.
London: Rivingtons, 1900.

3060. DONALDSON, AUGUSTUS BLAIR. Henry Parry Liddon.
London: Rivingtons, 1905.

3061. DONOHUE, G. J. "Oxford Movement." Commonweal 18
(June 30, 1933), 231-233.

3062. DONOVAN, MARCUS. After the Tractarians, from
the Recollections of Athelstan Riley. London: P.
Allan, 1933.

3063. DONOVAN, MARCUS. "John Mason Neale." Church
Quarterly Review 167 (1966), 317-322.

3064. DOUGHERTY, CHARLES T. AND HOMER C. WELSH. "Wiseman
and the Oxford Movement: An Early Report to the
Vatican." Victorian Studies 2 (1958-59), 149-154.
(See letter by Louis Allen, above)

3065. DRUMMOND, ANDREW LANDALE. "Fiction and the Oxford
Movement." Church Quarterly Review 140 (Apr.
1945), 29-50.

3066. DUGRE, ALEXANDRE, S.J. "Un Centenaire anglais."
Relations 5 (Mar. 1945), 65-68.

3067. DURNFORD, FRANCIS HENRY. "Richard William Church."
Expository Times 59, No. 9 (June 1948), 238-240.

3068. EAVES, ARTHUR J. "William Sewell: An Oxford
Humanist." Dissertation Abstracts International 34
(1974), 4194A (University of Notre Dame, 1973).

3069. ETHERINGTON, MICHAEL. "Hawker of Morwenstow."
Church Quarterly Review 143, No. 285 (Oct.-Dec.
1946), 76-81. (Robert Hawker)

3070. FABER, SIR GEOFFREY CUST. Oxford Apostles: A
Character Study of the Oxford Movement. London:
Faber and Faber, 1933.

3071. FAIRCHILD, HOXIE NEALE. "Romanticism and the
Religious Revival in England." Journal of the
History of Ideas 2 (1941), 330-338.

3072. FAIRWEATHER, E. R. "Apostolical Tradition and the
Defence of Dogma: An Episode in the Anglo-Catholic
Revival." Canadian Journal of Theology 11 (Oct.
1965), 277-289. (R. D. Hampden and the Tractarians)

3073. FAIRWEATHER, E. R. (Ed.) The Oxford Movement. New
York: Oxford University Press, 1964.

3074.   FIGGIS, JOHN NEVILLE.   Churches in the Modern
        State.   London, New York: Longmans, Green and Co.,
        1913.

3075.   FOAKES-JACKSON, FREDERICK JOHN.   "The Oxford
        Movement."   American Church Monthly 14 (1924),
        198-205.

3076.   FOX, ADAM.   "John Mason Neale."   Hymn Society
        Bulletin No. 36 (July 1946), 1-4.

3077.   FULWEILER, HOWARD WELLS.   "Heaven versus Utopia: A
        Study of the 'Tracts for the Times', 1833-1841."
        Dissertation Abstracts 21 (1961), 1939-40 (University
        of North Carolina, 1960).

3078.   FULWEILER, HOWARD WELLS.   "Tractarians and Philistines:
        The 'Tracts for the Times' versus Victorian Middle-
        Class Values."   Historical Magazine of the Protestant
        Episcopal Church 31 (Mar. 1962), 36-53.

3079.   GARBETT, CYRIL FORSTER.   "John Keble and the Oxford
        Movement."   Theology 27 (Sept. 1933), 126-130.

3080.   GORCE, DENYS.   Faber, un anglican à l'âme franciscaine
        (années d'avant sa conversion), 1814-1845.   Paris:
        Editions Franciscaines, 1967.

3081.   GREENFIELD, R. H.   "The Attitude of the Tractarians
        to the Roman Catholic Church, 1833-50."   Ph.D. Dis-
        sertation, University of Oxford, 1956.

3082.   GREY, FRANCIS W.   "Some Factors in the Oxford
        Movement."   The American Catholic Quarterly Review
        35 (1910), 193-215.

3083.   GRIFFIN, JOHN R.   "John Keble: Radical."   Anglican
        Theological Review 53 (July 1971), 167-173.

3084.   GRIFFIN, JOHN R.   "The Social Implications of the
        Oxford Movement."   Historical Magazine of the Protes-
        tant Episcopal Church 44 (1975), 155-165.

3085.   GUINEY, LOUISE IMOGEN.   Hurrell Froude: Memoranda
        and Comments.   London: Methuen and Co., 1904.

3086.   HÄRDELIN, ALF.   The Tractarian Understanding of
        the Eucharist.   Uppsala: Almqvist och Wiksell,
        1965.

3087. HALIFAX, EDWARD FREDERICK LINDLEY WOOD.   John
       Keble.   London: A. R. Mowbray, 1909.
3088. HALL, SAMUEL.   A Short History of the Oxford
       Movement.   London, New York: Longmans, Green, and
       Co., 1906.
3089. HALL-PATCH, W.   Father Faber: With a Foreword
       by His Eminence Cardinal Bourne.   New York: P. J.
       Kenedy and Sons, 1914.
3090. HAMMOND, T. C.   "The Evangelical Revival and the
       Oxford Movement."   Churchman: The Evangelical
       Quarterly (London) 47 (Apr. 1933), 79-86.
3091. HARDWICK, JOHN CHARLTON.   The Light That Failed:
       Reflections on the Oxford Movement.   Oxford: B.
       Blackwell, 1933.
3092. HARPER, GORDON HUNTINGTON.   Cardinal Newman and
       William Froude, F.R.S..   Baltimore: The Johns
       Hopkins Press, 1933.
3093. HARPER, GORDON HUNTINGTON.   The Froude Family
       in the Oxford Movement.   Baltimore: The Johns
       Hopkins Press, 1933.
3094. HARRIS, SILAS M.   The Oxford Movement and the
       Holy See.   Tractate 1.   The First Ten Years: The
       Witness of the Early Tractarians.   London: Talbot,
       1934.
3095. HARRIS, T. L.   "The Conception of Authority in the
       Oxford Movement."   Church History 3 (1934), 115-125.
3096. HARRISON, ARCHIBALD HAROLD WALTER.   "Romanticism in
       Religious Revivals."   Hibbert Journal 31 (July
       1933), 582-594.
3097. HARROLD, CHARLES FREDERICK.   "The Oxford Movement:
       A Reconsideration."   In Joseph Ellis Baker (Ed.),
       The Reinterpretation of Victorian Literature.
       Princeton, New Jersey: Princeton University Press,
       1950, 33-56.
3098. HEADLAM, A. C.   "Hugh James Rose and the Oxford
       Movement."   Church Quarterly Review 93 (1922),
       86-102.

3099. HEAZELL, FRANCIS NICHOLSON. The History of
S. Michael's Church, Croydon: A Chapter in the
Oxford Movement. London and Oxford: A. R. Mowbray
and Co., Ltd.; Milwaukee, U.S.A.: Morehouse
Publishing Co., 1934.

3100. HEENEY, BRIAN. Mission to the Middle Classes:
The Woodard Schools, 1848-1891. London: S.P.C.K.,
1969.

3101. HEENEY, BRIAN. "Tractarian Parson: Edward Monro
of Harrow Weald." Canadian Journal of Theology 13
(1967), 241-253; 14 (1968), 13-27.

3102. HEILSAM, INGEBORG. "Die Oxfordbewegung und ihr
Einfluss auf die englische Dichtung." Ph.D. Dis-
sertation, Universität Wien, 1950. (Includes G.
M. Hopkins, C. Patmore, F. Thompson, J. H. Newman)

3103. HENSON, HERBERT HENSLEY. The Oxford Groups: The
Charge Delivered at the Third Quadrennial Visita-
tion of His Diocese. London: H. Milford, Oxford
University Press, 1933.

3104. HEYWOOD, BERNARD OLIVER FRANCIS. Sermon Notes
for a Suggested Course on the Oxford Movement.
London: Society for the Promotion of Christian
Knowledge, 1933.

3105. HILL, A. G. "The Tractarian Challenge." Theology
64 (July 1963), 280-287.

3106. HOLLIS, CHRISTOPHER. "Newman and Dean Church: A
Friendship that Endured." Tablet 211 (June 7,
1958), 528.

3107. HOLLOWAY, O. E. "The Tractarian Movement in
Oxford." Bodleian Quarterly Record 7, No. 78
(1932-34), 213-232.

3108. HOLMES, J. DEREK. "Newman, Froude and Pattison:
Some Aspects of Their Relations." Journal of
Religious History 4 (June 1966), 23-38.

3109. HOPPEN, K. THEODORE. "The Oxford Movement."
History Today 17 (Mar. 1967), 145-152.

3110.   HOUGHTON, ESTHER RHOADS.   "The 'British Critic' and
        the Oxford Movement."   Studies in Bibliography 16
        (1963), 119-137.

3111.   HUELIN, GORDON.   "A Tractarian Clergyman and His
        Friends."   Church Quarterly Review 160 (1959),
        37-48.   (William Scott)

3112.   HUNKIN, JOSEPH WELLINGTON.   "The Flying Start of
        the Oxford Movement."   Contemporary Review 143
        (Jan.-June 1933), 561-569.

3113.   HUNT, R. W.   "Newman's Notes on Dean Church's
        'Oxford Movement'."   Bodleian Library Record 8
        (1969), 135-137.   (See Church, R. W., 3035)

3114.   HUNTER-BLAIR, SIR DAVID OSWALD.   "Memories of Old
        Oxford."   Dublin Review 195 (1934), 227-239.

3115.   HUTCHISON, WILLIAM GEORGE. (Ed.)   The Oxford
        Movement: Being a Selection from Tracts for the
        Times.   London, New York: W. Scott, 1906.

3116.   INGRAM, KENNETH.   "A Century of Anglo-Catholicism."
        Bookman (London) 84 (June 1933), 140-141.

3117.   INGRAM, KENNETH.   John Keble.   London: P. Allan,
        1933.

3118.   JACKSON, GEORGE.   "Dean Church: An Appreciation."
        London Quarterly Review 118 (1912), 24-47.

3119.   JAMES, LIONEL.   A Forgotten Genius: Sewell of
        St. Columba's and Radley.   London: Faber and Faber
        Limited, 1945.

3120.   JANSSENS, A.   Anglicaansche bekeerlingen: Newman-
        Faber-Manning-Benson-Knox-Kinsman-Chesterton.
        Anvers-Brux.: Standaard, 1928.

3121.   JANSSENS, A.   De Beweging van Oxford.   Louvain:
        Davidsfonds, 1930.

3122.   JOHNSON, HUMPHREY JOHN THEWLIS.   "The Church of
        England in 1845: The Ecclesiastical Background of
        Newman's Conversion."   Dublin Review 217 (Oct.
        1945), 147-155.

3123.   JOHNSTON, JOHN OCTAVIUS. Life and Letters of
        Henry Parry Liddon. London, New York and Bombay:
        Longmans, Green, and Co., 1904.

3124.   JONES, OWAIN W. Isaac Williams and His Circle.
        London: S.P.C.K., 1971.

3125.   JUDGE, RAYMOND J. "The Forgotten Priest of the
        Oxford Movement." Catholic World 161 (Sept. 1945),
        476-479. (Father Dominic Barberi who received J.
        H. Newman into the Roman Catholic Church)

3126.   KAULEN, LORE. "Die traktarianische Bewegung und
        ihre Beziehungen zur Literatur ihrer Zeit." Ph.D.
        Dissertation, Universität Göttingen, 1928.

3127.   KAYE-SMITH, SHEILA. "Ninety Years of Oxford
        Nonsense." Fortnightly 123 (June 1925), 762-772.

3128.   KELDANY, HERBERT. "Victorian Rector: Oakeley of
        Islington." Clergy Review 23 (Nov. 1943), 487-494.
        (Frederick Oakeley)

3129.   KELWAY, ALBERT CLIFTON. George Rundle Prynne:
        A Chapter in the Early History of the Catholic
        Revival. London, New York: Longmans, Green, and
        Co., 1905.

3130.   KENYON, RUTH. "Two Studies in the Social Outlook
        of the Tractarians: 1. Newman. 2. Keble and
        Pusey." Theology 24 (June 1932), 317-324; 25 (July
        1932), 24-34.

3131.   KNOX, EDMUND ARBUTHNOTT. The Tractarian Movement,
        1833-1845: The Oxford Movement as a Phase of the
        Religious Revival in Western Europe in the Second
        Quarter of the Nineteenth Century. London: Putnam,
        1933.

3132.   KNOX, EDMUND ARBUTHNOTT. "Tractarianism and the
        Episcopacy." Nineteenth Century and After 114
        (July 1933), 73-81.

3133.   KNOX, EDMUND ARBUTHNOTT. "Tractarianism and the
        National Life." Spectator (May 12, 1933), 673-674.

3134.   KNOX, RONALD. "Many Mansions: The Conversions of
        Newman and Faber." Tablet 185 (June 30, 1945), 310.

3135. KNOX, WILFRED LAWRENCE. The Catholic Movement
       in the Church of England. London: Allan, 1923.

3136. KNOX, WILFRED LAWRENCE AND ALEXANDER ROPER VIDLER.
       The Development of Modern Catholicism. London:
       Allan, 1933. (Anglo-Catholicism from the middle of
       the nineteenth century to the present)

3137. LANG, COSMO GORDON, ARCHBISHOP OF CANTERBURY. "Mr.
       Gladstone and the Oxford Movement." Nineteenth
       Century and After 114 (Sept. 1933), 374-384.

3138. LATHBURY, DANIEL C. Dean Church. London: Mowbray
       and Co., 1905.

3139. LA VERDONIE, J. L. DE. "Influence du 'Mouvement
       d'Oxford' sur les conversions au Catholicisme en
       Angleterre." Revue Apostolique 58 (1936), 199-223.

3140. LESLIE, SIR JOHN RANDOLPH SHANE. "Lewis Carroll
       and the Oxford Movement: A Paper Submitted to the
       Historical Theological School at Göttingen Univer-
       sity." London Mercury 28 (July 1933), 233-239.

3141. LESLIE, SIR JOHN RANDOLPH SHANE. The Oxford Move-
       ment, 1833-1933. London: Burns, Oates, and Wash-
       bourne, Ltd., 1933.

3142. LOCHHEAD, MARION. "Lockhart, the 'Quarterly', and
       the Tractarians." Quarterly Review 291 (1953),
       196-209.

3143. LOUGH, A. G. The Influence of John Mason Neale.
       London: S.P.C.K., 1962.

3144. LOUGH, A. G. John Mason Neale: Priest Extra-
       ordinary. Newton Abbot: Hennech Vicarage, 1975.

3145. LOVERA DI CASTIGLIONE, CARLO. Il Movimento di
       Oxford. Brescia: Morcelliana, 1935.

3146. LUCKOCK, HERBERT MORTIMER. The Beautiful Life
       of an Ideal Priest: Or Reminiscences of Thomas
       Thellusson Carter, Hon. Canon of Christ Church,
       Oxford, and Warden of the House of Mercy, Clewer.
       London: Simpkin, Marshall, Hamilton, Kent and Co.;
       Lichfield: A. C. Lomax's Successors, 1902.

3147. LUNN, SIR ARNOLD HENRY MOORE. Roman Converts.
       London: Chapman and Hall, 1924. (J. H. Newman, H.
       Manning, G. Tyrrell, Ronald Knox, G. K. Chesterton)
3148. LYNCH, M. J. "Gladstone and the Oxford Movement."
       M.A. Dissertation, University of Leicester, 1973.
3149. MCALLASTER, ELVA A. "The Oxford Movement and
       Victorian Poetry." Ph.D. Dissertation, University
       of Illinois at Urbana, 1948.
3150. MCCARTHY, JOHN ROBERT. "E. B. K. Fortescue: A
       Study in Some Nineteenth Century Ecclesiastical
       Problems in England and Scotland." Dissertation
       Abstracts International 33 (1973), 5097-98A (Case
       Western Reserve University, 1972).
3151. MCCLELLAN, WILLIAM H. "The Oxford Movement in
       Action and Reaction." Ecclesiastical Review 88
       (Mar. 1933), 225-240.
3152. MCGRATH, ALBERTUS M., SISTER. "The History of the
       Anglo-Catholic Movement, 1850-1875." Ph.D. Dis-
       sertation, University of Wisconsin, 1947.
3153. MCGREEVY, MICHAEL A. "John Keble on the Anglican
       Church and the Church Catholic." Heythrop Journal
       5 (1964), 27-35.
3154. MACKEAN, W. H. The Eucharistic Doctrine of the
       Oxford Movement: A Critical Study. London and New
       York: Putnam, 1933.
3155. MACKERNESS, ERIC DAVID. "Henry Parry Liddon: The
       Diadochus of the Tractarians." Church Quarterly
       Review 158 (Oct.-Dec. 1957), 466-477.
3156. MCNABB, V. "Le Mouvement tractarien: à propos
       d'un aperçu sur le mouvement philosophique et reli-
       gieux de la Haute Eglise anglicane." Revue Thomiste
       18 (1910), 308-320.
3157. MAHONEY, LEONARD, S.J. "The Genesis of the Oxford
       Movement." America 76 (Feb. 22, 1947), 575-576.
3158. MALLET, SIR CHARLES. "Theology and Religion."
       Contemporary Review 164 (Oct. 1943), 218-224.

3159.   MARE, MARGARET AND ALICIA C. PERCIVAL.   Victorian
        Best-Seller: The World of Charlotte M. Yonge.
        London: George G. Harrap and Co. Ltd., 1948.

3160.   MASTERMAN, J. H. B.   "The Oxford Movement."
        Spectator 150 (Feb. 17, 1933), 208-209; (Feb. 24,
        1933), 251.

3161.   MAY, JAMES LEWIS.   The Oxford Movement, Its History
        and Its Future: A Layman's Estimate.   London:
        John Lane, 1933.

3162.   MAY, JAMES LEWIS.   The Unchanging Witness: Some
        Detached Reflections on the Oxford Movement and Its
        Future.   London: Centenary Press, 1933.

3163.   MEAD, A. H.   "Richard Bagot, Bishop of Oxford, and
        the Oxford Movement, 1833-45."   B. Litt. Dissertation,
        University of Oxford, 1966.

3164.   MIDDLETON, ROBERT DUDLEY.   Magdalen Studies.
        London: Society for Promoting Christian Knowledge,
        1936.   (Martin Routh, J. R. Bloxam, Frederick
        Bulley, William Palmer, Roundell Palmer, J. B.
        Mozley, Henry Best, R. W. Sibthorp, Bernard Smith,
        Henry Bramley)

3165.   MIDDLETON, ROBERT DUDLEY.   "Tract Ninety."   Journal
        of Ecclesiastical History 2 (1951), 81-101.

3166.   MOORMAN, JOHN RICHARD HUMPIDGE.   "Forerunners of
        the Oxford Movement."   Theology 26 (Jan. 1933),
        2-15.

3167.   MORRISON, JOHN L.   "The Oxford Movement and the
        British Periodicals."   Catholic Historical Review
        45 (1959), 137-160.

3168.   MORSE-BOYCOTT, REV. DESMOND LIONEL.   Lead, Kindly
        Light: Studies of the Saints and Heroes of the
        Oxford Movement.   London: Centenary Press, 1932.
        (Includes J. H. Newman, H. J. Rose, R. H. Froude,
        I. Williams, J. Keble, E. B. Pusey, C. Marriott,
        F. Faber, H. Manning, C. Rossetti, Father Ignatius,

A. H. Stanton, Mother Kate, C. Lowder, R. Dolling,
H. P. Liddon, Mary Scharlieb, Frank Weston, Arthur
Tooth, Thomas Lacey)

3169. MORSE-BOYCOTT, REV. DESMOND LIONEL. The Secret
Story of the Oxford Movement. London: Skeffington
and Son, Ltd., 1933.

3170. MORTLOCK, CHARLES BERNARD. (Ed.) Oxford Movement
Centenary Sermons, by Eminent Preachers. London:
Skeffington, 1933.

3171. MORTLOCK, CHARLES BERNARD. The People's Book
of the Oxford Movement: With a Who's Who of the
Movement. London: Skeffington and Son, Ltd.,
1933.

3172. MOSS, CLAUDE BEAUFORT. The Orthodox Revival:
1833-1933. Oxford and London: A. R. Mowbray and
Co., 1933.

3173. MOWAT, JOHN DICKSON. Bishop A. P. Forbes. Edinburgh:
R. Grant and Son, 1925.

3174. MUNSON, J. E. B. "The Oxford Movement by the End
of the Nineteenth Century: The Anglo-Catholic
Clergy." Church History 44 (Sept. 1975), 382-395.

3175. NEWSOME, DAVID H. The Parting of Friends: A
Study of the Wilberforces and Henry Manning.
London: Murray, 1966.

3176. NICHOLLS, NORAH. "A Bibliography of the Oxford
Movement." Bookman (London) 84, No. 501 (June
1933), 143.

3177. NIKOL, JOHN. "The Oxford Movement in Decline:
Lord John Russell and the Tractarians 1846-1852."
Historical Magazine of the Protestant Episcopal
Church 43 (1974), 341-357.

3178. NIKOL, JOHN. "The Oxford Movement in Decline:
Lord John Russell, the Tractarians and the Church
of England, 1846-1852." Dissertation Abstracts International 33 (1972), 257A (Fordham University,
1972).

3179.   NYE, G. H. F.   The Story of the Oxford Movement.
        London: G. Allen, 1910.

3180.   O'CONNELL, MARVIN R.   The Oxford Conspirators:
        A History of the Oxford Movement, 1833-1845.   New
        York: Macmillan, 1969.

3181.   O'HALLORAN, BERNARD CHRISTOPHER.   "Richard Hurrell
        Froude: His Influence on John Henry Newman and the
        Oxford Movement."   Dissertation Abstracts 26 (1966),
        4636-37 (Columbia University, 1965).

3182.   O'HARE, CHARLES M.   "A Protestant Writer on the
        Oxford Movement."   Irish Ecclesiastical Record, 5th
        Series, 42 (1933), 561-573; 43 (1934), 36-47,
        271-284; 44 (1934), 225-240, 481-494; 45 (1935),
        27-38, 374-385.   (Walter Walsh's The Secret History
        of the Oxford Movement, 1897)

3183.   OLLARD, SIDNEY LESLIE AND FRANK LESLIE CROSS.
        The Anglo-Catholic Revival in Outline.   London:
        Society for Promoting Christian Knowledge, 1933.

3184.   OLLARD, SIDNEY LESLIE.   "A Famous Centenary.
        February 13th, 1845."   Oxford Magazine 63, No. 12
        (Feb. 15, 1945), 142-144.   (Meeting of Convocation
        to censure Tracts for the Times)

3185.   OLLARD, SIDNEY LESLIE.   "The Oxford Architectural
        and Historical Society and the Oxford Movement."
        Oxoniensia 5 (1940), 146-160.

3186.   OLLARD, SIDNEY LESLIE.   A Short History of the
        Oxford Movement: A Critical Survey.   London: A.
        R. Mowbray & Co., 1915.

3187.   OLLARD, SIDNEY LESLIE.   "Some Directions of the
        Oxford Movement."   Dublin Review 193 (1933), 74-84.

3188.   O'ROURKE, JAMES.   "Richard Hurrell Froude."   Irish
        Ecclesiastical Record 48 (Nov. 1936), 485-495.

3189.   OTTER, SIR JOHN LONSDALE.   Nathaniel Woodard:
        A Memoir of His Life.   London: John Lane, 1925.

3190.   PAGE, FREDERICK.   "Froude, Kingsley, and Arnold, on
        Newman."   Notes and Queries 184 (Apr. 10, 1943),
        220-221.

3191. PEARSALL, RONALD. "The Oxford Movement in Retrospect." Quarterly Review 304 (1966), 75-83.

3192. PECK, WILLIAM GEORGE. The Social Implications of the Oxford Movement. London, New York: C. Scribner's Sons, 1933.

3193. PECK, WINIFRED F. "The Ladies of the Oxford Movement." Cornhill Magazine 75 (July-Dec. 1933), 3-14.

3194. PERRY, WILLIAM. Alexander Penrose Forbes, Bishop of Brechin: The Scottish Pusey. London: S.P.C.K., 1939.

3195. PERRY, WILLIAM. The Oxford Movement in Scotland. Cambridge: Cambridge University Press, 1933.

3196. PFAFF, RICHARD W. "The Library of the Fathers: The Tractarians as Patristic Translators." Studies in Philology 70 (July 1973), 329-344.

3197. PLUS, RAOUL. "L'Oratorien Faber." Nouvelle Revue Théologique 72 (1950), 296-301.

3198. PRESTON, WILLIAM. Anglo-Catholicism and the Oxford Movement: Reviewed in the Light of the Holy Scriptures, the Book of Common Prayer, and Contemporary History. Revised and enlarged by G. E. A. Weeks. London: Protestant Reformation Society, 1933.

3199. RAMSEY, ARTHUR MICHAEL. "Hugh James Rose." Durham University Journal New Series 3, No. 1 (Dec. 1941), 50-58.

3200. REGINA, M. WILLIAMS, SISTER. "Richard Hurrell Froude." Ph.D. Dissertation, Loyola University of Chicago, 1961. (Also cited as in 3251)

3201. REXROTH, KENNETH. "Evolution of Anglo-Catholicism." Continuum 7 (1969), 345-360, 463-477.

3202. REYNOLDS, ERNEST EDWIN. Three Cardinals: Newman, Wiseman, Manning. New York: P. J. Kennedy, 1958.

3203. RICHARDS, GEORGE CHATTERTON. "Oriel College and the Oxford Movement." Nineteenth Century and After 113 (June 1933), 724-738.

3204.   RICHARDS, GEORGE CHATTERTON. "Pusey and the Oxford
        Movement." Durham University Journal 28 (1932-34),
        161-178, 245-257.
3205.   RICHEY, J. ARTHUR M. "The Oxford Movement Centennial."
        Catholic World 137 (1933), 158-165.
3206.   RUSSELL, GEORGE WILLIAM ERSKINE. Dr. Liddon.
        London: A. R. Mowbray, 1905.
3206a. ST. JOHN, H. "Hurrell Froude and the Beginnings
        of the Movement." Blackfriars 14 (July 1933),
        560-568.
3207.   SCHOLFIELD, J. F. "The Meaning of the Oxford
        Revival of 1833." Ave Maria 36 (July 30, 1932),
        129-134.
3208.   SCHWARZ, MARC L. "The Paradox of Commitment: John
        Keble and the Establishment, 1833-1850." Historical
        Magazine of the Protestant Episcopal Church 37
        (1968), 298-310.
3209.   SELWYN, EDWARD GORDON. "The Oxford Movement."
        Quarterly Review 260 (1933), 301-314.
3210.   SHAW, PLATO ERNEST. The Early Tractarians and
        the Eastern Church. Milwaukee: Morehouse
        Publishing Co.; London: A. R. Mowbray and Co.,
        1930.
3211.   SHEEN, H. E. "The Oxford Movement in a Manchester
        Parish." M.A. Dissertation, University of
        Manchester, 1971.
3212.   SHIPTON, I. A. M. "Christina Rossetti: The Poetess
        of the Oxford Movement." Church Quarterly Review
        116 (July 1933), 219-229.
3213.   SMITH, BASIL A. "The Anglicanism of Dean Church."
        Church Quarterly Review 156 (Jan.-Mar. 1955),
        70-81.
3214.   SMITH, BASIL A. Dean Church: The Anglican Response
        to Newman. London: Oxford University Press, 1958.
3215.   SNEAD-COX, J. C. [sic] "Hereditary Catholics and the
        Oxford Converts." Catholic World 126 (Jan. 1928),
        541-542. (Should be J. G.; extract from 4300)

3216.   SNOW, DOROTHY M. B.   "Hugh James Rose: Rector of
        Hadleigh, Suffolk."   B. Litt. Dissertation, Uni-
        versity of Oxford, 1960.

3217.   STEWARD, SAMUEL M.   "Provocatives of the Oxford
        Movement and Its Nexus with English Literary Romanti-
        cism."   Ph.D. Dissertation, Ohio State University,
        1934.   (Also see Ohio State University Abstracts
        No. 15, 1935, 219-228)

3218.   STEWART, HERBERT LESLIE.   A Century of Anglo-
        Catholicism.   London and Toronto: J. M. Dent and
        Sons Ltd., 1929.

3219.   STOCKLEY, WILLIAM FREDERICK PAUL.   "Keble and
        Newman."   Irish Ecclesiastical Record 35 (1930),
        40-57; 135-148.

3220.   STORR, V. H.   "Le Mouvement Oxford."   Oecumenica 1,
        No. 2 (juin 1934), 112-118.

3221.   STORR, VERNON FAITHFUL.   The Oxford Movement:
        A Liberal Evangelical View.   London: S.P.C.K.,
        1933.

3222.   STRONG, THOMAS BANKS et al.   Henry Parry Liddon,
        D.D., D.C.L., 1829-1929:   A Centenary Memoir.
        London, Oxford: A. R. Mowbray and Co., 1929.

3223.   STUNT, TIMOTHY C. F.   "Two Nineteenth-Century
        Movements."   Evangelical Quarterly 37 (Oct.-Dec.
        1965), 221-231.   (Oxford and the Brethren)

3224.   SYKES, CHRISTOPHER.   Two Studies in Virtue.
        London: Collins, 1953.   (Includes Richard Sibthorp)

3225.   TAYLOR, BRIAN.   "George Mason, Parish Priest."
        Historical Magazine of the Protestant Episcopal
        Church 41 (1972), 77-83.

3226.   THUREAU-DANGIN, PAUL.   The English Catholic Revival
        in the Nineteenth Century.   Revised and re-edited
        from a translation by Wilfred Wilberforce.   2 vols.
        London: Simpkin, Marshall & Co., 1914.

3227.   THUREAU-DANGIN, PAUL.   "La Renaissance catholique
        en Angleterre au XIXe siècle."   Le Correspondant
        166 (1901), 890-911, 1049-91; 168 (1901), 26-70,

981-1005. (I. Les Converts, 1845-1847. II. Pusey
et Manning au lendemain de conversion de Newman,
1845-1847. III. Les Mécomptes du Puseyisme, 1846-
1850.)

3228. TOWLE, ELEANOR A. John Mason Neale, D.D.: A
Memoir. London, New York: Longmans, Green, and
Co., 1906.

3229. TRISTRAM, HENRY. "Tractarians and Education."
Blackfriars 14 (Aug. 1933), 642-653.

3230. TYSZKIEWICZ, STANISLAS. "Un Episode du mouvement
d'Oxford: la mission de William Palmer: I. De
L'Anglicanisme aux confins de l'orthodoxie; II.
Des Confins de l'orthodoxie au seuil du catholic-
isme; III. Au Bercail." Etudes Publiées par des
PP. de la Compagnie de Jésus (Paris), 136 (1913),
43-63, 190-210, 329-347. (Published as a book,
Paris: 1913)

3231. UNDERHILL, EVELYN. "The Spiritual Significance of
the Oxford Movement." Hibbert Journal 31 (1933),
401-412.

3232. VOLL, DIETER. Catholic Evangelicalism: The Acceptance of Evangelical Traditions by the Oxford Movement
during the Second Half of the Nineteenth Century.
Trans. Veronica Ruffer. London: Faith Press,
1963.

3233. VROOM, F. W. "The Oxford Movement: 1833-1933."
Dalhousie Review 13 (1933-34), 152-164.

3234. WAGER, CHARLES HENRY ADAMS. "The Oxford Movement
and Its Results." Dial 61 (Nov. 16, 1916), 393-394.

3235. WALLER, JOHN O. "A Composite Anglo-Catholic Concept
of the Novel, 1841-1868." Bulletin of the New York
Public Library 70 (1966), 356-368.

3236. WALSH, LEO JOSEPH. "William G. Ward and the 'Dublin
Review'." Dissertation Abstracts 26 (1966), 6028
(Columbia University, 1962).

3237. WALSH, WALTER. The History of the Romeward Movement in the Church of England 1833-1864. London: J. Nisbet, 1900.

3238. WARD, MAISIE. "W. G. Ward and Wilfrid Ward." Dublin Review 198 (1936), 235-252.

3239. WARD, WILFRID PHILIP. (Ed.) "John Keble: An Unpublished Fragment." Dublin Review 143 (1908), 376-383. (By J. H. Newman)

3240. WARD, WILFRID PHILIP. William George Ward and the Catholic Revival. London and New York: Macmillan and Co., 1893.

3241. WARD, WILFRID PHILIP. William George Ward and the Oxford Movement. London: Macmillan and Co., 1889.

3242. WARD, WILLIAM REGINALD. "Oxford and the Origins of Liberal Catholicism in the Church of England." Studies in Church History 1 (1964), 233-252.

3243. WATKIN-JONES, HOWARD. "Two Oxford Movements: Wesley and Newman." Hibbert Journal 31 (Oct. 1932), 83-96.

3244. WEBB, CLEMENT CHARLES JULIAN. Religious Thought in the Oxford Movement. London: S.P.C.K., 1928.

3245. WEBB, CLEMENT CHARLES JULIAN. "The Significance of the Oxford Movement in the History of Anglicanism." Theology 26 (Jan. 1933), 25-36.

3246. WEBB, CLEMENT CHARLES JULIAN. "Two Philosophers of the Oxford Movement." Philosophy 8 (1933), 273-284. (J. H. Newman and W. G. Ward)

3247. WEBSTER, ALAN B. "The Ministry of Dean Church." Theology 75 (Aug. 1972), 405-411.

3248. WELLDON, J. E. C. "The Oxford Movement." Spectator 150 (Feb. 24, 1933), 251; 150 (Mar. 3, 1933), 288.

3249. WHITE, ROSEMARY A. "Women of the Oxford Movement." Catholic World 161 (1945), 255-257.

3250. WILLIAMS, NORMAN POWELL AND CHARLES HARRIS. (Eds.)
      Northern Catholicism: Centenary Studies in the
      Oxford and Parallel Movements. New York: The
      Macmillan Co., 1933.

3251. WILLIAMS, REGINA M., SISTER. "Richard Hurrell
      Froude." Ph.D. Dissertation, Loyola University of
      Chicago. 1961. (Also cited as in 3200)

3252. WILLIAMSON, HUGH R. "The Oxford Movement Cent-
      enary." Bookman (London), 84 (1933), 133-134.

3253. WILLOUGHBY, L. A. "On Some German Affinities with
      the Oxford Movement." Modern Language Review 29
      (1934), 52-66.

3254. WINDLE, SIR BERTRAM COGHILL ALAN. Who's Who of the
      Oxford Movement. New York and London: The Century
      Co., 1926.

3255. WOOD, EDWARD FREDERICK LINDLEY. John Keble.
      London and Oxford: A. R. Mowbray and Co., 1905.

3256. WOODS, CHARLOTTE E. Archdeacon Wilberforce: His
      Ideals and Teaching. London: Elliot Stock, 1917.
      (Robert Wilberforce)

3257. WRIGHT, CUTHBERT. "Second Spring: The Tractarian
      Movement, 1833-1933." Sewanee Review 41 (July
      1933), 268-285.

3258. WUENSCHEL, E. A. "The Character of the Oxford
      Movement." Homiletic and Pastoral Review 33 (July
      1933), 1026-35.

3259. WUENSCHEL, E. A. "The Oxford Movement and the
      Anglican Revival." Homiletic and Pastoral Review
      33 (Aug. 1933), 1140-52.

3260. YATES, NIGEL. The Oxford Movement and Parish
      Life: St. Saviour's Leeds, 1839-1929. York: St.
      Anthony's Press, 1975.

4. Broad Church

Works on the Broad Church and on the views and activities of individuals associated with it (except for those to be found in XIX. Varieties of Belief. . .) are listed in this section. Included here among others are Thomas Arnold, Julius Hare, Rowland Williams, Frederick Temple, Richard Holt Hutton, Arthur Penrhyn Stanley, Mark Pattison, Brooke Foss Westcott, and Baden Powell. See I. Science, 2. Method and Philosophy (Baden Powell); II. Religion; other sections in this category (XIV), particularly 5. Biblical Criticism and Essays and Reviews (1860).

3261. AITKEN, W. FRANCIS. Frederick Temple: Archbishop of Canterbury. London: S. W. Partridge, 1901.

3262. ANON. "John Llewelyn Davies: In Memoriam." Contemporary Review 109 (1916), 782-788.

3263. BAILEY, IVOR. "The Challenge of Change: A Study of Relevance versus Authority in the Victorian Pulpit." Expository Times 86 (Oct. 1974), 18-22. (F. W. Robertson)

3264. BAKER, WILLIAM J. "Julius Charles Hare: A Victorian Interpreter of Luther." South Atlantic Quarterly 70 (1971), 88-101.

3265. BAMFORD, THOMAS W. Thomas Arnold. London: Cresset Press, 1960.

3266. BAMFORD, THOMAS W. (Ed.) Thomas Arnold on Education. London and New York: Cambridge University Press, 1970.

3267. BLACKWOOD, JAMES RUSSELL. The Soul of Frederick W. Robertson, the Brighton Preacher. New York and London: Harper, 1947.

3268. BOLITHO, HECTOR AND THE DEAN OF WINDSOR. (Eds.) A Victorian Dean: A Memoir of Arthur Penrhyn Stanley, Dean of Westminster. London: Chatto and Windus, 1930.

3269. BOLTON, J. R. G. "Jowett of Balliol." Spectator
       171 (Oct. 1, 1943), 308-309.
3270. CAMPBELL, REGINALD JOHN. Thomas Arnold. London:
       Macmillan, 1927.
3271. CANNON, WALTER F. "Scientists and Broad Churchmen:
       An Early Victorian Intellectual Network." Journal
       of British Studies 4 (1964), 65-86.
3272. CHADWICK, HENRY. The Vindication of Christianity
       in Westcott's Thought. Cambridge: Cambridge
       University Press, 1961.
3273. CHADWICK, OWEN. Westcott and the University.
       London: Cambridge University Press, 1963.
3274. CHRISTENSEN, MERTON A. "Thomas Arnold's Debt to
       German Theologians: A Prelude to 'Literature and
       Dogma'." Modern Philology 55 (Aug. 1957), 14-20.
3275. CLAYTON, JOSEPH. Bishop Westcott. London: A. R.
       Mowbray and Co., Limited, 1906.
3276. COLEMAN, ARTHUR MAINWARING. Six Liberal Thinkers.
       Oxford: Blackwell, 1936. (Includes Mark Pattison,
       Edwin Hatch)
3277. CURNOW, A. G. "Robertson of Brighton: A Centenary
       Tribute." London Quarterly and Holborn Review 178
       (July 1953), 175-179.
3278. DANT, CHARLES HARRY. Archbishop Temple. London
       and Newcastle-on-Tyne: Walter Scott Publishing
       Co., Ltd., 1903.
3279. DARK, SIDNEY. Five Deans: John Colet, John Donne,
       Jonathan Swift, Arthur Penrhyn Stanley, William Ralph
       Inge. London: Jonathan Cape, 1928.
3280. FABER, SIR GEOFFREY CUST. "Doctor Jowett."
       National Review 131 (1948), 51-56.
3281. FABER, SIR GEOFFREY CUST. Jowett: A Portrait
       with Background. London: Faber and Faber, 1957.
3282. FAIRWEATHER, E. R. "Apostolical Tradition and the
       Defence of Dogma: An Episode in the Anglo-Catholic
       Revival." Canadian Journal of Theology 11 (Oct.
       1965), 277-289. (R. D. Hampden)

3283. FALLOWS, W. G. "Thomas Arnold: A Prophet for Today." Modern Churchman 6 (July 1963), 274-278.

3284. FORBES, DUNCAN. The Liberal Anglican Idea of History. Cambridge: The Cambridge University Press, 1952.

3285. GREEN, VIVIAN HUBERT HOWARD. Oxford Common Room: A Study of Lincoln College and Mark Pattison. London: E. Arnold, 1957.

3286. GUTSCHE, HUGO. "Thomas Arnold als Reformator des höheren englischen Schulwesens im 19. Jahrhundert." Ph.D. Dissertation, Universität Erlangen, 1914.

3287. GWYNN, STEPHEN LUCIUS. "Mark Pattison." Cornhill Magazine 62 (1927), 539-554.

3288. HENSON, HERBERT HENSLEY. Robertson of Brighton, 1816-1853. London: Smith, Elder and Co., 1916.

3289. HOLMES, J. DEREK. "Newman, Froude and Pattison: Some Aspects of Their Relations." Journal of Religious History 4 (June 1966), 23-38.

3290. HOWES, GRAHAM. "Dr. Arnold and Bishop Stanley." Studies in Church History 2 (1965), 320-337.

3291. JOHN, BRIAN. "Thomas Arnold as Educator of the Liberal Conscience." Journal of General Education 19 (July 1967), 132-140.

3292. JUDD, A. F. "The Victorian Compromise." Theology 35, No. 210 (Dec. 1937), 326-338.

3293. KNIGHT, DAVID M. "Professor Baden Powell and the Inductive Philosophy." Durham University Journal 60 (1968), 81-87.

3294. MCFARLAND, GEORGE FOSTER. "The Early Literary Career of Julius Charles Hare." Bulletin of the John Rylands Library 46 (1963), 42-83.

3295. MCFARLAND, GEORGE FOSTER. "The Early Literary Career of Julius Charles Hare, 1818 to 1834." Dissertation Abstracts 26 (1965), 371 (University of Pennsylvania, 1964).

3296.   MCFARLAND, GEORGE FOSTER.   "Julius Charles Hare:
        Coleridge, De Quincey, and German Literature."
        Bulletin of the John Rylands Library 47 (1964),
        165-197.

3297.   MCFARLAND, GEORGE FOSTER.   "Shelley and Julius
        Hare: A Review and a Response." Bulletin of the
        John Rylands Library 57 (Spring 1975), 406-429.

3298.   MCFARLAND, GEORGE FOSTER.   "Wordsworth and Julius
        Hare." Bulletin of the John Rylands Library 55
        (Spring 1973), 403-433.

3299.   MCGIFFERT, A. C., JR.   "James Marsh, 1794-1842:
        Philosophical Theologian, Evangelical Liberal."
        Church History 38 (Dec. 1969), 437-458.

3300.   MACKERNESS, ERIC DAVID.   "Benjamin Jowett: Preacher
        and Prophet." Modern Churchman 45, No. 4 (1955),
        319-326.

3301.   MACKERNESS, ERIC DAVID.   "R. H. Hutton and the
        Victorian Lay Sermon." Dalhousie Review 32 (1957),
        259-267.

3302.   MOMERIE, VEHIA. Dr. Momerie: His Life and Work.
        Edinburgh: W. Blackwood, 1905. (Alfred Momerie)

3303.   MONTAGUE, FRANCIS CHARLES. Some Early Letters of
        Mark Pattison . . . Reprinted from the "Bulletin of
        the John Rylands Library." Manchester: Manchester
        University Press; Librarian, John Rylands Library,
        1934. (Also in Bulletin of the John Rylands Library,
        18, 1934, 156-176)

3304.   OAKELEY, EDWARD MURRAY. Bishop Percival: A Brief
        Sketch of a Great Career. Oxford: B. H. Blackwell,
        1919.

3305.   PASKO, MICHAEL.   "Mark Pattison's Course through
        the Oxford Movement." Dissertation Abstracts 25
        (1965), 6633 (University of Illinois, 1964).

3306.   PEELE, R. DE C.   "Arnold of Rugby." Spectator 168
        (June 12, 1942), 552.

3307. PREYER, ROBERT OTTO. "Julius Hare and Coleridgean
       Criticism." Journal of Aesthetics and Art Criticism
       15 (1957), 449-460.

3308. PRICE, FANNY. "Jowett on Carlyle." Notes and
       Queries 185 (June-Dec. 1943), 45-46.

3309. PYKE, RICHARD. "Bishop Westcott: Scholar, Saint
       and Statesman." London Quarterly and Holborn Review
       169 (Apr. 1944), 116-120.

3310. RUPP, ERNEST GORDON. Hort and the Cambridge Tradition.
       London: Cambridge University Press, 1970.

3311. SAMBROOK, A. J. "A Welsh Heretic: Rowland Williams."
       Church Quarterly Review 166 (1965), 448-462.

3312. SANDERS, CHARLES RICHARD. Coleridge and the Broad
       Church Movement: Studies in S. T. Coleridge,
       Dr. Arnold of Rugby, J. C. Hare, Thomas Carlyle
       and F. D. Maurice. Durham, N.C.: Duke University
       Press, 1942.

3313. SANDFORD, ERNEST GREY. Frederick Temple: An
       Appreciation. London: Macmillan and Co., Limited;
       New York: Macmillan Co., 1907.

3314. SANDFORD, ERNEST GREY. (Ed.) Memoirs of Archbishop
       Temple by Seven Friends. 2 vols. London: Macmillan
       and Co., Limited; New York: Macmillan Co., 1906.

3315. SCOTT, D. L. "Rowland Williams, 1817-1870." Modern
       Churchman 45 (June 1955), 118-125.

3316. SIMPSON, F. S. W. "Frederick Temple's Contribution
       to English Modernism." Modern Churchman 45 (June
       1955), 126-129.

3317. SNELL, FREDERICK JOHN. Early Associations of
       Archbishop Temple: A Record of Blundell's School and
       Its Neighbourhood. London: Hutchinson and Co.,
       1904.

3318. SNIEGOWSKI, DONALD CHESTER. "The Early Career of
       Mark Pattison." Dissertation Abstracts 27 (1966),
       247A (Yale University, 1966).

3319. SPARROW, JOHN HANBURY ANGUS. Mark Pattison and the Idea of a University. London: Cambridge University Press, 1967.

3320. STACKHOUSE, REGINALD FRANCIS. "Thomas Arnold's Theory of Church and State." Ph.D. Dissertation, Yale University, 1962.

3321. STEPHENSON, ALAN MALCOLM GEORGE. "Liberal Anglicanism in the Nineteenth Century." Modern Churchman 13 (Oct. 1969), 87-102. (S. T. Coleridge, Essays and Reviews, F. D. Maurice)

3322. STEVENS, ALBERT KUNNEN. "Richard Holt Hutton, Theologian and Critic." Microfilm Abstracts 10, No. 2 (1950), 119-120 (University of Michigan, 1950).

3323. SWANSTON, HAMISH F. G. "Dean Stanley and the Enlargement of the Church." Theology 77 (Oct. 1974), 528-535.

3324. SYMES, JOHN E. "Broad Churchmanship: A.D. 1500-1900." Constructive Quarterly 2 (Sept. 1914), 617-626.

3325. TEMPLE, WILLIAM. Life of Bishop Percival. London: Macmillan and Co., Limited, 1921. (John Percival)

3326. TENER, ROBERT H. "R. H. Hutton and 'Agnostic'." Notes and Queries 11 (Nov. 1964), 429-431.

3327. THOMAS, GLYN NICHOLAS. "Richard Holt Hutton: A Biographical and Critical Study." Microfilm Abstracts 10, No. 1 (1950), 90-92 (University of Illinois, 1949).

3328. THUREAU-DANGIN, PAUL. "Une Page de l'histoire de l'Anglicanisme: les débuts du Broad Church 1845-1865." Revue des Deux Mondes 15, 5e pér. (1903), 77-111.

3329. TOLLEMACHE, LIONEL ARTHUR. "Jowett and Tennyson." Spectator 119 (Oct. 20, 1917), 411.

3330. TUELL, ANNE KIMBALL. John Sterling: A Representative Victorian. New York: The Macmillan Company, 1941.

3331. WEST, ARTHUR GEORGE BAINBRIDGE. Memoirs of Brooke
       Foss Westcott. Cambridge: Heffer, 1936.
3332. WHITRIDGE, ARNOLD. Dr. Arnold of Rugby. London:
       Constable and Co., Ltd., 1928.
3333. WIGMORE-BEDDOES, DENNIS G. "How The Unitarian
       Movement Paid Its Debt to Anglicanism." Transactions
       of the Unitarian Historical Society 13 (Oct. 1963),
       69-79. (The influence of Unitarianism on leaders
       of the Broad Church)
3334. WIGMORE-BEDDOES, DENNIS G. Yesterday's Radicals:
       A Study of the Affinity between Unitarianism and
       Broad Church Anglicanism in the Nineteenth Century.
       London: James Clarke, 1971.
3335. WILLEY, BASIL. "Rector and Master: Mark Pattison
       and Benjamin Jowett." Modern Churchman New Series
       2 (Mar. 1959), 135-139.
3336. WILLIAMSON, EUGENE LA COSTE. The Liberalism of
       Thomas Arnold: A Study of His Religious and Political
       Writings. University: University of Alabama
       Press, 1964.
3337. WINSLOW, DONALD F. "Francis W. Newman's Assessment
       of John Sterling: Two Letters." English Language
       Notes 11 (1974), 278-283. (To Moncure Conway, Oct.
       6, 1869; Oct. 16, 1869)
3338. WOODWARD, FRANCES J. The Doctor's Disciples:
       A Study of Four Pupils of Arnold of Rugby: Stanley,
       Gelt, Clough, William Arnold. New York: Oxford
       University Press, 1954.
3339. WRAGGE, REV. WALTER. "Broad Church Movement in the
       Last Two Decades of the 19th Century." Church
       Quarterly Review 114, No. 227 (Apr. 1932), 27-42.
3340. WYMER, NORMAN. Dr. Arnold of Rugby. London:
       Hale, 1953.
3341. YOUNG, IVAN. "Brooke Foss Westcott (1825-1906)."
       Oecumenica 4, No. 4 (Jan. 1938), 665-674.

5.   Biblical Criticism and Essays and Reviews (1860)

This section looks at studies of the Bible and
biblical criticism. It deals with the impact of Essays
and Reviews (written by Frederick Temple, Rowland Williams,
Baden Powell, Henry Bristow Wilson, G. W. Goodwin, Mark
Pattison, and Benjamin Jowett) and with the controversial
life of Bishop John William Colenso. See II. Religion;
other sections in this category (XIV), especially 4.
Broad Church; various sections in XIX. Varieties of
Belief. . ., particularly 1. Arnold, Matthew.

3342.   BADGER, KINGSBURY.   "Mark Pattison and the Victorian
        Scholar."   Modern Language Quarterly 6 (1945),
        423-447.

3343.   BAGNALL, A. G.   "William Colenso."   Times Literary
        Supplement (Oct. 26, 1946), 521.

3344.   BARRETT, CHARLES KINGSLEY.   "Joseph Barber Light-
        foot."   Durham University Journal 33 (1972), 193-204.

3345.   BARRETT, CHARLES KINGSLEY.   Westcott as Commentator.
        Cambridge: Cambridge University Press, 1959.

3346.   BROWN, JESSE HUNCHBERGER.   "The Contribution of
        William Robertson Smith to Old Testament Scholarship,
        with Special Emphasis on Higher Criticism."   Disserta-
        tion Abstracts 25 (1965), 4839-40 (Duke University,
        1964).

3347.   BURNETT, B. B.   "The Missionary Work of the First
        Anglican Bishop of Natal, the Rt. Reverend John
        William Colenso, D.D. between the Years, 1852-1873."
        M.A. Thesis, University of South Africa, 1947.

3348.   CHRISTENSEN, MERTON A.   "The Impact of Biblical
        Criticism upon English Literary Thought from 1800
        to 1875."   Ph.D. Dissertation, University of Maryland,
        1954.

3349.   CHRISTENSEN, MERTON A.   "Taylor of Norwich and the
        Higher Criticism." Journal of the History of Ideas
        20 (1959), 179-194. (William Taylor)

3350.   COLENSO, FRANCES BUNYON. Colenso Letters from
        Natal: Arranged with Comments by Wyn Rees.
        Pietermaritzburg: Shuter and Shooter, 1958.

3351.   COULLING, SIDNEY M. B. "The Background of 'The
        Function of Criticism at the Present Time'."
        Philological Quarterly 42 (1963), 36-54. (M.
        Arnold, J. Colenso)

3352.   CROWTHER, MARGARET ANNE. Church Embattled:
        Religious Controversy in Mid-Victorian England.
        Newton Abbot, England: David and Charles, 1970.

3353.   DE GROOT, H. B. "Baden Powell, Scientist and
        Theologian: A Bibliographical Note." Victorian
        Periodicals Newsletter Nos. 5-6 (1969), 16-18.
        (List of articles by Powell)

3354.   EAKER, J. GORDON. "Matthew Arnold's Biblical
        Criticism." Religion in Life 32 (1963), 257-266.

3355.   EDEN, GEORGE RODNEY AND FREDERICK CHARLES MACDONALD.
        (Eds.) Lightfoot of Durham; Memories and Apprecia-
        tions. Cambridge, England: Cambridge University
        Press, 1932.

3356.   ELLIS, IEUAN PRYCE. "'Essays and Reviews' Recon-
        sidered." Theology 74 (Sept. 1971), 396-404.

3357.   EMERTON, J. A. "Old Testament Scholarship and the
        Church: A Century after Colenso." Modern Churchman
        5 (July 1962), 266-271.

3358.   FRANCIS, MARK. "The Origins of 'Essays and Reviews':
        An Interpretation of Mark Pattison in the 1850's."
        Historical Journal 17 (Dec. 1974), 797-811.

3359.   FRASER, B. D. John William Colenso: A Bibliography.
        Cape Town: School of Librarianship, University of
        Cape Town, 1952.

3360.   FREI, HANS W.   The Eclipse of Biblical Narrative:
        A Study in Eighteenth and Nineteenth Century
        Hermeneutics.   New Haven, Conn.:   Yale University
        Press, 1974.

3361.   GLOVER, WILLIS BORDERS.   Evangelical Nonconformists
        and Higher Criticism in the Nineteenth Century.
        London: Independent Press, 1954.

3362.   GRYLLS, ROSALIE GLYNN.   "Bishop Colenso of Natal."
        Contemporary Review 184 (1953), 27-32.

3363.   HARVEY, VAN A.   "D. F. Strauss's 'Life of Jesus'
        Revisited."   Church History 30 (1961), 191-211.

3364.   HINCHLIFF, PETER BINGHAM.   The Anglican Church in
        South Africa.   London: Darton, Longman and Todd,
        1963.

3365.   HINCHLIFF, PETER BINGHAM.   "John William Colenso:
        A Fresh Appraisal."   Journal of Ecclesiastical
        History 13 (1962), 203-216.

3366.   HINCHLIFF, PETER BINGHAM.   John William Colenso:
        Bishop of Natal.   London: Nelson, 1964.

3367.   HOOKER, M. A.   "The Place of Bishop J. W. Colenso
        in the History of South Africa."   2 vols.   Ph.D.
        Thesis, University of Witwatersrand, 1953.

3368.   KORINKO, STEPHEN JOHN.   "Matthew Arnold and Biblical
        Higher Criticism."   Dissertation Abstracts Inter-
        national 31 (1971), 4124-25A (University of Nebraska,
        1970).

3369.   KRIEFALL, LUTHER HARRY.   "A Victorian Apocalypse:
        A Study of George Eliot's 'Daniel Deronda' and Its
        Relation to David F. Strauss' 'Das Leben Jesu'."
        Dissertation Abstracts 28 (1967), 234A (University
        of Michigan, 1966).

3370.   MARSH, PETER T.   "Prophecy and Concession: A
        Victorian Quandary over Biblical Criticism."
        Canadian Journal of Theology 12 (July 1966),
        172-183.

3371. NEILL, STEPHEN CHARLES. The Interpretation of
the New Testament 1861-1961. London, New York:
Oxford University Press, 1964.

3372. OWEN, RALPH ALBERT DORNFELD. Christian Bunsen and
Liberal English Theology. Montpelier, Vt.: Capital
City Press, 1924.

3373. POWELL, H. GORDON. "'Ecce Homo': The Historical
Jesus in 1865." London Quarterly and Holborn Review
35 (1966), 52-56.

3374. REARDON, BERNARD M. G. "'Essays and Reviews': A
Centenary of Liberal Anglicanism." Quarterly Review
298 (July 1960), 301-308.

3375. REID, JAMES EDWARD. "The Higher Criticism in
England and the Periodical Debate of the 1860's."
Dissertation Abstracts International 32 (1972),
6390A (University of Minnesota, 1971).

3376. SHAFFER, ELINOR S. "Kubla Khan" and "The Fall of
Jerusalem": The Mythological School in Biblical
Criticism and Secular Literature, 1770-1880.
Cambridge: Cambridge University Press, 1975.

3377. THRANE, JAMES ROBERT. "The Rise of Higher Criticism
in England, 1800-1870." Dissertation Abstracts 16
(1956), 1457 (Columbia University, 1956).

3378. VARNER, LEO BENTLEY. "The Literary Reception of
Bishop Colenso: Arnold, Kingsley, Newman and
Others." Dissertation Abstracts International 35
(1975) 4567-68A (University of Illinois at Urbana-
Champaign, 1974). (Includes M. Arnold, C. Kingsley,
J. H. Newman, F. Newman, F. D. Maurice, J. F.
Stephen, J. A. Froude, W. R. Greg, F. Harrison)

3379. WHITE, P. O. G. "The Colenso Controversy." Theology
65 (1962), 402-408.

3380. WHITE, P. O. G. "'Essays and Reviews'." Theology
63 (1960), 46-53.

3381. WILLEY, BASIL. More Nineteenth Century Studies:
A Group of Honest Doubters. London: Chatto &

Windus, 1956. (F. W. Newman, A. Tennyson, J. A.
Froude, Essays and Reviews, William Hale White, J.
Morley)

3382.   WILLIAMSON, EUGENE LA COSTE. "Significant Points
of Comparison between the Biblical Criticism of
Thomas and Matthew Arnold." PMLA 76 (1961), 539-543.

3383.   WILSON, D. J. "The Life of J. B. Lightfoot
(1829-89), with Special Reference to the Training
of the Ministry." Ph.D. Dissertation, University
of Edinburgh, 1956.

6.   Lux Mundi (1889)

Works dealing with Lux Mundi and the relationship
of its theological perspective to that of the Oxford
Movement and Anglo-Catholicism are listed in this section.
Studies focus mainly on Charles Gore and, to a lesser
extent, Henry Scott Holland. See II. Religion; other
sections in this category (XIV).

3384.   AVIS, P. D. L. "Gore and Theological Synthesis."
Scottish Journal of Theology 28, No. 5 (1975),
461-476.

3385.   BAYNES, ARTHUR HAMILTON. "From Newman to Gore."
Hibbert Journal 32 (1933), 1-8.

3386.   BELPAIRE, T. "Bishop Charles Gore." Irénikon 12
(1935), 486-498.

3387.   CARPENTER, JAMES A. Gore: A Study in Liberal
Catholic Thought. London: Faith Press, 1960.

3388.   CHADWICK, HENRY. "Charles Gore and Roman Catholic
Claims." Theology 78 (Feb. 1975), 68-75.

3389.   CHESHIRE, CHRISTOPHER. "Charles Gore: The Christian
Socialist." Christendom: A Journal of Christian
Sociology 2, No. 5 (Mar. 1932), 47-53.

3390. CHESHIRE, CHRISTOPHER. (Ed.) Henry Scott Holland:
       Some Appreciations. London: W. Gardner, Darton,
       1919.

3391. CROSSE, GORDON. Charles Gore: A Biographical
       Sketch. London: A. R. Mowbray and Co., Ltd.,
       1932.

3392. DILLISTONE, F. W. "Charles Gore and Charles Raven."
       Church Quarterly 4 (July 1971), 27-37.

3393. DREW, MARY. "Henry Scott Holland." Contemporary
       Review 114 (1918), 652-660.

3394. EKSTRÖM, RAGNAR. The Theology of Charles Gore: A
       Study in Modern Anglican Theology. Lund: C. W. K.
       Gleerup, 1944.

3395. FOSTER, J. H. "Henry Scott Holland: A Biography."
       Ph.D. Dissertation, University of Wales, 1970.

3396. FROST, FRANCIS. "Un Anglo-Catholique du XIX siècle:
       Charles Gore et la question du développement du
       dogme." Mélanges de Science Religieuse 38 (Dec.
       1971), 205-220.

3397. GORE, JOHN. Charles Gore, Father and Son: A Back-
       ground to the Early Years and Family Life of Bishop
       Gore. London: J. Murray, 1932.

3398. GRAYSTON, D. "Critical Orthodoxy and Social Change:
       Charles Gore." Ecumenist 14 (Nov.-Dec. 1975), 610.

3399. HOOD, A. FREDERIC. "Gore on the Incarnation."
       Church Quarterly Review 161, No. 340 (1960), 280-282.

3400. HUBBARD, H. E. "Charles Gore." Church Quarterly
       Review 155 (1954), 22-34.

3401. JASPER, RONALD CLAUD DUDLEY. "Gore on Liturgical
       Revision." Church Quarterly Review 166 (Jan.-Mar.
       1965), 21-36.

3402. KENYON, RUTH. "The 'Lux Mundi' School as Continuators
       of the Tractarian Sociology." Theology 26 (Jan.
       1933), 16-25.

3403. KNOX, WILFRED LAWRENCE AND ALEXANDER ROPER VIDLER.
       The Development of Modern Catholicism. London:
       Allan, 1933. (Anglo-Catholicism from the middle
       of the nineteenth century to the present)

3404. LOFTHOUSE, W. F. "Charles Gore." London Quarterly
       and Holborn Review 186 (Apr. 1961), 123-127.

3405. LYTTELTON, EDWARD. The Mind and Character of Henry
       Scott Holland. London: A. R. Mowbray, 1926.

3406. MANSBRIDGE, ALBERT. Edward Stuart Talbot and
       Charles Gore: Witnesses to and Interpreters of the
       Christian Faith in Church and State. London: J.
       M. Dent and Sons, Limited, 1935.

3407. MAYOR, STEPHEN H. "'Lux Mundi': A Reassessment."
       Church Quarterly Review 166 (Jan.-Mar. 1965),
       74-80.

3408. MIDDLETON, ROBERT DUDLEY. "Charles Gore." Theology
       56 (Jan. 1953), 6-13.

3409. MOORMAN, JOHN RICHARD HUMPIDGE. "Charles Gore and
       the Doctrine of the Church." Church Quarterly Review
       158 (Apr.-June 1957), 128-140.

3410. MORTIMER, R. C. "The Moral Emphasis in Gore's
       Theology." Church Quarterly Review 159 (Apr.-June
       1958), 231-245.

3411. NEWSOME, DAVID H. "The Assault on Mammon: Charles
       Gore and John Neville Figgis." Journal of Ecclesi-
       astical History 17 (1966), 227-241. (Anglo-
       Catholic theology and social reform)

3412. NUTTALL, GEOFFREY FILLINGHAM. "Charles Gore and
       the Solidarity of the Faith." Church Quarterly 1
       (July 1968), 52-64.

3413. OLIVER, R. "Scott Holland in 1945." London
       Quarterly and Holborn Review 171 (Apr. 1946),
       139-142.

3414. PAGE, ROBERT JEFFRESS. "Charles Gore: Anglican
       Apologist." Dissertation Abstracts 15 (1955),
       2321-22 (Columbia University, 1955).

3415. PAGET, STEPHEN AND J. M. C. CRUM. Francis Paget:
       Bishop of Oxford. London: Macmillan, 1912.
3416. PAGET, STEPHEN. (Ed.) Henry Scott Holland: Memoirs
       and Letters. London: J. Murray, 1921.
3417. PRESTIGE, GEORGE LEONARD. The Life of Charles Gore.
       London: Heinemann, 1935.
3418. RAMSEY, ARTHUR MICHAEL. Charles Gore and Anglican
       Theology. London: S.P.C.K., 1955.
3419. RAMSEY, ARTHUR MICHAEL. "Christian Faith and the
       Historical Jesus." Theology 75 (Mar. 1972), 118-126.
       (Includes C. Gore)
3420. REARDON, BERNARD M. G. (Ed.) Henry Scott Holland:
       A Selection from His Writings. London: S.P.C.K.,
       1962.
3421. SELWYN, EDWARD GORDON. "Gore the Liberator."
       Quarterly Review 297 (Jan. 1959), 28-38.
3422. SMITH, BARDWELL. "Liberal Catholicism: An Anglican
       Perspective." Anglican Theological Review 54 (July
       1972), 175-193. (Gore)
3423. STEPHENSON, GWENDOLEN. Edward Stuart Talbot,
       1844-1934. London: S.P.C.K.; New York: Macmillan
       Co., 1936.
3424. WARD, WILLIAM REGINALD. "Oxford and the Origins of
       Liberal Catholicism in the Church of England."
       Studies in Church History 1 (1964), 233-252.

XV.   Nonconformity

This category contains material on Nonconformity in general and on a variety of Dissenting churches and associations. It includes Presbyterians (mainly in England, where they were Dissenters, and occasionally in Scotland, where they were the established church), Methodists (of various kinds), Congregationalists, Baptists, Quakers, Unitarians, Plymouth Brethren, Irvingites, Millerites, Moravians, the Free Church of Scotland, Churches of Christ, and Mormons. There are many studies of individual Nonconformists, such as Hugh Bourne, Jabez Bunting, Thomas Chalmers, John Clifford, Robert William Dale, Henry Drummond, Andrew Martin Fairbairn, Peter Taylor Forsyth, Hugh Price Hughes, James Martineau, William Robertson Smith, and Charles Haddon Spurgeon. Attention is also given to the issue of church disestablishment. (The denomination of an individual is indicated where it is not obvious.) See II. Religion; XIV. Church of England, particularly 1. General and 2. Evangelicalism.

3425.   ADAMSON, WILLIAM.   The Life of the Rev. Joseph Parker. Glasgow: I. Ker and Co., 1902. (Congregationalist)

3426.   ADDISON, WILLIAM GEORGE.   Religious Equality in Modern England, 1714-1914. London: Society for Promoting Christian Knowledge; New York: The Macmillan Company, 1944.

3427.   ADDISON, WILLIAM GEORGE.   The Renewed Church of the United Brethren, 1722-1930. London: Society for the Promotion of Christian Knowledge, 1932. (Moravians)

3428.   ANDREWS, STUART.   Methodism and Society. Harlow: Longmans, 1970.

3429.   ANDREWS, STUART. "The Wesley Naturalist." History
        Today 21 (Nov. 1971), 810-817. (The Wesley Scientific
        Society and its journal, The Wesley Naturalist, in
        the last quarter of the 19th century)

3430.   ANON. Albert Spicer, 1847-1934: A Man of His
        Time. By One of His Family. London: Simpkin,
        Marshall, 1938. (Congregationalist)

3431.   ANON. "Alexander Fletcher, 1787-1860." Trans-
        actions of the Congregational Historical Society
        16, No. 2 (1949), 91-95. (Presbyterian)

3432.   ANON. "Henry Drummond." Parson 2, No. 1 (Jan.-
        Mar. 1947), 21-26. (Free Church of Scotland)

3433.   ANON. "Martineau Studies: 1. His Prayers; 2.
        His Christology." Hibbert Journal 61 (Apr. 1963),
        146-149. (Unitarian)

3434.   ANON. "William Robertson Smith." Parson 1, No. 3
        (Oct.-Dec. 1946), 21-27. (Free Church of Scotland)

3435.   ANSON, PETER FREDERICK. Bishops at Large. London:
        Faber and Faber, 1964. (Includes Dr. J. Joseph
        Overbeck, Arnold Harris Mathew, Vilatte Succession
        in England, Churches of the Mathew Succession,
        Catholic Apostolic Church)

3436.   ARMSTRONG, ANTHONY. The Church of England, The
        Methodists and Society, 1700-1850. New York:
        Rowman and Littlefield, 1973.

3437.   ARMSTRONG, GEORGE GILBERT. Richard Acland Armstrong:
        A Memoir. London: Philip Green, 1906. (Unitarian)

3438.   BACON, ERNEST WALLACE. Spurgeon: Heir of the
        Puritans. London: Allen and Unwin, 1967. (Baptist)

3439.   BAILEY, WARNER MCREYNOLDS. "Theology and Criticism
        in William Robertson Smith." Dissertation Abstracts
        International 31 (1971), 6697A (Yale University,
        1970). (Free Church of Scotland)

3440.   BAKER, ERIC. "William Cobbett and the Quakers."
        Friends' Quarterly New Series 2, No. 4 (Oct. 1948),
        249-252.

3441.   BAKER, FRANK. "The Bournes and the Primitive
        Methodist Dead Poll: Some Unpublished Documents."
        Proceedings of the Wesley Historical Society 28,
        No. 7 (1952), 138-142.
3442.   BAKER, FRANK. A Charge to Keep: An Introduc-
        tion to the People Called Methodists. London:
        Epworth Press, 1947.
3443.   BALLEINE, GEORGE REGINALD. Past Finding Out:
        The Tragic Story of Joanna Southcott and Her
        Successors. London: S.P.C.K., 1956.
        (Pentecostal)
3444.   BATEMAN, CHARLES THOMAS. John Clifford. London:
        S. W. Partridge and Co., 1902. (Baptist)
3445.   BATEMAN, CHARLES THOMAS. John Clifford, Free
        Church Leader and Preacher. London: National
        Council of the Evangelical Free Churches, 1904.
        (Baptist)
3446.   BATEMAN, CHARLES THOMAS. R. J. Campbell, M.A.:
        Pastor of the City Temple, London. London: S. W.
        Partridge Co., 1903. (Congregationalist)
3447.   BEATTIE, DAVID JOHNSTONE. Brethren: The Story
        of a Great Recovery. Kilmarnock: J. Ritchie,
        1939. (Plymouth Brethren)
3448.   BEBB, EVELYN DOUGLAS. Nonconformity and Social
        and Economic Life, 1660-1880: Some Problems of the
        Present As They Appeared in the Past. London:
        Epworth Press, 1935.
3449.   BEBBINGTON, D. W. "Gladstone and the Nonconform-
        ists: A Religious Affinity in Politics." In Derek
        Baker (Ed.), Church Society and Politics. Oxford:
        Basil Blackwell, 1975, 369-382.
3450.   BEBBINGTON, D. W. "The Nonconformist Conscience:
        A Study of the Political Attitudes and Activities
        of Evangelical Nonconformists, 1886-1902." Ph.D.
        Dissertation, University of Cambridge, 1975.

3451.   BECKERLEGGE, OLIVER AVEYARD.   The United Methodist
        Free Churches: A Study in Freedom.   London:
        Epworth Press, 1957.

3452.   BEIDELMAN, THOMAS O.   W. Robertson Smith and the
        Sociological Study of Religion.   Chicago:
        University of Chicago Press, 1974.   (Free Church
        of Scotland)

3453.   BEST, GEOFFREY FRANCIS ANDREW.   "The Whigs and the
        Church Establishment in the Age of Grey and Holland."
        History 45, No. 154 (1960), 103-118.   (Role of
        Dissent in parliamentary politics)

3454.   BIGGS, B. J.   "The Disciplined Society: Early
        Victorian Preachers in the Retford Wesleyan Circuit."
        Transactions of the Thoroton Society 75 (1971),
        98-102.

3455.   BILLINGTON, LOUIS.   "The Churches of Christ in
        Britain: A Study in XIXth-Century Sectarianism."
        Journal of Religious History 8 (1974-75), 21-48.

3456.   BILLINGTON, LOUIS.   "The Millerite Adventists in
        Great Britain, 1810-50."   Journal of American
        Studies 1 (1967), 191-212.

3457.   BINFIELD, CLYDE.   "Thomas Binney and Congrega-
        tionalism's 'Special Mission'."   Transactions of the
        Congregational Historical Society 21 (June 1971),
        1-10.

3458.   BLACK, JOHN SUTHERLAND AND GEORGE CHRYSTAL.   The
        Life of William Robertson Smith.   London: A. and
        C. Black, 1912.   (Free Church of Scotland)

3459.   BLACK, KENNETH MACLEOD.   The Scots Churches in
        England.   Edinburgh: W. Blackwood, 1906.
        (Presbyterian)

3460.   BLOCH-HOELL, NILS.   The Pentecostal Movement:
        Its Origin, Development, and Distinctive Character.
        London: Allen and Unwin; New York: Humanities
        Press, 1964.

3461.   BOGGS, W. ARTHUR.   "Reflections of Unitarianism in
        Mrs. Gaskell's Novels."   Ph.D. Dissertation, Uni-
        versity of California at Berkeley, 1951.

3462.   BOLAM, C. G. et al.   The English Presbyterians:
        From Elizabethan Puritanism to Modern Unitarianism.
        London: Allen and Unwin, 1968.

3463.   BOURNE, FREDERICK WILLIAM.   The Bible Christians:
        Their Origin and History, 1815-1900.   London:
        Bible Christian Book Room, 1905.

3464.   BOWMER, JOHN C.   "Church and Ministry in Wesleyan
        Methodism from the Death of John Wesley to the
        Death of Jabez Bunting (1791-1858)."   Ph.D
        Dissertation, University of Leeds, 1967.

3465.   BOWMER, JOHN C.   The Lord's Supper in Methodism,
        1791-1960.   London: Epworth Press, 1961.

3466.   BOWRAN, JOHN GEORGE.   The Life of Arthur Thomas
        Guttery.   London: Holborn Pub. House, 1922.
        (Primitive Methodist)

3467.   BOYD, THOMAS HUNTER.   Henry Drummond: Some
        Recollections.   London: Headley Bros., 1907.
        (Free Church of Scotland)

3468.   BRADLEY, WILLIAM LEE.   P. T. Forsyth: The Man
        and His Work.   London: Independent Press, 1952.
        (Congregationalist)

3469.   BRAILSFORD, MABEL R.   "Elizabeth Fry, Amelia Opie:
        Two Lives."   Journal of the Friends' Historical
        Society 34 (1938), 35-38.   (Quaker)

3470.   BRASH, WILLIAM BARDSLEY.   The Story of Our Col-
        leges, 1835-1935:   A Centenary Record of Ministerial
        Training in the Methodist Church.   London: Epworth
        Press, 1935.

3471.   BRAYSHAWE, ALFRED NEAVE.   The Quakers: Their
        Story and Message.   London: R. Davis, 1921.

3472.   BRIGGS, ASA.   Social Thought and Social Action:
        A Study of the Work of Seebohm Rowntree 1871-1954.
        London: Longmans, 1961.   (Quaker)

3473.   BRIGGS, JOHN HENRY YORK AND IAN SELLERS. (Eds.)
        Victorian Nonconformity. New York: St. Martin's
        Press, 1974.

3474.   BRIGHT, JOHN. "John Bright and the 'State of the
        Society' in 1851." Journal of the Friends'
        Historical Society 43 (1951), 23-28. (Quaker)

3475.   BRINTON, HOWARD HAINES. Friends for 300 Years:
        The History and Beliefs of the Society of Friends
        since George Fox Started the Quaker Movement. New
        York: Harper, 1952.

3476.   BROCKETT, ALAN. Nonconformity in Exeter, 1650-1875.
        Manchester: Published on behalf of the University
        of Exeter by Manchester University Press, 1962.

3477.   BROWN, JESSE HUNCHBERGER. "The Contribution of
        William Robertson Smith to Old Testament Scholarship,
        with Special Emphasis on Higher Criticism." Diss-
        ertation Abstracts 25 (1965), 4839-40 (Duke Uni-
        versity, 1964). (Free Church of Scotland)

3478.   BROWN, KENNETH D. "Non-conformity and the British
        Labour Movement: A Case Study." Journal of Social
        History 8, No. 2 (Winter 1974), 113-120.

3479.   BRUIJN, JACOB DE. Thomas Chalmers en zijn kerkelijk
        streven. Nijkerk: G. F. Callenbach, 1954. (Free
        Church of Scotland)

3480.   BUNN, LESLIE HENRY. Seventy Years of English
        Presbyterian Praise, 1857-1927. London: Presbyterian
        Historical Society of England, 1959. (Presbyterian
        worship)

3481.   BURLEY, ALFRED CUNNINGHAM. Charles Haddon Spurgeon.
        London: Religious Tract Society, 1928. (Baptist)

3482.   BURLEY, ALFRED CUNNINGHAM. Spurgeon and His Friends.
        London: Epworth Press, 1933. (Baptist)

3483.   BUSS, FREDERICK HAROLD AND R. G. BURNETT. A Goodly
        Fellowship: A History of the Hundred Years of the
        Methodist Local Preachers Mutual Aid Association,
        1849-1949. London: Epworth Press, 1949.

3484. BUTLER, P. "Irvingism as an Analogue of the Oxford Movement." Church History 6 (1937), 101-112.

3485. BYRT, GEORGE WILLIAM. John Clifford: A Fighting Free Churchman. London: Kingsgate Press, 1947. (Baptist)

3486. CANNON, WILLIAM RAGSDALE. "Methodism in a Philosophy of History." Methodist History 12 (July 1974), 27-43.

3487. CARLILE, JOHN CHARLES. Alexander Maclaren, D.D., the Man and His Message: A Character Sketch. London: S. W. Partridge and Co., 1901. (Baptist)

3488. CARLILE, JOHN CHARLES. C. H. Spurgeon: An Interpretive Biography. London: The Religious Tract Society and the Kingsgate Press, 1933. (Baptist)

3489. CARLILE, JOHN CHARLES. The Story of the English Baptists. London: J. Clarke and Co., 1905.

3490. CARPENTER, JOSEPH ESTLIN. James Martineau, Theologian and Teacher: A Study of His Life and Thought. London: P. Green, 1905. (Unitarian)

3491. CARRUTHERS, SAMUEL WILLIAM. Fifty Years, 1876-1926: Being a Brief Survey of the Work and Progress of the Presbyterian Church of England since the Union. London: Presbyterian Church of England, 1926.

3492. CAVE, SYDNEY. "Dr. P. T. Forsyth: The Man and His Writings." Congregational Quarterly 26, No. 2 (Apr. 1948), 107-119. (Congregationalist)

3493. CHAMBERLAYNE, JOHN H. "From Sect to Church in British Methodism." British Journal of Sociology 15 (June 1964), 139-149.

3494. CHAMPNESS, ELIZA M. The Life Story of Thomas Champness. London: C. H. Kelly, 1907. (Editor of the Methodist Recorder)

3495. CHRISTIAN, C. J. "Relation of Methodist Movement to Political Thought of England." M.A. Dissertation, University of Manchester, 1936.

3496.   CHURCH, LESLIE F.   "The Dutch Reformed Church in
        London (1550-1950)." London Quarterly and Holborn
        Review 20 (Jan. 1951), 5-10.

3497.   CLARE, ALBERT. The City Temple, 1640-1940: The
        Tercentenary Commemoration Volume. London:
        Independent Press, Ltd., 1940. (Important center
        for English Congregationalism)

3498.   CLARK, HENRY WILLIAM. History of English Non-
        conformity: From Wiclif to the Close of the Nine-
        teenth Century. 2 vols. London: Chapman and
        Hall, 1911-13.

3499.   CLELAND, I. D. "The Development of Wesleyan
        Methodist Principles and Ideas between 1791 and
        1914." M.Phil. Dissertation, University of
        Nottingham, 1970.

3500.   COAD, F. ROY. A History of the Brethren Movement:
        Its Origins, Its Worldwide Development and Its Signi-
        ficance for the Present Day. Exeter: Paternoster
        Press, 1968. (Plymouth Brethren)

3501.   COCKS, HARRY FRANCIS LOVELL. The Nonconformist
        Conscience. London: Independent Press, 1943.

3502.   COCKS, HARRY FRANCIS LOVELL. "The Social Thought
        of 19th Century English Nonconformity." Christianity
        and Society 10, No. 3 (Summer 1945), 16-24.

3503.   COLLYER, W. ISLIP. Robert Roberts: A Study of
        Life and Character. Birmingham: The Christadelphian,
        1948. (Editor of The Ambassador of the Coming Age
        which he renamed the Christadelphian)

3504.   COMPTON-RICKETT, ARTHUR. I Look Back: Memoirs
        of Fifty Years. London: Herbert, Jenkins, 1932.
        (Congregationalist; includes J. Martineau)

3505.   COMPTON-RICKETT, ARTHUR. (Ed.) Joseph Compton-
        Rickett: A Memoir. Bournemouth: E. Cooper, 1922.
        (Congregationalist)

3506.   COOPER, JAMES NICHOLL. "The Predecessors of
        Militant Dissent: English Evangelical Dissent in

Its Relations with Church and State, 1832-1841."
Ph.D. Dissertation, Harvard University, 1970.

3507. COWELL, H. J. "C. H. Spurgeon the Preacher."
Essex Review 60 (1951), 39-42. (Baptist)

3508. COWELL, H. J. Charles Haddon Spurgeon. London:
Lutterworth Press, 1950. (Baptist)

3509. COWELL, H. J. "John Clifford." Holborn Review 60
(1918), 433-451. (Baptist)

3510. COWHERD, RAYMOND GIBSON. "The Politics of English
Dissent, 1832-1848." Church History 23 (1954),
136-143.

3511. COWHERD, RAYMOND GIBSON. Politics of English
Dissent: The Religious Aspects of Liberal and Human-
itarian Reform Movements from 1815 to 1848. New
York: New York University Press, 1956.

3512. CRANFIELD, WALTER THOMAS [DENIS CRANE]. John
Clifford: God's Soldier and the People's Tribune.
London: Edwin Dalton, 1908. (Baptist)

3513. CRAUFURD, ALEXANDER HENRY GREGAN. Recollections
of James Martineau, with Some Letters from Him and an
Essay on His Religion. Edinburgh: G. A. Morton,
1903. (Unitarian)

3514. CRIPPS, ERNEST CHARLES. "William Allen: Quaker,
Humanitarian, Scientist." Hibbert Journal 42 (July
1944), 353-358.

3515. CROCKER, LIONEL. "Charles Haddon Spurgeon's Theory
of Preaching." Quarterly Journal of Speech 25
(1939), 214-224. (Baptist)

3516. CROOK, WILLIAM H. "The Contributive Factors in the
Life and Preaching of Charles Haddon Spurgeon."
Ph.D. Dissertation, Southwestern Baptist Theological
Seminary, 1957. (Baptist)

3517. CULLEN, M. J. "Making of the Civil Registration
Act of 1836." Journal of Ecclesiastical History 25
(Jan. 1974), 39-59.

3518. CUMBERS, FRANK HENRY. The Book Room: The Story
       of the Methodist Publishing House and Epworth Press.
       London: Epworth Press, 1956.

3519. CUNNINGHAM, VALENTINE D. Everywhere Spoken Against:
       Dissent in the Victorian Novel. Oxford: Clarendon
       Press, 1975.

3520. CURR, H. S. "Spurgeon and Gladstone." Baptist
       Quarterly 11 (1942), 46-54.

3521. CURRIE, ROBERT. Methodism Divided: A Study in
       the Sociology of Ecumenicalism. London: Faber &
       Faber, 1968.

3522. DAKIN, ARTHUR. The Baptist View of the Church
       and Ministry. London: Baptist Union Publ. Dept.,
       1944.

3523. DALE, ROBERT WILLIAM. History of English Congre-
       gationalism. Completed and edited by A. W. W.
       Dale. London: Hodder and Stoughton, 1907.

3523a. DARLOW, T. H. William Robertson Nicoll: Life and
       Letters. London: Hodder and Stoughton, 1925.
       (Free Church of Scotland)

3524. DARTON, L. "The Baptism of Maria Hack, 1837: An
       Episode of the Beacon Controversy." Journal of
       the Friends' Historical Society 46 (1954), 67-77.

3525. DAVIES, HORTON. The English Free Churches.
       London, New York: Oxford University Press, 1952.

3526. DAVIES, HORTON. "Liturgical Reform in 19th-century
       English Congregationalism." Transactions of the
       Congregational Historical Society 17, No. 3 (1954),
       73-82.

3527. DAVIES, RUPERT ERIC AND ERNEST GORDON RUPP. (Eds.)
       A History of the Methodist Church in Great Britain.
       London: Epworth Press, 1965.

3528. DAVIES, RUPERT ERIC. (Ed.) John Scott Lidgett:
       A Symposium. London: Epworth Press, 1957.
       (Wesleyan)

3529. DAVIES, RUPERT ERIC. Methodism. London: Epworth
       Press; Baltimore: Penguin Books, 1963.

3530. DAVIES, W. R.   "The Relation of Methodism and the
       Church of England between 1738 and 1850."   M.A.
       Dissertation, University of Manchester, 1959.

3531. DAVIS, RICHARD W.   Dissent in Politics, 1780-1830:
       The Political Life of William Smith, M.P.   London:
       Epworth Press, 1971.   (Unitarian)

3532. DAVIS, RICHARD W.   "The Strategy of 'Dissent' in
       the Repeal Campaign, 1820-1828."   Journal of Modern
       History 38 (1966), 374-393.

3533. DAWSON, ALBERT.   Joseph Parker, D.D.: His Life
       and Ministry.   London: S. W. Partridge and Co.,
       1901.   (Congregationalist)

3534. DEARING, TREVOR.   Wesleyan and Tractarian Worship:
       An Ecumenical Study.   London: Epworth Press;
       S.P.C.K., 1966.

3535. DONCASTER, L. HUGH.   Friends of Humanity: With
       Special Reference to the Quaker, William Allen,
       1770-1843.   London: Dr. Williams's Trust, 1965.

3536. DOUGHTY, W. L.   "George J. Stevenson: A Letter to
       Zechariah Taft."   Proceedings of the Wesley Histor-
       ical Society 28, Pt. 2 (1951), 33-38.   (A letter of
       1845)

3537. DRIVER, ARTHUR HARRY.   "On Certain Aspects of John
       Henry Newman and Robert William Dale."   Congrega-
       tional Quarterly 24, No. 1 (Jan. 1946), 31-40.

3538. DRUMMOND, ANDREW LANDALE.   Edward Irving and His
       Circle.   London: J. Clarke and Co., Ltd., 1937.

3539. DRUMMOND, ANDREW LANDALE.   "The Public Worship of
       the English Free Churches."   Church Service Society
       Annual for 1938-9 No. 11 (1939), 32-42.   (From the
       17th to the 20th century)

3540. DUNLAP, ELDEN DALE.   "Methodist Theology in Great
       Britain in the Nineteenth Century: With Special
       Reference to the Theology of Adam Clarke, Richard
       Watson, and William Burt Pope."   Dissertation
       Abstracts 28 (1968), 4700-01A (Yale University,
       1956).

3541. DUNNING, NORMAN GROVE. Samuel Chadwick. London:
       Hodder and Stoughton, 1933. (Wesleyan; editor of
       The Joyful News)

3542. EDWARDS, M. S. "S. E. Keeble and Nonconformist
       Social Thinking, 1880-1939." M.Litt. Dissertation,
       University of Bristol, 1969. (Methodist)

3543. EDWARDS, MALDWYN LLOYD. After Wesley: A Study
       of the Social and Political Influence of Methodism
       in the Middle Period (1791-1849). London: The
       Epworth Press (E. C. Barton), 1935.

3544. EDWARDS, MALDWYN LLOYD. Methodism and England:
       A Study of Methodism in Its Social and Political
       Aspects during the Period 1850-1932. London: The
       Epworth Press (E. C. Barton), 1943.

3545. EDWARDS, MALDWYN LLOYD. This Methodism: Eight
       Studies. London: Epworth Press, 1939.

3546. EDWARDS, MICHAEL S. Joseph Rayner Stephens, 1805-79:
       A Lecture. Hyde, Cheshire: Wesley Historical
       Society, 1968. (Wesleyan)

3547. EDWARDS, MICHAEL S. "Methodism and the Chartist
       Movement." London Quarterly and Holborn Review 35
       (1966), 301-310.

3548. EMBLEY, PETER L. "The Origins and Early Develop-
       ment of the Plymouth Brethren." Ph.D. Dissertation,
       University of Cambridge, 1967.

3549. EMDEN, PAUL HERMAN. Quakers In Commerce: A Record
       of Business Achievement. London: S. Low, Marston
       and Co., Ltd., 1940.

3550. EMMOTT, ELIZABETH BRAITHWAITE. The Story of Quakerism.
       London: Headley, 1908.

3551. ESCOTT, HARRY. Peter Taylor Forsyth, 1848-1921,
       Director of Souls: Selections from His Practical
       Writings. London: Epworth Press, 1948.
       (Congregationalist)

3552. ETTEN, HENRY VAN. Les Quakers: histoire de la
       Société Religieuse des Amis depuis sa fondation
       jusqu'à nos jours. Paris: Fischbacher, 1924.

3553.   EVANS, RICHARD LOUIS. A Century of 'Mormonism'
        in Great Britain: A Brief Summary of the Activities
        of the Church of Jesus Christ of the Latter-Day
        Saints in the United Kingdom, with Emphasis on Its
        Introduction One Hundred Years Ago. Salt Lake
        City, Utah: The Desert News Press, 1937.

3554.   EVANS, W. A. "A Statistical Study of the Develop-
        ment of Nonconformity in North Wales in the 19th
        Century, with Special Reference to the Period
        1850-1901." M.A. Dissertation, University of
        Liverpool, 1928.

3555.   EVERITT, ALAN. "Nonconformity in Country Parishes."
        In Joan Thirsk (Ed.), Land, Church and People: Essays
        Presented to Professor H. P. R. Finberg. Reading,
        Berks.: Museum of English Rural Life, 1970, 178-199.

3556.   EVERITT, ALAN. The Pattern of Rural Dissent:
        The XIXth Century. Leicester: Leicester University
        Press, 1972.

3557.   FARNDALE, W. E. "Hugh Bourne - and His Vital
        Message." London Quarterly and Holborn Review 177
        (1952), 161-166. (Primitive Methodist)

3558.   FARNDALE, W. E. "Hugh Bourne and the 'Spiritual
        Manifestation'." Proceedings of the Wesley Historical
        Society 28, No. 7 (1952), 131-137. (Primitive
        Methodist)

3559.   FIELD, C. D. "Methodism in Metropolitan London,
        1850-1920: A Social and Sociological Study."
        Ph.D. Dissertation, University of Oxford, 1975.

3560.   FINDLAY, GEORGE GILLANDERS. William F. Moulton:
        The Methodist Scholar. London: Robert Culley,
        1910.

3561.   FINDLAY, JAMES FRANKLIN, JR. "Dwight L. Moody,
        Evangelist of the Gilded Age, 1837-1899."
        Dissertation Abstracts 22 (1962), 2770-71
        (Northwestern University, 1961). (Congregationalist;
        American evangelist who made several trips to
        Britain with I. D. Sankey)

3562. FOSHEE, CHARLES NEWELL. "Andrew Martin Fairbairn: Philosopher of the Christian Religion." Dissertation Abstracts 19 (1958), 892-893 (Duke University, 1958). (Congregationalist)

3563. FOULDS, ELFRIDA VIPONT (BROWN). The Story of Quakerism, 1652-1952. By Elfrida Vipont [pseud.]. London: Bannisdale Press, 1954.

3564. FOWLER, WILLIAM STEWART. A Study in Radicalism and Dissent: The Life and Times of Henry Joseph Wilson, 1833-1914. London: Epworth Press, 1961. (Brought up as a Congregationalist but not associated with any one church)

3565. FULLERTON, WILLIAM YOUNG. Charles Haddon Spurgeon: A Biography. London: W. Williams and Norgate, 1920. (Baptist)

3566. FULLERTON, WILLIAM YOUNG. F. B. Meyer: A Biography. London: Marshall, Morgan and Scott, 1929. (Baptist)

3567. FULLERTON, WILLIAM YOUNG. Thomas Spurgeon: A Biography. London, New York: Hodder and Stoughton, 1919. (Baptist)

3568. GALLAGHER, J. R. "The Presbyterian Synod and Catholic Emancipation, 1825-9." M.A. Dissertation, National University of Ireland (Dublin), 1970.

3569. GARDINER, ALFRED GEORGE. Life of George Cadbury. London, New York: Cassell and Company, Limited, 1923. (Quaker)

3570. GASH, NORMAN. Reaction and Reconstruction in English Politics 1832-1852. Oxford: At the Clarendon Press, 1965. (Includes "Church and Dissent")

3571. GILBERT, A. D. "The Growth and Decline of Nonconformity in England and Wales, with Special Reference to the Period before 1850: An Historical Interpretation of Statistics of Religious Practice." Ph.D. Dissertation, University of Oxford, 1973.

3572. GLASER, JOHN F. "English Nonconformity and the Decline of Liberalism." American Historical Review 63 (1957-58), 352-363.

3573.   GLASER, JOHN F.   "Nonconformity and Liberalism,
        1868-1885:   A Study in English Party History."
        Ph.D. Dissertation, Harvard University, 1949.

3574.   GLASER, JOHN F.   "Parnell's Fall and the Noncon-
        formist Conscience."   Irish Historical Studies 12
        (1960), 119-138.

3575.   GLOVER, WILLIS BORDERS.   Evangelical Nonconform-
        ists and Higher Criticism in the Nineteenth Century.
        London: Independent Press, 1954.

3576.   GOSSE, SIR EDMUND WILLIAM.   Father and Son: A
        Study of Two Temperaments.   London: W. Heinemann,
        1907.   (Plymouth Brethren; the father is P. H.
        Gosse)

3577.   GOW, HENRY.   The Unitarians.   Garden City, N.Y.:
        Doubleday, Doran, 1928.

3578.   GRANT, JOHN WEBSTER.   Free Churchmanship in England,
        1870-1940, with Special Reference to Congregationalism.
        London: Independent Press, 1955.

3579.   GREGORY, BENJAMIN.   Side Lights on the Conflicts
        of Methodism during the Second Quarter of the Nine-
        teenth Century, 1827-1852.   London: Cassell, 1898.

3580.   GREGORY, JOHN ROBINSON.   A History of Methodism:
        Chiefly for the Use of Students.   2 vols.   London:
        C. H. Kelly, 1911.

3581.   GRIFFIN, A. R.   "Methodism and Trade Unionism in
        the Nottinghamshire-Derbyshire Coalfield 1844-90."
        Proceedings of the Wesley Historical Society 37
        (Feb. 1969), 1-9.

3582.   GRIFFITH, GWILYM OSWALD.   The Theology of P. T.
        Forsyth.   London: Lutterworth Press, 1948.
        (Congregationalist)

3583.   GRUBB, EDWARD.   "The Evangelical Movement and Its
        Impact on the Society of Friends."   Friends' Quarterly
        Examiner 58 (1924), 1-34.   (c. 1780-1880)

3584.   GWYTHER, C. E.   "Methodist Social and Political
        Theory and Practice, 1848 to 1914, with Particular

Reference to the Forward Movement." M.A. Dissertation, University of Liverpool, 1961.

3585.   HAGSTOTZ, GIDEON DAVID. The Seventh-Day Adventists
in the British Isles, 1878-1933. Lincoln, Nebraska:
Union College Press, 1936.

3586.   HAIG, CHARLES ANEURIN. John Angell James (1785-1859).
London: Independent Press, 1962. (Congregationalist)

3587.   HAINES, MEREDITH C. "The Nonconformists and the
Nonconformist Periodical Press in Mid-Nineteenth
Century England." Dissertation Abstracts 27 (1967),
2117-18A (Indiana University, 1966).

3588.   HALDANE, ELIZABETH S. "Edward Irving." Quarterly
Review 263 (1934), 111-126.

3589.   HALEVY, ELIE. The Birth of Methodism in England.
Trans. and Ed. Bernard Semmel. Chicago: Chicago
University Press, 1971.

3590.   HALL, ALFRED. James Martineau: The Story of
His Life. London: The Sunday School Association,
1906. (Unitarian)

3591.   HALL, B. "From Dissent to Free Churchmanship."
Journal of the Presbyterian Historical Society 10,
No. 2 (1953), 75-79.

3592.   HALL, DAVID J. "Historical Study of the Discipline
of the Society of Friends, 1738-1861." M.A.
Dissertation, University of Durham, 1972.

3593.   HALL, DAVID J. "Membership Statistics of the
Society of Friends, 1800-1850." Journal of the
Friends' Historical Society 52, No. 2 (1969),
97-100.

3594.   HALL, H. William Allen, 1770-1843: Member of
the Society of Friends. Haywards Heath: Charles
Clarke, 1953.

3595.   HAMILTON, JOHN TAYLOR. A History of the Church
Known as the Moravian Church, or the Unitas Fratrum,
or the Unity of the Brethren, during the Eighteenth
and Nineteenth Centuries. Bethlehem, Pa.: Times
Publishing Company, Printers, 1900.

3596. HANKINSON, FREDERICK. "Dissenters' Chapels Act,
       1844." Transactions of the Unitarian Historical
       Society 8, No. 2 (Oct. 1944), 52-57.

3597. HANNA, JEANNETTE CHALMERS. "Thomas Chalmers."
       Journal of the University of Edinburgh 3 (1930),
       97-103. (Free Church of Scotland)

3598. HARRIES, JOHN. G. Campbell Morgan: The Man and
       His Ministry. New York, Chicago: Fleming H.
       Revell Company, 1930. (Congregationalist)

3599. HARRIS, THOMAS ROBERTS. Dr. George Smith, 1800-68:
       Wesleyan Methodist Layman. Redruth: Cornish
       Methodist Historical Association, 1968.

3600. HARRIS, THOMAS ROBERTS. Samuel Dunn, Reformer,
       1798-1882. Redruth, Cornwall: Cornish Methodist
       Historical Association, 1963. (Wesleyan)

3601. HARRISON, ARCHIBALD HAROLD WALTER. "The Arminian
       Methodists." Proceedings of the Wesley Historical
       Society 23, Pt. 2 (June 1941), 25-26. (Derby Faith
       Movement, 1832)

3602. HARRISON, ARCHIBALD HAROLD WALTER, B. AQUILA BARBER,
       GEORGE C. HORNBY AND E. TEGLA DAVIES. The Methodist
       Church: Its Origin, Divisions, and Re-union.
       London: Methodist Publishing House, 1932.

3603. HARRISON, F. M. W. "The Nottinghamshire Baptists:
       Church Relations, Social Composition, Finance,
       Theology." Baptist Quarterly 26 (Oct. 1975),
       169-190.

3604. HARRISON, G. ELSIE. Methodist Good Companions.
       London: Epworth Press, 1935. (Includes the
       influence of Methodism on the Brontës)

3605. HARVEY, F. B. "Methodism and the Romantic Move-
       ment." London Quarterly Review 159 (July 1934),
       289-302.

3606. HAYDEN, ERIC W. Spurgeon on Revival. Grand Rapids,
       Michigan: Zondervan, 1962. (Baptist)

3607.   HAYDEN, ROGER.   "John Taylor (1831-1901) and the
        Records of Northants Nonconformity."   Baptist
        Quarterly 24 (July 1972), 342-344.   (Baptist)

3608.   HEALEY, FRANCIS G.   Rooted in Faith: Three
        Centuries of Nonconformity, 1662-1962.   London:
        Independent Press, 1961.

3609.   HELMSTADTER, RICHARD J.   "Voluntaryism, 1828-1860."
        Dissertation Abstracts 22 (1962), 3994-95 (Columbia
        University, 1961).

3610.   HERFORD, CHARLES HAROLD.   (Ed.)   Joseph Estlin
        Carpenter: A Memorial Volume.   Oxford: The
        Clarendon Press, 1929.   (Unitarian)

3611.   HERFORD, CHARLES HAROLD.   Philip Henry Wicksteed:
        His Life and Work.   London & Toronto: J. M. Dent &
        Sons Ltd., 1931.   (Unitarian)

3612.   HIMBURY, D. MERVYN.   British Baptists: A Short
        History.   London: Carey Kingsgate Press, 1962.

3613.   HIRST, MARGARET E.   The Quakers in Peace and War:
        An Account of Their Peace Principles and Practice.
        London: Swarthmore Press Ltd.; New York: George
        H. Doran Company, 1923.

3614.   HOBSBAWM, ERIC J.   "Methodism and the Threat of
        Revolution in Britain."   History Today 7 (Feb.
        1957), 115-124.

3615.   HOBSBAWM, ERIC J.   Primitive Rebels: Studies in
        Archaic Forms of Social Movement in the 19th and
        20th Centuries.   Manchester, England: Manchester
        University Press, 1959.   (Includes the "labour
        sects" and Methodism)

3616.   HOLDER, CHARLES FREDERICK.   The Quakers in Great
        Britain and America: The Religious and Political
        History of the Society of Friends from the
        Seventeenth to the Twentieth Century.   New York,
        Los Angeles: The Neuner Company, 1913.

3617.   HOLIFIELD, E. BROOKS.   "English Methodist Response
        to Darwin."   Methodist History 10 (Jan. 1972),
        14-22.

3618.   HOLLOWAY, JOHN. (Ed.)   The Journals of Two Poor
        Dissenters, 1786-1880. London: Routledge and K.
        Paul, 1970. (Baptist; William Thomas Swan and
        William Swan)

3619.   HOLT, FELIX. "The Hincks Family." Transactions
        of the Unitarian Historical Society 8, Pt. 2 (Oct.
        1944), 84-85. (Unitarian)

3620.   HOLT, RAYMOND VINCENT. The Unitarian Contribution
        to Social Progress in England. London: G. Allen
        and Unwin, Ltd., 1938.

3621.   HOPKINS, JAMES KIRKLAND. "Joanna Southcott: A
        Study of Popular Religion and Radical Politics,
        1789-1814." Dissertation Abstracts International
        34 (1974), 5870-71A (University of Texas at
        Austin, 1972). (Pentecostal)

3622.   HORNE, CHARLES SILVESTER. Nonconformity in the
        XIXth Century. London: National Council of
        Evangelical Free Churches, 1905.

3623.   HORNE, CHARLES SILVESTER. A Popular History of
        the Free Churches. London: Clarke, 1903.

3624.   HORWILL, HERBERT WILLIAM. "Hugh Price Hughes."
        Methodist Review 85 (1903), 345-359. (Methodist)

3625.   HOUGH, L. H. "Robert William Dale." Congrega-
        tional Quarterly 7, No. 4 (Oct. 1929), 417-424.
        (Congregationalist)

3626.   HUBBARD, GEOFFREY. Quaker by Convincement. London:
        Penguin, 1974. (Quakers in the nineteenth and
        twentieth centuries)

3627.   HUGH, R. L. "The Theological Background of Noncon-
        formist Social Influence in Wales, 1800-1850."
        Ph.D. Dissertation, University of London, 1951.

3628.   HUGHES, DOROTHEA PRICE. The Life of Hugh Price
        Hughes. London: Hodder and Stoughton, 1904.
        (Methodist)

3629.   HUNTER, F. "The Influence of the Church of England
        and Dissent upon Methodism in the 19th Century."
        M.A. Dissertation, University of Manchester, 1939.

3630.   HURST, JOHN FLETCHER.   The History of Methodism.   7
        vols. New York: Eaton and Mains, 1902-04.

3631.   HUTSON, HARRY M.   "Methodist Concern with Social
        Problems in England, 1848-1873." Ph.D. Disser-
        tation, University of Iowa, 1952.

3632.   HUTTON, JOSEPH EDMUND.   A History of the Moravian
        Church. (2d ed., rev. and enl.) London: Moravian
        Publication Office, 1909. (First edition, A Short
        History of the Moravian Church, 1895)

3633.   INGHAM, S. M.   "The Disestablishment Movement in
        England, 1868-74." Journal of Religious History 3
        (1964-65), 38-60.

3634.   INGLIS, KENNETH STANLEY.   "English Nonconformity
        and Social Reform, 1880-1900." Past and Present 13
        (1958), 73-88.

3635.   IRONSIDE, HENRY ALLAN.   A Historical Sketch of
        the Brethren Movement: An Account of Its Inception,
        Progress and Failures, and Its Lessons for Present
        Day Believers. Grand Rapids, Michigan: Zondervan
        Publishing House, 1942. (Plymouth Brethren)

3636.   ISICHEI, ELIZABETH.   "From Sect to Denomination in
        English Quakerism." British Journal of Sociology
        15 (Sept. 1964), 207-222.

3637.   ISICHEI, ELIZABETH.   "Organization and Power in the
        Society of Friends (1852-1859)." Archives de
        Sociologie des Religions 19 (Jan.-June 1965),
        31-49.

3638.   ISICHEI, ELIZABETH.   Victorian Quakers. London:
        Oxford University Press, 1970.

3639.   JACKS, LAWRENCE PEARSALL. (Ed.)   From Authority
        to Freedom: The Spiritual Pilgrimage of Charles
        Hargrove. London: Williams and Norgate, 1920.
        (Unitarian)

3640.   JACKSON, ABRAHAM WILLARD.   James Martineau: A
        Biography and Study. Boston: Little, Brown, and
        Company, 1900. (Unitarian)

3641.   JACKSON, GEORGE.   "R. W. Dale."   London Quarterly
        and Holborn Review 167 (Apr. 1942), 133-142.
        (Congregationalist)

3642.   JENKINS, JAMES HEALD.   Ebenezer E. Jenkins: A
        Memoir.   London: Charles H. Kelly, 1906.   (Wesleyan)

3643.   JEWSON, C. B.   "Joseph Kinghorn and His Friends."
        Baptist Quarterly 8, No. 8 (Oct. 1937), 440-443.
        (Baptist)

3644.   JOHNSON, DALE A.   "Anticipations of the Future:
        Some Early Letters of R. W. Dale."   Journal of the
        United Reformed Church Historical Society 1 (Nov.
        1973), 56-61.   (Congregationalist)

3645.   JOHNSON, DALE A.   "Between Evangelicalism and a
        Social Gospel: The Case of Joseph Rayner Stephens."
        Church History 42 (1973), 229-242.   (Wesleyan)

3646.   JOHNSON, WILLIAM CHARLES.   Encounter in London:
        The Story of the London Baptist Association,
        1865-1965.   London: Carey Kingsgate Press, 1965.

3647.   JONES, E. K.   "Some Old Association Reports."
        Baptist Quarterly 13, No. 8 (1950), 355-359.   (New
        Connexion of General Baptists, 1838-49)

3648.   JONES, HUGH.   Hanes Wesleyaeth Gymreig.   3 vols.
        Bangor: Llyfrfa Wesleyaidd, 1911-12.   (Welsh
        Wesleyanism)

3649.   JONES, IEUAN GWYNEDD.   "Dr. Thomas Price and the
        Election of 1868 in Merthyr Tydfil: A Study in
        Nonconformist Politics."   Welsh History Review 2
        (1964-65), 147-172, 251-270.   (Baptist)

3650.   JONES, ROBERT TUDUR.   Congregationalism in England,
        1662-1962.   London: Independent Press, 1962.

3651.   JONES, ROBERT TUDUR.   "The Origins of the Non-
        conformist Disestablishment Campaign, 1830-1840."
        Journal of the Historical Society of the Church in
        Wales 20 (1970), 39-76.

3652.   JONES, RUFUS MATTHEW.   The Later Periods of
        Quakerism.   2 vols.   London: Macmillan and Co.,
        Limited, 1921.

3653.   JONES, WILLIAM HAROLD. History of the Wesleyan
        Reform Union. London: Epworth Press, 1952.
3654.   JORNS, AUGUSTE. The Quakers as Pioneers in Social
        Work. Trans. Thomas Kite Brown. New York: The
        Macmillan Company, 1931.
3655.   KEEP, H. F. "Dale of Birmingham." Transactions
        of the Congregational Historical Society 10, No. 6
        (Sept. 1929), 243-249. (Congregationalist)
3656.   KENDALL, HOLLIDAY BICKERSTAFFE. The History of the
        Primitive Methodist Church. Revised and enlarged
        edition. London: Joseph Johnson, 1919.
3657.   KENDALL, HOLLIDAY BICKERSTAFFE. History of the
        Primitive Methodist Connexion. 2nd edition, revised
        and enlarged. London: R. Bryant, 1902.
3658.   KENDALL, HOLLIDAY BICKERSTAFFE. The Origin and
        History of the Primitive Methodist Church. 2 vols.
        London: E. Dalton, n.d.
3659.   KENNEDY, J. R. "The Life, Work and Thought of John
        Angell James (1785-1859)." Ph.D. Dissertation,
        University of Edinburgh, 1957. (Congregationalist)
3660.   KENT, JOHN HENRY SOMERSET. The Age of Disunity.
        London: Epworth Press, 1966. (Methodism)
3661.   KENT, JOHN HENRY SOMERSET. "The Clash between
        Radicalism and Conservatism in Methodism, 1815-1848."
        Ph.D. Dissertation, University of Cambridge, 1951.
3662.   KENT, JOHN HENRY SOMERSET. Jabez Bunting, the
        Last Wesleyan: A Study in the Methodist Ministry
        After the Death of John Wesley. London: Epworth
        Press, 1955.
3663.   KENT, JOHN HENRY SOMERSET. "Methodism and Revolution."
        Methodist History 12 (July 1974), 136-144.
3664.   KENWORTHY, F. "The Unitarian Tradition in Liberal
        Christianity." Transactions of the Unitarian
        Historical Society 8, No. 2 (1944), 58-67. (Includes
        J. Martineau)

3665.   KIRBY, THOMAS A.   "Carlyle and Irving."   ELH: A
        Magazine of English Literary History 13, No. 1
        (Mar. 1946), 59-63.

3666.   KITSON CLARK, GEORGE SIDNEY ROBERTS.   The English
        Inheritance: An Historical Essay.   London: SCM
        Press, 1950.   (Places Dissent in its social and
        political setting)

3667.   KNOX, ROBERT BUICK.   Voices from the Past: History
        of the English Conference of the Presbyterian Church
        of Wales, 1889-1938.   Llandyssul: J. D. Lewis,
        1969.

3668.   KOLDE, T.   "Edward Irving."   Neue Kirchliche
        Zeitschrift 11 (1900), 468-506, 518-560.

3669.   KOSS, STEPHEN.   Nonconformity in Modern British
        Politics.   Hamden, Conn.:   Shoe String Press/
        Archon; London: Batsford, 1975.

3670.   KOSS, STEPHEN.   "Wesleyanism and Empire."
        Historical Journal 18 (1975), 105-118.

3671.   KRUPPA, PATRICIA STALLINGS.   "Charles Haddon
        Spurgeon: A Preacher's Progress."   Dissertation
        Abstracts International 32 (1971), 360-361A
        (Columbia University, 1968).   (Baptist)

3672.   LANDELS, THOMAS DURLEY.   William Landels, D.D.:
        A Memoir.   London, New York: Cassell and Company,
        Limited, 1900.   (Baptist)

3673.   LANGLEY, ARTHUR SWAINSON.   Birmingham Baptists:
        Past and Present.   London: The Kingsgate Press,
        1939.

3674.   LAWRENCE, G. W.   "William Robertson Nicoll (1851-
        1923) and Religious Journalism in the Nineteenth
        Century."   Ph.D. Dissertation, University of
        Edinburgh, 1954.   (Free Church of Scotland;
        founder of the Bookman)

3675.   LAWTON, GEORGE.   "Methodist Statistics, 1838."
        Proceedings of the Wesley Historical Society 26,
        Pt. 1 (Mar. 1947), 10-13.

3676.   LEA, JOHN.   "The Davidson Controversy, 1856-1857."
        Durham University Journal 68 (Dec. 1975), 15-32.
        (Samuel Davidson, Congregationalist)

3677.   LEA, JOHN.   "The Growth of the Baptist Denomination
        in Mid-Victorian Lancashire and Cheshire."
        Transactions of the Historical Society of Lancashire
        and Cheshire 124 (1972), 128-153.

3678.   LEA, JOHN.   "Historical Source Materials on Congre-
        gationalism in Nineteenth Century Lancashire."
        Journal of the United Reformed Church Historical
        Society 1 (Dec. 1974), 106-112.

3679.   LEA, JOHN.   "The Journal of Rev. D. T. Carnson,
        Secretary to the Executive Committee of the
        Lancashire Congregational Union, 1847-1854."
        Transactions of the Historical Society of Lancashire
        and Cheshire 125 (1974), 119-148.

3680.   LEE, S. G.   "Robert Hibbert and His Religious
        Background."   Hibbert Journal 51, No. 4 (1953),
        319-328. (Unitarian)

3681.   LENNOX, CUTHBERT.   Henry Drummond: A Biographical
        Sketch.   London: Andrew Melrose, 1901.   (Free
        Church of Scotland)

3682.   LEWIS, HOWELL ELVET.   Nonconformity in Wales.
        London: National Council of Evangelical Free
        Churches, 1904.

3683.   LIDGETT, JOHN SCOTT.   "The Theological Institution:
        Some Noted Tutors of Yesterday."   London Quarterly
        and Holborn Review 161 (Jan. 1936), 1-13.
        (Methodist ministerial colleges and the remini-
        scences of Methodist ministers, including John
        Scott, George Osborn, William Moulton, William
        Pope, Marshall Randles)

3684.   LINDEBOOM, J.   Austin Friars: History of the
        Dutch Reformed Church in London, 1550-1950.   Trans.
        from the Dutch by D. de Iongh.   The Hague: M.
        Nijhoff, 1950.

3685.   LLOYD, E. C.   "The Influence of the Methodist
        Movement on Social Life in Wales."   B.Litt.
        Dissertation, University of Oxford, 1921.

3686.   LLOYD, HUMPHREY.   The Quaker Lloyds in the Indus-
        trial Revolution.   London: Hutchinson, 1975.

3687.   LLOYD-JONES, DAVID MARTYN.   1662-1962:   From
        Puritanism to Non-conformity.   London: The
        Evangelical Library, 1962.

3688.   LORD, FRED TOWNLEY.   Baptist World Fellowship:
        A Short History of the Baptist World Alliance.
        London: Carey Kingsgate Press, 1955.

3688a. LOWERSON, J. R.   "The Political Career of Sir
        Edward Baines, 1800-90."   M.A. Dissertation,
        University of Leeds, 1965.   (Congregationalist)

3689.   LUCAS, GEORGE A.   "Maclaren of Manchester."
        Holborn Review 58 (1916), 262-269.   (Alexander
        Maclaren, Baptist)

3690.   LUCCOCK, HALFORD EDWARD AND PAUL HUTCHINSON.
        The Story of Methodism.   New York, Cincinnati:
        Methodist Book Concern, 1926.

3691.   LUKE, WILLIAM BALKWILL.   Memorials of Frederick
        William Bourne.   London: W. H. Gregory, 1906.
        (Bible Christian Methodist)

3692.   MACARTNEY, CLARENCE EDWARD.   "Edward Irving."
        Princeton Theological Review 20 (1922), 232-262.

3693.   MACARTNEY, CLARENCE EDWARD.   "Thomas Chalmers."
        Princeton Theological Review 17 (1919), 365-400.
        (Free Church of Scotland)

3694.   MACHIN, G. I. T.   "Gladstone and Nonconformity in
        the 1860's: The Formation of an Alliance."
        Historical Journal 17 (1974), 347-364.

3695.   MCKENZIE, ISABEL.   Social Activities of the English
        Friends in the First Half of the Nineteenth Century.
        New York: Privately printed for the author, 1935.

3696.   MACKINTOSH, WILLIAM H.   Disestablishment and
        Liberation.   London: Epworth, 1972.   (History of
        the Liberation Society)

3697.   MCLACHLAN, HERBERT.   Alexander Gordon (9 June
        1841-21 February 1931):   A Biography with a
        Bibliography.   Manchester: Manchester University
        Press, 1932.   (Unitarian)

3698.   MCLACHLAN, HERBERT.   "Cross Street Chapel in the
        Life of Manchester."   Memoirs and Proceedings of
        the Manchester Literary and Philosophical Society
        84, No. 3 (1939-41), 29-41.   (Unitarian)

3699.   MCLACHLAN, HERBERT.   Essays and Addresses.
        Manchester: Manchester University Press, 1950.
        (Includes Unitarianism and liberal Nonconformity
        from the seventeenth century onwards)

3700.   MCLACHLAN, HERBERT.   The Methodist Unitarian
        Movement.   Manchester: Manchester University
        Press; London, New York: Longmans, Green and Co.,
        1919.

3701.   MCLACHLAN, HERBERT.   Records of a Family, 1800-1933:
        Pioneers in Education, Social Service, and Liberal
        Religion.   Manchester: Manchester University
        Press, 1935.   (The Beard family, Unitarians)

3702.   MCLACHLAN, HERBERT.   The Story of a Nonconformist
        Library.   Manchester: Manchester University Press;
        London, New York: Longmans, Green and Co., 1923.
        (Unitarian literature and the library of Unitarian
        College, Manchester)

3703.   MCLACHLAN, HERBERT.   The Unitarian Movement in
        the Religious Life of England.   London: G. Allen
        and Unwin, Ltd., 1934.

3704.   MACLAREN, A. ALLAN.   "Presbyterianism and the
        Working Class in a Mid-Nineteenth Century City."
        Scottish Historical Review 46 (1967), 115-139.
        (Aberdeen)

3705.   MCLAREN, ELIZABETH T.   Dr. McLaren of Manchester:
        A Sketch.   London: Hodder and Stoughton, 1911.
        (Baptist)

3706.   MANNING, BERNARD LORD. The Protestant Dissenting
        Deputies. Edited by Ormerod Greenwood. Cambridge:
        Cambridge University Press, 1952. (Two members
        chosen from each congregation of Presbyterians,
        Independents, and Baptists, located within twelve
        miles of London, to protect civil rights of Dis-
        senters)

3707.   MANTLE, JOHN GREGORY. Hugh Price Hughes. London:
        S. W. Partridge and Co., 1901. (Methodist)

3708.   MARCHANT, SIR JAMES. Dr. John Clifford, C.H.
        London: Cassell and Company, Ltd., 1924.
        (Baptist)

3709.   MARCHANT, SIR JAMES. J. B. Paton, M.A., D.D.,
        Educational and Social Pioneer. London: J. Clarke,
        1909. (Congregationalist)

3710.   MARQUIS, JOHN FRANCIS. English Presbyterian Or-
        dination Vows, 1882-1935. London: Presbyterian
        Historical Society of England, 1955.

3711.   MARSH, H. G. "The Cultivation of the Spiritual
        Life in Early Primitive Methodism." London Quarterly
        and Holborn Review 177 (1952), 180-184.

3712.   MARSH, J. T. "Old Quaker Dalton." Memoirs and
        Proceedings of the Manchester Literary and
        Philosophical Society 111 (1968-69), 27-47.

3713.   MARTIN, H. R. "The Politics of the Congregational-
        ists, 1830-56." Ph.D. Dissertation, University of
        Durham, 1972.

3714.   MARTINEAU, MISS E. I. J. "Quakerism and Public
        Service, Chiefly between 1832 and 1867." B.Litt.
        Dissertation, University of Oxford, 1938.

3715.   MARWICK, WILLIAM H. "Quakers in Victorian Scot-
        land." Journal of the Friends' Historical Society
        52, No. 2 (1969), 67-77.

3716.   MASON, B. J. "The Part Played in British Social
        and Political Life by the Protestant Nonconformists
        between the Years 1832-1859, with Special Reference

to the Disestablishment of the Church of England."
M.A. Dissertation, University of Southampton, 1958.

3717. MATTHEWS, RONALD. English Messiahs: Six English
Religious Pretenders, 1656-1927. London: Methuen,
1936. (James Nayler, Joanna Southcott, Richard
Brothers, John Nichols Tom, Henry James Prince,
John Hugh Smyth-Pigott)

3718. MAXWELL, WILLIAM DELBERT. The Book of Common
Prayer and the Worship of Non-Anglican Churches.
London: Oxford University Press, 1950.

3719. MAYNARD, W. B. "The Constitutional Authority of
Jabez Bunting over Wesleyan Methodism as Seen
through His Correspondence." M.A. Dissertation,
University of Durham, 1970.

3720. MAYOR, STEPHEN H. "R. W. Dale and Nineteenth
Century Thought." Transactions of the Congregational
Historical Society 20 (1965), 4-18. (Congregational-
ist)

3721. MELLONE, SYDNEY HERBERT. Liberty and Religion:
The First Century of the British and Foreign Unitarian
Association. London: Lindsey Press, 1925.

3722. MEREDITH, ALBERT ROGER. "The Social and Political
Views of Charles Haddon Spurgeon, 1834-1892."
Dissertation Abstracts International 34 (1974),
5879A (Michigan State University, 1973). (Baptist)

3723. MICHELL, WILLIAM JOHN. Brief Biographical Sketches
of Bible Christian Ministers and Laymen, with
Portraits. 2 vols. Jersey: Beresford Press,
1906.

3724. MICKLEWRIGHT, FREDERICK HENRY AMPHLETT. "Dissenters'
Chapels Act." Notes and Queries 187 (1944), 33-34,
107. (July 15, 1844)

3725. MICKLEWRIGHT, FREDERICK HENRY AMPHLETT. "A Nineteenth-Century Religious Reformer." London Quarterly
and Holborn Review 169 (July 1944), 230-235.
(Robert Suffield, Unitarian convert from Roman
Catholicism)

3726.   MIKOLASKI, S. J.   "R. W. Dale on the Atonement."
        Evangelical Quarterly 35 (Jan.-Mar. 1963), 23-29.
        (Congregationalist)

3727.   MIKOLASKI, S. J.   "The Theology of R. W. Dale."
        Evangelical Quarterly 34 (July-Sept. 1962), 131-143;
        (Oct.-Dec. 1962), 196-205. (Congregationalist)

3728.   MILLS, JOSEPH TRAVIS. John Bright and the Quakers.
        2 vols. London: Methuen and Co., Ltd., 1935.

3729.   MINEKA, FRANCIS EDWARD. The Dissidence of Dissent:
        The Monthly Repository 1806-1838. Chapel Hill:
        The University of North Carolina Press, 1944.
        (Includes religious periodicals 1700-1825)

3730.   MINET, SUSAN. "Hommage à l'église de Londres,
        1550-1950: A Résumé of the Story of London's First
        Huguenot Church." Huguenot Society of London
        Proceedings 18, No. 3 (1949), 232-242.

3731.   MONTGOMERY, R. MORTIMER. "The Significance of the
        Dissenters' Chapels Act of 1844." Transactions
        of the Unitarian Historical Society 8, No. 2 (Oct.
        1944), 45-51.

3732.   MOORE, ROBERT SAMUEL. Pit-Men, Preachers and
        Politics: The Effects of Methodism in a Durham
        Mining Community. New York and London: Cambridge
        University Press, 1974.

3733.   MORGAN, GEORGE ERNEST. "A Veteran in Revival":
        R. C. Morgan: His Life and Times. London: Morgan
        and Scott, 1909. (Evangelical publisher not
        affiliated with any particular denomination)

3734.   MORGAN, JILL. A Man of the Word: Life of G.
        Campbell Morgan. London: Pickering and Inglis,
        1951. (Congregationalist)

3735.   MORGAN, JILL. (Ed.) This Was His Faith: The
        Expository Letters of G. Campbell Morgan. London:
        Pickering and Inglis, 1954. (Congregationalist)

3736.   MORRIS, R. J. "The Unitarian View of Fifteen
        Periodicals in 1834." Victorian Periodicals
        Newsletter No. 18 (Dec. 1972), 31-32.

3737.   MOSS, RICHARD WADDY.   The Rev. W. B. Pope, D.D.,
        Theologian and Saint.   London: R. Culley, 1909.
        (Wesleyan)

3738.   MULLETT, CHARLES FREDERIC.   "Protestant Dissent as
        Crime (1660-1828)."   Review of Religion 13 (May
        1949), 339-353.

3739.   MUMFORD, NORMAN W.   "The Administration of the
        Sacrament of Baptism in the Methodist Church."
        London Quarterly and Holborn Review 16 (1947),
        113-119.

3740.   MUMFORD, NORMAN W.   "The Administration of the
        Sacrament of the Lord's Supper in the Methodist
        Church after the Death of John Wesley."   London
        Quarterly and Holborn Review 20 (Jan. 1951), 61-70.

3741.   MURRAY, HAROLD.   Campbell Morgan, Bible Teacher:
        A Sketch of the Great Expositor and Evangelist.
        London, Edinburgh: Marshall, Morgan and Scott,
        Ltd., 1938.   (Congregationalist)

3742.   MURRAY, HAROLD.   Gipsy Smith: An Intimate Memoir.
        Exeter: H. Murray, 1947.   (Rodney Smith,
        Methodist)

3743.   MURRAY, HAROLD.   Sixty Years an Evangelist: An
        Intimate Study of Gipsy Smith.   London and
        Edinburgh: Marshall, Morgan and Scott Ltd., 1937.
        (Rodney Smith, Methodist)

3744.   MYERS, SYDNEY.   R. W. Dale (1829-1895).   London:
        Independent Press, 1962.   (Congregationalist)

3745.   NEATBY, WILLIAM BLAIR.   A History of the Plymouth
        Brethren.   London: Hodder and Stoughton, 1901.

3746.   NEWSON, MRS.   John Thain Davidson: Reminiscences.
        By His Daughter.   London: Hodder and Stoughton,
        1906.   (Presbyterian)

3747.   NEWTON, DAVID.   Sir Halley Stewart.   London: Allen
        and Unwin, 1968.   (Congregationalist)

3748.   NEWTON, J. S.   "The Political Career of Edward
        Miall, Editor of the 'Nonconformist', and Founder

of the Liberation Society." Ph.D. Dissertation,
University of Durham, 1975. (Congregationalist)

3749.   NICOLL, SIR WILLIAM ROBERTSON. "Ian Maclaren":
Life of the Rev. John Watson, D.D. London: Hodder
and Stoughton, 1908. (Presbyterian)

3750.   NOEL, NAPOLEON. The History of the Brethren. Ed.
William F. Knapp. 2 vols. Denver, Colo.: W. F.
Knapp, 1936. (Plymouth Brethren)

3751.   NUELSEN, JOHN LOUIS, THEOPHIL MANN AND J. J. SOMMER.
Kurzgefasste Geschichte des Methodismus von seinen
Anfängen bis zur Gegenwart. Bremen: Trakthaus,
1920.

3752.   NUTTALL, GEOFFREY FILLINGHAM. "Early Quakerism and
Early Primitive Methodism." Friends' Quarterly 7
(1953), 179-187.

3753.   NUTTALL, GEOFFREY FILLINGHAM AND OWEN CHADWICK.
(Eds.) From Uniformity to Unity, 1662-1962.
London: S.P.C.K., 1962. (Studies of Dissent)

3754.   O'NEILL, THOMAS P. "The Society of Friends and the
Great Famine (1845-48)." Studies: An Irish
Quarterly Review 39 (1951), 203-213.

3755.   OWENS, ERNEST SIBLEY, JR. "The Role of John
Clifford in the Life and Thought of English Baptists."
Ph.D. Dissertation, New Orleans Baptist Theological
Seminary, 1960.

3756.   PARRY, R. I. "The Attitude of Welsh Independents
towards Working-class Movements, Including Public
Education, from 1815 to 1870." M.A. Dissertation,
University College of North Wales (Bangor), 1931.

3757.   PATON, JOHN LEWIS ALEXANDER. John Brown Paton:
A Biography. London, New York: Hodder and
Stoughton, 1914. (Congregationalist)

3758.   PATTERSON, W. M. Northern Primitive Methodism.
London: E. Dalton, 1909.

3759.   PAUL, SYDNEY FRANK. Further History of the Gospel
Standard Baptists. 4 vols. Brighton: S. F. Paul,
1951-1962.

3760.   PAUL, SYDNEY FRANK.   The Seceders, 1829-69.   Volume
        Three: Continuing and Concluding Life and Letters
        of Joseph Charles Philpot M.A.; with a Further
        Relation of the Progress of "The Gospel Standard"
        and of the Strict Baptist Churches Connected there-
        with during the Period 1849-1869.   Brighton:
        S. F. Paul, 1960.   (See Philpot, Joseph Henry,
        3788)

3761.   PAUL, SYDNEY FRANK.   Story of the Gospel in England.
        4 vols.   Ilfracombe: Arthur H. Stockwell, 1948-50.

3762.   PAYNE, ERNEST ALEXANDER.   "The Baptist Connections
        of George Dyer: A Postscript to E. V. Lucas's
        'Life of Charles Lamb'."   Baptist Quarterly 10, No.
        5 (Jan. 1941), 260-267; 11, Nos. 8-9 (1944), 237-238.

3763.   PAYNE, ERNEST ALEXANDER.   The Baptist Union: A Short
        History.   Greenwood, S.C.:   Attic Press, 1959.

3764.   PAYNE, ERNEST ALEXANDER.   The Fellowship of Believers:
        Baptist Thought and Practice Yesterday and Today.
        London: Kingsgate Press, 1944.

3765.   PAYNE, ERNEST ALEXANDER.   The Free Church Tradi-
        tion in the Life of England.   London: Student
        Christian Movement Press, 1944.

3766.   PAYNE, ERNEST ALEXANDER.   "Great Preachers, 9:
        Spurgeon."   Theology 54 (1951), 376-380.   (Baptist)

3767.   PAYNE, ERNEST ALEXANDER.   "The Period of Establish-
        ment, 1828-1910."   London Quarterly and Holborn
        Review 187 (July 1962), 193-197.   (Growing strength
        of Nonconformity during this period)

3768.   PEACOCK, ARTHUR.   "Social Factors in British
        Unitarian History."   Faith and Freedom 22 (1969),
        64-74.

3769.   PEAKE, ARTHUR SAMUEL.   The Life of Sir William
        Hartley.   London: Hodder and Stoughton, 1926.
        (Primitive Methodist)

3770.   PEAKE, LESLIE SILLMAN.   Arthur Samuel Peake:
        A Memoir.   London: Hodder and Stoughton, Limited,
        1930.   (Methodist)

3771.   PEARCE, EDWARD GEORGE.   The Story of the Lutheran
        Church in Britain through Four Centuries of History.
        London: Evangelical Lutheran Church of England,
        1969.

3772.   PEARCE, GORDON JAMES MARTIN.   Charles Haddon
        Spurgeon (1834-1892).   London: Independent Press,
        1962. (Baptist)

3773.   PEARSALL, RONALD.   "The Case for Spurgeon."
        Quarterly Review 304 (July 1966), 322-328.
        (Baptist)

3774.   PEARSON, P. C.   "Wesleyan Methodism from 1850 to
        1900 in Relation to the Life and Thought of the
        Victorian Age."   M.A. Dissertation, Victoria Uni-
        versity of Manchester, 1965.

3775.   PEASTON, ALEXANDER ELLIOTT.   "Dr. Martineau and the
        'Ten Services'."   Transactions of the Unitarian
        Historical Society 7, No. 4 (Oct. 1942), 290-293.
        (Unitarian; services of Baptism, Confirmation,
        Matrimony and Holy Communion, among others, in use
        in the Church of England)

3776.   PEASTON, ALEXANDER ELLIOTT.   "Nineteenth Century
        Liturgies."   Transactions of the Unitarian Historical
        Society 7, No. 3 (Oct. 1941), 215-225.   (Unitarian)

3777.   PEEL, ALBERT.   A Brief History of English Con-
        gregationalism.   London: Independent Press, 1931.

3778.   PEEL, ALBERT.   The Congregational Two Hundred,
        1530-1948.   London: Independent Press, 1948.

3779.   PEEL, ALBERT, A. G. MATTHEWS, BERNARD L. MANNING
        AND ALEXANDER JAMES GRIEVE.   Congregationalism
        through the Centuries.   Cambridge: Independent
        Press, 1937.

3780.   PEEL, ALBERT. (Ed.)   Essays Congregational and
        Catholic.   London: Congregational Union of England
        and Wales, 1931.

3781.   PEEL, ALBERT.   A Hundred Eminent Congregational-
        ists, 1530-1924.   London: Independent Press, 1927.

3782.   PEEL, ALBERT AND DOUGLAS HORTON. International
        Congregationalism. London: Independent Press,
        1949.
3783.   PEEL, ALBERT. (Ed.) Letters to a Victorian Editor,
        Henry Allon, Editor of the British Quarterly Review.
        London: Independent Press, 1929. (Congregationalist)
3784.   PEEL, ALBERT AND J. A. R. MARRIOTT. Robert Forman
        Horton. London: G. Allen and Unwin, Ltd., 1937.
        (Congregationalist)
3785.   PEEL, ALBERT. These Hundred Years: A History
        of the Congregational Union of England and Wales,
        1831-1931. London: Congregational Union of
        England and Wales, 1931.
3786.   PERKIN, J. R. C. "Baptism in Nonconformist Theology,
        1820-1920, with Special Reference to the Baptists."
        Ph.D. Dissertation, University of Oxford, 1955.
3787.   PHILIP, ADAM. Thomas Chalmers: Apostle of Union.
        London: Clarke, 1929. (Free Church of Scotland)
3788.   PHILPOT, JOSEPH HENRY. (Ed.) The Seceders, 1829-
        1869: The Story of a Spiritual Awakening as Told
        in the Letters of Joseph Charles Philpot and of
        William Tiptaft. 2 vols. London: C. J. Farncombe
        and Sons, Ltd., 1930-32. (Baptist; see Paul,
        Sydney Frank, 3760)
3789.   PIKE, GODFREY HOLDEN. Dr. Parker and His Friends.
        London: T. Fisher Unwin, 1904. (Joseph Parker,
        Congregationalist)
3790.   POLLARD, C. "Notes from City Road Minute Books: A
        Chartist at Wesley's Chapel." Proceedings of the
        Wesley Historical Society 23, Pt. 6 (June 1942),
        121-122. (Expulsion of James Ardrey)
3791.   POLLOCK, JOHN CHARLES. Moody without Sankey:
        A New Biographical Portrait. London: Hodder and
        Stoughton, 1963. (D. L. Moody, Congregationalist,
        and I. D. Sankey, Methodist; American evangelists
        who toured Britain)

3792.   PORRITT, ARTHUR.   J. D. Jones of Bournemouth.
        London: Independent Press Ltd., 1942.   (John
        Daniel Jones, Congregationalist)

3793.   POWELL, W. R.   "Bibliographical Aids to Research
        XII: The Sources for the History of Protestant
        Nonconformist Churches in England."   Bulletin of
        the Institute of Historical Research 25 (Nov.
        1952), 213-227.

3794.   PRESSLY, H. E.   "Evangelicalism in England in the
        First Half of the Nineteenth Century as Exemplified
        in the Life and Works of William Jay (1769-1853)."
        Ph.D. Dissertation, University of Edinburgh, 1950.
        (Congregationalist)

3795.   PRICE, E. J.   "Dr. Fairbairn and Airedale College:
        The Hour and the Man."   Transactions of the Congrega-
        tional Historical Society 13, No. 3 (Apr. 1939),
        131-139.   (Congregationalist)

3796.   PRICE, SEYMOUR JAMES.   "The Clifford Centenary."
        Baptist Quarterly 8, No. 4 (Oct. 1936), 189-190.
        (John Clifford, Baptist)

3797.   PRICE, SEYMOUR JAMES.   "London Strict Baptist
        Association, 1846-1853."   Baptist Quarterly 9, No.
        2 (Apr. 1938), 109-120.

3798.   PRICE, SEYMOUR JAMES.   A Popular History of the
        Baptist Building Fund: The Centenary Volume, 1824-
        1924.   London: Kingsgate Press, 1927.

3799.   PYKE, RICHARD.   The Early Bible Christians.   London:
        Epworth Press, 1941.

3800.   PYKE, RICHARD.   The Golden Chain: The Story of
        the Bible Christian Methodists from the Formation of
        the First Society in 1815 to the Union of the Denomi-
        nation in 1907 with the Methodist New Connexion
        and the United Methodist Free Churches in Forming
        the United Methodist Church.   London: Henry Hooks,
        1915.

3801. RAISTRICK, ARTHUR. Two Centuries of Industrial
      Welfare: The London (Quaker) Lead Company, 1692-
      1905. London: Friends' Historical Society, 1938.

3802. RAM, R. W. "Dissent in Urban Yorkshire, 1800-1850."
      Baptist Quarterly 22 (Jan. 1967), 3-22.

3803. RAY, CHARLES. The Life of Charles Haddon Spurgeon.
      With an Introduction by Pastor Thomas Spurgeon.
      London: Isbister and Company, Limited, 1903.
      (Baptist)

3804. RAY, CHARLES. A Marvellous Ministry: The Story of
      C. H. Spurgeon's Sermons, 1855 to 1905. London:
      Passmore and Alabaster, 1905. (Baptist)

3805. REDFERN, WILLIAM. Modern Developments in Methodism.
      London: National Council of Evangelical Free
      Churches, 1906.

3806. REES, FREDERICK ABIJAH. (Ed.) Alexander Maclaren:
      Appreciations of His Ministry. Manchester: R.
      Hayward, 1906. (Baptist)

3807. REYNOLDS, R. "The Catholic Boys at Ackworth: A
      Footnote to Quaker and Catholic History in the 19th
      Century." Journal of the Friends' Historical
      Society 43 (1951), 57-71.

3808. RICE, DANIEL FREDERICK. "Natural Theology and the
      Scottish Philosophy in the Thought of Thomas
      Chalmers." Scottish Journal of Theology 24 (Feb.
      1971), 23-46. (Free Church of Scotland)

3809. RICE, DANIEL FREDERICK. "The Theology of Thomas
      Chalmers." Dissertation Abstracts 27 (1967),
      3114-15A (Drew University, 1966). (Free Church
      of Scotland)

3810. RICHARDS, NOEL JUDD. "The Political and Social
      Impact of British Nonconformity in the Late Nine-
      teenth Century 1870-1902." Dissertation Abstracts
      29 (1968), 1851-52A (University of Wisconsin,
      1968).

3811. RICHEY, RUSSELL E. "Did the English Presbyterians
      Become Unitarians?" Church History 42 (Mar. 1973),
      58-72.

3812. RITSON, JOSEPH. "The Date of the Centenary of
      Primitive Methodism." Primitive Methodist Quarterly
      Review 47 (1905), 1-8.

3813. RITSON, JOSEPH. The Romance of Primitive Methodism.
      London: Edwin Dalton, 1909.

3814. ROBERTS, JOHN. The Calvinistic Methodism of Wales.
      Caernarvon: Calvinistic Methodist Book Agency,
      1934.

3815. ROBERTS, PHILIP ILOTT (A. CHESTER MANN). F. B. Meyer,
      Preacher, Teacher, Man of God. New York, Chicago:
      Fleming H. Revell Company, 1929. (Baptist)

3816. ROBINSON, HENRY WHEELER. Baptist Principles.
      London: Kingsgate Press, 1925.

3817. ROBINSON, HENRY WHEELER. The Life and Faith
      of the Baptists. London: Methuen and Co., Ltd.,
      1927.

3818. ROBINSON, WILLIAM GORDON. A History of the
      Lancashire Congregational Union, 1806-1951.
      Manchester: Lancashire Congregational Union, 1955.

3819. ROBSON, R. S. "Sir Henry Irving - Presbyterian:
      A Period Piece." Presbyterian Historical Society
      of England Journal 10, No. 1 (1952), 19-23.

3820. RODGERS, JOHN HEWITT. The Theology of P. T. Forsyth:
      The Cross of Christ and the Revelation of God.
      London: Independent Press, 1965. (Congregationalist)

3821. ROGERS, J. G. "Dissent in the Victorian Era."
      Nineteenth Century 50 (July 1901), 114-126. (Also
      in Living Age 230, Sept. 21, 1901, 729-739)

3822. ROGERS, PHILIP GEORGE. The Sixth Trumpeter: The
      Story of Jezreel and His Tower. London, New York:
      Oxford University Press, 1963. (James Jezreel, New
      and Latter House of Israel)

3823.   ROSE, EDWARD ALAN. (Ed.)   A Register of Methodist
        Circuit Plans, 1777-1860.   3 Pts. Manchester:
        Society of Cirplanologists, 1961-65.

3824.   ROSEBERY, ARCHIBALD PHILIP PRIMROSE.   "Dr. Chalmers."
        In Miscellanies Literary and Historical.   London:
        Hodder and Stoughton, Limited, 1921, Vol. 1, 238-254.
        (Free Church of Scotland)

3825.   ROUTLEY, ERIK.   English Religious Dissent.   Cambridge,
        England: Cambridge University Press, 1960.

3826.   ROUTLEY, ERIK.   The Story of Congregationalism,
        Briefly Told.   London: Independent Press, 1961.

3827.   ROWDON, HAROLD HAMLYN.   The Origins of the Brethren,
        1825-1850.   London: Pickering and Inglis, 1967.
        (Plymouth Brethren)

3828.   ROWELL, GEOFFREY.   "The Origins and History of
        Universalist Societies in Britain, 1750-1850."
        Journal of Ecclesiastical History 22 (1971), 35-56.

3829.   RUPP, ERNEST GORDON.   Methodism in Relation to
        Protestant Tradition.   London: Epworth Press,
        1954.

3830.   RUPP, ERNEST GORDON.   "Some Reflections on the
        Origin and Development of the English Methodist
        Tradition, 1738-1898."   London Quarterly and Holborn
        Review 178 (July 1953), 166-175.

3831.   RUPP, ERNEST GORDON.   Thomas Jackson: Methodist
        Patriarch.   London: Epworth Press, 1954.

3832.   RUPP, ERNEST GORDON.   "Victorian Humanity: The
        Influence of Victorian Nonconformity."   Listener 53
        (Mar. 17, 1955), 469-471.

3833.   RUSLING, G. W.   "The Nonconformist Conscience."
        Baptist Quarterly 22 (July 1967), 126-142.

3834.   RUSSELL, ELBERT.   The History of Quakerism.   New
        York: The Macmillan Company, 1942.

3835.   SAILLENS, EMILE.   La Vie et l'oeuvre de C. H.
        Spurgeon.   Lyon: E. Bichsel, 1902.   (Baptist)

3836.   SALTER, FRANK REYNER.   Dissenters and Public
        Affairs in Mid-Victorian England.   London:
        Dr. Williams's Trust, 1967.

3837.   SALTER, FRANK REYNER.   "Political Nonconformity in
        the Eighteen-Thirties."   Transactions of the Royal
        Historical Society 3 (1953), 125-143.

3838.   SANDEEN, ERNEST ROBERT.   The Roots of Fundamen-
        talism: British and American Millenarianism,
        1800-1930.   Chicago: University of Chicago Press,
        1970.

3839.   SANDEEN, ERNEST ROBERT.   "Towards a Historical
        Interpretation of the Origins of Fundamentalism."
        Church History 36 (Mar. 1967), 66-83.   (John Darby
        and the origin of the Plymouth Brethren in the
        1820's)

3840.   SCHOTT, KENNETH RONALD.   "An Analysis of Henry
        Drummond and His Rhetoric of Reconciliation."
        Ph.D. Dissertation, Ohio State University, 1972.
        (Free Church of Scotland)

3841.   SCOTT, PATRICK GREIG.   "'Zion's Trumpet':
        Evangelical Enterprise and Rivalry, 1830-1835."
        Victorian Studies 13 (Dec. 1969), 199-203.
        (Journal edited by Calvinistic Independent laymen
        in London)

3842.   SCOTT, RICHENDA C.   "Authority or Experience: John
        Wilhelm Rowntree and the Dilemma of 19th Century
        British Quakerism."   Journal of the Friends' Historical
        Society 49, No. 1 (1959), 75-95.

3843.   SCOTT, WILLIAM.   "Preface of 'Memoirs of Dissenting
        Churches' (1831)."   Transcribed by F. A. Homer.
        Transactions of the Unitarian Historical Society 6,
        No. 1 (Oct. 1935), 62-66.

3844.   SELBIE, WILLIAM BOOTHBY.   "Andrew Martin Fairbairn,
        1838-1938."   Congregational Quarterly 16 (1938),
        395-404.   (Congregationalist)

3845. SELBIE, WILLIAM BOOTHBY. The Life of Andrew Martin Fairbairn. London, New York: Hodder and Stoughton, 1914. (Congregationalist)

3846. SELBIE, WILLIAM BOOTHBY. (Ed.) The Life of Charles Sylvester Horne. London: Hodder and Stoughton, 1920. (Congregationalist)

3847. SELLERS, IAN. "Congregationalists and Presbyterians in Nineteenth Century Liverpool." Transactions of the Congregational Historical Society 20 (1965), 74-85.

3848. SELLERS, IAN. "Nonconformist Attitudes in Later Nineteenth-Century Liverpool." Transactions of the Historic Society of Lancashire and Cheshire 114 (1962), 215-239.

3849. SELLERS, IAN. "Political and Social Attitudes of Representative English Unitarians (1795-1850)." B.Litt. Dissertation, University of Oxford, 1956.

3850. SELLERS, IAN. "Unitarians and Social Change." Hibbert Journal 61 (Jan., Apr., July 1963), 16-22, 76-80, 122-127, 177-180.

3851. SEMMEL, BERNARD. "The Halévy Thesis." Encounter 37 (July 1971), 44-55. (Methodism)

3852. SEMMEL, BERNARD. The Methodist Revolution. New York: Basic Books, 1973. (The impact of Methodism on the shaping of nineteenth-century England)

3853. SHAW, PLATO ERNEST. The Catholic Apostolic Church Sometimes Called Irvingite: A Historical Study. New York: King's Crown, 1946.

3854. SHAW, THOMAS. The Bible Christians, 1815-1907. London: Epworth Press, 1965.

3855. SHAW, THOMAS. Foolish Dick and His Chapel: The Story of Porthtowan Methodism, 1796-1966. Redruth (Corn.): T. Shaw, 1966. (Richard Hampton)

3856. SHAW, THOMAS. A History of Cornish Methodism. Truro: Barton, 1967.

3857.   SHORT, H. L.   "From 'First Cause' to 'Indwelling
        Life': The Influence of American Hymns on English
        Unitarian Worship in the 19th Century." Transactions
        of the Unitarian Historical Society 10, No. 2
        (1952), 53-65.

3858.   SHORT, H. L.   "The Later History of the English
        Presbyterians." Hibbert Journal 65 (1966-67),
        32-37, 60-66, 117-122, 157-162; 66 (1967-68),
        31-35, 70-73, 131-136.

3859.   SHORT, H. L.   "Priestley and Martineau." Hibbert
        Journal 60 (1962), 211-219.   (Unitarian)

3860.   SHORT, KENNETH RICHARD M.   "The English Indemnity
        Acts, 1726-1867." Church History 42 (Sept. 1973),
        366-376.

3861.   SHORT, KENNETH RICHARD M.   "London's General Body
        of Protestant Ministers: Its Disruption in 1836."
        Journal of Ecclesiastical History 24 (Oct. 1973),
        377-393.

3862.   SHORT, KENNETH RICHARD M.   "A Study in Political
        Nonconformity: The Baptists, 1827-45, with Particular
        Reference to Slavery." Ph.D. Dissertation, University
        of Oxford, 1972.

3863.   SIDER, EARL MORRIS.   "Dissent and the Religious
        Issue in British Politics: 1840-1868." Dissertation
        Abstracts 27 (1966), 1771-72A (State University of
        New York at Buffalo, 1966).

3864.   SIL, NARASINGHA PROSAD.   "Influence of Methodism on
        the English Working Class, 1740-1819." Quarterly
        Review of Historical Studies 14, No. 1 (1974-75),
        19-34.

3865.   SIMPSON, JAMES YOUNG.   Henry Drummond.   Edinburgh
        and London: Oliphant, Anderson and Ferrier, 1901.
        (Free Church of Scotland)

3866.   SKEATS, HERBERT S. AND CHARLES S. MIALL.   History
        of the Free Churches of England, 1618-1891.
        London: Alexander and Shepheard, 1894.

3867. SMALLBONE, JEAN A. "Matthew Arnold and the
       Bicentenary of 1862." Baptist Quarterly 14 (1952),
       222-226. (Commemoration of the ejection of clergy,
       1662)

3868. SMALLBONE, JEAN A. "Matthew Arnold and the Noncon-
       formists." Baptist Quarterly 14 (1952), 345-355.

3869. SMITH, HENRY, JOHN E. SWALLOW AND WILLIAM TREFFRY.
       (Eds.) The Story of the United Methodist Church.
       London: Henry Hooks, 1932.

3870. SMITH, WARREN SYLVESTER. "London Quakers at the
       Turn of the Century." Quaker History 53 (1964),
       94-108.

3871. SPARKES, DOUGLAS C. "James Bradford of Hill Cliffe."
       Baptist Quarterly 26 (1975), 106-108. (Baptist)

3872. STALKER, JAMES. "A Notable Leader of the Scottish
       Church: Professor Henry Drummond." Review and
       Expositor 20 (1923), 423-433. (Free Church of
       Scotland)

3873. STEVENS, W. "Oxford's Attitude to Dissenters,
       1646-1946." Baptist Quarterly 13, No. 1 (1949),
       4-17.

3874. STIGANT, P. "Wesleyan Methodism and Working-class
       Radicalism in the North, 1792-1821." Northern
       History 6 (1971), 98-116.

3875. STONE, S. M. "A Survey of Baptist Expansion in
       England from 1785 to 1850, with Special Reference
       to the Emergence of Permanent Structures of Organ-
       isation." M.A. Dissertation, University of Bristol,
       1959.

3876. STREET, M. JENNIE. F. B. Meyer: His Life and
       Work. London: S. W. Partridge and Co., 1902.
       (Baptist)

3877. STUNT, TIMOTHY C. F. "Early Brethren and the
       Society of Friends." C.B.R.F. Occasional Paper No.
       3 (1970), 5-27. (Plymouth Brethren and Quakers)

3878. SWIFT, DAVID E. "Charles Simeon and J. J. Gurney: A Chapter in Anglican-Quaker Relations." Church History 29 (1960), 167-186.

3879. SWIFT, DAVID E. Joseph John Gurney: Banker, Reformer and Quaker. Middletown, Conn.: Wesleyan University Press, 1962.

3880. SWIFT, W. F. "The Women Itinerant Preachers of Early Methodism." Proceedings of the Wesley Historical Society 28, No. 5 (1952), 89-94; 29, No. 4 (1953), 76-83.

3881. SYKES, JOHN. The Quakers. London: Wingate, 1959.

3882. SYMONS, ALPHONSE JAMES ALBERT. "Irving and the Irvingites." Horizon 12, No. 71 (Nov. 1945), 310-324.

3883. TARRANT, WILLIAM GEORGE. Unitarianism. London: Constable and Company Ltd., 1912.

3884. TAYLOR, ERNEST R. Methodism and Politics 1791-1851. Cambridge: Cambridge University Press, 1935.

3885. TAYLOR, JOHN H. "London Congregational Churches since 1850." Transactions of the Congregational Historical Society 20 (1965), 22-41.

3886. TAYLOR, PHILIP A. M. Expectations Westward: The Mormons and the Emigration of Their British Converts in the Nineteenth Century. Edinburgh: Oliver and Boyd, 1965.

3887. TELFORD, JOHN. The Life of James Harrison Rigg, D.D., 1821-1909. London: Robert Culley, 1909. (Wesleyan)

3888. THOMAS, ALLEN CLAPP. "Caroline Emelia Stephen (1835-1909)." Bulletin of the Friends' Historical Society 3 (1910), 95-98. (Sister of Sir Leslie Stephen, who joined the Society of Friends)

3889. THOMAS, FRANK HOPKINS, JR. "The Development of Denominational Consciousness in Baptist Historical Writings, 1738-1886." Dissertation Abstracts International 36 (1976), 6161-62A (Southern Baptist

Theological Seminary, 1975). (Both in England and
the United States)

3890. THOMPSON, DAVID MICHAEL. "Churches of Christ in
the British Isles 1842-1972: A Historical Sketch."
Journal of the United Reformed Church History Society
1 (May 1973), 23-34.

3891. THOMPSON, DAVID MICHAEL. (Ed.) Nonconformity
in the Nineteenth Century. London: Routledge and
Kegan Paul, 1972.

3892. THORBURN, DONALD B. "The Effects of the Wesleyan
Movement on the Brontë Sisters, As Evidenced by an
Examination of Certain of Their Novels." Microfilm
Abstracts 8, No. 2 (1948), 109-111 (New York
University, 1947).

3893. TICE, FRANK. The History of Methodism in Cambridge. London: Epworth Press, 1966.

3894. TILLYARD, FRANK. "The Distribution of the Free
Churches in England." Sociological Review 27 (Jan.
1935), 1-18.

3895. TORBET, ROBERT GEORGE. A History of the Baptists.
Philadelphia: Judson Press, 1950.

3896. TOWNSEND, WILLIAM JOHN, HERBERT BROOK WORKMAN AND
GEORGE EAYRS. (Eds.) A New History of Methodism.
2 vols. London: Hodder and Stoughton, 1909.

3897. TOWNSEND, WILLIAM JOHN. The Story of Methodist
Union. Halifax, London: Milner and Co., Ltd.,
1906.

3898. TRISTRAM, HENRY. "Repeal of the Corporation and
Test Acts, 1828." Dublin Review 183 (1928), 105-
115, 201-207.

3899. TURNER, J. M. "Methodism and Anglicanism, 1791-1850."
M.A. Dissertation, University of Bristol, 1957.

3900. TURNER, W. G. John Nelson Darby. London: C. A.
Hammond, 1926. (Plymouth Brethren)

3901. TWADDLE, M. "Oxford and Cambridge Admissions
Controversy of 1834." British Journal of Educational
Studies 14 (Nov. 1966), 45-58.

3902.   UNDERWOOD, ALFRED CLAIR. A History of the English
        Baptists. London: Baptist Union Publ. Dept.,
        1947.

3903.   UNRUH, FRED PAUL. "A Historical Study of Robert
        Vaughan and His Views on Politics, Education,
        Religion and History As Reflected in the 'British
        Quarterly Review'." Dissertation Abstracts 23
        (1962), 1005-06 (University of Missouri, 1962).
        (Congregationalist)

3904.   UNWIN, FRANCES MABELLE AND JOHN TELFORD. Mark
        Guy Pearse: Preacher, Author, Artist, 1842-1930.
        London: Epworth Press, 1930. (Methodist)

3905.   UPTON, CHARLES BARNES. Dr. Martineau's Philosophy:
        A Survey. London: J. Nisbet and Co., Limited,
        1905. (Unitarian)

3906.   URWIN, EVELYN CLIFFORD. The Significance of 1849:
        Methodism's Greatest Upheaval. London: Epworth
        Press, 1949.

3907.   VARLEY, HENRY. Henry Varley's Life Story. London:
        Alfred Holness, 1916. (Born into a Baptist family,
        headed an undenominational church, refused ordin-
        ation)

3908.   VAUGHAN, FRANK. A History of the Free Church of
        England, Otherwise Called the Reformed Episcopal
        Church. Bath: H. Sharp and Sons, 1938.

3909.   VEITCH, THOMAS STEWART. The Story of the Brethren
        Movement: A Simple and Straightforward Account of
        the Features and Failures of a Sincere Attempt to
        Carry Out the Principles of Scripture during the
        Last 100 Years. London: Pickering and Inglis,
        1933.

3910.   VERNON, ANNE. A Quaker Businessman: The Life of
        Joseph Rowntree, 1836-1925. London: Allen and
        Unwin, 1958.

3911.   VINCENT, JOHN. The Formation of the Liberal Party
        1857-1868. New York: Scribner, 1966. (Includes
        the role of Nonconformity)

3912.   VLUGT, W. VAN DER.   "James Martineau, 1805-1900."
        Theologisch Tijdschrift 35 (1901), 289-319, 385-419,
        481-499.   (Unitarian; written in Dutch)

3913.   VOGELER, ALBERT RICHARD.   "Disestablishmentarianism:
        The Liberationist Crusade against the Church of
        England, 1868-1886."   Dissertation Abstracts
        International 34 (1973), 3326-27A (Columbia
        University, 1973).

3914.   WAKEFIELD, GORDON STEVENS.   Methodist Devotion:
        The Spiritual Life in the Methodist Tradition,
        1791-1945.   London: Epworth Press, 1966.

3915.   WALKER, J. D.   "Methodist Discipline, 1750-1900."
        M.A. Dissertation, University of Manchester, 1972.

3916.   WALKER, R. B.   "The Growth of Wesleyan Methodism in
        Victorian England and Wales."   Journal of Ecclesiastical
        History 24 (July 1973), 267-284.

3917.   WALTON, ROBERT CLIFFORD.   The Gathered Community.
        London: Carey Press, 1946.   (Baptist)

3918.   WARD, WILLIAM REGINALD.   "The Baptists and the
        Transformation of the Church, 1780-1830."   Baptist
        Quarterly 25 (Oct. 1973), 167-184.

3919.   WARD, WILLIAM REGINALD. (Ed.)   The Early Correspon-
        dence of Jabez Bunting, 1820-1829.   London: Royal
        Historical Society, 1972.   (Wesleyan)

3920.   WARD, WILLIAM REGINALD.   "The Religion of the
        People and the Problem of Control, 1790-1830."   In
        G. J. Cuming and Derek Baker (Eds.), Popular Belief
        and Practice.   Cambridge: Cambridge University
        Press, 1972, 237-257.   (Methodism)

3921.   WATERS, A. C.   "History of the British 'Churches of
        Christ'."   Ph.D. Dissertation, University of
        Edinburgh, 1940.

3922.   WATT, HUGH.   Thomas Chalmers and the Disruption.
        London: Nelson, 1943.   (Free Church of Scotland)

3923.   WATT, MARGARET HEWITT.   "Carlyle and Irving: The
        Story of a Friendship."   Scots Magazine 12 (Oct.
        1934-Mar. 1935), 202-208.

3924.   WATTS, M. R.   "John Clifford and Radical
        Nonconformity."   Ph.D. Dissertation, University
        of Oxford, 1966.   (Baptist)

3925.   WEARMOUTH, ROBERT FEATHERSTONE.   "The Background of
        Hugh Bourne's Achievements."   London Quarterly and
        Holborn Review 21 (1952), 167-170.   (Primitive
        Methodist)

3926.   WEARMOUTH, ROBERT FEATHERSTONE.   Methodism and the
        Struggle of the Working Classes, 1850-1900.
        Leicester: Edgar Backus, 1954.

3927.   WEARMOUTH, ROBERT FEATHERSTONE.   Methodism and the
        Trade Unions.   London: Epworth Press, 1959.

3928.   WEARMOUTH, ROBERT FEATHERSTONE.   Methodism and the
        Working Class Movements of England 1800-1850.
        London: The Epworth Press, 1937.

3929.   WEARMOUTH, ROBERT FEATHERSTONE.   Some Working-class
        Movements of the Nineteenth Century.   London:
        Epworth Press, 1948.   (Methodism)

3930.   WELCH, ALLEN HOWARD.   "John Carvell Williams: The
        Nonconformist Watchdog (1821-1907)."   Dissertation
        Abstracts International 30 (1969), 868-869A
        (University of Kansas, 1968).   (Congregationalist)

3931.   WHITBY, A. C.   "Matthew Arnold and the Nonconformists:
        A Study in Social and Political Attitudes."   B.Litt.
        Dissertation, University of Oxford, 1955.

3932.   WHITE, J. W.   "The Influence of North American
        Evangelism in Great Britain between 1830 and 1914
        on the Origin and Development of the Ecumenical
        Movement."   Ph.D. Dissertation, University of
        Oxford, 1963.

3933.   WHITE, JOHN HARROP.   The Story of the Old Meeting
        House, Mansfield.   London: Lindsey Press, 1959.
        (Drift from Presbyterianism to Unitarianism in a
        Midlands town)

3934.   WHITLEY, HENRY CHARLES.   Blinded Eagle: An Intro-
        duction to the Life and Teaching of Edward Irving.
        London: S.C.M. Press, 1955.

3935. WHITLEY, HENRY CHARLES. Laughter in Heaven.
       London: Hutchinson, 1962. (Edward Irving and the
       Irvingites)
3936. WHITLEY, WILLIAM THOMAS. The Baptists of London,
       1612-1928. London: Kingsgate Press, 1928.
3937. WHITLEY, WILLIAM THOMAS. A History of British
       Baptists. London: C. Griffen and Company, Limited,
       1923.
3938. WHITNEY, JANET PAYNE. Elizabeth Fry: Quaker
       Heroine. Boston: Little, Brown and Company, 1936.
3939. WHITTAKER, M. B. "The Revival of Dissent, 1800-35."
       M.Litt. Dissertation, University of Cambridge,
       1959.
3940. WIGMORE-BEDDOES, DENNIS G. Yesterday's Radicals:
       A Study of the Affinity between Unitarianism and
       Broad Church Anglicanism in the Nineteenth Century.
       London: James Clarke, 1971.
3941. WILBUR, EARL MORSE. A History of Unitarianism in
       Transylvania, England and America. 2 vols.
       Cambridge: Harvard University Press, 1945-52.
3942. WILBUR, EARL MORSE. A History of Unitarianism:
       Socinianism and Its Antecedents. Cambridge, Mass.:
       Harvard University Press, 1945.
3943. WILKERSON, ALBERT H. The Rev. R. J. Campbell:
       The Man and His Message. London: Francis Griffiths,
       1907. (Congregationalist)
3944. WILKINSON, JOHN THOMAS. (Ed.) Arthur Samuel Peake,
       1865-1929: Essays in Commemoration. By Elsie Cann
       [and others] and Selections from His Writings.
       London: Epworth Press, 1958. (Primitive Methodist)
3945. WILKINSON, JOHN THOMAS. Hugh Bourne, 1772-1852.
       London: Epworth Press, 1952. (Primitive Methodist)
3946. WILKINSON, JOHN THOMAS. "Hugh Bourne, 1772-1852: A
       Centenary Tribute." Proceedings of the Wesley
       Historical Society 28, No. 7 (1952), 126-130.
       (Primitive Methodist)

3947.   WILKINSON, JOHN THOMAS.   Samuel Drew, 1765-1833.
        Redruth, Cornwall: Cornish Methodist Historical
        Association, 1963.   (Wesleyan)

3948.   WILKINSON, JOHN THOMAS.   1662 and After: Three
        Centuries of English Nonconformity.   London:
        Epworth Press, 1962.

3949.   WILKINSON, JOHN THOMAS.   William Clowes, 1780-1851.
        London: Epworth Press, 1951.   (Primitive Methodist)

3950.   WILKINSON, JOHN THOMAS.   "William Clowes, 1780-1851:
        A Centenary Tribute."   Proceedings of the Wesley
        Historical Society 28, Pt. 1 (1951), 812.   (Primitive
        Methodist)

3951.   WILLIAMS, ALBERT HUGHES.   Welsh Wesleyan Methodism,
        1800-1858:   Its Origins, Growth and Secessions.
        Bangor, Wales: Cyhoeddwyd Gan Lyfrfa'r Methodistiaid,
        1935.

3952.   WILLIAMS, WILLIAM.   Charles Haddon Spurgeon: Personal
        Reminiscences.   Revised and edited by his daughter,
        Marguerite Williams.   Foreword by Dr. John Charles
        Carlile.   London: The Religious Tract Soc., 1933.
        (Baptist)

3953.   WILLIAMSON, DAVID.   The Life of Alexander Maclaren.
        London: James Clarke and Co., 1910.   (Baptist)

3954.   WILSON, BRYAN R. (Ed.)   Patterns of Sectarianism:
        Organisation and Ideology in Social and Religious
        Movements.   London: Heinemann, 1967.   (Includes
        the Salvation Army, Quakers, Plymouth Brethren)

3955.   WILSON, BRYAN R.   Religious Sects: A Sociological
        Study.   London: Weidenfeld and Nicolson, 1970.

3956.   WILSON, BRYAN R.   Sects and Society: A Sociological
        Study of the Elim Tabernacle, Christian Science, and
        Christadelphians.   Berkeley: University of California
        Press, 1961.

3957.   WILSON, H. G.   "The Ministers' Benevolent Society."
        Transactions of the Unitarian Historical Society
        10, No. 1 (1951), 19-23.   (Founded 1852, for
        Unitarian ministers and their families)

3958.   WINGATE, S. D.   "The Ghost of John Nelson Darby."
        Contemporary Review 207 (1965), 91-95. (Plymouth
        Brethren)
3959.   WINNIFRITH, T. J.   "Charlotte Brontë and Calvinism."
        Notes and Queries 215 (Jan. 1970), 17-18.
3960.   WOFFORD, MILTON GENE.   "The Theology of Robert
        William Dale." Ph.D. Dissertation, Southwestern
        Baptist Theological Seminary, 1961. (Congregationalist)
3961.   WOLFENDEN, J. W.   "English Non-conformity and the
        Social Conscience, 1880-1906." Ph.D. Dissertation,
        Yale University, 1954.
3962.   WOODFIN, YANDALL C.   "The Apologetic Method of
        Andrew Martin Fairbairn." Ph.D. Dissertation,
        Southwestern Baptist Theological Seminary, 1957.
        (Congregationalist)
3963.   WORKMAN, HERBERT BROOK.   Methodism. Cambridge:
        Cambridge University Press; New York: G. P. Putnam's
        Sons, 1912.
3964.   WRIGHT, EDGAR.   Mrs. Gaskell: The Basis for
        Reassessment. London, New York: Oxford University
        Press, 1965. (Includes Unitarianism)
3965.   YOUNG, JESSE BOWMAN.   "Hugh Price Hughes: Evangelist."
        Methodist Review 87 (1905), 419-428. (Methodist)
3966.   YOUNG, KENNETH.   Chapel: The Joyful Days and
        Prayerful Nights of the Non-conformists in Their
        Heyday, c. 1850-1950. London: Eyre Methuen, 1972.
3967.   ZAREK, OTTO.   The Quakers. Trans. E. W. Dickes.
        London: Dakers, 1943.

XVI.   Catholicism

        Under this heading are listed works on English
Catholicism and its leaders. A significant portion of
the material deals with Catholic Emancipation and the
restoration of the hierarchy in England. Included,
too, are works on orders, social thought, modernism,
periodicals, the reaction to new scientific ideas, and
the attitudes of non-Catholics towards Catholics. Studies
of Catholic thinkers and literary figures, e.g., Lord
Acton, Gerard Manley Hopkins, Henry Edward Manning,
Francis Thompson, George Tyrrell, Herbert Vaughan, the
Wards, and Nicholas Wiseman are also assembled here.
See II. Religion; XIV. Church of England, particularly
1. General and 3. Oxford Movement; appropriate sections
in XIX. Varieties of Belief. . . .

3968.   ABERCROMBIE, NIGEL J.   "Edmund Bishop and the Roman
        Breviary." Clergy Review 38 (1953), 75-86, 129-139.
3969.   ABERCROMBIE, NIGEL J.   The Life and Work of Edmund
        Bishop. London: Longmans, 1959.
3970.   AHAUS, H. Kardinaal Herbert Vaughan, stichter
        van Mill-Hill. Louvain: Xaveriana, 1939.
3971.   ALATRI, P. "Lord Acton e il suo cattolicesimo
        liberale." Belfagor 4 (1949), 21-35, 137-148.
3972.   ALBION, G. "Papal Aggression on England." Irish
        Ecclesiastical Record 74 (Oct. 1950), 350-357.
3973.   ALEXANDER, CALVERT. The Catholic Literary Revival:
        Three Phases in its Development from 1845 to the
        Present. Milwaukee: Bruce, 1935.
3974.   ALLEN, LOUIS. "Ambrose Phillipps de Lisle, 1809-78."
        Catholic Historical Review 40, No. 1 (1954), 1-26.
3975.   ALTHOLZ, JOSEF LEWIS AND DAMIAN MCELRATH. (Eds.)
        The Correspondence of Lord Acton and Richard Simpson.

3 vols. Cambridge: Cambridge University Press, 1971.

3976. ALTHOLZ, JOSEF LEWIS. The Liberal Catholic Movement in England: The 'Rambler' and its Contributors 1848-1864. London: Burns and Oates, 1962.

3977. ALTHOLZ, JOSEF LEWIS. "The Political Behaviour of the English Catholics 1850-1867." Journal of British Studies 4, No. 1 (Nov. 1964), 89-103.

3978. ALTHOLZ, JOSEF LEWIS. "Some Observations on Victorian Religious Biography: Newman and Manning." Worship 43 (Aug.-Sept. 1969), 407-415.

3979. ALTHOLZ, JOSEF LEWIS. "The Vatican Decrees Controversy, 1874-1875." Catholic Historical Review 57 (1971-72), 593-605.

3980. ANON. "The Five-Fold Growth of Theology: Catholicism and Liberalism (1837-1937)." Times Literary Supplement No. 1839 (May 1, 1937), 330-331.

3981. ANON. "Forgotten Passages in the Life of Cardinal Wiseman." Dublin Review 163 (1918), 145-168.

3982. ANON. "Newman and Manning." Quarterly Review 206 (Apr. 1907), 354-383.

3983. ANSON, PETER FREDERICK. The Benedictines of Caldey: The Story of the Anglican Benedictines of Caldey and Their Submission to the Catholic Church. London: Burns, Oates, 1940.

3984. ANSTRUTHER, GEORGE ELLIOT. "Francis Cardinal Bourne." Dublin Review 196 (Apr. 1935), 177-195.

3985. ANSTRUTHER, GEORGE ELLIOT. "A Great Idea Extinct." Blackfriars 31 (1950), 416-421. (Dominicans)

3986. ANSTRUTHER, GEORGE ELLIOT. A Hundred Years of Catholic Progress: Being a Short Account of the Church's Fortunes in Great Britain since the Time of the Emancipation Act. London: Burns, Oates and Washbourne, 1929.

3987. ARNSTEIN, WALTER L. "Victorian Prejudice Reexamined." Victorian Studies 12 (1969), 452-457. (Anti-Catholicism)

3988. ATTWATER, DONALD. The Catholic Church in Modern
       Wales: A Record of the Past Century. London:
       Burns, Oates and Washbourne, 1935.

3989. BARCUS, JAMES EDGAR. "Structuring the Rage Within:
       The Spiritual Autobiographies of Newman and Orestes
       Brownson." Cithara 15 (Nov. 1975), 45-57.

3990. BARMANN, LAWRENCE F. Baron Friedrich von Hügel
       and the Modernist Crisis in England. Cambridge:
       Cambridge University Press, 1972.

3991. BARRY, WILLIAM FRANCIS et al. Catholic Emanci-
       pation, 1829-1929: Essays by Various Writers.
       London: Longmans, 1929.

3992. BARRY, WILLIAM FRANCIS. "Emancipation." Dublin
       Review 152 (Apr. 1913), 382-398.

3993. BASSET, BERNARD. The English Jesuits: From Campion
       to Martindale. London: Burns and Oates, 1967.

3994. BECK, GEORGE ANDREW. (Ed.) The English Catholics,
       1850-1950. London: Burns, 1950.

3995. BECK, GEORGE ANDREW. "English Spiritual Writers:
       XXI. Cardinal Manning." Clergy Review 45 (Sept.
       1960), 513-524.

3996. BECQUET, T. "Les Catholiques anglais 1850-1950."
       Revue Générale Belge (juillet 1951), 427-443.

3997. BELLESHEIM, A. "Herbert Kardinal Vaughan, dritter
       Erzbischof von Westminster 1832-1903." Historisch-
       politische Blätter für das katholische Deutschland
       146 (1910), 36-47; 105-116.

3998. BENNETT, SCOTT. "Catholic Emancipation, the
       'Quarterly Review', and Britain's Constitutional
       Revolution." Victorian Studies 12 (Mar. 1969),
       283-304.

3999. BERGONZI, BERNARD. "The English Catholics: Forward
       from 'The Chesterbelloc'." Encounter 24 (Jan.
       1965), 19-30. (19th-century origins of present-day
       Catholicism; "Chesterbelloc" is a compound of
       G. K. Chesterton and Hilaire Belloc)

4000. BERKELEY, HUMPHRY. "Catholics in English Politics."
Wiseman Review 236, No. 494 (1962), 300-308.

4001. BERRY, MARGARET ANN, D.C., SISTER. "Literary
Theory and Criticism in the English Catholic Revival
(1845-1900)." Ph.D. Dissertation, St. John's
University, 1956.

4002. BIRRELL, T. A. Non-Catholic Writers and Catholic
Emancipation: An Aspect of Sidney Smith, Shelley,
Coleridge and Cobbett. Nijmegen: Dekker and Van
de Vegt, 1953.

4003. BLAKISTON, NOEL. (Ed.) The Roman Question: Extracts
from the Dispatches of Odo Russell from Rome,
1858-1870. London: Chapman and Hall, 1962.

4004. BLÖTZER, JOSEPH. Die Katholikenemanzipation in
Grossbritannien und Irland: Ein Beitrag zur
Geschichte religiöser Toleranz. Freiburg im
Breisgau, St. Louis, Mo.: Herder, 1905.

4005. BLUNT, HUGH F. "Aubrey Beardsley: A Study in
Conversion." Catholic World 134 (1932), 641-650.

4006. BODLEY, JOHN EDWARD COURTENAY. Cardinal Manning,
and Other Essays. London, New York: Longmans,
Green and Co., 1912.

4007. BOLTON, CHARLES ANSELM. "Cardinal Vaughan as
Educator." Clergy Review 28, No. 4 (Oct. 1947),
237-245.

4008. BONNEY, E. "Cardinal Hinsley at Ushaw (1876-1898)."
Ushaw Magazine 53, No. 158 (July 1943), 34-37.

4009. BOSSY, JOHN. "Four Catholic Congregations in Rural
Northumberland, 1750-1850." Recusant History 9
(Apr. 1967), 88-119.

4010. BOTTINO, EDWARD JOSEPH. "The 'Rambler' Controversy:
Positions of Simpson, Newman and Acton, 1856-1862."
Dissertation Abstracts International 31 (1970),
2295A (St. John's University, 1970).

4011. BOWMAN, LEONARD JOSEPH. "The Religious Tradition
behind the Imagery of Gerard Manley Hopkins' Poetry."

Dissertation Abstracts International 34 (1973),
402A (Fordham University, 1973).

4012. BRAYBROOKE, PATRICK. Some Victorian and Georgian
Catholics. London: Burns, Oates and Washbourne,
1932.

4013. BRINLEE, ROBERT WASHINGTON, JR. "Hopkins' Reconciliation of Religion and Poetry: Its Critical
History and Its Implications for the Christian
Imagination." Dissertation Abstracts 29 (1969),
4479A (University of Missouri, 1968).

4014. BROWN, STEPHEN J. "Some Catholic Periodicals."
Studies 25 (Sept. 1936), 428-442.

4015. BUMP, JEROME. "Art and Religion: Hopkins and
Savonarola." Thought 50 (1975), 132-147.

4016. BURKE, EDWARD J. "Ruskin and Catholicism:
1819-1858." Ph.D. Dissertation, St. John's
University, 1955.

4017. BURTCHAELL, JAMES TUNSTEAD. "The Biblical Question
and the English Liberal Catholics." Review of
Politics 31 (Jan. 1969), 108-120.

4018. BURTCHAELL, JAMES TUNSTEAD. Catholic Theories
of Biblical Inspiration since 1810: A Review and
Critique. London: Cambridge University Press,
1969.

4019. BURY, JOHN BAGNELL. History of the Papacy in
the Nineteenth Century: Liberty and Authority in
the Catholic Church. Edited, with a memoir, by the
Rev. Robert Henry Murray. London: Macmillan and
Co., Limited, 1930.

4020. BUTLER, EDWARD CUTHBERT. The Life and Times of
Bishop Ullathorne, 1806-1889. 2 vols. London:
Burns, Oates and Washbourne, Ltd., 1926.

4021. BUTLER, EDWARD CUTHBERT. The Vatican Council:
The Story Told from the Inside in Bishop Ullathorne's
Letters. 2 vols. London, New York: Longmans,
Green and Co., 1930.

4022.   BUTLER, WILLIAM FRANCIS THOMAS. "What Catholic
        Emancipation Meant." Dublin Review 184 (Apr.
        1929), 194-205.

4023.   BUTTERFIELD, HERBERT. Lord Acton. London:
        Published for the Historical Assn. by G. Philip,
        1948.

4024.   CAHILL, GILBERT ALOYSIUS. "Irish Catholicism and
        English Toryism." Review of Politics 19 (1957),
        62-76. (Anti-Catholicism)

4025.   CAHILL, GILBERT ALOYSIUS. "Irish Catholicism and
        English Toryism, 1832-1848: A Study in Ideology."
        Dissertation Abstracts 14 (1954), 2323-24 (State
        University of Iowa, 1954).

4026.   CAHILL, GILBERT ALOYSIUS. "The Protestant Associa-
        tion and the Anti-Maynooth Agitation of 1845."
        Catholic Historical Review 43 (1957), 273-308.

4027.   CARSON, R. L. "Multiplication Tables: The Progress
        of Catholicism in England and Wales, 1702-1949."
        Clergy Review 32 (1949), 21-30.

4028.   CATCHESIDE, P. H. "Catholicism in London in 1847:
        Fr Gentili Reports to the Holy See." Tablet 197
        (Mar. 24, 1951), 225-226.

4029.   CHAPEAU, ALPHONSE-LOUIS-EUGENE. "Les Lettres de
        Manning à Gladstone (1837-1851)." Thèse
        Complémentaire, Université de Paris, 1955.

4030.   CHAPEAU, ALPHONSE-LOUIS-EUGENE. "La Vie anglicane
        de Manning." Dissertation, Université de Paris,
        1955.

4031.   CHARLES, CONRAD. "The Origins of the Parish Mission
        in England and the Early Passionist Apostolate,
        1840-1850." Journal of Ecclesiastical History 15
        (1964), 60-75.

4032.   CLARK, EUGENE V. "Catholic Liberalism and Ultra-
        montanism: Freedom and Duty: A Study of the
        Quarrel over the Control of Catholic Affairs in
        England, 1858-1866." Dissertation Abstracts 26
        (1966), 5395-96 (University of Notre Dame, 1965).

4033. CLARKE, EGERTON. "Gerard Hopkins, Jesuit." Dublin Review 198 (1936), 127-141.

4034. CLERKE, ELLEN MARY. "Catholic Progress in the Reign of Victoria." Dublin Review 128 (1901), 227-255.

4035. CLIFFORD, WILLIAM. "Catholic Allegiance: A Pastoral Letter of 1874." Buckfast Abbey Chronicle 8, No. 4 (Dec. 1938), 211-224. (Bishop Clifford on W. E. Gladstone's criticism of English Catholics)

4036. COHEN, SELMA J. "The Poems of Gerard Manley Hopkins in Relation to His Religious Thought." Ph.D. Dissertation, University of Chicago, 1947.

4037. COLDRICK, HELEN F., R.D.C., SISTER. "Daniel O'Connell and Religious Freedom." Dissertation Abstracts International 35 (1974), 2892A (Fordham University, 1974).

4038. COLLINGWOOD, C. "The Catholic Truth Society." Clergy Review 37 (1952), 641-658.

4039. CONNELL, JOAN. "The Roman Catholic Church in England, 1553-1850: A Study in Internal Politics." Ph.D. Dissertation, University of Chicago, 1969.

4040. CONSTANT, G. "Les Progrès du catholicisme en Angleterre." Revue des Deux Mondes, 8th Series, 28 (1935), 519-538.

4041. CONZEMIUS, VICTOR. (Ed.) Briefwechsel, 1820-1890 [von] Ignaz von Döllinger. München: Beck, 1963-65. (Döllinger's correspondence with Lord Acton, 1850-70)

4042. CONZEMIUS, VICTOR. "Lord Acton and the First Vatican Council." Journal of Ecclesiastical History 20, (Oct. 1969), 267-294.

4043. COOLEN, GEORGES. "L'Emancipation catholique en Angleterre." Revue Apologétique 48 (1929), 704-716.

4044. CORISH, P. J. "Restoration of the English Catholic Hierarchy." Irish Ecclesiastical Record 74 (Oct. 1950), 289-307.

4045. CORR, GERARD M. Servites in London: An Account of the Coming of the Servite Fathers to England and of the Founding of the English Province of the Order. Newbury, Berks.: Servite Fathers, 1952.

4046. CORRIGAN, RAYMOND. The Church and the Nineteenth Century. Milwaukee: Bruce, 1938. (Catholicism in Europe and England)

4047. COTTER, JAMES FINN. Inscape: The Christology and Poetry of Gerard Manley Hopkins. Pittsburgh: University of Pittsburgh Press, 1972.

4048. COURSON, BARBARA FRANCES MARY DE. "Herbert, Cardinal Vaughan." Revue Générale 93 (1911), 706-728.

4049. Coverdale, Philip F. Some Notes on the 1851 Religious Census: With a Summary of Roman Catholic Returns. London, [1966].

4050. COYNE, J. J. "The First Westminster Synod." Clergy Review 38 (1953), 269-280. (1852)

4051. CRANE, PAUL. "The Catholic Minority in Great Britain, 1850-1950." Thought 31 (Winter 1956), 509-541.

4052. CULKIN, G. "John Lingard and Ushaw." Clergy Review 35 (1951), 361-371.

4053. CURTIS, LEWIS PERRY. Anglo-Saxons and Celts: A Study of Anti-Irish Prejudice in Victorian England. Bridgeport, Conn.: Published by the Conference on British Studies at the University of Bridgeport, 1968. (Anti-Catholicism)

4054. CWIEKOWSKI, FREDERICK J. The English Bishops and the First Vatican Council. Louvain: Bureaux de la R.H.E., Bibliothèque de l'Université, Publications Universitaires de Louvain, 1971.

4055. DAKIN, ARTHUR HAZARD. Von Hügel and the Supernatural. London: Society for Promoting Christian Knowledge; New York: The Macmillan Company, 1934.

4056. DANIEL-ROPS, HENRY. The Church in an Age of Revolution, 1789-1870. Trans. John Warrington. London: J. M. Dent and Sons Ltd.; New York: E. P. Dutton and Co. Inc., 1965.

4057. DARK, SIDNEY. Manning. London: Duckworth, 1936.

4058. DAVIS, CHARLES. (Ed.) English Spiritual Writers. London: Burns and Oates, 1961. (Includes J. H. Newman, H. Manning, Father Faber, and John Chapman)

4059. DAWSON, CHRISTOPHER HENRY. "English Catholics, 1850-1950." Dublin Review 224 (Winter 1950), 1-12.

4060. DE LA BEDOYERE, MICHAEL. The Life of Baron von Hügel. London: Dent, 1951.

4061. DELL, ROBERT EDWARD. "George Tyrrell." Cornhill Magazine 27 (1909), 665-675.

4062. DENHOLM, ANTHONY F. "The Conversion of Lord Ripon in 1874." Recusant History 10 (Apr. 1969), 111-118. (Ripon's conversion to Catholicism)

4063. DERKS, K. J. "Het eeuwfeest der Katholieke Emancipatie in Engeland." Studiën 112 (1929), 442-463.

4064. DEVLIN, CHRISTOPHER. "The Ignatian Inspiration of Gerard Manley Hopkins." Blackfriars 16 (1935), 887-900.

4065. DI CICCO, REV. MARIO M. "Gerard Manley Hopkins and the Mystery of Christ." Dissertation Abstracts International 32 (1971), 425A (Case Western Reserve University, 1970).

4066. DOCKERY, J. B. Collingridge: A Franciscan Contribution to Catholic Emancipation. Newport: Johns, 1954.

4067. DOMMERSEN, HAROLD. "Aston Hall and the Passionists." Clergy Review 22 (Sept. 1942), 400-407. (Fr. Dominic)

4068. DONALD, GERTRUDE. "By Way of Brighton Pier." Buckfast Abbey Chronicle 8, No. 3 (Sept. 1938), 158-173. (Richard Sibthorp)

4069. DORAN, P. "The Restoration of the Hierarchy."
Clergy Review 34 (1950), 161-178.

4070. DORLODOT, HENRI DE. Darwinism and Catholic Thought.
Trans. The Rev. Ernest Wessenger. London: Burns,
Oates and Washbourne, Ltd., 1922.

4071. DOUGHERTY, CHARLES T. "Manning, Ollivier and Papal
Infallibility." Manuscripta 15 (1971), 96-99.
(Emile Ollivier, author of a history of the first
Vatican Council)

4072. DOUGHERTY, CHARLES T. AND HOMER C. WELSH. "Wiseman
and the Oxford Movement: An Early Report to the
Vatican." Victorian Studies 2 (1958-59), 149-154.

4073. DOWNES, DAVID ANTHONY. Victorian Portraits: Hopkins
and Pater. New York: Bookman Associates, 1965.

4074. EATON, ROBERT ORMSTON. The Benedictines of Colwich,
1829-1929: England's First House of Perpetual
Adoration. London: Sands, 1929.

4075. EDWARDS, T. CHARLES. "Papal Aggression: 1851."
History Today 1 (Dec. 1951), 42-49.

4076. FITZGERALD, PERCY HETHERINGTON. Fifty Years of
Catholic Life and Social Progress under Cardinals
Wiseman, Manning, Vaughan, and Newman: With an
Account of the Various Personages, Events and
Movements, during the Era. 2 vols. London: T.
Fisher Unwin, 1901.

4077. FITZSIMMONS, JOHN. "Cardinal Manning, Friend of
the People." Clergy Review 32 (1949), 145-156.

4078. FITZSIMMONS, JOHN. (Ed.) Manning: Anglican and
Catholic. London: Burns & Oates, 1951.

4079. FLETCHER, JOHN R. "Early Catholic Periodicals in
England." Dublin Review 198 (1936), 284-310.
(Lists periodicals up to 1876)

4080. FLINDALL, R. P. "Anglican and Roman Attitudes:
1825-1875." Church Quarterly Review 169 (Apr.-June
1968), 206-215.

4081.   FOTHERGILL, BRIAN. Nicholas Wiseman. Garden City,
        New York: Doubleday, 1963.

4082.   FOTHERGILL, BRIAN. "Wiseman: The Man and His
        Mission." Wiseman Review 236 (Fall 1962), 236-246.

4083.   FRANKE, L. "Herbert Vaughan, kardinaal-aartsbisschop
        van Westminster." Studiën 80 (1913), 59-84.

4084.   GARDNER, W. H. "The Religious Problem in Gerard
        Manley Hopkins." Scrutiny 6 (1937), 32-42.

4085.   GASQUET, FRANCIS AIDAN. "Letters of Cardinal
        Wiseman: Communicated, with a Commentary."
        Dublin Review 164 (Jan. 1919), 1-25.

4086.   GASQUET, FRANCIS AIDAN. (Ed.) Lord Acton and
        His Circle. London: G. Allen, 1906.

4087.   GATELY, MARY JOSEPHINE, SISTER. The Sisters of
        Mercy: Historical Sketches, 1831-1931. New York:
        The Macmillan Company, 1931.

4088.   GILL, G. "Un secolo di cattolicismo in Inghilterra,
        1850-1950." La civiltà cattolica 4, anno 101
        (1950), 679-692.

4089.   GILLEY, SHERIDAN. "Papists, Protestants and the
        Irish in London, 1835-70." In G. J. Cuming and
        Derek Baker (Eds.), Popular Belief and Practice.
        Cambridge: Cambridge University Press, 1972,
        259-266.

4090.   GILLEY, SHERIDAN. "Protestant London, No-Popery
        and the Irish Poor." Recusant History 10 (1969-70),
        210-230; 11 (1971), 24-46.

4091.   GLASER, JOHN F. "Parnell's Fall and the Nonconformist
        Conscience." Irish Historical Studies 12 (1960),
        119-138.

4092.   GOETZ, JOSEPH, S.J. "Coleridge, Newman and Tyrrell:
        A Note." Heythrop Journal 14 (Oct. 1973), 431-436.

4093.   GORDON-GORMAN, WILLIAM JAMES. Converts to Rome:
        A Biographical List of the More Notable Converts to
        the Catholic Church in the United Kingdom during the
        Last Sixty Years. London: Sands and Co., 1910.

4094. GREAVES, R. W. "Roman Catholic Relief and the Leicester Election of 1826." Transactions of the Royal Historical Society Series 4, 22 (1940), 199-223.

4095. GREENE, GWENDOLEN MAUD. Two Witnesses: A Personal Recollection of Hubert Parry and Friedrich von Hügel. London and Toronto: J. M. Dent and Sons, Ltd.; New York: E. P. Dutton and Co., Inc., 1930.

4096. GUIBERT, JEAN. Le Réveil du catholicisme en Angleterre au XIXe siècle: conférences prêchées dans l'église Saint-Sulpice. 1901-1906. Paris: Poussielgue, 1907.

4097. GUMBLEY, WALTER. "The English Dominicans from 1555 to 1950." Dominican Studies 5 (1952), 103-133.

4098. GUMBLEY, WALTER. (Ed.) Obituary Notices of the English Dominicans, 1555-1952. London: Blackfriars Publications, 1955.

4099. GWATKIN, J. "Freedom's Advocate: Lord Acton and the Liberal Catholic Movement in England." Dublin Review 239 (Sept. 1965), 90-99.

4100. GWYNN, DENIS ROLLESTON. Cardinal Wiseman. London: Burns, Oates and Washbourne Ltd., 1929.

4101. GWYNN, DENIS ROLLESTON. "Dominic Barberi and the 'Cambridge Converts'." Clergy Review 22, No. 6 (June 1942), 241-249.

4102. GWYNN, DENIS ROLLESTON. "The Famine and the Church in England." Irish Ecclesiastical Record 69, No. 10 (Oct. 1947), 896-909. (Impact of Irish immigration on the Roman Catholic Church in England)

4103. GWYNN, DENIS ROLLESTON. Father Dominic Barberi. London: Burns, Oates, 1947.

4104. GWYNN, DENIS ROLLESTON. "Father Dominic Barberi and the English." Month 181, No. 948 (Nov.-Dec. 1945), 384-392.

4105. GWYNN, DENIS ROLLESTON. "Father Gentili, 1801-1848." Irish Ecclesiastical Record 70 (1948), 769-784.

4106.   GWYNN, DENIS ROLLESTON.   Father Luigi Gentili
        and His Mission, 1801-1848.   London: Burns and
        Oates, 1951.
4107.   GWYNN, DENIS ROLLESTON.   "Fr Paul Pakenham,
        Passionist."   Clergy Review 42 (July 1957),
        400-419.
4108.   GWYNN, DENIS ROLLESTON.   "Heralds of the Second
        Spring."   Clergy Review 29 (1948), 94-109, 251-265,
        395-412; 30 (1948), 39-53, 100-118, 176-196, 248-265,
        310-327, 389-406.   (Bishop Baines, Ambrose Phillipps,
        Father Gentili, Augustus Pugin, Bishop Walsh,
        Father Dominic Barberi, Cardinal Acton, Frederick
        Lucas, Bishop Ullathorne)
4109.   GWYNN, DENIS ROLLESTON.   A Hundred Years of Catholic
        Emancipation, 1829-1929.   London, New York: Longmans,
        Green and Co., 1929.
4110.   GWYNN, DENIS ROLLESTON.   "Lingard and Cardinal
        Wiseman."   Clergy Review 35 (1951), 372-386.
4111.   GWYNN, DENIS ROLLESTON.   Lord Shrewsbury, Pugin,
        and the Catholic Revival.   London: Hollis & Carter,
        1946.
4112.   GWYNN, DENIS ROLLESTON.   "Newman, Wiseman and Dr.
        Russell."   Irish Ecclesiastical Record 58 (Sept.
        1941), 275-286.   (Charles William Russell)
4113.   GWYNN, DENIS ROLLESTON.   "The Paradox of Wiseman."
        Clergy Review 34 (Sept. 1950), 187-202.
4114.   GWYNN, DENIS ROLLESTON.   "The Radicals and Emanci-
        pation."   Dublin Review 184 (Apr. 1929), 258-267.
4115.   GWYNN, DENIS ROLLESTON.   The Second Spring, 1818-
        1852:   A Study of the Catholic Revival in England.
        London: Burns, Oates, 1942.
4116.   GWYNN, DENIS ROLLESTON.   The Struggle for Catholic
        Emancipation (1750-1829).   London, New York:
        Longmans, Green and Co., 1928.
4117.   GWYNN, DENIS ROLLESTON.   "Sydney Smith and Catholic
        Emancipation."   Edinburgh Review 249 (1929), 232-246.

4118.   GWYNN, DENIS ROLLESTON.   "The Venerable Dominic and
        London."   Westminster Cathedral Chronicle 41, No. 2
        (Feb. 1947), 35-37.   (Dominic Barberi, 1840-49)

4119.   GWYNN, DENIS ROLLESTON.   "Wiseman and Daniel O'Connell."
        Clergy Review 28 (1947), 361-373.

4120.   GWYNN, DENIS ROLLESTON.   "Wiseman's Return to
        England in 1840."   Clergy Review 19 (Sept. 1940),
        189-204.

4121.   HALES, E. E. Y.   Revolution and Papacy, 1769-1846.
        Garden City, N.Y.: Hanover House, 1960.

4122.   HANLEY, MARY INEZ, R.S.M., SISTER.   "Religious
        Symbolism in the Poetry of the Catholic Revival."
        Ph.D. Dissertation, Marquette University, 1944.

4123.   HANRAHAN, N.   "Cardinal Vaughan and the Secular
        Clergy."   Clergy Review 46 (1961), 715-733.

4124.   HARRINGTON, HENRY.   "Catholic Emancipation in
        England."   Thought 4 (1929), 480-499.

4125.   HEENAN, JOHN CARMEL.   Cardinal Hinsley.   London:
        Burns, Oates, 1944.

4126.   HEGARTY, W. J.   "Gladstone's Attitude to Catholicism."
        Irish Ecclesiastical Record 86 (1956), 26-42.

4127.   HEGARTY, W. J.   "Was Lingard a Cardinal?"   Irish
        Ecclesiastical Record 79 (1953), 81-93.

4128.   HERRMANN, IRMGARD.   Benjamin Disraelis Stellung
        zur katholischen Kirche.   Schramberg (Württemberg):
        Gatzer & Hahn, 1932.

4129.   HESELTINE, GEORGE COULEHAN.   The English Cardinals.
        London: Burns, Oates and Washbourne, Ltd., 1931.

4130.   HEXTER, J. H.   "The Protestant Revival and the
        Catholic Question in England, 1778-1829."   Journal
        of Modern History 8 (1936), 297-319.

4131.   HICKEY, JOHN.   Urban Catholics: Urban Catholicism
        in England and Wales from 1829 to the Present Day.
        London: G. Chapman, 1967.

4132.   HILL, ROLAND.   "Lord Acton and the Catholic Reviews."
        Blackfriars 36 (1955), 469-482.   (Lord Acton's
        connection with Catholic periodicals)

4133.   HILL, ROLAND.   "Newman und Manning."   Stimmen der Zeit
        150 (Sept. 1952), 435-439.

4134.   HIMMELFARB, GERTRUDE.   Lord Acton: A Study in
        Conscience and Politics.   Chicago: University of
        Chicago Press, 1952.

4135.   HOHL, CLARENCE L., JR.   "Lord Acton and the Vatican
        Council."   Historical Bulletin 28 (Nov. 1949),
        7-11.

4136.   HOLLIS, CHRISTOPHER.   A History of the Jesuits.
        London: Weidenfeld and Nicolson, 1968.

4137.   HOLMES, J. DEREK.   "Cardinal Manning's Letters to
        Father Lawless and Mrs. King."   Downside Review 92
        (Jan. 1974), 19-24.

4138.   HOLMES, J. DEREK.   "The Character of English
        Catholicism: A Legacy of the Nineteenth Century."
        Month 8 (Dec. 1975), 352-354.

4139.   HOLMES, J. DEREK.   "English Catholicism from
        Wiseman to Bourne."   Clergy Review 61 (Feb. 1976),
        57-69; 61 (Mar. 1976), 107-116.

4140.   HOLMES, J. DEREK.   "Liberal Catholicism and Newman's
        Letter to the Duke of Norfolk."   Clergy Review 60
        (Aug. 1975), 498-511.

4141.   HOLMES, J. DEREK.   "Some Unpublished Passages from
        Cardinal Wiseman's Correspondence."   Downside
        Review 90 (1972), 41-52.

4142.   HOPPEN, K. THEODORE.   "Tories, Catholics, and the
        General Election of 1859."   Historical Journal 13
        (1970), 48-67.

4143.   HOPPEN, K. THEODORE.   "W. G. Ward and Liberal
        Catholicism."   Journal of Ecclesiastical History 23
        (1972), 323-344.

4144.   HOPPEN, K. THEODORE.   "William George Ward and
        Nineteenth-Century Catholicism."   Ph.D. Dissertation,
        University of Cambridge, 1966.

4145.   HOUSE, HUMPHRY.   "A Note on Hopkins' Religious
        Life."   New Verse No. 14 (Apr. 1935), 3-5.

4146. HOUTIN, ALBERT. "Le Père Tyrrell et la Société de
       Jésus." Révolution Française 59 (1910), 122-128.
4147. HUDSON, DEREK. "Lewis Carroll and G .M. Hopkins:
       Clergymen on a Victorian See-Saw." Dalhousie Review
       50 (1970), 83-87.
4148. HUGHES, JOHN JAY. Absolutely Null and Utterly
       Void: The Papal Condemnation of Anglican Orders,
       1896. London: Sheed and Ward, 1968.
4149. HUGHES, JOHN JAY. "The Papal Condemnation of
       Anglican Orders, 1896." Journal of Ecumenical
       Studies 4 (Spring 1967), 235-267.
4150. HUGHES, PHILIP. "Bishops of 1850: Wiseman's
       Colleagues in the Restored Hierarchy." Tablet 196
       (Sept. 23, 1950), 253-255.
4151. HUGHES, PHILIP. A Popular History of the Catholic
       Church. London: Burns & Oates, 1958.
4152. HUGHES, PHILIP. "Uproar of 1850: Who Raised the
       Cry of Papal Aggression?" Tablet 196 (Sept. 30,
       1950), 284.
4153. HURLEY, E. P. "George Tyrrell: A Character Study."
       Queen's Quarterly 35 (1928), 408-420.
4154. HUTTON, EDWARD. Catholicism and English Literature.
       London: Muhler, 1942.
4155. ISAACSON, CHARLES STUTEVILLE. The Story of the
       English Cardinals. London: E. Stock, 1907.
4156. JACKMAN, ARTHUR. "Some Memories of the Cardinal."
       Dublin Review 196 (Apr. 1935), 196-211. (Herbert
       Vaughan)
4157. JANSSENS, A. Anglicaansche bekeerlingen: Newman-
       Faber-Manning-Benson-Knox-Kinsman-Chesterton.
       Anvers-Brux.: Standaard, 1928.
4158. JOHNSON, HUMPHREY JOHN THEWLIS. "Parliament and
       the Restored Hierarchy." Dublin Review No. 448
       (1950), 1-16.
4159. JONES, ANNE MARIAN. "The Theological Vision of
       Gerard Manley Hopkins: Seeing as 'Kenosis'."

Dissertation Abstracts International 36 (1976),
7485A (University of Notre Dame, 1976).

4160.   JOYCE, T. P.   *The Restoration of the Catholic
Hierarchy in England and Wales, 1850: A Study of
Certain Public Reactions*. Rome: Officium Libri
Catholici, 1966.

4161.   KEENAN, MARJORIE GAILHAC, LA MERE.   "La Pensée
religieuse de Coventry Patmore." Ph.D. Dissertation, Université de Paris, 1953.

4162.   KEHOE, MONICA G.   "The Influence of Roman
Catholicism on Francis Thompson's Poetry." Ph.D.
Dissertation, Ohio State, 1936.

4163.   KELLY, BERNARD WILLIAM.   *The Mind and Poetry of
Gerard Manley Hopkins*. Ditchling, Sussex, and
London: Pepler and Sewell, 1935.

4164.   KENYON, RUTH.   "Ideal Ward and a Catholic Sociology."
Christendom: A Journal of Christian Sociology 2,
No. 5 (Mar. 1932), 35-46. (W. G. Ward)

4165.   KERNS, V.   "Wiseman and the 'Second Spring'. A
Centenary." *Irish Ecclesiastical Record* 74 (1950),
321-337.

4166.   KLAUSE, ROBERT JAMES.   "The Pope, the Protestants,
and the Irish: Papal Aggression and Anti-Catholicism
in Mid-Nineteenth-Century England." *Dissertation
Abstracts International* 35 (1974), 371A (University
of Iowa, 1973).

4167.   KNOX, WILFRED LAWRENCE AND ALEXANDER ROPER VIDLER.
The Development of Modern Catholicism. London:
Allan, 1933.

4168.   KRÄMER, KONRAD W.   "Die religiöse Dichtung Francis
Thompsons: Eine Untersuchung der religiösen
Sensibilität des Dichters und Versuch einer
Darstellung seiner Seinkonzeption." Ph.D. Dis-
sertation, Universität Münster, 1954.

4169.   LA GORCE, AGNES DE.   *Francis Thompson*. Trans. H.
F. Kynaston-Snell. London: Burns, Oates and
Washbourne Ltd., 1933.

4170.   LARKIN, EMMET.   "Launching the Counterattack: Part
        II of the Roman Catholic Hierarchy and the Destruc-
        tion of Parnellism."   Review of Politics 28 (1966),
        359-383.

4171.   LARKIN, EMMET.   "Mounting the Counter-attack: The
        Roman Catholic Hierarchy and the Destruction of
        Parnellism."   Review of Politics 25 (Apr. 1963),
        157-182.

4172.   LARKIN, EMMET.   "The Roman Catholic Hierarchy and
        the Fall of Parnell."   Victorian Studies 4 (June
        1961), 315-336.

4173.   LE BUFFE, FRANCIS PETER.   "The Hound of Heaven":
        An Interpretation.   New York: Macmillan Company,
        1921.   (Francis Thompson)

4174.   LEETHAM, CLAUDE RICHARD HARBORD.   "Gentili's
        Reports to Rome."   Wiseman Review 237 (1964),
        395-415.

4175.   LEETHAM, CLAUDE RICHARD HARBORD.   Luigi Gentili:
        A Sower of the Second Spring.   London: Burns and
        Oates, 1965.

4176.   LEITCH, VINCENT BARRY.   "Religious Desolation in
        the Poetry of Southwell, Herbert, Hopkins and
        Eliot."   Dissertation Abstracts International 34
        (1973), 278A (University of Florida, 1972).

4177.   LESLIE, SIR JOHN RANDOLPH SHANE.   Cardinal Gasquet:
        A Memoir.   London: Burns, Oates, 1953.

4178.   LESLIE, SIR JOHN RANDOLPH SHANE.   "Cardinal Manning."
        Month 5 (1951), 134-141.

4179.   LESLIE, SIR JOHN RANDOLPH SHANE.   "Cardinal Manning
        and the London Strike of 1889."   Dublin Review 167
        (1920), 219-231.

4180.   LESLIE, SIR JOHN RANDOLPH SHANE.   "Cardinal Manning
        in Mrs. Crawford's Journals."   Quarterly Review 289
        (1951), 68-83.

4181.   LESLIE, SIR JOHN RANDOLPH SHANE.   Henry Edward
        Manning: His Life and Labours.   London: Burns,
        Oates and Washbourne Limited, 1921.

4182.   LESLIE, SIR JOHN RANDOLPH SHANE. (Ed.) "More
        Letters of Wiseman and Manning." Dublin Review 172
        (1923), 106-129.
4183.   LESLIE, SIR JOHN RANDOLPH SHANE. "Some Birmingham
        Bygones: Illustrated by Letters from the Postbags
        of Ullathorne and Manning." Dublin Review 166
        (1920), 203-221.
4184.   LESLIE, SIR JOHN RANDOLPH SHANE. (Ed.) "Unpub-
        lished Letters of Cardinal Wiseman to Dr. Manning."
        Dublin Review 169 (1921), 161-191.
4185.   LESTER-GARLAND, LESTER VALLIS. The Religious
        Philosophy of Baron F. von Hügel. London: J. M.
        Dent and Sons, Ltd., 1933.
4186.   LEWIS, C. J. "The Disintegration of the Tory-
        Anglican Alliance in the Struggle for Catholic
        Emancipation." Church History 29 (1960), 25-43.
4187.   LILLEY, ALFRED LESLIE. Modernism: A Record and
        a Review. New York: C. Scribner's Sons, 1908.
4188.   LLOINCH, J. "La emancipación del catolicismo en
        Inglaterra." Razón y Fe 87 (1929), 217-232.
4189.   LOISY, ALFRED FIRMIN. George Tyrrell et Henri
        Brémond. Paris: E. Nourry, 1936.
4190.   LUNN, SIR ARNOLD HENRY MOORE. Roman Converts.
        London: Chapman and Hall, 1924. (J. H. Newman, H.
        Manning, G. Tyrrell, Ronald Knox, G. K. Chesterton)
4191.   LYON, JOHN JOSEPH. "Immediate Reactions to Darwin:
        The English Catholic Press' First Reviews of the
        'Origin of the [sic] Species'." Church History 41
        (Mar. 1972), 78-93.
4192.   LYON, JOHN JOSEPH. "The Reaction of English
        Catholics to the Developments in Earth and Life
        Sciences, 1825-1864." Dissertation Abstracts 27
        (1967), 2120A (University of Pittsburgh, 1966).
4193.   MACCAFFREY, JAMES. History of the Catholic Church
        in the Nineteenth Century (1789-1908). Dublin and
        Waterford: M. H. Gill and Son, Ltd.; St. Louis,
        Mo.: B. Herder, 1909.

4194. MCCANN, JUSTIN AND COLUMBA CARY-ELWES. (Eds.)
Ampleforth and Its Origins: Essays on a Living Tradi-
tion by Members of the Ampleforth Community.
London: Burns, Oates and Washbourne, 1952.

4195. MCCLELLAND, VINCENT ALAN. "Archbishop Ullathorne
and Religious Education." Pax 54 (Fall 1964),
124-130.

4196. MCCLELLAND, VINCENT ALAN. Cardinal Manning, His
Public Life and Influence, 1865-1892. London, New
York: Oxford University Press, 1962.

4197. MCCLELLAND, VINCENT ALAN. "The Liberal Training of
England's Catholic Youth: William Joseph Petre
(1847-93) and Educational Reform." Victorian Studies
15 (1972), 257-278.

4198. MCCORMACK, ARTHUR. "Cardinal Vaughan: The IVth
Cardinal." Dublin Review No. 506 (1965), 295-336.

4199. MCCORMACK, ARTHUR. Cardinal Vaughan: The Life
of the Third Archbishop of Westminster. London:
Burns and Oates, 1966.

4200. MACDONAGH, M. Daniel O'Connell and the Story
of Catholic Emancipation. London: Burns, Oates
and Washbourne, 1929.

4201. MACDONAGH, M. "The Eve of Catholic Emancipation."
Catholic World 124 (Jan. 1927), 489-498.

4202. MACDOUGALL, HUGH A. The Acton-Newman Relations:
The Dilemma of Christian Liberalism. New York:
Fordham University Press, 1962.

4203. MACDOUGALL, HUGH A. Lord Acton on Papal Power.
London: Sheed and Ward, 1973.

4204. MCELRATH, DAMIAN. "Richard Simpson and John Henry
Newman: The 'Rambler', Laymen, and Theology."
Catholic Historical Review 52 (Jan. 1967), 509-533.

4205. MCELRATH, DAMIAN. Richard Simpson, 1820-1876:
A Study in XIXth-Century English Liberal Catholicism.
Louvain: Bureau de la R.H.E. Bibliothèque de
l'Université, Publications Universitaires de Louvain,
1972.

4206.   MCELRATH, DAMIAN.   The Syllabus of Pius IX: Some
        Reactions in England.   Louvain: Bibliothèque de
        l'Université, Bureau de la Revue, 1964.
4207.   MCENTEE, GEORGIANA PUTNAM.   The Social Catholic
        Movement in Great Britain.   New York: The Macmillan
        Company, 1927.   (Includes H. Manning)
4208.   MACHIN, G. I. T.   "The Catholic Emancipation Crisis
        of 1825."   English Historical Review 78 (1963),
        458-482.
4209.   MACHIN, G. I. T.   The Catholic Question in English
        Politics 1820 to 1830.   Oxford: Clarendon Press,
        1964.
4210.   MACHIN, G. I. T.   "The Duke of Wellington and
        Catholic Emancipation."   Journal of Ecclesiastical
        History 14 (1963), 190-208.
4211.   MACHIN, G. I. T.   "Lord John Russell and the Prelude
        to the Ecclesiastical Titles Bill, 1846-1851."
        Journal of Ecclesiastical History 25 (1974), 277-295.
4212.   MACHIN, G. I. T.   "The Maynooth Grant, the Dis-
        senters and Disestablishment."   English Historical
        Review 82 (1967), 61-85.
4213.   MACHIN, G. I. T.   "The No-Popery Movement in Britain
        in 1828-1829."   Historical Journal 6 (1963), 193-211.
4214.   MCIVER, A.   "Priest of the Midland District:
        Robert Richmond."   Clergy Review 43 (June 1958),
        348-358.
4215.   MCNABB, V.   "Cardinal Wiseman and Cardinal Mercier
        on Reunion."   Dublin Review 204 (Jan. 1939), 160-173.
4216.   MARTINDALE, CYRIL CHARLES.   Bernard Vaughan, S.J.
        London, New York: Longmans, Green and Co., 1923.
        (Brother of Cardinal Vaughan)
4217.   MARTINDALE, CYRIL CHARLES.   Catholics in Oxford:
        Being a Sketch of Their Struggles and Fortunes from
        the Martyrdom of Edmund Campion in 1581 Down to the
        Present Day.   Oxford: B. Blackwell, 1925.

4218. MARTINDALE, CYRIL CHARLES. "Le Centenaire de
       l'émancipation du catholicisme en Angleterre."
       Etudes publiées par les PP. de la Compagnie de Jésus
       200 (1929), 67-76.

4219. MARTINDALE, CYRIL CHARLES. "Sibyl and Sphinx:
       Newman and Manning in the '80's." Contemporary Review
       138 (1930), 470-479.

4220. MATHEW, DAVID. Catholicism in England, 1535-1935:
       A Portrait of a Minority, Its Culture and Tradition.
       London, New York: Longmans, Green and Co., 1936.

4221. MATHEW, DAVID. Lord Acton and His Times. London:
       Eyre and Spottiswoode, 1968.

4222. MAY, JAMES LEWIS. Father Tyrrell and the Modernist
       Movement. London: Eyre and Spottiswoode, 1932.

4223. MENCZER, BELA. (Ed.) Catholic Political Thought,
       1789-1848: Texts Selected with an Introduction and
       Biographical Notes. London: Burns, Oates, 1952.

4224. MESSENGER, E. C. "Wiseman, the Donatists, and
       Newman: A Dublin Centenary." Dublin Review 205
       (July 1939), 110-119. (Wiseman's article on Tracts
       for the Times)

4225. MICKLEWRIGHT, FREDERICK HENRY AMPHLETT. "G. M.
       Hopkins and Provost Fortescue." Notes and Queries
       197 (1952), 169-172, 174, 365-366.

4226. MIETTA, LUIGI. L'Inghilterra cattolica e la crisi
       anglicana. Milano: Alacer, 1934.

4227. MILBURN, D. "Impressions of an English Bishop at
       the First Vatican Council: Letters of Bishop
       Chadwick of Hexham and Newcastle to the President
       of Ushaw." Wiseman Review No. 493 (Autumn 1962),
       217-235.

4228. MORRISON, JOHN L. "Orestes Brownson and the Catholic
       Reaction to Darwinism." Duquesne Review 6 (Spring
       1961), 75-87.

4229. MOURRET, FERNAND. A History of the Catholic Church.
       Trans. by Rev. Newton Thompson. St. Louis, Mo.,
       and London: B. Herder Book Co., 1930.

4230. MOWAN, O.   "The First Bishop of Plymouth."   Buckfast
       Abbey Chronicle 20 (1950), 127-139.   (George Errington)

4231. MUCKERMAN, H.   Attitude of Catholics to Darwin-
       ism and Evolution.   St. Louis, Mo.:   B. Herder,
       1928.

4232. MURPHY, JOHN F.   "Francis Thompson: A Catholic
       Poet of Nature."   Ph.D. Dissertation, University of
       Ottawa, 1944.

4233. MURRAY, J. M.   "Manning: His Anglican Career and
       Conversion."   Ave Maria 42 (Nov. 16, 1935), 624-626.

4234. NEDONCELLE, MAURICE GUSTAVE.   Baron Friedrich
       von Hügel: A Study of His Life and Thought.
       Trans. Marjorie Vernon.   London, New York:
       Longmans, Green and Co., 1937.

4235. NEWTON, WILFRID DOUGLAS.   Catholic London.   London:
       Robert Hale, 1950.   (From the earliest times to
       1850)

4236. NOACK, ULRICH.   "Liberale Ideen auf dem ersten
       vatikanischen Konzil: Lord Acton in Rom 1869-70."
       Historische Zeitschrift 205 (Aug. 1967), 81-100.

4237. NOACK, ULRICH.   "Lord Acton: Der Lebensweg eines
       Kämpfers für Christentum und Freiheit."   Hochland
       33 Jg., Bd. 1, Heft 5 (Feb. 1936), 385-397.

4238. NORMAN, E. R.   Anti-Catholicism in Victorian England.
       London: Allen and Unwin, 1968.

4239. NORMAN, E. R.   The Catholic Church and Ireland
       in the Age of Rebellion.   London: Longmans, Green
       and Co., Ltd., 1965.

4240. NORMAN, E. R.   "The Maynooth Question of 1845."
       Irish Historical Studies 15 (1967), 407-437.

4241. NORRIS, DAVID.   "Cardinal Wiseman: The Diocesan
       Bishop."   Wiseman Review 237 (1963), 158-167.

4242. NURSER, JOHN S.   "The Religious Conscience in Lord
       Acton's Political Thought."   Journal of the History
       of Ideas 22 (1961), 47-62.

4243.   OATES, AUSTIN.   "Le Progrès du catholicisme dans
        l'empire britannique sous le règne de Victoria."
        Revue Générale 73 (1901), 902-923.

4244.   O'CONNOR, JOHN JOSEPH.   The Catholic Revival in
        England. New York: The Macmillan Co., 1942.
        (1770-1892)

4245.   O'CONOR, REV. JOHN FRANCIS XAVIER, S.J.   A Study
        of Francis Thompson's "Hound of Heaven." New York:
        John Lane Company, 1912.

4246.   O'HERLIHY, TIMOTHY C. M.   Catholic Emancipation
        Reviewed a Century After. Dublin: M. H. Gill and
        Son, Ltd., 1928.

4247.   OLDMEADOW, E. J.   Francis Cardinal Bourne. 2 vols.
        London: Burns, Oates, 1940-44.

4248.   OLNEY, JAMES L.   "George Herbert and Gerard Manley
        Hopkins: A Comparative Study in Two Religious
        Poets." Dissertation Abstracts 25 (1964), 1895-96
        (Columbia University, 1963).

4249.   O'NEILL, JAMES E.   "The British Quarterlies and the
        Religious Question, 1802-1829." Catholic Historical
        Review 52 (Oct. 1966), 350-371. (Catholic
        Emancipation in the Edinburgh Review, the Quarterly
        Review, and the Westminster Review)

4250.   O'ROURKE, JAMES.   "Manning and Newman." Irish
        Ecclesiastical Record 52 (Nov. 1938), 459-469.

4251.   OSBORNE, JOHN W.   "William Cobbett's Role in the
        Catholic Emancipation Crisis, 1823-1829." Catholic
        Historical Review 49 (Oct. 1963), 382-389.

4252.   PAINTING, DAVID E.   "Disraeli and the Roman Catholic
        Church." Quarterly Review 304 (Jan. 1966), 17-25.

4253.   PATERSON, GARY HUME.   "The Place of the Roman
        Catholic Church in the Literature of the Decadence
        in England." Dissertation Abstracts International
        32 (1971), 928A (University of Toronto, 1969).

4254.   PATERSON, GARY HUME.   "The Religious Thought of
        Lionel Johnson." Antigonish Review 13 (Spring
        1973), 95-109.

4255.   PEARSALL, RONALD.   "The Earl of Shaftesbury and the
        Papal Aggression, 1850."   New Blackfriars 52 (Oct.
        1971), 466-471.

4256.   PETRE, MAUDE DOMINICA MARY.   Autobiography and
        Life of George Tyrrell.   2 vols.   New York:
        Longmans, Green and Co.; London: E. Arnold, 1912.

4257.   PETRE, MAUDE DOMINICA MARY. (Ed.)   George Tyrrell's
        Letters.   London: T. F. Unwin, Ltd., 1920.

4258.   PETRE, MAUDE DOMINICA MARY.   Modernism: Its Failure
        and Its Fruits.   London and Edinburgh: T. C. and
        E. C. Jack, Ltd., 1918.   (G. Tyrrell)

4259.   PETRE, MAUDE DOMINICA MARY.   Von Hügel and Tyrrell.
        London: J. M. Dent and Sons, Ltd., 1937.

4260.   PICHON, CHARLES.   "Le 'Cardinal diable', Herbert
        Vaughan, archevêque de Westminster."   Correspondant
        327 (1932), 558-571.

4261.   PICK, JOHN F.   Gerard Manley Hopkins: Priest
        and Poet.   London, New York: Oxford University
        Press, 1942.

4262.   PIUS IX, POPE.   "The Apostolic Letter of Pope Pius
        IX Re-establishing the Catholic Hierarchy in England."
        Clergy Review 34 (1950), 153-160.

4263.   PLANQUE, GABRIEL.   Questions historiques: histoire
        du catholicisme en Angleterre.   Paris: Bloud,
        1909.

4264.   POLLEN, JOHN HUNGERFORD.   "The Recognition of the
        Jesuits in England."   Month 116 (1910), 23-36.

4265.   POLLEN, JOHN HUNGERFORD.   "The Restoration of the
        English Jesuits 1803-1817."   Month 115 (1910),
        585-597.

4266.   POLLEN, JOHN HUNGERFORD.   "An Unobserved Centenary."
        Month 115 (1910), 449-461, 585-597; 116 (1910),
        23-36.   (Restoration of the Society of Jesus in
        England)

4267.   PRINCE, J. F. T.   "The Catholic Emergence: A
        Contemporary Diary: 1823-1840."   Tablet 209
        (Mar. 9, 1957), 225-226.   (Diary of John Prince)

4268. QUATRE-SOLZ DE MAROLLES, VICTOR. Le Cardinal Manning. Préface de Ferdinand Brunetière. Paris: Librairie des Saints-Pères, 1905.

4269. QUIRK, ANASTASIA. "The Movement for Catholic Emancipation." M.A. Dissertation, University of Liverpool, 1913.

4270. RAICO, RALPH. "The Place of Religion in the Liberal Philosophy of Constant, Tocqueville, and Lord Acton." Ph.D. Dissertation, University of Chicago, 1971.

4271. RALLS, WALTER. "The Papal Aggression of 1850: A Study in Victorian Anti-Catholicism." Church History 43 (June 1974), 242-256.

4272. RALLS, WALTER. "The Papal Aggression of 1850: Its Background and Meaning." Dissertation Abstracts 21 (1960), 867-868 (Columbia University, 1960).

4273. RANCHETTI, MICHELE. The Catholic Modernists: A Study of the Religious Reform Movement 1864-1907. Trans. Isabel Quigly. London: Oxford University Press, 1969.

4274. RANDALL, SIR ALEC. "Papal Infallibility and the Politicians." Times Literary Supplement (Sept. 11, 1970), 1001-02.

4275. RATTE, JOHN. Three Modernists: Alfred Loisy, George Tyrrell, William L. Sullivan. New York: Sheed and Ward, 1967.

4276. READ, SIR HERBERT. "Gerard Manley Hopkins." In Phyllis Maude Jones (Ed.), English Critical Essays, Twentieth Century. London: Oxford University Press, H. Milford, 1933, 351-374.

4277. REYNOLDS, ERNEST EDWIN. "Cardinal Bourne: A Centenary Tribute." Wiseman Review 235 (1961), 4-14.

4278. REYNOLDS, ERNEST EDWIN. The Roman Catholic Church in England and Wales: A Short History. Wheathampstead: Anthony Clarke Books, 1973.

4279.   REYNOLDS, ERNEST EDWIN. Three Cardinals: Newman,
        Wiseman, Manning. New York: P. J. Kenedy, 1958.

4280.   REYNOLDS, R. "The Catholic Boys at Ackworth: A
        Footnote to Quaker and Catholic History in the 19th
        Century." Journal of the Friends' Historical Society
        43 (1951), 57-71.

4281.   ROCHE, KENNEDY F. "The Relations of the Catholic
        Church and the State in England and Ireland,
        1800-1852." Historical Studies 3 (1961), 9-24.

4282.   ROOT, JOHN. "Catholics and Science in Mid-Victorian
        England." Dissertation Abstracts International 35
        (1974), 2191A (Indiana University, 1974).

4283.   ROSKELL, MARY FRANCIS. Memoirs of Francis Kerril
        Amherst, D.D., Lord Bishop of Northampton. Ed.
        Henry F. J. Vaughan. London: Art and Book Company,
        1903.

4284.   ROWLEY, NORBERT, F.S.C., BROTHER. "Kenelm H. Digby
        and the English Catholic Literary Revival." Ph.D.
        Dissertation, St. John's University, 1942.

4285.   RYAN, ALVAN SHERMAN. "Catholic Social Thought and
        the Great Victorians." Thought 23 (1948), 641-656.

4286.   RYAN, GUY. "The Acton Circle and the 'Chronicle'
        1867-68." Victorian Periodicals Newsletter 7, No.
        2 (June 1974), 10-24.

4287.   RYDER, CYRIL. Life of Thomas Edward Bridgett:
        Priest of the Congregation of the Most Holy Redeemer,
        With Characteristics from His Writings. London:
        Burns and Oates, 1906.

4288.   SAUER, MARY OF THE ANGELS, SISTER. "Lionel Johnson,
        Catholic Humanist." Ph.D. Dissertation, University
        of Michigan, 1944.

4289.   SCHIEFEN, R. J. "The Organisation and Administra-
        tion of Roman Catholic Dioceses in England and
        Wales in the Mid Nineteenth Century." Ph.D. Dis-
        sertation, University of London, 1970.

4290.   SCHIEFEN, R. J.   "Some Aspects of the Controversy
        between Wiseman and the Westminster Chapter."
        Journal of Ecclesiastical History 21 (Apr. 1970),
        125-148.

4291.   SCHOECK, R. J.   "The Historian as Dissenter: The
        Function of Criticism in Lord Acton's 'Inaugural
        Lecture on the Study of History'."   In C. Robert
        Cole and Michael E. Moody (Eds.), The Dissenting
        Tradition.   Athens: Ohio University Press, 1975,
        262-269.

4292.   SCHOENL, WILLIAM JAMES.   "The Intellectual Crisis
        in English Catholicism, 1890-1907:   Liberals,
        Modernists, and the Vatican."   Dissertation Abstracts
        29 (1969), 2657A (Columbia University, 1968).

4293.   SCHOLFIELD, J. F.   "Herbert, Cardinal Vaughan."
        American Catholic Quarterly Review 36 (Jan. 1911),
        1-17.

4294.   SCOTT, NATHAN A., JR.   "The Poetry and Theology of
        Earth: Reflections on the Testimony of Joseph
        Sittler and Gerard Manley Hopkins."   Journal of
        Religion 54 (1974), 102-118.

4295.   SEITZ, MARGARET LYNN.   "Catholic Symbol and Ritual
        in Minor British Poetry of the Later Nineteenth
        Century."   Dissertation Abstracts International 35
        (1974), 1634A (Arizona State University, 1974).

4296.   SIDNEY, PHILIP.   Modern Rome in Modern England:
        Being Some Account of the Roman Catholic Revival in
        England during the Nineteenth Century.   London:
        Religious Tract Society, 1906.

4297.   SIMPSON, RICHARD.   "The Catholic Church in England
        in 1859."   Downside Review 88 (Apr. 1966), 171-192.

4298.   SINCLAIR, WILLIAM MACDONALD.   "A Glimpse of Manning
        and Wilberforce at Lavington."   Chambers's Journal,
        7th Series, 8 (1918), 49-53.

4299.   SMITH, SYDNEY FENN.   "The Life of Cardinal Vaughan."
        Month 116 (1910), 1-16, 113-127.

4300.   SNEAD-COX, JOHN GEORGE. The Life of Cardinal
        Vaughan. 2 vols. London: Herbert and Daniel,
        1910.

4301.   SOMERVILLE, H. "Disraeli and Catholicism." Month
        159 (1932), 114-124.

4302.   STAM, JOHANNES JACOBUS. George Tyrrell, 1861-1909.
        Utrecht: H. Honig, 1938.

4303.   STAPLETON, MARY WINIFRED FRIDESWIDE. Reminiscences:
        Life Story of a Catholic Victorian Family. By
        "Frideswide" O.S.B. East Bergholt: Abbey Press,
        1938.

4304.   STEINMANN, JEAN. Friedrich von Hügel: sa vie,
        son oeuvre et ses amitiés. Paris: Editions
        Montaigne, 1962.

4305.   STEPHAN, J. "Seventy Years Ago: The Colonization
        of Buckfast Abbey by French Monks." Buckfast
        Abbey Chronicle 22 (1952), 171-181.

4306.   STEUERT, HILARY. "The Catholic Tradition."
        Catholic World 145 (May 1937), 229-231. (Influence
        of Catholic thought on English art and literature)

4307.   STEWART, HERBERT LESLIE. Modernism, Past and
        Present. London: J. Murray, 1932.

4308.   TAYLOR, IDA ASHWORTH. The Cardinal Democrat:
        Henry Edward Manning. London: K. Paul, Trench,
        Trübner and Co., Ltd., 1908.

4309.   THOMAS, ALFRED. Hopkins the Jesuit. London:
        Oxford University Press, 1969.

4310.   THOMAS, GEORGE STEPHEN. "Wordsworth, Scott,
        Coleridge, Southey, and De Quincy on Catholic
        Emancipation, 1800-1828: The Conservative Reaction."
        Dissertation Abstracts 25 (1964), 487-488 (New York
        University, 1963).

4311.   THOMPSON, CYPRIAN. "The Venerable Dominic Barberi."
        Buckfast Abbey Chronicle 15, No. 2 (Summer 1945),
        48-53.

4312. THUREAU-DANGIN, PAUL. Le Cardinal Vaughan. Paris:
       Bloud et Cie, 1911. (First published in Correspondant
       241, 1910, 865-888, 1070-1104)

4313. TONSOR, STEPHEN J. "Lord Acton on Döllinger's
       Historical Theology." Journal of the History of Ideas
       20 (1959), 329-352.

4314. TRACY, CLARENCE R. "Bishop Blougram." Modern
       Language Review 34 (July 1939), 422-425. (R.
       Browning's poem as a comment on J. H. Newman and N.
       Wiseman)

4315. TRISTRAM, HENRY. "Catholic Emancipation, Mr. Peel,
       and the University of Oxford." Cornhill Magazine
       66 (Apr. 1929), 406-418.

4316. ULLATHORNE, WILLIAM BERNARD, ARCHBISHOP. From
       Cabin-Boy to Archbishop: The Autobiography of
       Archbishop Ullathorne: Printed from the Original
       Draft. With an introduction by Shane Leslie.
       London: Burns, Oates, 1941.

4317. USHERWOOD, STEPHEN. "'No Popery' under Queen
       Victoria." History Today 23 (1973), 274-279.

4318. VAUGHAN, HENRY FRANCIS JOHN. (Ed.) Memoirs of
       Francis Kerrill Amherst, D.D., Lord Bishop of
       Northampton. London: Art and Book Co., 1903.

4319. VIDLER, ALEXANDER ROPER. A Century of Social
       Catholicism, 1820-1920. London: S.P.C.K., 1964.

4320. VIDLER, ALEXANDER ROPER. The Modernist Movement
       in the Roman Church. Cambridge: Cambridge
       University Press, 1934.

4321. VIDLER, ALEXANDER ROPER. A Variety of Catholic
       Modernists. London: Cambridge University Press,
       1970.

4322. WADHAM, JULIANA. The Case of Cornelia Connelly.
       London: Collins, 1956.

4323. WARD, BERNARD NICOLAS. The Sequel to Catholic
       Emancipation: The Story of the English Catholics
       Continued Down to the Re-establishment of Their

Hierarchy in 1850. 2 vols. London, New York:
Longmans, Green, and Co., 1915.

4324. WARD, MAISIE. The Wilfrid Wards and the Transition. 2 vols. London: Sheed and Ward, 1934-37.
(See Tristram, Henry, 5925)

4325. WARD, WILFRID PHILIP. "Cardinal Vaughan." Dublin
Review 147 (July-Oct. 1910), 6-24, 217-238. (Also
in Men and Matters. New York, London: Longmans,
Green, and Co., 1914, 201-250)

4326. WARD, WILFRID PHILIP. Life and Times of Cardinal
Wiseman. 2 vols. London, New York: Longmans,
Green, and Co., 1897.

4327. WARD, WILFRID PHILIP. "Some Recollections, 1882-
1887." Life and Letters 10 (1934), 677-690.

4328. WARD, WILFRID PHILIP. "Ushaw Centenary and English
Catholicism." Dublin Review 143 (Oct. 1908),
217-243.

4329. WATKIN, EDWARD INGRAM. Roman Catholicism in
England: From the Reformation to 1950. London,
New York: Oxford University Press, 1957.

4330. WATT, E. D. "Rome and Lord Acton: A Reinter-
pretation." Review of Politics 28 (Oct. 1966),
493-507.

4331. WEAVER, CATHARINE CAROLIN. "Francis Thompson's
Philosophical Poetry: An Evaluation." Microfilm
Abstracts 11, No. 3 (1951), 691-692 (University of
Michigan, 1951).

4332. WEYAND, NORMAN T. (Ed.) Immortal Diamond: Studies
in Gerard Manley Hopkins. London: Sheed and Ward,
1949.

4333. WHELAN, D. B. "Behind the Scenes of Catholic
Emancipation." Dublin Review 184 (Apr. 1929),
295-328.

4334. WHELAN, JOSEPH P. The Spirituality of Friedrich
von Hügel. London: Collins, 1971.

4335.   WHITE, GERTRUDE M.   "Hopkins' 'God's Grandeur':   A
        Poetic Statement of Christian Doctrine."   Victorian
        Poetry 4 (1966), 284-287.

4336.   WHYTE, JOHN HENRY.   "The Appointment of Catholic
        Bishops in Nineteenth Century Ireland."   Catholic
        Historical Review 48 (Apr. 1962), 12-32.

4337.   WILLIAMSON, CLAUDE. (Ed.)   Great Catholics.   London:
        Nicholson and Watson, Limited, 1938.   (Includes D.
        O'Connell, J. H. Newman, A. Pugin, G. Hopkins, F.
        Thompson, F. Bourne)

4338.   WILSON, JOSEPH ANSELM.   The Life of Bishop Hedley.
        New York: P. J. Kenedy and Sons, 1930.

4339.   WISKEMANN, ELIZABETH M.   "Great Britain and the
        Roman Question (1861-4)."   M.Litt. Dissertation,
        University of Cambridge, 1928.

4340.   WOODRUFF, D.   "The Early Writings of Lord Acton:
        An Introduction."   Dublin Review 226, No. 455
        (1952), 1-25.

4341.   WOODS, C. J.   "Ireland and Anglo-Papal Relations,
        1880-85."   Irish Historical Studies 18 (1972),
        29-60.

4342.   YOUNG, FATHER URBAN. (Ed.)   Dominic Barberi in
        England: A New Series of Letters.   Trans. and Ed.
        Father Urban Young.   With an introduction by Denis
        Rolleston Gwynn.   London: Burns, Oates and Co.,
        1935.

XVII.   Christian Socialism

The meaning of Christian Socialism and the leaders
of the Christian Socialist movement, including John Malcolm
Ludlow, Stewart Headlam, Charles Kingsley, Thomas Hughes,
and Frederick Denison Maurice, are the subjects of this
category. See II. Religion; XIV. Church of England,
particularly 1. General and 3. Broad Church; appropriate
sections under XIX. Varieties of Belief. . . .

4343.   ALLEN, PETER R.   "Christian Socialism and the Broad
        Church Circle." Dalhousie Review 49 (1969-70),
        58-68.

4344.   ALLEN, PETER R.   "F. D. Maurice and J. M. Ludlow:
        A Reassessment of the Leaders of Christian Socialism."
        Victorian Studies 11 (1968), 461-482.

4345.   ALTRINCHAM, JOHN EDWARD POYNDER GRIGG.   Two Anglican
        Essays. London: Secker and Warburg, 1958.
        (Includes F. D. Maurice)

4346.   BACKSTROM, PHILIP N., JR.   Christian Socialism
        and Co-operation in Victorian England: Edward
        Vansittart Neale and the Co-operative Movement.
        London: Croom Helm, 1975.

4347.   BACKSTROM, PHILIP N., JR.   "John Malcolm Forbes
        Ludlow: A Little Known Contributor to the Cause of
        the British Working Man in the 19th Century." Dis-
        sertation Abstracts 21 (1960), 859-860 (Boston Uni-
        versity Graduate School, 1960).

4348.   BACKSTROM, PHILIP N., JR.   "The Practical Side of
        Christian Socialism in Victorian England." Vic-
        torian Studies 6 (1962-63), 305-324.

4349.   BEST, GEOFFREY FRANCIS ANDREW.   Bishop Westcott
        and the Miners. London: Cambridge University
        Press, 1967.

4350.   BETTANY, FREDERICK GEORGE.   Stewart Headlam: A
        Biography.   London: J. Murray, 1926.

4351.   BINYON, GILBERT CLIVE.   The Christian Socialist
        Movement in England: An Introduction to the Study
        of Its History.   London: Society for Promoting
        Christian Knowledge; New York and Toronto: The
        Macmillan Co., 1931.

4352.   BRUNNER, KARL.   "Charles Kingsley als christlich-
        sozialer Dichter."   Anglia 46 (1922), 289-322; 47
        (1923), 1-33.

4353.   BUTLER, BARBARA J.   "Frederick Lewis Donaldson and
        the Christian Socialist Movement."   M.Phil. Dis-
        sertation, University of Leeds, 1970.

4354.   CHRISTENSEN, TORBEN.   Origin and History of
        Christian Socialism, 1848-1854.   Copenhagen:
        Universitetsforlaget I Aarhus, 1962.

4355.   COOMBS, NORMAN RUSSELL.   "The Doctrine of the
        Incarnation and Christian Socialism."   Dissertation
        Abstracts 22 (1961), 851 (University of Wisconsin,
        1961).   (F. D. Maurice, C. Kingsley, S. Headlam)

4356.   DOTTIN, FRANCOISE.   "Chartism and Christian Socialism
        in 'Alton Locke'."   In Janie Teissedou et al.,
        Politics in Literature in the Nineteenth Century.
        Lille: Université de Lille, 1974, 31-59.

4357.   EVANS, I.   "Christian Socialism, Its Rise and
        Development, Its Economic and Social Results, and
        Its Relation to Other Working-class Movements."
        M.A. Dissertation, University College of Wales
        (Aberystwyth), 1912.

4358.   FRASER, F.   "Robert Owen and Christian Socialism."
        Ph.D. Dissertation, University of Edinburgh, 1927.

4359.   HARRINGTON, HENRY RANDOLPH.   "Muscular Christianity:
        A Study of the Development of a Victorian Idea."
        Dissertation Abstracts International 32 (1972),
        5738-39A (Stanford University, 1971).   (Includes C.
        Kingsley, T. Hughes)

4360.   HARTLEY, A. J.   "Christian Socialism and Victorian
        Morality: The Inner Meaning of 'Tom Brown's School
        Days'."   Dalhousie Review 49 (1969), 216-223.

4361.   HARTLEY, A. J.   "Literary Aspects of Christian
        Socialism in the Work of F. D. Maurice and Charles
        Kingsley."   Ph.D. Dissertation, King's College,
        University of London, 1963.

4362.   JONES, PETER D'ALROY.   The Christian Socialist
        Revival: 1877-1914.   Princeton: Princeton
        University Press, 1968.

4363.   LANDIS, B. Y.   "Christian Socialism of 1848."
        Christian Century 65 (July 21, 1948), 730-732.

4364.   MACK, EDWARD CLARENCE AND WALTER H. G. ARMYTAGE.
        Thomas Hughes: The Life of the Author of "Tom
        Brown's School Days."   London: Benn, 1952.

4365.   MARTIN, HUGH. (Ed.)   Christian Social Reformers
        of the Nineteenth Century.   London: Student
        Christian Movement, 1933.

4366.   MASTERMAN, NEVILLE CHARLES.   "Christian Socialists
        of 1848-54."   Theology 73 (Jan. 1970), 15-23.

4367.   MASTERMAN, NEVILLE CHARLES.   "J. M. Ludlow's
        Criticism of F. D. Maurice's Theology."   Theology
        56 (Oct. 1953), 372-378.

4368.   MASTERMAN, NEVILLE CHARLES.   John Malcolm Ludlow:
        The Builder of Christian Socialism.   Cambridge,
        England: University Press, 1963.

4369.   NEWMAN, E. V.   "The Relation of Theology to Social
        Theory and Action in the Christian Social Movement
        in England from 1877 to 1914."   B.Litt. Dissertation,
        University of Oxford, 1936.

4370.   PARKYN, JOYCE.   "The Political Thought of Some of
        the Founders of Christian Socialism."   M.A. Dis-
        sertation, University of London, 1962.

4371.   RAVEN, CHARLES EARLE.   Christian Socialism, 1848-1854.
        London: Macmillan & Co., 1920.

4372.   RECKITT, MAURICE BENINGTON.   For Christ and the
        People: Studies of Four Socialist Priests and
        Prophets of the Church of England between 1870
        and 1930.   London: S.P.C.K., 1968.   (Thomas
        Hancock, S. Headlam, Charles Marson, Conrad Noel)

4373.   RECKITT, MAURICE BENINGTON.   Maurice to Temple:
        A Century of the Social Movement in the Church of
        England.   London: Faber & Faber, 1947.   (Includes
        C. Kingsley)

4374.   SAVILLE, JOHN.   "The Christian Socialists of 1848."
        In John Saville (Ed.), Democracy and the Labour
        Movement: Essays in Honour of Dona Torr.   London:
        Lawrence & Wishart, 1954, 135-159.

4375.   SEDDING, EDWARD HAROLD.   Charles William Stubbs,
        D.D., Fourth Bishop of Truro.   Plymouth: William
        Brendon and Son, Limited, 1914.

4376.   THOMAS, ALAN.   "A 'Missing' Christian Socialist
        Letter."   Notes and Queries 217 (1972), 254-255.
        (Morning Chronicle, Dec. 28, 1849; the letter is
        signed by Charles Mansfield, Archibald Campbell,
        J. Ludlow and T. Hughes)

4377.   VULLIAMY, COLWYN EDWARD.   Charles Kingsley and
        Christian Socialism.   London: The Fabian Society,
        1914.

4378.   WINN, WILLIAM E.   "'Tom Brown's School Days' and
        the Development of 'Muscular Christianity'."
        Church History 29 (Mar. 1960), 64-73.

4379.   WOODWORTH, ARTHUR V.   Christian Socialism in England.
        London: S. Sonnenschein and Co., Limited; New
        York: C. Scribner's Sons, 1903.

XVIII. Spiritualism, Theosophy, Psychical Research

     This category, which was only casually investigated,
deals with the varieties of spiritualism, with some atten-
tion given to organizations, practitioners (such as Madame
Helena Petrovna Blavatsky), and scientists who were
involved in psychical research, e.g., William Crookes,
Michael Faraday, F. W. H. Myers, Henry Sidgwick, and
Oliver Lodge. For writings on Annie Besant, see XXI.
Atheism. . . . See also VII. Evolution, 3. Chambers,
Robert and 5. Wallace, Alfred Russel.

4380. ANON. "Faraday and the Psychics." Scientific
        American 232 (Jan. 1975) 52-53.
4381. ANON. "Michael Faraday's Researches in Spiritualism."
        Scientific Monthly 83 (1956), 145-150.
4382. BARRINGTON, M. R. Crookes and the Spirit World:
        A Collection of Writings by or concerning the Work of
        Sir William Crookes, O.M.F.R.S., in the Field of
        Psychical Research. Collected by R. G. Medhurst.
        General introduction by K. M. Goldney. New York:
        Taplinger, 1972.
4383. BAYLEN, JOSEPH O. "W. T. Stead's 'Borderland: A
        Quarterly Review and Index of Psychic Phenomena',
        1893-97." Victorian Periodicals Newsletter No. 4
        (Apr. 1969), 30-35.
4384. BESTERMAN, THEODORE. Crystal-gazing: A Study
        in the History, Distribution, Theory and Practice
        of Scrying. London: W. Rider and Son, Limited,
        1924.
4385. BROAD, CHARLIE DUNBAR. "Henry Sidgwick and Psychical
        Research." In Religion, Philosophy, and Psychical
        Research: Selected Essays. London: Routledge and
        K. Paul, 1953, 86-115.

4386. BROAD, CHARLIE DUNBAR. Lectures on Psychical
       Research: Incorporating the Perrott Lectures Given
       in Cambridge University in 1959 and 1960. London:
       Routledge and K. Paul; New York: Humanities Press,
       1962.

4387. BURTON, JEAN. Heyday of a Wizard: Daniel Home,
       the Medium. New York: Knopf, 1944.

4388. CAMPBELL, JOHN LORNE AND TREVOR HENRY HALL.
       Strange Things: The Story of Fr. Allan McDonald, Ada
       Goodrich Freer, and the Society for Psychical Research's
       Inquiry into Highland Second Sight. London:
       Routledge and K. Paul, 1968.

4389. CARRINGTON, HEREWARD AND NANDOR FODOR. The Story
       of the Poltergeist down the Centuries. London:
       Rider, 1953.

4390. DOYLE, SIR ARTHUR CONAN. The History of Spirit-
       ualism. 2 vols. London, New York: Cassell and
       Company, Ltd., 1926.

4391. GAULD, ALAN. The Founders of Psychical Research.
       New York: Schocken Books, 1968. (Includes H.
       Sidgwick, F. W. H. Myers)

4392. GRIFFITH, FREDA G. The Swedenborg Society, 1810-1960.
       London: Swedenborg Society, 1960.

4393. HALL, TREVOR HENRY. The Spiritualists: The Story
       of Florence Cook and William Crookes. London: G.
       Duckworth, 1962.

4394. HALL, TREVOR HENRY. The Strange Case of Edmund
       Gurney. London: G. Duckworth, 1964.

4395. HILL, JOHN ARTHUR. (Ed.) Letters from Sir Oliver
       Lodge: Psychical, Religious, Scientific and Personal.
       London: Cassell and Company, Ltd., 1932.

4396. JOLLY, W. P. Sir Oliver Lodge: Psychical
       Researcher and Scientist. London: Constable,
       1974.

4397. KINGSLAND, WILLIAM. The Real H. P. Blavatsky.
       London: J. M. Watkins, 1928.

4398.   LANDEFELD, CHARLES S.   "Science and Spiritism in
        the Psychical Research of Oliver Lodge." Synthesis
        (Cambridge) 2, No. 2 (1974), 20-27.
4399.   MACARTHUR, J. S.   "A Believer in the Future Life:
        F. W. H. Myers, 1843-1901." Hibbert Journal 41
        (1943), 122-130.
4400.   MCCABE, JOSEPH. Spiritualism: A Popular History
        from 1847. London: T. F. Unwin, Ltd., 1920.
4401.   MAN OF KENT, A.   "Mark Rutherford among the
        Spiritualists." British Weekly 54 (Apr. 3, 1913)
        13.
4402.   MEDHURST, R. G. AND K. M. GOLDNEY.   "William Crookes
        and the Physical Phenomena of Mediumship."
        Proceedings of the Society for Psychical Research
        54 (1964), 25-157.
4403.   MILLER, JONATHAN.   "Spiritualism: The Victorian
        Cult." Sunday Times Magazine (Jan. 16, 1966),
        26-27.
4404.   NELSON, GEOFFREY K.   Spiritualism and Society.
        London: Routledge and K. Paul, 1969.
4405.   NICOL, FRASER.   "The Founders of the S.P.R."
        Proceedings of the Society for Psychical Research
        55 (1972), 341-367.
4406.   PEARSALL, RONALD. The Table-Rappers. London:
        Michael Joseph, 1972.
4407.   PHELPS, WILLIAM LYON.   "Robert Browning on Spirit-
        ualism." Yale Review 23 (1933), 125-138.
4408.   PODMORE, FRANK. Modern Spiritualism: A History
        and a Criticism. 2 vols. London: Methuen, 1902.
4409.   PORTER, KATHERINE H.   Through A Glass Darkly:
        Spiritualism in the Browning Circle. Lawrence:
        University of Kansas Press, 1958.
4410.   RANSOM, JOSEPHINE MARIA. Madame Blavatsky as Occultist.
        London: Theosophical Publishing House, 1931.
4411.   RANSOM, JOSEPHINE MARIA. A Short History of the
        Theosophical Society. Adyar, Madras, India: Theo-
        sophical Pub. House, 1938.

4412.   ROBERTS, CARL ERIC BECHHOFER. The Mysterious
        Madame: Helena Petrovna Blavatsky; the Life and
        Work of the Founder of the Theosophical Society with
        a Note on Her Successor, Annie Besant. New York:
        Brewer and Warren, Inc., 1931.

4413.   RYAN, CHARLES JAMES. H. P. Blavatsky and the
        Theosophical Movement. Point Loma, Calif.: Theo-
        sophical University Press, 1937.

4414.   SALTER, WILLIAM HENRY. The Society for Psychical
        Research: An Outline of Its History. London:
        Society for Psychical Research, 1948.

4415.   SMITH, WARREN SYLVESTER. The London Heretics,
        1870-1914. London: Constable, 1967. (Includes
        Spiritualism, Catholic Modernism, Positivism,
        Secularism)

4416.   STEIN, ROBERT D. "The Impact of the Psychical
        Research Movement on the Literary Theory and
        Literary Criticism of Frederic W. H. Myers."
        Dissertation Abstracts 29 (1969), 2229-30A
        (Northwestern University, 1968).

4417.   SYMONDS, JOHN. Madame Blavatsky: Medium and
        Magician. London: Odhams Press, 1959.

4418.   WILLIAMS, GERTRUDE LEAVENWORTH (MARVIN). Priestess
        of the Occult: Madame Blavatsky. New York:
        A. A. Knopf, 1946.

4419.   WILSON, DAVID B. "The Thought of Late Victorian
        Physicists: Oliver Lodge's Ethereal Body."
        Victorian Studies 15 (1971) 29-48.

XIX. Varieties of Belief, Doubt, and Unbelief

       To a greater or lesser degree all writers in this
category were engaged in personal struggles of conscience,
or involved in battles with others, on matters related
to religious belief. While the reasons for allocating
separate sections to the majority of these individuals is
obvious, the overall choice may be related more to the
classroom origins of this bibliography than to any
principle of priority (see Introduction). Some of the
writers in this category have been the subjects of a great
volume of commentary, and it has been difficult to choose,
for our purposes, those studies which touch on science or
religion. Where material has been plentiful, we have
included only a few biographies or memoirs; where it has
been sparse, we have tended to list these and whatever
else we could find, even when the relationship of a
particular study to our major theme is somewhat tenuous.
The research for sections 1-23 was more exhaustive than
for the last section ("Others"), which also houses those
authors for whom only one or two items have turned up.
       There are a number of cross-listings within this
category and there is a certain degree of overlap between
this one and others. There are several complementary
categories: III. Ideas . . ., which includes general
and comparative studies dealing with belief and unbelief;
VII. Evolution, 4. Darwin, Charles Robert, which
includes studies on the impact of biology on religion;
VIII. Evolution and . . . Thought, 6. Spencer, Herbert,
which includes studies of Spencer's views on religion
(and other sections in VIII); XXI. Atheism . . .; XXII.
Positivism. For further cross-references, see introductions to individual sections.

1. Arnold, Matthew

See introduction to this category; XIV. Church of
England, 4. Broad Church, 5. Biblical Criticism and Essays
and Reviews (1860).

4420.   AIMEE, SISTER. "The Religious Beliefs of Three
Victorian Poets: Tennyson, Browning, and Arnold,
and Their Influence on English Literature." Ph.D.
Dissertation, University of Ottawa, 1942.

4421.   ALEXANDER, EDWARD. Matthew Arnold and John Stuart
Mill. New York: Columbia University Press, 1965.

4422.   ALLOTT, KENNETH. "Matthew Arnold's Reading-Lists
in Three Early Diaries." Victorian Studies 2
(1958-59), 254-266. (See note by Eugene L. Williamson,
Jr., and reply by Allott, Victorian Studies 3, Mar.
1960, 317-320)

4423.   ANGELL, J. W. "Matthew Arnold's Indebtedness to
Renan's 'Essais de morale et de critique'." Revue
de Littérature Comparée 14 (1934), 714-733.

4424.   ARMYTAGE, WALTER H. G. "Matthew Arnold and T. H.
Huxley: Some New Letters: 1870-80." Review
of English Studies 4 (1953), 346-353.

4425.   BACHEM, ROSE. "Arnold's and Renan's Views on
Perfection." Revue de Littérature Comparée 41
(1967), 228-237.

4426.   BAMBROUGH, RENFORD. Reason, Truth, and God.
London: Methuen and Co., Ltd., 1969.

4427.   BARTON, ROBERT E. "Saving Religion: A Comparison
of Matthew Arnold and George Eliot." Dissertation
Abstracts International 34 (1974), 4240-41A (University of Washington, 1973).

4428.   BATTARBEE, KEITH J. "Secularization of Christianity
in the Work of Matthew Arnold." Ph.D. Dissertation,
University of Cambridge, 1971.

4429. BLACKBURN, WILLIAM MAXWELL. "The Background of
       Arnold's 'Literature and Dogma'." Modern Philology
       43 (1945), 130-139.

4430. BLACKBURN, WILLIAM MAXWELL. "Bishop Butler and the
       Design of Arnold's 'Literature and Dogma'." Modern
       Language Quarterly 9 (1948), 199-207.

4431. BLACKBURN, WILLIAM MAXWELL. "Matthew Arnold and
       the Oriel Noetics." Philological Quarterly 25
       (1946), 70-78.

4432. BLACKBURN, WILLIAM MAXWELL. "Matthew Arnold's
       'Literature and Dogma': An Essay towards a Better
       Appreciation of the Bible." Ph.D. Dissertation,
       Yale University, 1943.

4433. BURGUM, EDWIN B. "The Humanism of Matthew Arnold."
       Symposium 2 (1931), 85-112.

4434. BUTTS, DENIS. "Newman's Influence on Matthew
       Arnold's Theory of Poetry." Notes and Queries 203
       (1958), 255-256.

4435. CAMPBELL, H. M. "Arnold's Religion and the Theory
       of Fictions." Religion in Life 36 (Summer 1967),
       223-232.

4436. CHAMBERS, SIR EDMUND KERCHEVER. Matthew Arnold.
       London: H. Milford, 1932.

4437. CHERRY, DOUGLAS. "The Two Cultures of Matthew
       Arnold and T. H. Huxley." Wascana Review 1 (1966),
       53-61.

4438. CONNELL, WILLIAM FRASER. The Educational Thought
       and Influence of Matthew Arnold. London: Routledge
       and K. Paul, 1950.

4439. COOPER, F. B. K. "The Civil Theology of Matthew
       Arnold." Eglise et Théologie 6 (1975), 365-385.

4440. COULLING, SIDNEY M. B. "The Background of 'The
       Function of Criticism at the Present Time'."
       Philological Quarterly 42 (1963), 36-54. (Includes
       J. W. Colenso)

4441. COULLING, SIDNEY M. B.   "The Evolution of 'Culture
       and Anarchy'."   Studies in Philology 60 (1963),
       637-668.

4442. COULLING, SIDNEY M. B.   "Renan's Influence on
       Arnold's Literary and Social Criticism."   Florida
       State University Studies 5 (1952), 95-112.

4443. DAWSON, WILLIAM HARBUTT.   Matthew Arnold and His
       Relation to the Thought of Our Time.   New York and
       London: G. P. Putnam's Sons, 1904.

4444. DELAURA, DAVID JOSEPH.   "Matthew Arnold and John
       Henry Newman: The 'Oxford Sentiment' and the
       Religion of the Future."   Texas Studies in Literature
       and Language 6 (1965), 573-702.   (Reprinted in
       Hebrew and Hellene in Victorian England: Newman,
       Arnold and Pater.   Austin, London: University of
       Texas Press, 1969, 5-161)

4445. DELAURA, DAVID JOSEPH.   "Matthew Arnold's Religious
       and Historical Vision."   Dissertation Abstracts 21
       (1961), 3088 (University of Wisconsin, 1960).

4446. DUDLEY, FRED A.   "Matthew Arnold and Science."
       Ph.D. Dissertation, University of Iowa, 1939.

4447. DUDLEY, FRED A.   "Matthew Arnold and Science."
       Publications of the Modern Language Association of
       America 57 (1942), 275-294.

4448. EATON, GERTRUDE MCBRIDE.   "A Critical Study of
       Matthew Arnold's Theological Works."   Dissertation
       Abstracts International 33 (1973), 6868-69A (Uni-
       versity of Pennsylvania, 1972).

4449. FORSYTH, R. A.   "'The Buried Life' - The Contrasting
       Views of Arnold and Clough in the Context of
       Dr. Arnold's Historiography."   ELH: Journal of
       English Literary History 35 (1968), 218-253.

4450. FRANCIS, NELLE TREW.   "The Critical Reception of
       Arnold's Poetry: The Religious Issue."   Dissertation
       Abstracts 22 (1961), 259 (University of Texas,
       1961).

4451. FULLER, E. "Arnold, Newman, and Rossetti." Critic
       45 (Sept. 1904), 273-276.

4452. GARROD, HEATHCOTE WILLIAM. "The Theology of
       Matthew Arnold." Oxford and Cambridge Review No. 6
       (1909), 17-31. (Also in Living Age 264, Feb. 5,
       1910, 349-356)

4453. GOLLIN, RICHARD M. "'Dover Beach': The Background
       of Its Imagery." English Studies 48 (1967), 493-511.

4454. GOODSTEIN, JACK. "Poetry, Religion, and Fact:
       Matthew Arnold." Costerus 1 (1972), 115-122.

4455. GUDAS, FABIAN. "The Debate on Matthew Arnold's
       Religious Writings." Ph.D. Dissertation, University
       of Chicago, 1953.

4456. HANSEN, JOHN A., JR. "Ernest Renan and Matthew
       Arnold: A Study in Nineteenth Century Religious
       Thought." Ph.D. Dissertation, Yale University,
       1947.

4457. HARDING, JOAN NAUNTON. "Renan and Matthew Arnold:
       Two Saddened Searchers." Hibbert Journal 57 (1959),
       361-367.

4458. HARPER, GEORGE MCLEAN. "Matthew Arnold and the
       Zeit-Geist." Virginia Quarterly Review 2 (1926),
       415-431.

4459. HEINEGG, PETER DAMIAN. "Hebrew or Hellene?
       Religious Ambivalence in Heine, Renan, and Arnold."
       Ph.D. Dissertation, Harvard University, 1972.

4460. HICKS, JOHN, ERNEST E. SANDEEN AND ALVAN S. RYAN.
       Critical Studies in Arnold, Emerson, and Newman.
       Iowa City, Ia.: University of Iowa, 1942.

4461. JOHNSON, WENDELL STACY. "Matthew Arnold's Sea of
       Life." Philological Quarterly 31 (1952), 195-207.

4462. KENOSIAN, CHARLES KENNETH. "The Position of Matthew
       Arnold in the Religious Dilemma of His Time."
       Dissertation Abstracts 21 (1960), 897-898 (Boston
       University Graduate School, 1960).

4463.   KORINKO, STEPHEN JOHN. "Matthew Arnold and Biblical
        Higher Criticism." Dissertation Abstracts Inter-
        national 31 (1971), 4124-25A (University of Nebraska,
        1970).

4464.   KROOK, DOROTHEA. "Christian Humanism: Matthew
        Arnold's 'Literature and Dogma'." In Three Traditions
        of Moral Thought. Cambridge: University Press,
        1959, 202-225.

4465.   LOWRY, HOWARD FOSTER. (Ed.) The Letters of Matthew
        Arnold to Arthur Hugh Clough. London and New York:
        Oxford University Press, H. Milford, 1932.

4466.   LUBELL, ALBERT. "Matthew Arnold: Between Two
        Worlds." Modern Language Quarterly 22 (1961),
        248-263. (Worlds of Christian orthodoxy and
        social ideals)

4467.   LYTTLETON, ARTHUR TEMPLE. "The Poetry of Doubt:
        Arnold and Clough." In Modern Poets of Faith,
        Doubt & Paganism and Other Essays. London: Murray,
        1904, 73-105.

4468.   MCCORMICK, MARY ELLEN. "Arnold and Clough: A
        Comparative Study of Religious Thought." Masters
        Essay, Cornell University, 1929.

4469.   MACKIE, ALEXANDER. Nature Knowledge in Modern
        Poetry: Being Chapters on Tennyson, Wordsworth,
        Matthew Arnold, and Lowell as Exponents of Nature-
        Study. London: Longmans, Green, and Co., 1906.

4470.   MADDEN, WILLIAM ANTHONY. Matthew Arnold: A Study
        of the Aesthetic Temperament in Victorian England.
        Bloomington: Indiana University Press, 1967.

4471.   MADDEN, WILLIAM ANTHONY. "The Religious and Aesthetic
        Ideas of Matthew Arnold." Dissertation Abstracts
        15 (1955), 182-183 (University of Michigan, 1955).

4472.   MIDDLEBROOK, JONATHAN. "'Resignation', 'Rugby
        Chapel', and Thomas Arnold." Victorian Poetry 8
        (1970), 291-297.

4473.   MOORE, T. STURGE.   "Matthew Arnold."   Essays and
        Studies by Members of the English Association 24
        (1938), 7-27.   (Arnold's theology)

4474.   MOTT, LEWIS FREEMAN.   "Renan and Matthew Arnold."
        Modern Language Notes 33 (1918), 65-73.

4475.   MULLER, HERBERT J.   "Matthew Arnold: A Parable for
        Partisans."   Southern Review 5, No. 3 (1940),
        551-558.

4476.   NEIMAN, FRASER.   Matthew Arnold. New York: Twayne
        Publishers, 1968.

4477.   NEIMAN, FRASER.   "The Zeitgeist of Matthew Arnold."
        PMLA 72 (1957), 977-996.

4478.   NEWTON, J. M.   "Some Notes on Religion, Irreligion
        and Matthew Arnold."   Cambridge Quarterly 4 (Spring
        1969), 115-124.

4479.   NOLAND, RICHARD WELLS.   "Matthew Arnold's Religion
        of the Future."   Dissertation Abstracts International
        30 (1970), 4420-21A (Columbia University, 1969).

4480.   OSBORNE, DAVID GORDON.   "Matthew Arnold, 1843-1849:
        A Study of the Yale Manuscript."   Dissertation
        Abstracts 24 (1963), 2018 (University of Rochester,
        1963).

4481.   PAGE, FREDERICK.   "'Balder Dead'."   Essays and
        Studies by Members of the English Association 28
        (1942), 60-68.   (Arnold and the future of Christianity)

4482.   PAGE, FREDERICK.   "Froude, Kingsley, and Arnold, on
        Newman."   Notes and Queries 184 (Apr. 10, 1943)
        220-221.

4483.   REEVES, PASCHAL.   "'Neither Saint Nor Sophist-Led':
        Matthew Arnold's Christology."   Mississippi Quarterly
        16 (1963), 57-66.

4484.   ROBBINS, WILLIAM M.   The Ethical Idealism of Matthew
        Arnold: A Study of the Nature and Sources of His
        Moral and Religious Ideas. London: W. Heinemann,
        1959.

4485.   ROBBINS, WILLIAM M.   "The Religious Thought of
        Matthew Arnold."   Ph.D. Dissertation, University of
        Toronto, 1942.

4486.   RYALS, CLYDE DE L.   "Arnold's 'Balder Dead'."
        Victorian Poetry 4 (1966), 67-81.   (Arnold and the
        decline of Christian faith)

4487.   SAN JUAN, EPIFANIO, JR.   "Matthew Arnold and the
        Poetics of Belief: Some Implications of 'Literature
        and Dogma'."   Harvard Theological Review 57 (1964),
        97-118.

4488.   SAVORY, JEROLD JAMES.   "Matthew Arnold and the
        Higher Criticism of the Bible: A Study of 'God and
        the Bible' and 'Last Essays on Church and Religion'."
        Dissertation Abstracts International 32 (1971),
        1528A (University of Michigan, 1971).

4489.   SHELTON, H. S.   "Matthew Arnold and the Modern
        Church: A Possible Key to a Difficult Problem."
        Hibbert Journal 44 (1946), 119-124.

4490.   SMALLBONE, JEAN A.   "Matthew Arnold and the Bicen-
        tenary of 1862."   Baptist Quarterly 14 (1952),
        222-226.   (Commemoration of ejection of the clergy,
        1662)

4491.   SMALLBONE, JEAN A.   "Matthew Arnold's Theology."
        Baptist Quarterly 14 (Oct. 1951), 177-182.

4492.   STARZYK, LAWRENCE J.   "Arnold and Carlyle."
        Criticism 12 (1970), 281-300.

4493.   STEVENS, DAVID R.   "Matthew Arnold and Some Anglican
        Divines: The Influence on Arnold of John Smith,
        Thomas Wilson, and Joseph Butler."   Ph.D. Dissertation,
        University of Texas at Austin, 1955.

4494.   STOCKING, GEORGE W., JR.   "Matthew Arnold, E. B. Tylor,
        and the Uses of Invention."   American Anthropologist
        65 (1963), 783-799.   (The definition of "culture")

4495.   SUPER, ROBERT HENRY.   The Time-Spirit of Matthew
        Arnold.   Ann Arbor: University of Michigan Press,
        1970.

4496.   THOMPSON, CHARLES WILLIS. "A Page of Matthew
        Arnold." Catholic World 144 (Feb. 1937), 542-546.
4497.   TOEPFER, RAYMOND GRANT. "The Study of Perfection:
        Matthew Arnold as Religious Humanist." Dissertation
        Abstracts International 32 (1971), 3273-74A (City
        University of New York, 1971).
4498.   TOWNSEND, FRANCIS G. "'Literature and Dogma':
        Matthew Arnold's Letters to George Smith."
        Philological Quarterly 35 (1956), 195-198.
4499.   TOWNSEND, FRANCIS G. "A Neglected Edition of
        Arnold's 'St. Paul and Protestantism'." Review of
        English Studies 5 (1954), 66-69.
4500.   TOWNSEND, FRANCIS G. "The Third Instalment of
        Arnold's 'Literature and Dogma'." Modern Philology
        50 (1953), 195-200.
4501.   TRAWICK, BUCKNER B. "The Sea of Faith and the
        Battle by Night in 'Dover Beach'." PMLA 65 (1950),
        1282-83. (Finds a source in A. H. Clough).
4502.   TRILLING, LIONEL. Matthew Arnold. New York: W.
        W. Norton and Co., 1939.
4503.   UNIKEL, GRAHAM. "The Religious Arnoldism of Mrs.
        Humphry Ward." Ph.D. Dissertation, University of
        California, Berkeley, 1951.
4504.   VOGELER, MARTHA SALMON. "Matthew Arnold and Frederic
        Harrison: The Prophet of Culture and the Prophet
        of Positivism." Studies in English Literature 2
        (1962), 441-462.
4505.   WALLER, JOHN O. "Doctor Arnold's Sermons and
        Matthew Arnold's 'Rugby Chapel'." Studies in English
        Literature 9 (1969), 633-646.
4506.   WALLER, JOHN O. "Matthew and Thomas Arnold:
        Soteriology." Anglican Theological Review 44 (Jan.
        1962), 57-70.
4507.   WEAR, RICHARD ARLEN. "The Pursuit of Folly:
        Matthew Arnold and the Foes of Religion."
        Dissertation Abstracts International 32 (1971),
        2107A (Florida State, 1968).

4508. WHITBY, A. C.   "Matthew Arnold and the Nonconform-
       ists: A Study in Social and Political Attitudes."
       B.Litt. Dissertation, University of Oxford, 1955.
4509. WILLIAMSON, EUGENE LA COSTE.   "Matthew Arnold and
       the Archbishops."   Modern Language Quarterly 24
       (1963), 245-252.
4510. WILLIAMSON, EUGENE LA COSTE.   "Matthew Arnold's
       'Eternal Not Ourselves . . .'."   Modern Language
       Notes 75 (1960), 309-312. (Quotation from Arnold's
       God and the Bible)
4511. WILLIAMSON, EUGENE LA COSTE.   "Words from West-
       minster Abbey: Matthew Arnold and Arthur Stanley."
       Studies in English Literature 11 (1971), 749-761.
4512. WILLS, ANTONY ALDWIN.   "Matthew Arnold's Literary
       and Religious Thought."   Dissertation Abstracts 29
       (1969), 4028A (Stanford University, 1968).
4513. WRAGGE, REV. WALTER.   "The Religion of Matthew
       Arnold."   Hibbert Journal 30 (1932), 504-513.

                2. Balfour, Arthur James

       See introduction to this category; XIV. Church of
England, 1. General.

4514. ANON.   "Lord Balfour and Scientific Thought."
       Contemporary Review 139 (Feb. 1931), 261-264.
4515. BEARDSLEE, CLAUDE GILLETTE.   "Arthur James Balfour's
       Contribution to Philosophy."   Ph.D. Dissertation,
       Brown University, 1931.
4516. DUGDALE, BLANCHE ELIZABETH CAMPBELL (BALFOUR).
       Arthur James Balfour: First Earl of Balfour. 2
       vols. London: Hutchinson & Co., Ltd., 1936.
4517. NAAMANI, ISRAEL T.   "The Theism of Lord Balfour."
       History Today 17 (1967), 660-666.

4518.   RAYLEIGH, LORD.   Lord Balfour in His Relation
        to Science.   Cambridge: Cambridge University
        Press, 1930.
4519.   YOUNG, KENNETH.   Arthur James Balfour: The Happy
        Life of the Politician, Prime Minister, Statesman,
        and Philosopher, 1848-1930.   London: G. Bell &
        Sons, Ltd., 1963.

3.   Butler, Samuel

See introduction to this category.

4520.   ACKLOM, MOREBY.   "Samuel Butler the Third: A
        Theological Rebel."   The Constructive Quarterly 5
        (1917), 182-200.
4521.   BRALLIER, VIRGIL VICTOR.   "Authoritarianism in the
        Religious Experience of Samuel Butler."   Disserta-
        tion Abstracts 17 (1957), 1610 (Boston University
        Graduate School, 1957).
4522.   BREUER, HANS-PETER.   "Samuel Butler's 'The Book of
        the Machines' and the Argument from Design."
        Modern Philology 72 (1975), 365-383.
4523.   BREUER, HANS-PETER.   "Sources of Morality in Butler's
        'Erewhon'."   Victorian Studies 16 (Mar. 1973),
        317-328.
4524.   BRYAN, DANIEL VANCE.   "Samuel Butler: Creative
        Evolution in Literature."   Dissertation Abstracts
        13 (1953), 1191-92 (University of Minnesota, 1953).
4525.   CANNAN, GILBERT.   Samuel Butler: A Critical Study.
        London: M. Secker, 1915.
4526.   CAREY, GLENN O.   "Samuel Butler's Theory of Evolution:
        A Summary."   English Literature in Transition 7
        (1964), 230-233.

4527.   COLEMAN, BRIAN.   "A Bibliography of the Evolu-
        tionary Texts Read by Samuel Butler (1835-1902)."
        Journal of the Society for the Bibliography of
        Natural History 7 (1974), 107-110.

4528.   COLEMAN, BRIAN.   "Samuel Butler, Darwin and Darwinism."
        Journal of the Society for the Bibliography of Natural
        History 7 (1974), 93-105.

4529.   COLEMAN, BRIAN.   "The Writings of Samuel Butler
        with Reference to the Relations between Science and
        Humanism in the Nineteenth Century."   Ph.D. Dissert-
        ation, University of London, 1971.

4530.   DANIELS, R. BALFOUR.   "God and Samuel Butler (1835-
        1902)."   In Martin Shockley (Ed.), 1967 Proceed-
        ings of the Conference of College Teachers of English
        of Texas.   Volume 32.   Lubbock: Texas Tech. Coll.,
        1967, 11-17.

4531.   DONALD, J. R.   "Ydgrunism and the Church."   Quest
        21 (Jan. 1930), 170-177.   (Religion in Butler's
        Erewhon)

4532.   DYSON, A. E.   "The Honest Sceptic: Samuel Butler."
        Listener 68 (1962), 383-384.

4533.   FURBANK, PHILIP NICHOLAS.   Samuel Butler, 1835-1902.
        Cambridge: Cambridge University Press, 1948.

4534.   GRENDON, FELIX.   "Samuel Butler's God."   North
        American Review 20 (1918), 277-288.

4535.   JONES, HENRY FESTING.   Charles Darwin and Samuel
        Butler: A Step towards Reconciliation.   London:
        A. C. Fifield, 1911.

4536.   KNOEPFLMACHER, U. C.   Religious Humanism and the
        Victorian Novel: George Eliot, Walter Pater,
        and Samuel Butler.   Princeton, N.J.:   Princeton
        University Press, 1965.

4537.   LOTHAMER, EILEEN EVELYN.   "The Religious Evolution
        of Samuel Butler."   Dissertation Abstracts 25
        (1965), 5260 (University of California, Los Angeles,
        1964).

4538. MACDONALD, W. L. "Samuel Butler and Evolution."
North American Review 223 (1926), 626-637.

4539. MUDFORD, PETER G. "The Impact of the Theory of
Evolution on the Late Nineteenth Century, with
Special Reference to Thomas Henry Huxley and Samuel
Butler." B.Litt. Thesis, University of Oxford,
1966.

4540. MUGGERIDGE, MALCOLM. The Earnest Atheist: A
Study of Samuel Butler. London: Eyre and Spottis-
woode, 1936.

4541. RATTRAY, PAUL EDWARDS. "The Philosophy of Samuel
Butler." Mind 23 (1914), 371-385.

4542. RATTRAY, ROBERT FLEMING. Samuel Butler: A Chronicle
and an Introduction. London: Duckworth, 1935.

4543. SHORB, ELLIS. "Samuel Butler's Concept of a Vital
Principle and Its Use in His Novels." Ph.D. Dissertation, University of North Carolina at Chapel Hill,
1956.

4544. STAFF, RUDOLF. "Die Philosophie des Organischen
bei Samuel Butler." Dissertation, Universität
München, 1932.

4545. STILLMAN, CLARA GRUENING. Samuel Butler: A Mid-
Victorian Modern. New York: The Viking Press,
1932.

4546. WEE, DAVID LUTHER. "The Forms of Apostasy: The
Rejection of Orthodox Christianity in the British
Novel, 1880-1900." Dissertation Abstracts 28
(1967), 205-206A (Stanford University, 1967).
(Includes W. H. White, O. Schreiner, Mrs. Humphry
Ward)

4547. WILLEY, BASIL. Darwin and Butler: Two Versions
of Evolution. London: Chatto and Windus, 1960.

4548. WOLF, P. PLACIDUS J. "Die sittlich-religiöse
Persönlichkeit Samuel Butlers (1835-1902)." Ph.D.
Dissertation, Universität Graz, 1952.

4. Carlyle, Thomas

See introduction to this category.

4549. BARNES, SAMUEL G. "Formula for Faith: The New-
        tonian Pattern in the Transcendentalism of Thomas
        Carlyle." Ph.D. Dissertation, University of North
        Carolina at Chapel Hill, 1953.

4550. BEHNKEN, ELOISE MARJORIE. "Thomas Carlyle:
        'Calvinist without the Theology'." Dissertation
        Abstracts International 33 (1972), 2360A (Florida
        State University, 1972).

4551. BLACKBURN, WILLIAM MAXWELL. "Carlyle and the
        Composition of 'The Life of John Sterling'."
        Studies in Philology 44 (1947), 672-687.

4552. CALDER, GRACE J. "Carlyle and 'Irving's London
        Circle': Some Unpublished Letters by Thomas Carlyle
        and Mrs. Edward Strachey." PMLA 69 (1954), 1135-49.

4553. CAMPBELL, ROBERT ALLAN. "Victorian Pegasus in
        Harness: A Study of Charles Kingsley's Debt to
        Thomas Carlyle and F. D. Maurice." Dissertation
        Abstracts International 30 (1970), 3902-03A
        (University of Wisconsin, 1969).

4554. COLEMAN, A. M. "Keble: A Phrase from Carlyle."
        Notes and Queries 175 (Sept. 3, 1938), 173.

4555. ELANDER, P. H. "Thomas Carlyles 'religiöse Krise'
        und deren Darstellung im selbstbiographischen
        Roman." Studia Neophilologica 15 (1942-43), 49-70.

4556. FLETCHER, JEFFERSON B. "Newman and Carlyle."
        Atlantic Monthly 95 (1905), 669-679.

4557. GREY, WILHELM. "Carlyle und das Puritanertum."
        Ph.D. Dissertation, Universität Halle, 1937.

4558. GWILLIAM, STANFORD. "Thomas Carlyle, Reluctant
        Calvinist." Dissertation Abstracts 26 (1966), 4628
        (Columbia University, 1965).

4559. HARROLD, CHARLES FREDERICK. "The Nature of Carlyle's
      Calvinism." Studies in Philology 33 (1936), 475-486.
4560. HUXLEY, LEONARD. "Carlyle and Huxley." Cornhill
      Magazine 72 (1932), 290-302.
4561. IKELER, A. ABBOTT. Puritan Temper and Transcen-
      dental Faith: Carlyle's Literary Vision. Columbus:
      Ohio State Press, 1972.
4562. IRVINE, WILLIAM. "Carlyle and T. H. Huxley." In
      Hill Shine (Ed.), Booker Memorial Studies: Eight
      Essays in Victorian Literature. Chapel Hill: Uni-
      versity of North Carolina Press, 1950, 104-121.
      (Reprinted in Austin Wright, Ed., Victorian Liter-
      ature: Modern Essays in Criticism. New York:
      Oxford University Press, 1961, 193-207)
4563. JACOBUSSE, K. DON. "Carlyle's Christianity."
      Dissertation Abstracts International 34 (1973),
      1915A (University of Michigan, 1973).
4564. KIRBY, THOMAS A. "Carlyle and Irving." ELH: A
      Magazine of English Literary History 13, No. 1
      (Mar. 1946), 59-63.
4565. KLEIN, ALBERT. "Die Weltanschauung Carlyles."
      Neue Jahrbücher für das klassische Altertum 38
      (1916), 241-260.
4566. LEHMANN, EDVARD. "Die Religion Thomas Carlyles."
      Deutsche Rundschau 143 (1910), 214-226.
4567. LINDLEY, DWIGHT N. "The Saint-Simonians, Carlyle,
      and Mill: A Study in the History of Ideas."
      Dissertation Abstracts 19 (1958), 320 (Columbia
      University, 1958).
4568. MARWICK, WILLIAM H. "Carlyle and Quakerism."
      Friends' Quarterly 16 (1968), 37-45.
4569. MEYER, MARIA. Carlyles Einfluss auf Kingsley
      in sozial-politischer und religiös-ethischer Hinsicht.
      Weimar: R. Wagner, 1914.
4570. MOORE, CARLISLE. "'Sartor Resartus' and the Problem
      of Carlyle's Conversion." PMLA 70 (1955), 662-681.

4571.   PANKHURST, RICHARD KEIR PETHICK.   The Saint Simonians,
        Mill and Carlyle: A Preface to Modern Thought.
        London: Sidgwick and Jackson, 1957.
4572.   PINNOW, JERRY LYLE. "Carlyle's Influence on Charles
        Kingsley's Life and Writings." Dissertation
        Abstracts International 34 (1973), 1865-66A (Uni-
        versity of Colorado, 1973).
4573.   ROBERTS, D. E. "The Religious Thought of Thomas
        Carlyle." Ph.D. Dissertation, University of
        Edinburgh, 1947.
4574.   RYALS, CLYDE DE L. "The 'Heavenly Friend': The
        'New Mythus' of 'In Memoriam'." Personalist 43
        (1962), 383-402. (A. Tennyson's spiritual crisis
        in "In Memoriam" parallels Carlyle's in Sartor
        Resartus)
4575.   SANDERS, CHARLES RICHARD. "The Background of
        Carlyle's Portrait of Coleridge in 'The Life of
        John Sterling'." Bulletin of the John Rylands
        Library 55 (1973), 434-458.
4576.   SANDERS, CHARLES RICHARD. "Carlyle and Tennyson."
        PMLA 76 (1961), 82-97.
4577.   SANDERS, CHARLES RICHARD. "The Question of Carlyle's
        Conversion." Victorian Newsletter No. 10 (1956),
        10-12.
4578.   SHINE, HILL. Carlyle's Fusion of Poetry, History,
        and Religion by 1834. Chapel Hill: The University
        of North Carolina Press, 1938.
4579.   SHINE, HILL. "Carlyle's Fusion of Poetry, History
        and Religion by 1834." Studies in Philology 34
        (1937), 438-466.
4580.   SHINE, HILL. "Carlyle's Views on the Relation
        between Religion and Poetry up to 1832." Studies
        in Philology 33 (1936), 57-92.
4581.   SIMPSON, DWIGHT J. "Carlyle and the Natural Law."
        History of Ideas News Letter 1, No. 3 (1955),
        10-12.

4582.   SINGER, HERMAN RIDDLE.   "Thomas Carlyle's Religion:
        Its Sources, and Its Influence upon His Work."
        Ph.D. Dissertation, University of California,
        Berkeley, 1942.

4583.   STARZYK, LAWRENCE J.   "Arnold and Carlyle."
        Criticism 12 (1970), 281-300.

4584.   STEWART, HERBERT LESLIE.   "Carlyle's Conception of
        Religion."   American Journal of Theology 21 (1917),
        43-57.

4585.   TRISTRAM, HENRY.   "Two Leaders: Newman and
        Carlyle."   Cornhill Magazine 65 (1928), 367-382.

4586.   TURNER, FRANK MILLER.   "Victorian Scientific
        Naturalism and Thomas Carlyle."   Victorian Studies
        18 (1975), 325-343.

4587.   WATT, MARGARET HEWITT.   "Carlyle and Irving: The
        Story of a Friendship."   Scots Magazine 22 (Oct.
        1934-Mar. 1935), 202-208.

4588.   WILLIAMS, S. T.   "Carlyle's 'Life of John Sterling'."
        South Atlantic Quarterly 19 (1920), 341-349.

4589.   WILSON, JOHN ROBERT.   "Religious Existentialism in
        the Early Works of Thomas Carlyle 1820-1832."
        Dissertation Abstracts International 31 (1971),
        6028A (University of Kansas, 1969).

4590.   YANDELL, F. F.   "Thomas Carlyle: sa métaphysique,
        sa morale, sa conception religieuse."   Ph.D. Disserta-
        tion, Université de Lille, 1906.

5.   Clough, Arthur Hugh

See introduction to this category.

4591.   ARMSTRONG, ISOBEL.   Arthur Hugh Clough.   London:
        Longmans & Green, 1962.

4592. BADGER, KINGSBURY. "Arthur Hugh Clough as Dipsychus."
Modern Language Quarterly 12 (1951), 39-56. (Clough's
religious difficulties)

4593. BAKER, CLARENCE POTTER. "A Study of the Influence
of German Higher Criticism on the Poetry of Arthur
Hugh Clough." Ph.D. Dissertation, University of
California, Los Angeles, 1962.

4594. BARISH, EVELYN. "The Morals of Intellect: A Study
of Arthur Hugh Clough's Political and Religious
Prose, 1837-1853." Dissertation Abstracts 27
(1967), 3831A (New York University, 1966).

4595. BAUM, PAULL F. "Clough and Arnold." Modern Language
Notes 67 (1952), 546-547. (Argues that "Dover
Beach" was written after "Say Not the Struggle
Nought Availeth"; see Robertson, David A., Jr.,
below)

4596. BISWAS, ROBINDRA KUMAR. Arthur Hugh Clough: Towards
a Reconsideration. Oxford: Clarendon Press, 1972.

4597. CHORLEY, LADY KATHERINE. Arthur Hugh Clough: The
Uncommitted Mind: A Study of His Life and Poetry.
Oxford and New York: Clarendon Press, 1962.

4598. FORSYTH, R. A. "'The Buried Life': The Contrasting
Views of Arnold and Clough in the Context of
Dr. Arnold's Historiography." ELH: Journal of
English Literary History 35 (1968), 218-253.

4599. GOLLIN, RICHARD M. "Arthur Hugh Clough's Formative
Years: 1819-1841." Dissertation Abstracts 20
(1959), 2276 (University of Minnesota, 1959).

4600. GREENBERGER, EVELYN BARISH. Arthur Hugh Clough:
The Growth of a Poet's Mind. Cambridge, Mass.:
Harvard University Press; London: Oxford University
Press, 1970.

4601. HARRIS, WENDELL V. Arthur Hugh Clough. New York:
Twayne, 1970.

4602. JOHARI, G. P. "Arthur Hugh Clough at Oriel and at
University Hall." PMLA 66 (1951), 405-425.

4603.   JOHNSON, RALPH R.   "The Conflict of Faith and Doubt
        in the Life and Works of Arthur Hugh Clough."
        Ph.D. Dissertation, Drew University, 1948.

4604.   LUCAS, FRANK LAURENCE.   "Clough."   In Eight Victorian
        Poets.   Cambridge: Cambridge University Press,
        1930, 55-74.

4605.   LYTTLETON, ARTHUR TEMPLE.   "The Poetry of Doubt:
        Arnold and Clough."   In Modern Poets of Faith, Doubt,
        and Other Essays.   London: Murray, 1904, 73-105.

4606.   MCCORMICK, MARY ELLEN.   "Arnold and Clough: A
        Comparative Study of Religious Thought."   Masters
        Essay, Cornell University, 1929.

4607.   MULHAUSER, FREDERICK LUDWIG, JR.   "Clough's 'Love
        and Reason'."   Modern Philology 42 (Feb. 1945),
        174-186.

4608.   PALMER, FRANCIS W.   "The Bearing of Science on the
        Thought of Arthur Hugh Clough."   PMLA 59 (Mar.
        1944), 212-225.

4609.   PEAKE, LESLIE SILLMAN.   "A. H. Clough as a Religious
        Teacher."   Modern Churchman 22 (July 1932), 191-199.

4610.   ROBERTSON, DAVID A., JR.   "'Dover Beach' and 'Say
        Not the Struggle Nought Availeth'."   PMLA 66 (1951),
        919-926.   (Argues that Clough's poem is a reply to
        M. Arnold's; see Baum, Paull F., above)

4611.   SCOTT, PATRICK GREIG.   "A. H. Clough: A Case-study
        in Victorian Doubt."   In Derek Baker (Ed.), Heresy,
        Schism and Religious Protest.   London: Cambridge
        University Press, 1972, 383-389.

4612.   TIMKO, MICHAEL.   "Arthur Hugh Clough: A Portrait
        Retouched."   Victorian Newsletter No. 15 (Spring
        1959), 24-28.

4613.   TIMKO, MICHAEL.   "The 'True Creed' of Arthur Hugh
        Clough."   Modern Language Quarterly 21 (1960),
        208-222.   ("True Creed," a phrase from Clough's
        "Easter Day")

4614.   VEYRIRAS, PAUL.   Arthur Hugh Clough, 1819-1861.
        Paris: Didier, 1964.

4615. WILLIAMS, DAVID. Too Quick Despairer: The Life
       and Work of Arthur Hugh Clough. London: Hart-Davis,
       1969.

               6. Coleridge, Samuel Taylor

       See introduction to this category; XIV. Church of
England, 4. Broad Church.

4616. ABRAMS, MEYER HOWARD. "Coleridge's 'A Light in
       Sound': Science, Metascience, and Poetic Imagin-
       ation." Proceedings of the American Philosophical
       Society 116 (1972), 458-476. (Discusses "The
       Eolian Harp" and Coleridge's thoughts on science)
4617. ADAMS, MAURIANNE SCHIFREEN. "Coleridge and the
       Victorians: Studies in the Interpretation of
       Poetry, Scripture, and Myth." Dissertation Abstracts
       28 (1968), 3662-63A (Indiana University, 1967).
4618. BARCUS, JAMES EDGAR. "The Homogeneity of Structure
       and Idea of Coleridge's 'Biographia Literaria',
       'Philosophical Lectures', and 'Aids to Reflection'."
       Dissertation Abstracts 29 (1969), 2205-06A (University
       of Pennsylvania, 1968). (The organic unity of
       Coleridge's literary theory, philosophy, and theology)
4619. BARFIELD, OWEN. What Coleridge Thought. Mid-
       dleton, Conn.: Wesleyan University Press, 1971.
4620. BARNES, S. G. "Was 'Theory of Life' Coleridge's
       Opus Maximum?" Studies in Philology 55 (July
       1958), 494-514.
4621. BARTH, J. ROBERT. Coleridge and Christian Doctrine.
       Cambridge, Mass.: Harvard University Press, 1969.
4622. BARTH, J. ROBERT. "Symbol as Sacrament in Coleridge's
       Thought." Studies in Romanticism 11 (Fall 1972),
       320-331.

4623. BATE, WALTER JACKSON. Coleridge. New York:
       Macmillan, 1968.

4624. BILLINGS, MILDRED KITTO. "The Theology of Horace
       Bushnell Considered in Relation to That of Samuel
       Taylor Coleridge." Ph.D. Dissertation, University
       of Chicago, 1960.

4625. BLUNDEN, EDMUND CHARLES AND EARL LESLIE GRIGGS.
       (Eds.) Coleridge: Studies by Several Hands on
       the Hundredth Anniversary of His Death. London:
       Constable, 1934.

4626. BOULGER, JAMES D. Coleridge as a Religious Thinker.
       New Haven: Yale University Press, 1961.

4627. BROICHER, CHARLOTTE. "Anglikanische Kirche und
       deutsche Philosophie." Preussische Jahrbücher 142
       (1910), 205-233, 457-498. (Includes J. H. Newman)

4628. CARPENTER, JAMES A. "Samuel Taylor Coleridge,
       Theologian." St. Luke's Journal of Theology 16
       (Sept. 1973), 13-21.

4629. CASTLE, WILLIAM RICHARDS. "Newman and Coleridge."
       Sewanee Review 17 (Apr. 1909), 139-152.

4630. CHAMBERS, SIR EDMUND KERCHEVER. Samuel Taylor
       Coleridge: A Biographical Study. Oxford: Clarendon
       Press, 1938.

4631. COBURN, KATHLEEN. "Coleridge, a Bridge between
       Science and Poetry: Reflections on the Bicentenary
       of His Birth." Proceedings of the Royal Institution
       of Great Britain 46 (1973), 45-63.

4632. COOPER, P. "An Unlikely Chemist: Samuel Taylor
       Coleridge." Pharmaceutical Journal and Pharmacist
       189 (1962), 591-592.

4633. COULSON, JOHN. Newman and the Common Tradition.
       Oxford: Clarendon Press, 1970.

4634. DEEN, LEONARD W. "Coleridge and the Radicalism of
       Religious Dissent." Journal of English and Germanic
       Philology 61 (1962), 496-510.

4635. DUNSTAN, A. C. "The German Influence on Coleridge."
       Modern Language Review 17 (1922), 272-281; 18
       (1923), 183-201.

4636. EISELEY, LOREN COREY. "Darwin, Coleridge, and the
       Theory of Unconscious Creation." Library Chronicle
       (University of Pennsylvania) 31 (1965), 7-22. (Re-
       printed in Daedalus, 94, Summer 1965, 588-602)

4637. ELLIOTT, JOHN WESLEY, JR. "A Critical Index to the
       Letters of Samuel Taylor Coleridge from 1785 to
       1801 on the Subjects of 'Philosophy and Religion'
       and 'Literature and Literary Theory': A Present-
       ation of His Thought." Dissertation Abstracts
       International 34 (1974), 6587-88A (Columbia Uni-
       versity, 1971).

4638. FLOTHOW, RUDOLPH CARL. "The Ecclesiastical Polity
       of Richard Hooker and Samuel Taylor Coleridge: A
       Study of the Continuity of Historical Issues."
       Ph.D. Dissertation, University of Southern California,
       1959.

4639. FORSTMAN, H. JACKSON. "Samuel Taylor Coleridge's
       Notes toward the Understanding of Doctrine."
       Journal of Religion 44 (1964), 310-327. (The
       doctrine of election)

4640. FRAPPELL, L. O. "Coleridge and the 'Coleridgeans'
       on Luther." Journal of Religious History 7 (June
       1973), 307-323.

4641. FROTHINGHAM, RICHARD. "The Unitarianism of Samuel
       Taylor Coleridge." Dissertation Abstracts 27
       (1967), 4250A (Columbia University, 1964).

4642. GLOYN, CYRIL KENNARD. "Coleridge's Theory of the
       Church in the Social Order." Church History 3
       (1934), 285-299.

4643. GOETZ, J. "Coleridge, Newman and Tyrrell: A
       Note." Heythrop Journal 14 (Oct. 1973), 431-436.

4644. GOLD, V. "Samuel Taylor Coleridge and the Appoint-
       ment of J. F. Daniell, F.R.S., as Professor of

Chemistry at King's College London." Notes and
Records of the Royal Society of London 28 (1973),
25-29.

4645.   GRUBB, GERALD GILES. "Coleridge the Metaphysician."
Review and Expositor 42, No. 2 (Apr. 1945), 192-212.

4646.   GUTTERIDGE, J. D. "Coleridge and Descartes's
'Meditations'." Notes and Queries 218 (1973),
45-46.

4647.   HAEGER, JACK H. "The Scientific Speculation of
Samuel Taylor Coleridge: Manuscript Transcriptions
and a Commentary." Dissertation Abstracts International 32 (1971), 2642A (University of Washington,
1970).

4648.   HALL, WILLIAM THOMAS. "Coleridge's Religious
Doctrines and Significant Parallels in Calvinism."
Dissertation Abstracts 23 (1962), 224-225 (University
of Texas, 1962).

4649.   HANEY, JOHN LOUIS. The German Influence on Samuel
Taylor Coleridge. Philadelphia: 1902.

4650.   HARTLEY, A. J. "Frederick Denison Maurice, Disciple
and Interpreter of Coleridge: 'Constancy to an
Ideal Object'." Ariel 3, No. 2 (1972), 5-16.

4651.   HAVEN, RICHARD. "Coleridge, Hartley, and the
Mystics." Journal of the History of Ideas 20
(1959), 477-494.

4652.   HENRY, WILLIAM HARLEY. "Coleridge's Meditative
Poems and His Early Religious Thought: The Theology
of 'The Eolian Harp', 'This Lime Tree Bower My
Prison' and 'Frost at Midnight'." Dissertation
Abstracts International 31 (1970), 2345A (Johns
Hopkins University, 1970).

4653.   HOWARD, CLAUD. Coleridge's Idealism: A Study of
Its Relationship to Kant and to the Cambridge
Platonists. Boston: R. G. Badger, 1924.

4654.   KAUVAR, GERALD B. "Coleridge, Hawkesworth, and the
Willing Suspension of Disbelief." Papers on Language
and Literature 5 (1969), 91-94.

4655.   KIRKWOOD, JAMES JOHNSTON.   "Coleridge on Nature."
        Dissertation Abstracts 29 (1969), 3614A (Duke
        University, 1968).

4656.   KNIGHT, DAVID M.   "Davy, Coleridge, and Chemical
        Nomenclature."   Journal of Chemical Education 52
        (1975), 54-55.

4657.   LACEY, PAUL ALVIN.   "Samuel Taylor Coleridge's
        Political and Religious Development: 1795-1810."
        Ph.D. Dissertation, Harvard University, 1966.

4658.   LINK, ARTHUR S.   "Samuel Taylor Coleridge and the
        Economic and Political Crisis in Great Britain,
        1816-1820."   Journal of the History of Ideas 9
        (1948), 323-338.

4659.   LYON, JUDSON S.   "Romantic Psychology and the Inner
        Senses: Coleridge."   PMLA 81 (1966), 246-260.

4660.   MCFARLAND, THOMAS.   Coleridge and the Pantheist
        Tradition.   Oxford: Clarendon, 1969.

4661.   MCKENZIE, GORDON.   Organic Unity in Coleridge.
        Berkeley, California: University of California
        Press, 1939.

4662.   MALE, R.   "The Background of Coleridge's 'Theory of
        Life'."   University of Texas Studies in English 33
        (1954), 60-68.

4663.   MANN, PETER.   "Annotations by Coleridge in a Copy
        of 'The Friend' (1818)."   Studies in Bibliography
        26 (1973), 243-254.   (Coleridge's debt to Count
        Rumford's physico-theological argument)

4664.   MARCEL, GABRIEL.   Coleridge et Schelling.   Paris:
        Aubier-Montaigne, 1971.

4665.   MARCOUX, HERVE.   "The Philosophy of Coleridge."
        Revue de l'Université d'Ottawa, Section Spéciale,
        18 (1948), 38-52, 104-112, 150-169, 235-249.

4666.   MEISSNER, WOLFGANG R.   "Samuel Taylor Coleridge:
        Eine Deutung der relig. Persönlichkeit und ihrer
        Entwicklung."   Ph.D. Dissertation, Universität
        Freiburg, 1954.

4667. METZGER, LORE. "Coleridge's Vindication of Spinoza: An Unpublished Note." Journal of the History of Ideas 21 (1960), 279-293.

4668. MILLER, CRAIG W. "Coleridge's Concept of Nature." Journal of the History of Ideas 25 (1964), 77-96.

4669. MOSSNER, ERNEST C. "Coleridge and Bishop Butler." Philosophical Review 45 (1936), 206-208.

4670. MUIRHEAD, JOHN HENRY. Coleridge as Philosopher. London: G. Allen & Unwin Ltd., 1930.

4671. NEEDHAM, J. "S. T. Coleridge as a Philosophical Biologist." Science Progress 20 (Apr. 1926), 692-702.

4672. NETTESHEIM, JOSEPHINE. "Die Religiöse Umkehr von S. T. Coleridge." Ph.D. Dissertation, Universität Bonn, 1923.

4673. ORSINI, GIAN NAPOLEONE GIORDANO. Coleridge and German Idealism: A Study in the History of Philosophy with Unpublished Materials from Coleridge's Manuscripts. Carbondale: Southern Illinois University Press, 1969.

4674. OWEN, HUW PARRY. "The Theology of Coleridge." Critical Quarterly 4 (1962), 59-67.

4675. PIPER, HERBERT. "Pantheistic Sources of Coleridge's Early Poetry." Journal of the History of Ideas 20 (1959), 47-59.

4676. POTTER, GEORGE REUBEN. "Coleridge and the Idea of Evolution." Publications of the Modern Language Association of America 40 (June 1925), 379-397.

4677. PRICKETT, STEPHEN. "Coleridge, Newman and F. D. Maurice: Development of Doctrine and Growth of Mind." Theology 76 (July 1973), 340-349. (J. H. Newman)

4678. ROBBINS, DEREK M. "Literature and Natural Philosophy, 1770-1800: The Relation between Scientific Systems and Literary Fictions with Special Reference to Joseph Priestly and Samuel Taylor Coleridge." Ph.D. Dissertation, University of Cambridge, 1973.

4679. RULE, PHILIP CHARLES. "Coleridge's Reputation as a
       Religious Thinker: 1816-1972." Harvard Theological
       Review 67 (July 1974), 289-320. (Bibliographical
       survey)

4680. SANDERS, CHARLES RICHARD. Coleridge and the Broad
       Church Movement: Studies in S. T. Coleridge,
       Dr. Arnold of Rugby, J. C. Hare, Thomas Carlyle
       and F. D. Maurice. Durham, N.C.: Duke University
       Press, 1942.

4681. SANDERS, CHARLES RICHARD. "Coleridge as a Champion
       of Liberty." Studies in Philology 32 (1935),
       618-631.

4682. SANDERS, CHARLES RICHARD. "Coleridge, F. D. Maurice,
       and the Distinction between the Reason and the
       Understanding." PMLA 51 (1936), 459-475.

4683. SANDERS, CHARLES RICHARD. "Coleridge, Maurice, and
       the Church Universal." Journal of Religion 21
       (Jan. 1941), 31-45.

4684. SANDERS, CHARLES RICHARD. "Maurice as a Commen-
       tator on Coleridge." PMLA 53 (1938), 230-243.

4685. SANDERS, CHARLES RICHARD. "The Relation of Frederick
       Denison Maurice to Coleridge." Ph.D. Dissertation,
       University of Chicago, 1934.

4686. SANDERS, CHARLES RICHARD. "Sir Leslie Stephen,
       Coleridge, and Two Coleridgeans." PMLA 55 (1940),
       795-801. (F. D. Maurice, J. Dykes Campbell)

4687. SANDERSON, DAVID R. "Coleridge's Political 'Sermons':
       Discursive Language and the Voice of God." Modern
       Philology 70 (May 1973), 319-330.

4688. SANKEY, BENJAMIN. "Coleridge and the Visible
       World." Texas Studies in Literature and Language 6
       (1964), 59-67.

4689. SCHRICKX, W. "Coleridge and the Cambridge Platonists."
       Review of English Literature 7, No. 1 (1966),
       71-91.

4690.   SHAFFER, ELINOR S.   "Coleridge and Natural Philosophy:
        A Review of Recent Literary and Historical Research."
        History of Science 12 (1974), 284-298.
4691.   SHAFFER, ELINOR S.   "Metaphysics of Culture: Kant
        and Coleridge's 'Aids to Reflection'." Journal
        of the History of Ideas 31 (1970), 199-218.
4692.   SNYDER, A. D.   "Coleridge's 'Theory of Life'."
        Modern Language Notes 47 (May 1932), 299-301.
4693.   STEPHENSON, H. W.   "Unitarianism." Transactions
        of the Unitarian Historical Society 5, No. 2 (1932),
        165-184.
4694.   STEWART, HERBERT LESLIE.   "The Place of Coleridge
        in English Theology." Harvard Theological Review
        11 (1918), 1-31.
4695.   STROEBER, RUDOLF.   "Die Idee der Kirche von Coleridge
        bis Newman." Ph.D. Dissertation, Universität
        Erlangen, 1952.
4696.   STUART, J. A.   "Augustine and the Ancient Mariner."
        Modern Language Notes 76 (Feb. 1961), 116-120.
4697.   STUART, J. A.   "Augustinian Cause of Action in
        Coleridge's 'Rime of the Ancient Mariner'." Harvard
        Theological Review 60 (Apr. 1967), 177-211.
4698.   TAYLOR, HARRY M.   "Coleridge's Statement and Defense
        of the Evangelical Faith as Ultimate Metaphysics."
        Ph.D. Dissertation, Drew University, 1938.
4699.   WALSH, W.   "Coleridge and Education." Journal
        of Education (London) 88 (May 1956), 201-203.
4700.   WAPLES, DOROTHY.   "David Hartley in 'The Ancient
        Mariner'." Journal of English and Germanic Philology
        35 (July 1936), 337-351.
4701.   WARE, MALCOLM.   "'Rime of the Ancient Mariner': A
        Discourse on Prayer?" Review of English Studies 11
        (Aug. 1960), 303-304.
4702.   WELLS, GEORGE A.   "Coleridge and Goethe on Scien-
        tific Method in the Light of Some Unpublished
        Coleridge Marginalia." German Life and Letters 4
        (Jan. 1951), 101-114.

4703.   WELLS, GEORGE A.   "Man and Nature: An Elucidation
        of Coleridge's Rejection of Herder's Thought."
        Journal of English and Germanic Philology 51 (1952),
        314-325.

4704.   WERKMEISTER, LUCYLE.   "Coleridge on Science, Philos-
        ophy, and Poetry: Their Relation to Religion."
        Harvard Theological Review 52 (1959), 85-118.

4705.   WERKMEISTER, LUCYLE.   "Early Coleridge: His Rage
        for Metaphysics."   Harvard Theological Review 54
        (Apr. 1961), 99-123.

4706.   WILLEY, BASIL.   "Coleridge and Religion."   In R. L.
        Brett (Ed.), Writers and Their Background: Samuel
        Taylor Coleridge.   Athens: Ohio University Press,
        1972, 221-243.

4707.   WILLOUGHBY, L. A.   "Coleridge and His German Contemp-
        oraries."   Publications of the English Goethe
        Society 10 (1934), 43-62.

4708.   WILLOUGHBY, L. A.   "Coleridge as a Philologist."
        Modern Language Review 31 (1936), 176-201.

4709.   WILLOUGHBY, L. A.   "Coleridge und Deutschland."
        Germanisch-Romanische Monatsschrift 24 (1936),
        112-127.

4710.   WINKELMANN, ELISABETH.   Coleridge und die Kantische
        Philosophie: Erste Einwirkungen des deutschen
        Idealismus in England.   Leipzig: Mayer und Müller,
        1933.

4711.   WOJCIK, MANFRED.   "Coleridge and the Problem of
        Transcendentalism."   Zeitschrift für Anglistik und
        Amerikanistik (East Berlin) 18 (1970), 30-58.

4712.   WRIGHT, D.   "Coleridge, Opium and Theology."
        Open Court 38 (Jan. 1924), 37-45.

7.   Eliot, George

See introduction to this category; XX.   The Agnostic
Novel; XXII.   Positivism.

4713.   ANDERSON, ROLAND FRANK.   "Formative Influences on
         George Eliot with Special Reference to George Henry
         Lewes."   Dissertation Abstracts 25 (1964), 1205-06
         (University of Toronto, 1963).
4714.   ANON.   "George Eliot as a Religious Factor."
         Literary Digest 63 (Dec. 20, 1919), 36.
4715.   APPLEMAN, PHILIP.   "The Dread Factor: Eliot,
         Tennyson, and the Shaping of Science."   Columbia
         Forum 3 (1974), 32-38.
4716.   BAKER, WILLIAM J.   "The Kabbalah, Mordecai, and
         George Eliot's Religion of Humanity."   Yearbook of
         English Studies 3 (1973), 216-221.
4717.   BARTON, ROBERT E.   "Saving Religion: A Comparison
         of Matthew Arnold and George Eliot."   Dissertation
         Abstracts International 34 (1974), 4240-41A (Uni-
         versity of Washington, 1973).
4718.   BENNETT, JOAN.   George Eliot, Her Mind and Her Art.
         Cambridge, England: Cambridge University Press,
         1948.
4719.   BRAMLEY, J. A.   "Religion and the Novelists."
         Contemporary Review 180 (1951), 348-453.   (Includes
         T. Hardy)
4720.   BREMOND, HENRI.   "La Religion de George Eliot."
         Revue des Deux Mondes, 5th Series, 36 (1906),
         787-822.
4721.   BULLETT, GERALD.   George Eliot.   London: Collins,
         1947.
4722.   CHANDER, JAGDISH.   "Religious and Moral Ideas in
         the Novels of George Eliot."   Dissertation Abstracts
         24 (1964), 2905 (University of Wisconsin, 1963).

4723. CLARK, ISABEL C. Six Portraits. London: Hutchinson,
       1935. (Includes Eliot and Mrs. Oliphant considered
       from a religious point of view)

4724. DAVIS, J. S. "George Eliot and Education."
       Educational Forum 1 (Jan. 1937), 201-206.

4725. FLEISSNER, ROBERT F. "'Middlemarch' and Idealism:
       Newman, Naumann, Goethe." Greyfriar 14 (1973),
       21-26. (Adolph Naumann)

4726. FORRESTER, J. M. "George Eliot and Physiology."
       Proceedings of the Royal Society of Medicine 64
       (1971), 724-726.

4727. GEIBEL, JAMES WAYNE. "An Annotated Bibliography of
       British Criticism of George Eliot, 1858-1900."
       Dissertation Abstracts International 30 (1970),
       4450A (Ohio State University, 1969).

4728. HAIGHT, GORDON SHERMAN. George Eliot: A Biography.
       London: Clarendon Press, 1968.

4729. HAIGHT, GORDON SHERMAN. (Ed.) The George Eliot
       Letters. 7 vols. New Haven: Yale University
       Press, 1954-56.

4730. HARRIS, MASON D., JR. "George Eliot and the
       Problems of Agnosticism: A Study of Philosophical
       Psychology." Dissertation Abstracts International
       32 (1971), 1513A (State University of New York,
       Buffalo, 1971).

4731. HOUSE, HUMPHRY. "Qualities of George Eliot's
       Unbelief." Listener (Mar. 25, 1948), 496-498.

4732. HUDSON, STEWART M. "George Henry Lewes' Evolutionism
       in the Fiction of George Eliot." Dissertation
       Abstracts International 31 (1971), 6059A (University
       of Southern California, 1970).

4733. JOHNSON, M. L. "George Combe and George Eliot."
       Westminster Review 166 (Nov. 1906), 557-568.

4734. JONES, JESSE C. "The Use of the Bible in George
       Eliot's Fiction." Dissertation Abstracts Inter-
       national 36 (1975), 2846A (North Texas State
       University, 1975).

4735.   JUNG, WALTER. "Der Einfluss des Positivismus auf
        George Eliot." Ph.D. Dissertation, Universität
        Leipzig, 1923.

4736.   KAMINSKY, ALICE R. "George Eliot, George Henry
        Lewes, and the Novel." PMLA 70 (1955), 977-1013.

4737.   KITCHEL, ANNA THERESA. George Lewes and George
        Eliot. New York: John Day Company, 1933.

4738.   KNOEPFLMACHER, U. C. "George Eliot, Feuerbach, and
        the Question of Criticism." Victorian Studies 3
        (1964), 306-309.

4739.   KNOEPFLMACHER, U. C. Religious Humanism and the
        Victorian Novel: George Eliot, Walter Pater, and
        Samuel Butler. Princeton, N.J.: Princeton University
        Press, 1965.

4740.   KRIEFALL, LUTHER HARRY. "A Victorian Apocalypse:
        A Study of George Eliot's 'Daniel Deronda' and Its
        Relation to David F. Strauss' 'Das Leben Jesu'."
        Dissertation Abstracts 28 (1967), 234A (University
        of Michigan, 1966).

4741.   LEVINE, GEORGE. "Intelligence as Deception: 'The
        Mill on the Floss'." PMLA 80 (1965), 402-409.
        (Includes L. Feuerbach, A. Comte)

4742.   MAHEU, PLACIDE GUSTAVE. La Pensée religieuse
        et morale de George Eliot: essai d'interprétation.
        Paris: M. Didier, 1958.

4743.   MANSELL, DARREL, JR. "A Note on Hegel and George
        Eliot." Victorian Newsletter No. 27 (1965), 12-15.

4744.   MASON, MICHAEL YORK. "'Middlemarch' and Science:
        Problems of Life and Mind." Review of English Studies
        22 (1971), 151-169. (Includes J.S. Mill, W. Whewell)

4745.   MASTERS, DONALD C. "George Eliot and the Evangel-
        icals." Dalhousie Review 41 (1961-62), 505-512.

4746.   MERTON, EGON STEPHEN. "George Eliot and William
        Hale White." Victorian Newsletter No. 25 (1964),
        13-15.

4747. MOLDSTAD, DAVID FRANKLIN. "Evangelical Influences
      on George Eliot." Summaries of Doctoral Dissert-
      ations, University of Wisconsin 15 (1955), 620-621
      (Ph.D. Dissertation, University of Wisconsin,
      1954).

4748. NEWTON, K. M. "George Eliot, George Henry Lewes,
      and Darwinism." Durham University Journal 66
      (1974), 278-293.

4749. PARIS, BERNARD J. Experiments in Life: George
      Eliot's Quest for Values. Detroit: Wayne State
      University Press, 1965.

4750. PARIS, BERNARD J. "George Eliot and the Higher
      Criticism." Anglia 84 (1966), 59-73.

4751. PARIS, BERNARD J. "George Eliot's Religion of
      Humanity." English Literary History 29 (Dec.
      1962), 418-443.

4752. PARLETT, MATHILDE. "George Eliot and Humanism."
      Studies in Philology 27 (1930), 25-46.

4753. PERZL, ANTON. "Der Methodismus im englischen Roman
      von George Eliot bis Arnold Bennett." Ph.D.
      Dissertation, Universität Wien, 1928.

4754. POND, E. J. "Les Idées morales et religieuses de
      George Eliot." Ph.D. Dissertation, Université de
      Lille, 1927.

4755. ROBERTS, NEIL. George Eliot: Her Beliefs and
      Her Art. London: Elek, 1975.

4756. SCOTT, JAMES F. "George Eliot, Positivism, and the
      Social Vision of 'Middlemarch'." Victorian Studies
      16 (Sept. 1972), 59-76.

4757. SIMPSON, WILLIAM JOHN SPARROW. "The Religion of
      George Eliot." Church Quarterly Review 112, No.
      224 (July 1931), 233-247.

4758. SVAGLIC, MARTIN JAMES. "Religion in the Novels of
      George Eliot." Journal of English and Germanic
      Philology 53 (1954), 145-159.

4759. SZIROTNY, JUNE MARJORIE SKYE. "The Religious
      Background of George Eliot's Novels." Dissertation
      Abstracts 27 (1967), 2547A (Stanford University,
      1966).

4760. TEGNER, INGEBORG E. T. "George Eliot: en studie i
      hennes religiösa och filosofiska utveckling."
      Ph.D. Dissertation, Lunds Universitet, 1929.

4761. THUENTE, DAVID R. "Channels of Feeling: George
      Eliot's Search for the Natural Bases of Religion."
      Dissertation Abstracts International 34 (1974),
      5207-08A (University of Kentucky, 1973).

4762. WOLF, EMILY VAUGHAN. "George Eliot's Liberal
      Menagerie: Natural History, Biology, and Value in
      the Early Novels." Ph.D. Dissertation, Harvard
      University, 1969.

4763. WOLFF, MICHAEL. "Marian Evans to George Eliot:
      The Moral and Intellectual Foundations of Her
      Career." Dissertation Abstracts 19 (1959), 2350
      (Princeton University, 1958).

8.   Froude, James Anthony

See introduction to this category; XIV. Church of
England, 1. General; XX. The Agnostic Novel.

4764. BADGER, KINGSBURY. "The Ordeal of Anthony Froude,
      Protestant Historian." Modern Language Quarterly
      13 (1952) 41-55.

4765. BENNETT, RAYMOND M. (Ed.) "Letters of James Anthony
      Froude." Rutgers Library Journal 11 (1947), 1-15.
      (Six letters to Gen. Gustave Paul Cluseret)

4766. DUNN, WALDO HILARY. James Anthony Froude: A
      Biography 1818-1894. 2 vols. Oxford: Clarendon
      Press, 1961-63.

4767. DUNN, WALDO HILARY. "A Valiant Professorship."
       South Atlantic Quarterly 50 (1951), 519-529.
       (Froude and the Regius Chair of Modern History
       at Oxford)

4768. FISH, ANDREW. "The Reputation of James Anthony
       Froude." Pacific Historical Review 1 (1932),
       179-192.

4769. HARRISON, ARCHIBALD HAROLD WALTER. "Mark Ruther-
       ford and J. A. Froude." London Quarterly and Holborn
       Review 164 (1939), 40-44.

4770. MCCRAW, HARRY WELLS. "Two Novelists of Despair:
       James Anthony Froude and William Hale White."
       Southern Quarterly 13 (1974), 21-51.

4771. MAURER, OSCAR. "Froude and 'Fraser's Magazine',
       1860-1874." University of Texas Studies in English
       28 (1949), 213-243.

4772. NYGARD, MARGARET CRUDEN. "James Anthony Froude's
       Protestantism and the 'History of England'." Ph.D.
       Dissertation, University of California, Berkeley,
       1960.

4773. PAUL, HERBERT WOODFIELD. Life of Froude. London:
       Sir Isaac Pitman & Sons, 1905.

4774. STEWART, HERBERT LESLIE. "James Anthony Froude and
       Anglo-Catholicism." American Journal of Theology
       22 (1918), 253-273.

4775. V., M. E. "An Injustice to Froude." Notes and
       Queries 188 (June 2, 1945), 233.

9. Huxley, Thomas Henry

See introduction to this category; I. Science, 1. General; IV. Education; X. Evolution and Ethics.

4776. AINSWORTH-DAVIS, JAMES RICHARD. Thomas H. Huxley. London: J. M. Dent & Co.; New York: E. P. Dutton & Co., 1907.

4777. ANNAN, NOEL GILROY. "Thomas Henry Huxley." Times Educational Supplement 2086 (May 13, 1955), 476.

4778. ANON. "Apes and Bishops." Scientific American 190 (Mar. 1954), 52.

4779. ANON. "High Victorian Science: The Recipes for Survival of Three Scientific Cinderellas." Times Literary Supplement 71 (Nov. 3, 1972), 1301-02. (Includes M. Faraday)

4780. ANON. "Huxley and Natural Selection." Scientific American Supplement 59 (Apr. 29, 1905), 24515-16.

4781. ANON. "Thomas Huxley and the Victorian Mind." Nation 120 (May 1925), 509-510.

4782. ARMSTRONG, A. MACC. "Samuel Wilberforce vs. T. H. Huxley: A Retrospect." Quarterly Review 296 (1958), 426-437.

4783. ARMSTRONG, H. E. "Huxley's Message in Education." Nature 115 (May 9, 1925), 743-747.

4784. ARMYTAGE, WALTER H. G. "Matthew Arnold and T. H. Huxley: Some New Letters: 1870-80." Review of English Studies 4 (1953), 346-353.

4785. ASHFORTH, ALBERT. "Spokesman for Darwin and for Science." New York Times Magazine (Apr. 7, 1963), 64, 74, 76.

4786. ASHFORTH, ALBERT. Thomas Henry Huxley. New York: Twayne Publishers, 1969.

4787. AYRES, CLARENCE EDWIN. Huxley. New York: Norton, 1932.

4788.   BAKER, WILLIAM J.   "Thomas Huxley in Tennessee."
        South Atlantic Quarterly 73 (Autumn 1974), 475-486.

4789.   BARTHOLOMEW, MICHAEL J.   "Huxley's Defence of
        Darwin."   Annals of Science 32 (Nov. 1975), 525-535.

4790.   BATESON, W.   "Huxley and Evolution."   Nature 115
        (May 9, 1925), 715-717.

4791.   BIBBY, HAROLD CYRIL.   The Essence of T. H. Huxley.
        London: Macmillan, 1967.

4792.   BIBBY, HAROLD CYRIL.   "Huxley and the Reception of
        the 'Origin'."   Victorian Studies 3 (1959-60),
        76-86.

4793.   BIBBY, HAROLD CYRIL.   "The Huxley-Wilberforce
        Debate: A Postscript."   Nature 176 (1955), 363.

4794.   BIBBY, HAROLD CYRIL.   "The Prince of Controver-
        sialists."   Twentieth Century 161 (1957), 268-276.

4795.   BIBBY, HAROLD CYRIL.   Scientist Extraordinary:
        The Life and Scientific Work of Thomas Henry Huxley,
        1825-1895.   Oxford, New York: Pergamon Press,
        1972.

4796.   BIBBY, HAROLD CYRIL.   "T. H. Huxley and Education."
        Journal of Education (London) 87 (Dec. 1955),
        535-538.

4797.   BIBBY, HAROLD CYRIL.   "T. H. Huxley and the
        Universities of Scotland."   Aberdeen University
        Review 37 (Autumn 1957), 134-149.

4798.   BIBBY, HAROLD CYRIL. (Ed.)   T. H. Huxley on
        Education: A Selection from His Writings with an
        Introductory Essay and Notes by Cyril Bibby.
        Cambridge, England: Cambridge University Press,
        1971.

4799.   BIBBY, HAROLD CYRIL.   T. H. Huxley, Scientist,
        Humanist and Educator.   London: Watts, 1959.

4800.   BIBBY, HAROLD CYRIL.   "T. H. Huxley's Idea of a
        University."   Universities Quarterly 10 (1956),
        377-390.

4801.   BIBBY, HAROLD CYRIL.   "Thomas Henry Huxley and
        University Development." Victorian Studies 2
        (1958), 97-116.

4802.   BICKNELL, JOHN W.   "Neologizing." Times Literary
        Supplement (June 29, 1973), 749. (Huxley and the
        origin of the word "agnosticism")

4803.   BLINDERMAN, CHARLES S.   "The Great Bone Case."
        Perspectives in Biology and Medicine 14 (Spring
        1971), 370-393. (R. Owen and Huxley on the brain)

4804.   BLINDERMAN, CHARLES S.   "Huxley and Kingsley."
        Victorian Newsletter No. 20 (Fall 1961), 25-28.

4805.   BLINDERMAN, CHARLES S.   "The Oxford Debate and
        After." Notes and Queries 202 (Mar. 1957), 126-128.

4806.   BLINDERMAN, CHARLES S.   "T. H. Huxley." Scientific
        Monthly 84 (Apr. 1957), 171-182.

4807.   BLINDERMAN, CHARLES S.   "T. H. Huxley: A Re-eval-
        uation of His Philosophy." Rationalist Annual
        (1966), 50-62.

4808.   BLINDERMAN, CHARLES S.   "T. H. Huxley's Populari-
        zation of Darwinism." Dissertation Abstracts 17
        (1957), 2287-88 (Indiana University, 1957).

4809.   BOND, F. D.   "Huxley, Theologian." Christian
        Century 43 (Feb. 25, 1926), 250-252.

4810.   BOWER, FREDERICK ORPEN.   "Teaching of Biological
        Science." Nature 115 (May 9, 1925), 712-714.

4811.   BOYER, JAMES ALEXANDER.   "Thomas Henry Huxley and
        His Relation to the Recognition of Science in
        English Education." Microfilm Abstracts 9, No. 3
        (1950), 58-59 (University of Michigan, 1949).

4812.   BURKE, J. B.   "The Agnostic's Insufficiency."
        Dublin Review 178 (Jan. 1926), 13-26.

4813.   CHERRY, DOUGLAS.   "The Two Cultures of Matthew
        Arnold and T. H. Huxley." Wascana Review 1 (1966),
        53-61.

4814.   CHESTERTON, CECIL.   "The Art of Controversy:
        Macaulay, Huxley, and Newman." Catholic World 105
        (July 1917), 446-456. (J. H. Newman)

4815.   CHESTERTON, G. K.   "An Agnostic Defeat."   Dublin
        Review 150 (Jan. 1912), 162-172.   (An extract is
        reprinted in Living Age 242, No. 3534, Mar. 30,
        1912, 777-783; Huxley versus W. G. Ward)

4816.   CLARK, RONALD WILLIAM.   The Huxleys.   London:
        Heinemann, 1968.

4817.   CLODD, EDWARD.   "Thomas Henry Huxley."   Century 110
        (May 1925), 33-41.

4818.   CLODD, EDWARD.   Thomas Henry Huxley.   Edinburgh:
        W. Blackwood and Sons, 1902.

4819.   CROWE, M. B.   "Huxley and Humanism."   Studies 49
        (Autumn 1960), 249-260.

4820.   DAVIES, J. R. A.   Thomas Henry Huxley.   London:
        Dent, 1907.

4821.   DAVITASHVILI, L. SH.   ["V. O. Kovalevskii and T.
        Huxley as Naturalist-Evolutionists."]   Trudy Institut
        Istorii Estestvoznaniia 3 (1949), 351-367.   (In
        Russian)

4822.   DAWSON, WARREN ROYAL.   The Huxley Papers: A Des-
        criptive Catalogue of the Correspondence, Manuscripts
        and Miscellaneous Papers of the Right Hon. Thomas
        Henry Huxley, P.C., D.C.L., F.R.S., Preserved in the
        Imperial College of Science and Technology.   London:
        Published for the Imperial College of Science and
        Technology by Macmillan and Co., Ltd., 1946.

4823.   DOCKRILL, D. W.   "T. H. Huxley and the Meaning of
        'Agnosticism'."   Theology 74 (1971), 461-477.

4824.   EISEN, SYDNEY.   "Huxley and the Positivists."
        Victorian Studies 7 (June 1964), 337-358.

4825.   FOLEY, LOUIS.   "The Huxley Tradition of Language
        Study."   Modern Language Journal 26 (1942), 14-20.

4826.   FOSKETT, D. J.   "Wilberforce and Huxley on Evolu-
        tion."   Nature 172 (1953), 920.

4827.   FRIDAY, JAMES R.   "A Microscopic Incident in a
        Monumental Struggle: Huxley and Antibiosis in
        1875."   British Journal of the History of Science 7
        (1974), 61-71.

4828. GATES, R. R. "Huxley as a Mutationist." American
       Naturalist 50 (Feb. 1916), 126-128.

4829. GEDDES, P. "Huxley as Teacher." Nature 115 (May
       9, 1925), 740-743.

4830. GEISON, GERALD L. "The Protoplasmic Theory of Life
       and the Vitalist-Mechanist Debate." Isis 60 (1969),
       273-292. (Huxley versus Lionel Smith Beale)

4831. GRIER, N. M. "Huxley and General Science." School
       and Society 6 (Aug. 4, 1917), 141-142.

4832. GRUSENDORF, A. A. "Huxley on Higher Education."
       School and Society 53 (Mar. 15, 1941), 326-329.

4833. HALLAM, GEORGE W. "Source of the Word 'Agnostic'."
       Modern Language Notes 70 (Apr. 1955), 265-269.

4834. HOLMES, SAMUEL JACKSON. "Life, Morals and Huxley's
       'Evolution and Ethics'." In Science in the University.
       Berkeley and Los Angeles: University of California
       Press, 1944, 319-332.

4835. HOUGHTON, WALTER EDWARDS. "The Rhetoric of T. H.
       Huxley." University of Toronto Quarterly 18 (1948-
       49), 159-175.

4836. HUXLEY, JULIAN SORELL. "The Voyage of the Rattle-
       snake: An Unpublished Diary of Thomas Henry Huxley."
       Fortnightly 137 (1935), 661-673.

4837. HUXLEY, LEONARD. "Carlyle and Huxley." Cornhill
       Magazine 72 (1932), 290-302.

4838. HUXLEY, LEONARD. (Ed.) "Huxley and Agassiz:
       Unpublished Letters." Yale Review 13 (1924),
       63-80.

4839. HUXLEY, LEONARD. Thomas Henry Huxley: A Character
       Sketch. London: Watts and Co., 1920.

4840. IRVINE, WILLIAM. Apes, Angels and Victorians:
       Darwin, Huxley and Evolution. New York: McGraw-Hill
       Book Co. Inc., 1955. (Reprinted, Cleveland: The
       World Publishing Company, 1959)

4841. IRVINE, WILLIAM. "Carlyle and T. H. Huxley." In
       Hill Shine (Ed.), Booker Memorial Studies: Eight
       Essays in Victorian Literature. Chapel Hill:

University of North Carolina Press, 1950, 104-121.
(Reprinted in Austin Wright, Ed., Victorian Liter-
ature: Modern Essays in Criticism. New York:
Oxford University Press, 1961, 193-207)

4842. IRVINE, WILLIAM. Thomas Henry Huxley. London:
Published for the British Council by Longmans,
Green, 1960.

4843. JENSEN, JOHN VERNON. "The Rhetoric of Thomas H.
Huxley and Robert G. Ingersoll in Relation to the
Conflict between Science and Theology." Dissertation
Abstracts 20 (1960), 4215-16 (University of Minnesota,
1959).

4844. JENSEN, JOHN VERNON. "The Rhetorical Strategy of
Thomas H. Huxley and Robert G. Ingersoll: Agnostics
and Roadblock Removers." Speech Monographs 32
(1965), 59-68.

4845. KARNOUTSOS, GEORGE. "Agnosticism." Journal of
Critical Analysis 2 (July 1970), 1-12.

4846. KARNOUTSOS, GEORGE. "Thomas Henry Huxley: His
Educational Theory and Campaign of Enlightenment."
Dissertation Abstracts 27 (1966), 416A (Rutgers -
The State University, 1966).

4847. KEITH, SIR ARTHUR. "Huxley as Anthropologist."
Nature 115 (May 9, 1925), 719-723.

4848. LAYTON, DAVID. "Huxley as Educator: A Reappraisal."
History of Education Quarterly 15 (Summer 1975),
219-225.

4849. LESOURD, GILBERT QUINN. "Huxley and the Preacher."
Methodist Review 108 (Nov. 1925), 922-929.

4850. MACBRIDE, ERNEST WILLIAM. Huxley. London:
Duckworth, 1934.

4851. MARSHALL, ALAN J. Darwin and Huxley in Australia.
London: Hodder and Stoughton, 1970.

4852. MILNER, PHILIP A. "The Moral Vision of Thomas H.
Huxley." Dissertation Abstracts International 32
(1972), 6385A (University of Notre Dame, 1972).

4853.   MINNICK, WAYNE C.   "Thomas H. Huxley's American
        Lectures on Evolution."   Southern Speech Journal 17
        (1952), 225-233.

4854.   MOON, T. C.   "Thomas Huxley's Contributions to
        Science Education."   Science Education 55 (Jan.
        1971), 39-43.

4855.   MORGAN, C. L.   "The Garden of Ethics."   International
        Journal of Ethics 21 (July 1911), 377-407.

4856.   MUDFORD, PETER G.   "The Impact of the Theory of
        Evolution on the Late Nineteenth Century, with
        Special Reference to Thomas Henry Huxley and
        Samuel Butler."   B.Litt. Thesis, University of
        Oxford, 1966.

4857.   MURPHY, BRUCE G.   "Thomas Huxley and His New Reform-
        ation."   Dissertation Abstracts International 34
        (1973), 2527A (Northern Illinois University, 1973).

4858.   NOLAND, RICHARD WELLS.   "T. H. Huxley on Culture."
        Personalist 45 (1964), 94-111.

4859.   ONIONS, C. T.   "Agnostic."   Times Literary Supplement
        (May 10, 1947), 225; (Sept. 6, 1947), 451.

4860.   OSBORN, HENRY FAIRFIELD.   Huxley and Education.
        New York: Charles Scribner's Sons, 1910.

4861.   OSBORN, HENRY FAIRFIELD.   "Huxley on Education."
        Columbia University Quarterly 13 (Dec. 1910),
        25-38.   (Also in Science 32, Oct. 28, 1910, 569-
        578)

4862.   PETERSON, HOUSTON.   Huxley: Prophet of Science.
        London, New York: Longmans, Green and Co., 1932.

4863.   PHELPS, LYNN A. AND EDWIN COHEN.   "The Wilber-
        force-Huxley Debate."   Western Speech 37 (Winter
        1973), 56-64.

4864.   PINGREE, JEANNE.   Thomas Henry Huxley: A List
        of His Scientific Notebooks, Drawings and Other
        Papers, Preserved in the College Archives.   London:
        Imperial College of Science and Technology, 1968.

4865.   REHBOCK, PHILIP F.   "Huxley, Haeckel, and the
         Oceanographers: The Case of 'Bathybius haeckelii'."
         Isis 66 (1975), 504-533.

4866.   RENNER, STANLEY WILLIAM.   "The Garden of Civili-
         zation: Conrad, Huxley, and the Ethics of Evolution."
         Conradiana 7 (1975), 109-120.

4867.   STANLEY, OMA.   "T. H. Huxley's Treatment of 'Nature'."
         Journal of the History of Ideas 18 (1957), 120-127.

4868.   STRAUS, WILLIAM L., JR.   "Huxley's 'Evidence as to
         Man's Place in Nature' a Century Later."   In L. G.
         Stevenson and R. Multhauf (Eds.), Medicine, Science
         and Culture.   Baltimore: Johns Hopkins Press,
         1968, 160-167.

4869.   TELLER, JAMES D.   "Great Teachers of Science. 1.
         Thomas Henry Huxley."   Science Education 25, No. 5
         (Oct. 1941), 239-247.

4870.   TELLER, JAMES D.   "Huxley on the Aims of Educa-
         tion."   Educational Forum 8 (1944), 317-323.

4871.   TELLER, JAMES D.   "Huxley's Evil Influence: Many
         Tributes Refute Criticisms."   Scientific Monthly 56
         (Feb. 1943), 173-178.

4872.   TELLER, JAMES D.   "A Pioneer Propagandist for
         Science in the Elementary School."   Elementary
         School Journal 43 (Dec. 1942), 239-241.

4873.   TELLER, JAMES D.   "The Social Responsibility of a
         Scientist."   Education 63 (Sept. 1942), 3-10.

4874.   THOMPSON, WILLIAM HALLIDAY.   Professor Huxley
         and Religion.   London: H. R. Allenson, 1905.

4875.   THOMSON, JOHN ARTHUR.   "Huxley as Evolutionist."
         Nature 115 (May 9, 1925), 717-718.

4876.   TILBY, A. W.   "Huxley's Problem of Justice."
         Nineteenth Century 106 (Jan. 1929), 58-69.

4877.   VOORHEES, I. W.   "Thomas Henry Huxley: Crusader of
         Science."   Science Education 16 (Oct. 1931), 66-72.

4878.   WASERMAN, MANFRED J.   "Thomas H. Huxley and Religion
         and Immortality: A Letter from T. Lauder Brunton

to Fielding H. Garrison." Medical History 13
(1969), 71-73.

4879. WATTS, WILLIAM WHITEHEAD. "Huxley's Geological
Thought and Teaching." Nature 115 (May 9, 1925),
732-734.

4880. WERSKEY, PAUL GARY. "Haldane and Huxley: The
First Appraisals." Journal of the History of Biology
4 (1971), 171-183.

4881. WEST, ANTHONY. Principles and Persuasions: The
Literary Essays of Anthony West. New York: Harcourt
and Brace, 1957. (Includes Darwin)

4882. WHITE, P. O. G. "Three Victorians and the New
Reformation." Theology 69 (Aug. 1966), 352-358.
(Huxley, Francis Power Cobbe, Mrs. Humphry Ward)

4883. WILSON, DAVID. "Huxley and Wilberforce at Oxford
and Elsewhere." Westminster Review 167 (Mar.
1907), 311-316.

10. Kingsley, Charles

See introduction to this category; XIV. Church of
England, 4. Broad Church; XVII. Christian Socialism.

4884. ALLEN, PETER R. "Charles Kingsley: The Broad
Church Background to His Thought." Dissertation
Abstracts 28 (1967), 662A (University of Toronto,
1965).

4885. ANNAN, NOEL GILROY. "Famous Controversies: Kingsley
and Cardinal Newman." New Statesman and Nation 27
(Mar. 25, 1944), 209.

4886. ANON. "Charles Kingsley: Sermons, 1850-1862."
Bodleian Library Record 1 (June 1940), 174.

4887. BALDWIN, STANLEY EVERETT. Charles Kingsley, Novelist
and Reformer. Ithaca, N.Y.: Cornell University
Press, 1934.

4888.   BEER, GILLIAN. "Kingsley: 'Pebbles on the Shore'."
        Listener 93 (Apr. 17, 1975), 506-507.
4889.   BLINDERMAN, CHARLES S. "Huxley and Kingsley."
        Victorian Newsletter No. 20 (Fall 1961), 25-28.
4890.   BRIGGS, ASA. "Charles Kingsley." Times Educa-
        tional Supplement 2091 (June 17, 1955), 650.
4891.   BROWN, WILLIAM HENRY. Charles Kingsley: The
        Work and Influence of Parson Lot. Manchester: The
        Co-operative Union Ltd., 1924.
4892.   BRUNNER, KARL. "Charles Kingsley als christlich-
        sozialer Dichter." Anglia 46 (1922), 289-322; 47
        (1923), 1-33.
4893.   CAMPBELL, ROBERT ALLAN. "Victorian Pegasus in
        Harness: A Study of Charles Kingsley's Debt to
        Thomas Carlyle and F. D. Maurice." Dissertation
        Abstracts International 30 (1970), 3902-03A (Uni-
        versity of Wisconsin, 1969).
4894.   CHADWICK, OWEN. "Charles Kingsley at Cambridge."
        Historical Journal 18 (June 1975), 303-325.
4895.   CHADWICK, OWEN. "Kingsley's Chair." Theology 78
        (Jan. 1975), 2-8.
4896.   CHITTY, SUSAN. The Beast and the Monk: A Life
        of Charles Kingsley. London: Hodder and Stoughton,
        1974.
4897.   COLEMAN, DOROTHY. "Rabelais and 'The Water-Babies'."
        Modern Language Review 66 (July 1971), 511-521.
4898.   COLLOMS, BRENDA. Charles Kingsley: The Lion
        of Eversley. London: Constable, 1975.
4899.   DORRILL, JAMES FRANK. "The Sermons of Charles
        Kingsley." Ph.D. Dissertation, Harvard University,
        1969.
4900.   DOTTIN, FRANCOISE. "Chartism and Christian Socialism
        in 'Alton Locke'." In Janie Teissedou et al.,
        Politics in Literature in the Nineteenth Century.
        Lille: Université de Lille, 1974, 31-59.
4901.   DOWNES, DAVID ANTHONY. The Temper of Victorian
        Belief: Studies in the Religious Novels of Pater,

Kingsley, and Newman. New York: Twayne Publishers, 1972.

4902.   GOLDBERG, F. S. "Kingsley and the Social Problems of His Day." Westminster Review 167 (Jan. 1907), 41-49.

4903.   HANAWALT, MARY WHEAT. "The Attitude of Charles Kingsley toward Science." Ph.D. Dissertation, University of Iowa, 1936.

4904.   HANAWALT, MARY WHEAT. "Charles Kingsley and Science." Studies in Philology 34 (Oct. 1937), 589-611.

4905.   HANBURY, M. "The Enigma of Charles Kingsley." Pax 39 (1949), 25-33. (Attitude towards Rome)

4906.   HEILMAN, ROBERT B. "Muscular Christianity." Notes and Queries 185 (1943), 44-45. (Credits Kingsley with originating the phrase)

4907.   HOPE, N. V. "Issue between Newman and Kingsley: A Reconciliation and Rejoinder." Theology Today 6 (Apr. 1949), 77-90.

4908.   HOUGHTON, WALTER EDWARDS. "The Issue between Kingsley and Newman." Theology Today 4 (1947), 80-101. (Reprinted in Robert O. Preyer, Ed., Victorian Literature: Selected Essays. New York: Harper and Row, 1967, 13-36, and in David J. DeLaura, Ed., Apologia Pro Vita Sua. New York: Norton, 1968, 390-409)

4909.   JAMES, S. B. "Kingsley vs. Newman." Catholic Digest 3 (Nov. 1938), 77-80.

4910.   JOHNSTON, ARTHUR. "'The Water Babies': Kingsley's Debt to Darwin." English 12 (Autumn 1959), 215-219.

4911.   JUHNKE, ELLA. "Charles Kingsley als sozialreformator-ischer Schriftsteller." Anglia 49 (1925), 32-79.

4912.   JULIAN, MRS. HESTER F. "The Scientific Correspond-ence of Charles Kingsley and William Pengelly." Journal of the Torquay Natural History Society 2 (1920), 361-373.

4913.   KELLER, LUDWIG.   Charles Kingsley und die religiös-
        sozialen Kämpfe in England im 19. Jahrhundert.
        Jena: E. Diederichs, 1911.

4914.   KENDALL, GUY.   Charles Kingsley and His Ideas.
        London: Hutchinson, 1947.

4915.   KÖHLER, FRITZ.   Charles Kingsley als religiöser
        Tendenz-Schriftsteller.   Borna-Leipzig: Noske,
        1912.   (Ph.D. Dissertation, Universität Marburg,
        1912)

4916.   LEWIS, W. DAVID.   "Three Religious Orators and the
        Chartist Movement."   Quarterly Journal of Speech 43
        (1957), 62-68.   (Joseph Rayner Stephens, Kingsley,
        F. W. Robertson)

4917.   MARTIN, ROBERT BERNARD.   "Charles Kingsley."
        Princeton University Library Chronicle 13 (1952),
        168.

4918.   MARTIN, ROBERT BERNARD.   The Dust of Combat: A
        Life of Charles Kingsley.   London: Faber and
        Faber, 1959.

4919.   MARTIN, ROBERT BERNARD.   "Manuscript Sermons of
        Charles Kingsley."   Princeton University Library
        Chronicle 23 (1961-62), 181.

4920.   MEADOWS, ARTHUR JACK.   "Kingsley's Attitude to
        Science."   Theology 78 (Jan. 1975), 15-22.

4921.   MEYER, MARIA.   Carlyles Einfluss auf Kingsley
        in sozial-politischer und religiös-ethischer Hinsicht.
        Weimar: R. Wagner, 1914.

4922.   PAGE, FREDERICK.   "Froude, Kingsley, and Arnold, on
        Newman."   Notes and Queries 184 (Apr. 10, 1943),
        220-221.

4923.   PERKINS, JOCELYN.   "Charles Kingsley at the Abbey:
        Installed Canon April 12, 1873; Died January 23,
        1875."   Westminster Abbey Quarterly 1, No. 1 (Jan.
        1939), 28-32.

4924.   PINNOW, JERRY LYLE.   "Carlyle's Influence on Charles
        Kingsley's Life and Writings."   Dissertation Abstracts

International 34 (1973), 1865-66A (University of
Colorado, 1973).

4925. POPE-HENNESSY, DAME UNA. _Canon Charles Kingsley:_
A Biography. London: Chatto & Windus, 1947.

4926. PRANGE, AUGUST. "Charles Kingsley als Verkörperer
des Broad-Church-Geistes unter besonderer Berücksichtigung der 'Letters and Memories of his Life' und
seiner nichtromanhaften Schriften sowie seiner
Predigten." Ph.D. Dissertation, Universität Halle,
1924.

4927. REBOUL, MARC. _Charles Kingsley: la formation_
d'une personnalité et son affirmation littéraire,
_1819-1850_. Paris: Presses Universitaires de
France, 1973.

4928. ROBERTSON, THOMAS LUTHER. "The Kingsley-Newman
Controversy and the 'Apologia'." _Modern Language_
Notes 69 (Dec. 1954), 564-569.

4929. ROWSE, A. L. "Charles Kingsley at Eversley."
Contemporary Review 221 (1972), 234-238, 322-326;
222 (1973), 7-12.

4930. RULE, PHILIP CHARLES. "Victorian Apostle: The
Religious Thought of Charles Kingsley." Ph.D.
Dissertation, Harvard University, 1968.

4931. SAXBY, DOUGLAS LLOYD. "Charles Kingsley, His
Religious and Social Ideas: The Critique of an
Age." _Dissertation Abstracts_ 26 (1966), 4100
(Princeton University, 1965).

4932. SCOTT, PATRICK GREIG. "Tennyson and Charles Kingsley."
Tennyson Research Bulletin 2, No. 3 (1974), 135-136.

4933. SEDGWICK, J. H. "Mid-Victorian Nordic." _North_
America 225 (Jan. 1928), 86-93.

4934. SPENCE, M. "Charles Kingsley and Education." M.A.
Dissertation, University of Bristol, 1940.

4935. STEVENS, THOMAS PRIMMITT. _Father Adderley_. London:
T. Werner Laurie, Ltd., 1943. (Includes the Christian
humanism of Kingsley and F. D. Maurice)

4936.   THORP, MARGARET FARRAND. Charles Kingsley 1819-1875.
        Princeton: Princeton University Press, 1937.
4937.   VANCE, NORMAN. "'Anythingarianism'." Notes and
        Queries 220 (1975), 108-109. (To Kingsley the word
        connotes moral and metaphysical, rather than
        ecclesiastical vagueness)
4938.   VANCE, NORMAN. "Kingsley's Christian Manliness."
        Theology 78 (Jan. 1975), 30-38.
4939.   VULLIAMY, COLWYN EDWARD. Charles Kingsley and
        Christian Socialism. London: The Fabian Society,
        1914.
4940.   WEST, H. W. "The Social and Religious Thought of
        Charles Kingsley, and His Place in the Christian
        Socialist School of 1848-54." Ph.D. Dissertation,
        University of Edinburgh, 1947.
4941.   WRIGHT, CUTHBERT. "Newman and Kingsley." Harvard
        Graduate Magazine 40 (1931), 127-134.
4942.   YOUNG, GEORGE MALCOLM. "Sophist and Swashbuckler."
        In Daylight and Champaign. London: Jonathan Cape,
        1937, 102-111. (Kingsley and J. H. Newman)

11. Lewes, George Henry

See introduction to this category.

4943.   ANDERSON, ROLAND FRANK. "Formative Influences on
        George Eliot with Special Reference to George Henry
        Lewes." Dissertation Abstracts 25 (1964), 1205-06
        (University of Toronto, 1963).
4944.   BAKER, WILLIAM J. "'A Problematical Thinker' to a
        'Sagacious Philosopher': Some Unpublished George
        Henry Lewes-Herbert Spencer Correspondence."
        English Studies 56 (1975), 217-221.

4945. DOREMUS, ROBERT BARNARD. "George Henry Lewes: A
       Descriptive Biography." Harvard University Sum-
       maries of Theses, 1940 (published 1942), 337-340.
       (Ph.D. Dissertation, Harvard University, 1940)

4946. EVERETT, EDWIN MALLARD. The Party of Humanity:
       The Fortnightly Review and Its Contributors, 1865-
       1874. Chapel Hill: University of North Carolina
       Press, 1939. (Includes J. Morley, J. S. Mill)

4947. FISHMAN, STEPHEN MICHAEL. "The Epistemology of G.
       H. Lewes." Dissertation Abstracts 28 (1967),
       722-723A (Columbia University, 1967).

4948. FISHMAN, STEPHEN MICHAEL. "James and Lewes on
       Unconscious Judgement." Journal of the History of
       Behavioral Sciences 4 (1968), 335-348. (William
       James)

4949. GOLDBERG, HANNAH FRIEDMAN. "George Henry Lewes and
       the Secular Revelation." Ph.D. Dissertation, Johns
       Hopkins University, 1966.

4950. GRASSI-BERTAZZI, GIAMBATTISTA. Esame critico
       della filosofia di George Henry Lewes: Parte 1. Le
       idee metodologiche e metafisiche. Messina: 1906.

4951. HIRSHBERG, EDGAR W. George Henry Lewes. New York:
       Twayne, 1972.

4952. HUDSON, STEWART M. "George Henry Lewes' Evolutionism
       in the Fiction of George Eliot." Dissertation
       Abstracts International 31 (1971), 6059A (University
       of Southern California, 1970).

4953. JONES, HELEN ELIZABETH. "George Henry Lewes's Use
       in His Literary Criticism of Ideas from His Empirical
       Metaphysics." Dissertation Abstracts International
       30 (1969), 2530A (University of Colorado, 1969).

4954. KAMINSKY, JACK. "The Empirical Metaphysics of
       George Henry Lewes." Journal of the History of
       Ideas 13 (1952), 314-332.

4955. KAMINSKY, JACK. "The Philosophy of George Henry
       Lewes." Ph.D. Dissertation, New York University,
       1950.

4956.   KITCHEL, ANNA THERESA.   George Lewes and George
        Eliot. New York: John Day Company, 1933.

4957.   MILES, F. R. AND R. L. MORETON.   "A Portrait of G.
        H. Lewes."   Notes and Queries 194 (1949), 368-369;
        459.

4958.   NEWTON, K. M.   "George Eliot, George Henry Lewes,
        and Darwinism."   Durham University Journal 66
        (1974), 278-293.

4959.   OCKENDEN, R. E.   "George Henry Lewes (1817-1878)."
        Isis 32 (1940), 70-86.

4960.   SMITH, R. E.   "George Henry Lewes and His 'Physiology
        of Common Life', 1859."   Proceedings of the Royal
        Society of Medicine 53 (1960), 569-574.

4961.   TJOA, HOCK GUAN.   "George Henry Lewes: A Victorian
        Mind."   Ph.D. Dissertation, Harvard University,
        1972.   (Published as a monograph by Harvard Univer-
        sity Press, 1977)

                12.   Mansel, Henry Longueville

        See introduction to this category; XIV.   Church of
England, 1. General.

4962.   BEVAN, EDWYN ROBERT.   "Mansel and Pragmatism."   In
        Symbolism and Belief. London: George Allen &
        Unwin Ltd., 1938, 318-340.

4963.   CUPITT, DON.   "Mansel and Maurice on Our Knowledge
        of God."   Theology 73 (July 1970), 301-311.

4964.   CUPITT, DON.   "Mansel's Theory of Regulative Truth."
        Journal of Theological Studies 18 (Apr. 1967),
        104-126.

4965.   CUPITT, DON.   "What Was Mansel Trying to Do?"
        Journal of Theological Studies 22 (Oct. 1971),
        544-547.

4966. DOCKRILL, D. W. "Doctrine of Regulative Truth and
       Mansel's Intentions." Journal of Theological
       Studies 25 (Oct. 1974), 453-465.
4967. DOCKRILL, D. W. "The Limits of Thought and Regulative
       Truths." Journal of Theological Studies 21 (1970),
       370-387.
4968. FREEMAN, KENNETH D. "Mansel's Religious Positiv-
       ism." Southern Journal of Philosophy 5 (Summer
       1967), 91-102.
4969. FREEMAN, KENNETH D. The Role of Reason in Religion:
       A Study of Henry Mansel. The Hague: Martinus
       Nijhoff, 1969.
4970. KANE, G. STANLEY. "The Concept of Divine Goodness
       and the Problem of Evil." Religious Studies 11
       (Mar. 1975), 49-71.
4971. KNOX, B. A. "Filling the Oxford Chair of Ecclesias-
       tical History, 1866: The Nomination of H. L.
       Mansel." Journal of Religious History 5 (June
       1968), 62-70.
4972. MARCUCCI, SILVESTRO. Henry L. Mansel: filosofia
       della coscienza ed epistemologia della religione.
       Firenze: P. Le Monnier, 1969.
4973. MATTHEWS, WALTER ROBERT. The Religious Philosophy
       of Dean Mansel. London: Oxford University Press,
       1956.
4974. NEDONCELLE, MAURICE GUSTAVE. La Philosophie
       religieuse en Grande-Bretagne de 1850 à nos jours.
       Paris: Bloud & Gay, 1934. (Includes Mansel's
       notion of God as unknowable)

13. Maurice, Frederick Denison

See introduction to this category; IV. Education;
XIV. Church of England, 4. Broad Church; XVII. Christian
Socialism.

4975. AHLERS, ROLF. "Die Vermittlungstheologie des
        Frederick Denison Maurice." Dissertation, Universität
        Hamburg, 1966.

4976. ALLCHIN, ARTHUR MACDONALD. "F. D. Maurice as
        Theologian." Theology 76 (Oct. 1973), 513-525.

4977. ALLEN, PETER R. "F. D. Maurice and J. M. Ludlow:
        A Reassessment of the Leaders of Christian Socialism."
        Victorian Studies 11 (1968), 461-482.

4978. ANON. "Amends to F. D. Maurice." Spectator 142
        (Apr. 20, 1929), 609-610.

4979. ANON. "Maurice and the Workingmen's College."
        Nation 82 (Feb. 22, 1906), 154-155.

4980. BALLARD, A. W. "F. D. Maurice: A Retrospect."
        Church Quarterly Review 140, No. 279 (Apr. 1945),
        51-60.

4981. BARTH, EUGENE HOWARD. "Rooted and Grounded in
        Love, F. D. Maurice's Relational Ethic of Reconcilia-
        tion and Transformation." Dissertation Abstracts
        26 (1966), 4097 (Princeton University, 1965).

4982. BEST, ERNEST E. "F. D. Maurice in Perspective."
        Encounter 34 (Winter 1973), 16-26.

4983. BICKNELL, PERCY F. "Frederick Denison Maurice."
        Dial 39 (Aug. 1, 1905), 53-56.

4984. BIRLEY, ROBERT. "Frederick Denison Maurice, 1805-
        1872." Times Educational Supplement 2987 (Aug. 18,
        1972), 30.

4985. BIRLEY, ROBERT. "Maurice and Education." Theology
        76 (Sept. 1973), 449-462.

4986.   BLAAUW, PIETER.   F. D. Maurice, zijn leven en
        werken. Amsterdam: Drukkerij "Concordia,"
        1908.

4987.   BOOTH, HARRY FEHR.   "The Knowledge of God and the
        Practice of Society in Frederick Denison Maurice."
        Dissertation Abstracts 24 (1963), 863 (Boston
        University Graduate School, 1963).

4988.   BROSE, OLIVE J.   "Frederick Denison Maurice and the
        Victorian Crisis of Belief." Victorian Studies 3
        (1959-60), 227-248.

4989.   BROSE, OLIVE J.   Frederick Denison Maurice: Rebel-
        lious Conformist. Athens, Ohio: Ohio University
        Press, 1971.

4990.   BROSE, OLIVE J.   "The Nature of Belief: In Commemora-
        tion of F. D. Maurice, August 29, 1805-April 1,
        1872." Christian Century 89 (Mar. 29, 1972),
        360-363.

4991.   CAMPBELL, ROBERT ALLAN.   "Victorian Pegasus in
        Harness: A Study of Charles Kingsley's Debt to
        Thomas Carlyle and F. D. Maurice." Dissertation
        Abstracts International 30 (1970), 3902-03A
        (University of Wisconsin, 1969).

4992.   CHRISTENSEN, TORBEN.   The Divine Order: A Study
        in F. D. Maurice's Theology. Leiden: E. J. Brill,
        1973.

4993.   CHRISTENSEN, TORBEN.   "F. D. Maurice and the
        Contemporary Religious World." Studies in Church
        History 3 (1966), 69-90.

4994.   CLAYTON, JAMES W.   "Reason and Society: An Approach
        to F. D. Maurice." Harvard Theological Review 65
        (1972), 305-335.

4995.   COULSON, JOHN.   "Criticism: Maurice's Idea of the
        Kingdom." In Newman and the Common Tradition:
        A Study in the Language of Church and Society.
        Oxford: Clarendon Press, 1970, 187-224.

4996. COX, JAMES WILLIAM. "God Manifesting Himself: A
       Study of Some Central Elements in the Theology of
       F. D. Maurice." Ph.D. Dissertation, Cambridge
       University, 1960.

4997. CUNLIFFE-JONES, H. "A New Assessment of F. D.
       Maurice's 'The Kingdom of Christ'." Church
       Quarterly 4 (July 1971), 38-50.

4998. CUPITT, DON. "Mansel and Maurice on Our Knowledge
       of God." Theology 73 (July 1970), 301-311.

4999. DAVIES, WALTER MERLIN. An Introduction to F. D.
       Maurice's Theology. London: S.P.C.K., 1964.

5000. DIMAN, LOUISE. "F. D. Maurice." Outlook 80 (1905),
       967-974.

5001. DRING, TOM. "Frederick Denison Maurice: The
       Greatest Prophet of the Nineteenth Century."
       London Quarterly and Holborn Review 173 (Jan.
       1948), 33-46.

5002. DRING, TOM. "The Philosophy of Frederick Denison
       Maurice." London Quarterly and Holborn Review 187
       (Oct. 1962), 276-284.

5003. EDE, W. MOORE. "What We Owe to F. D. Maurice and
       His Disciples." Modern Churchman 23 (Dec. 1933),
       527-534.

5004. FLESSEMAN-VAN LEER, ELLEN. De overmacht van de
       liefde: inleiding in de theologie van F. D. Maurice.
       Wageningen: N. Veenman, 1968.

5005. GARDNER, C. "Frederick Denison Maurice, 1805-1872."
       Hibbert Journal 28 (Jan. 1930), 311-319.

5006. GOVAN, THOMAS P. "The Task and Purpose of the
       University: A Report on the Educational Writings
       of Frederick Denison Maurice." Anglican Theological
       Review 47 (Oct. 1965), 395-409.

5007. GRYLLS, ROSALIE GLYNN. Queen's College, 1848-1948.
       London: Routledge, 1948.

5008. HALL, R. O. "Revised Reviews. 5. F. D. Maurice's
       'The Doctrine of Sacrifice'." Theology 64 (May
       1961), 189-193.

5009. HALL, ROBERT TOM.   "Autonomy and the Social Order:
       The Moral Philosophy of F. D. Maurice."   Monist 55
       (July 1971), 504-519.

5010. HALL, ROBERT TOM.   "The Unity of Philosophy, Theology
       and Ethics in the Thought of Frederick Denison
       Maurice."   Dissertation Abstracts 28 (1968), 3756A
       (Drew University, 1967).

5011. HARTLEY, A. J.   "Frederick Denison Maurice, Disciple
       and Interpreter of Coleridge: 'Constancy to an
       Ideal Object'."   Ariel 3, No. 2 (1972), 5-16.

5012. HARTLEY, A. J.   "The Way to Unity: Maurice's
       Exegesis for Society."   Canadian Journal of Theology
       16 (1970), 95-99.

5013. HAUGHEY, REVEREND JOHN C., S.J.   "The Ecclesiology
       of Frederick Denison Maurice."   Dissertation Abstracts
       28 (1967), 767A (Catholic University of America,
       1967).

5014. HIGHAM, FLORENCE.   Frederick Denison Maurice.
       London: Student Christian Movement Press, 1948.

5015. HODKIN, H.   "The Theological Teaching of F. D.
       Maurice."   Theology 34 (Feb. 1937), 97-107.

5016. JENKINS, CLAUDE.   Frederick Denison Maurice and
       the New Reformation.   London: Society for Promoting
       Christian Knowledge; New York: The Macmillan Company,
       1938.

5017. KETCHUM, ROBERT HOLYOKE.   "Frederick Denison Maurice:
       An Assessment of His Contributions to Nineteenth
       Century English Education."   Dissertation Abstracts
       International 30 (1970), 5262A (Syracuse University,
       1969).

5018. LIDGETT, JOHN SCOTT.   "Frederick Denison Maurice."
       Contemporary Review 135 (1929), 584-586.

5019. MCCLAIN, FRANK MAULDIN.   Maurice: Man and Moralist.
       London: S.P.C.K., 1972.

5020. MCNAB, JOHN INGRAM.   "Towards a Theology of Social
       Concern: A Comparative Study of the Elements for

Social Concern in the Writings of Frederick D.
Maurice and Walter Rauschenbusch." Dissertation
Abstracts International 33 (1973), 3761A (McGill
University, 1972).

5021.   MAJOR, H. D. A.   "Renaissance of F. D. Maurice."
Modern Churchman 41 (June 1951), 99-103.

5022.   MASTERMAN, CHARLES FREDERICK GURNEY.   Frederick
Denison Maurice. London: A. R. Mowbray and Co.,
Limited, 1907.

5023.   MASTERMAN, J. H. B.   "A Great Anglican." Spectator
150 (1933), 330-331.

5024.   MASTERMAN, NEVILLE CHARLES.   "F. D. Maurice:
Progressive or Reactionary?" Theology 76
(Nov. 1973), 575-585.

5025.   MASTERMAN, NEVILLE CHARLES.   "F. D. Maurice: The
Philosopher Theologian." Hibbert Journal 52 (July
1954), 381-388.

5026.   MASTERMAN, NEVILLE CHARLES.   "The Mental Processes
of the Reverend F. D. Maurice." Theology 68 (Jan.
1965), 46-53.

5027.   MURPHY, DAVID M.   "Irenic Method in the Ecumenical
Theology of F. D. Maurice." Foundations 15 (Apr.-
June 1972), 126-145.

5028.   MURPHY, DAVID M.   "Mission of the Church and Secularity
in the Life and Work of F. D. Maurice." Foundations
15 (July-Sept. 1972), 251-265.

5029.   NIEBUHR, HELMUT RICHARD.   "Christ the Transformer
of Culture: The Views of F. D. Maurice." In
Christ and Culture. New York: Harper and Brothers,
1951, 218-229.

5030.   OGDEN, S. M. et al. "American Maurice Centenary."
Anglican Theological Review 54 (Oct. 1972), 260-342.
(Includes articles on reason and unity in Maurice,
Maurice's treatment of heaven and hell, and Maurice
and Eustace Conway)

5031.   ORAM, EANNE.   "Emily and F. D. Maurice: Some
        Parallels of Thought."   Brontë Society Transactions
        13 (1957), 131-140.   (Emily Brontë)

5032.   PAKENHAM, F., EARL OF LONGFORD.   "Humanity a Holy
        Thing."   Theology 76 (Dec. 1973), 631-638.

5033.   POLLOCK, SIR FREDERICK.   "Frederick Denison Maurice
        and the Working Men's College."   Cambridge Review
        22 (1901), 132-133.

5034.   PORTER, JOHN FRANCIS.   "The Place of Christ in the
        Thought of F. D. Maurice."   Dissertation Abstracts
        20 (1960), 3869-70 (Columbia University, 1959).

5035.   PORTER, JOHN FRANCIS AND WILLIAM J. WOLFF. (Eds.)
        Toward the Recovery of Unity: The Thought of
        Frederick Denison Maurice, Edited from His Letters.
        New York: Seabury Press, 1964.

5036.   POWICKE, FREDERICK JAMES.   "Frederick Denison
        Maurice (1805-1872):   A Personal Reminiscence."
        Congregational Quarterly 8 (1920), 169-184.

5037.   PRICKETT, STEPHEN.   "Coleridge, Newman and F. D.
        Maurice: Development of Doctrine and Growth of
        Mind."   Theology 76 (July 1973), 340-349.
        (J. H. Newman)

5038.   RAMSEY, ARTHUR MICHAEL.   F. D. Maurice and the
        Conflicts of Modern Theology. London: Cambridge
        University Press, 1951.

5039.   RAMSEY, ARTHUR MICHAEL.   "Frederick Denison Maurice."
        Oecumenica 2, No. 4 (Jan. 1936), 308-317.

5040.   RANSON, GUY H.   "F. D. Maurice on the Social Nature
        of Man."   Canadian Journal of Theology 11 (Oct.
        1965), 265-276.

5041.   RANSON, GUY H.   "F. D. Maurice's Theology of Society:
        A Critical Study."   Dissertation Abstracts 25
        (1964), 3720 (Yale University, 1956).

5042.   RANSON, GUY H.   "The Kingdom of God as the Design
        of Society: An Important Aspect of F. D. Maurice's
        Theology."   Church History 30 (1961), 458-472.

5043.   RANSON, GUY H.   "Trinity and Society: A Unique
         Dimension of F. D. Maurice's Theology."   Religion
         in Life 29 (Winter 1959-60), 64-74.
5044.   RICE, JESSIE FOLSOM.   "The Influence of Frederick
         Dennison [sic] Maurice on Tennyson."   Dissertation,
         University of Chicago, 1913.
5045.   SANDERS, CHARLES RICHARD.   "Coleridge, F. D. Maurice,
         and the Distinction between the Reason and the
         Understanding."   PMLA 51 (1936), 459-475.
5046.   SANDERS, CHARLES RICHARD.   "Coleridge, Maurice, and
         the Church Universal."   Journal of Religion 21
         (1941), 31-45.
5047.   SANDERS, CHARLES RICHARD.   "Maurice as a Commentator
         on Coleridge."   PMLA 53 (1938), 230-243.
5048.   SANDERS, CHARLES RICHARD.   "The Relation of Frederick
         Denison Maurice to Coleridge."   Ph.D. Dissertation,
         University of Chicago, 1934.
5049.   SANDERS, CHARLES RICHARD.   "Sir Leslie Stephen,
         Coleridge and Two Coleridgeans."   PMLA 55 (1940),
         795-801.   (Maurice and J. Dykes Campbell)
5050.   SANDERS, CHARLES RICHARD.   "Was F. D. Maurice a
         Broad Churchman?"   Church History 3 (1934), 222-231.
5051.   SMITH, A. C.   "A Study of Frederick Denison Maurice."
         M.A. Dissertation, University of Sheffield, 1950.
5052.   SOKOLOFF, NANCY BOYD.   "Revelation as Education in
         the Thought of F. D. Maurice."   Dissertation Abstracts
         International 32 (1971), 3415-16A (Columbia University,
         1971).
5053.   STEVENS, THOMAS PRIMMITT.   Father Adderly.   London:
         T. Werner Laurie, 1943.   (Includes the Christian
         humanism of C. Kingsley and Maurice)
5054.   TERWILLIGER, ROBERT E.   "The Doctrine of the Church
         in the Works of Frederick Denison Maurice."   Ph.D.
         Dissertation, Yale University, 1948.
5055.   TURNBULL, JOHN WINTER.   "Catholic with a Lower-case
         c."   Ecumenical Review 12 (Oct. 1959), 114-119.

John Stuart Mill (1806-1873)   475

5056.   TURNBULL, JOHN WINTER.   "The Idea of a National
         Church in the Theology of Frederick Denison Maurice."
         Ph.D. Dissertation, Yale University, 1957.

5057.   VAUGHAN, MICHELINA AND MARGARET SCOTFORD ARCHER.
         "F. D. Maurice and the Educational Role of the
         National Church."   A Sociological Yearbook of
         Religion in Britain 5 (1972), 48-59.

5058.   VIDLER, ALEXANDER ROPER.   The Theology of F. D.
         Maurice.   London: Student Christian Movement
         Press, 1948.   (Published as Witness to the Light.
         New York: Scribner's, 1948; enlarged edition
         published as F. D. Maurice and Company.   London:
         S.C.M. Press, 1966)

5059.   WINN, WILLIAM E.   "F. D. Maurice and Contemporary
         Theology."   South East Asia Journal of Theology 2
         (Oct. 1960), 33-36.

5060.   WOOD, HERBERT GEORGE.   Frederick Denison Maurice.
         London: Cambridge University Press, 1950.

14.   Mill, John Stuart

See introduction to this category; I.   Science, 2.
Method and Philosophy; XIII.   Evolution and Psychology;
XXII.   Positivism.

5061.   ADELMAN, PAUL.   "Frederic Harrison on Mill."
         Mill News Letter 5, No. 2 (Spring 1970), 2-4.

5062.   ALEXANDER, EDWARD.   "John Stuart Mill on Dogmatism,
         'Liberticide' and Revolution."   Victorian News-
         letter No. 37 (Spring 1970), 12-18.

5063.   ALEXANDER, EDWARD.   Matthew Arnold and John Stuart
         Mill.   New York: Columbia University Press, 1965.

5064.   ANON.   "The Testimony of John Stuart Mill to
         Mysticism."   Outlook (New York) 95 (1910), 818-820.

5065.   AUGUST, EUGENE R.   John Stuart Mill: A Mind at
        Large. New York: Scribners, 1975.

5066.   BERLIN, ISAIAH.   John Stuart Mill and the Ends
        of Life. London: Council of Christians and Jews,
        1959.

5067.   BILL, THOMAS LEE.   "The Theory of Nature in John
        Stuart Mill." Dissertation Abstracts 25 (1964),
        535 (St. Louis University, 1963).

5068.   BRITTON, KARL W.   John Stuart Mill. London:
        Penguin Books, 1953.

5069.   BRITTON, KARL W.   "Perpetuating a Mistake about
        Mill's Three Essays on Religion." Mill News Letter
        5 (Spring 1970), 6-7.

5070.   CARR, ROBERT.   "The Religious Thought of John
        Stuart Mill: A Study in Reluctant Scepticism."
        Journal of the History of Ideas 23 (1961), 475-495.

5071.   ELLERY, JOHN B.   John Stuart Mill. New York:
        Twayne, 1964.

5072.   EVERETT, EDWIN MALLARD.   The Party of Humanity:
        The Fortnightly Review and Its Contributors, 1865-1874.
        Chapel Hill: University of North Carolina Press,
        1939. (Includes J. Morley, G. H. Lewes)

5073.   FABBRICOTTI, C. A.   Positivismo? John Stuart Mill.
        Firenze: Lumachi, 1910.

5074.   HAINDS, J. R.   "John Stuart Mill and the Saint-
        Simonians." Journal of the History of Ideas 6
        (1946), 103-112.

5075.   HARRIS, ROBERT T.   "Nature: Emerson and Mill."
        Western Humanities Review 6 (1952), 1-13.

5076.   HAYEK, FRIEDRICH AUGUST VON.   John Stuart Mill and
        Harriet Taylor: Their Correspondence and Subsequent
        Marriage. Chicago: University of Chicago Press,
        1951.

5077.   HESSLER, C. A.   "John Stuart Mill och Religions-
        Friheten." In Civibus et Rei Publicae. Uppsala:
        Amgrist and Wiksell, 1960, 176-185.

5078. HIMMELFARB, GERTRUDE. On Liberty and Liberalism:
       The Case of John Stuart Mill. New York: Knopf,
       1974.
5079. JANES, GEORGE M. "John Stuart Mill's Religion."
       Quarterly Journal of the University of North Dakota
       20 (1930), 294-299.
5080. KANTZER, E. MARCEL. La Religion de J. Stuart Mill.
       Caen: C. Valin, 1906.
5081. KROOK, DOROTHEA. "Rationalist Humanism: J. S.
       Mill's 'Three Essays on Religion'." In Three
       Traditions of Moral Thought. Cambridge: University
       Press, 1959, 181-201.
5082. LEVIN, RUDOLF. Der Geschichtsbegriff des Posi-
       tivismus unter besonderer Berücksichtigung Mills und
       der rechtsphilosophischen Anschauungen John Austins.
       Leipzig: Buchdruckerei Joh. Moltzen, 1935.
5083. LEWELS, MAXIMILIAN. John Stuart Mill: Die Stellung
       eines Empiristen zur Religion. Münster: Theis-
       sing'sche Buchhandlung, 1903.
5084. LINDLEY, DWIGHT N. "John Stuart Mill: The Second
       Greatest Influence." Victorian Newsletter No. 11
       (1957), 25-26. (The influence of French thought)
5085. LINDLEY, DWIGHT N. "The Saint-Simonians, Carlyle,
       and Mill: A Study in the History of Ideas."
       Dissertation Abstracts 19 (1958), 320 (Columbia
       University, 1958).
5086. LLOYD, WALTER. "J. S. Mill's Letters to Auguste
       Comte." Westminster Review 153 (1900), 421-426.
5087. MCCLOSKEY, HENRY JOHN. John Stuart Mill: A Critical
       Study. London: Macmillan, 1971.
5088. MACMINN, NEY, J. R. HAINDS AND JAMES MCNAB MCCRIMMON.
       Bibliography of the Published Writings of John Stuart
       Mill. Evanston: Northwestern University, 1945.
5089. MAN, GLENN K. S. "John Stuart Mill on Bentham and
       Coleridge." Revue de l'Université d'Ottawa 45
       (1975), 320-332.

5090.   MAZLISH, BRUCE.   James and John Stuart Mill: Father
        and Son in the Nineteenth Century.   London:
        Hutchinson; New York: Basic Books, 1975.

5091.   MEGILL, ALLAN D.   "J. S. Mill's Religion of Humanity
        and the Second Justification for the Writing of 'On
        Liberty'."   Journal of Politics 34 (1972), 612-629.

5092.   MINEKA, FRANCIS EDWARD. (Ed.)   The Earlier Letters
        of John Stuart Mill, 1812-1848.   2 vols. In Collected
        Works of John Stuart Mill.   Volumes 12 and 13.
        Toronto: University of Toronto Press; London:
        Routledge & Kegan Paul, 1963.

5093.   MINEKA, FRANCIS EDWARD AND DWIGHT N. LINDLEY.
        (Eds.)   The Later Letters of John Stuart Mill,
        1849-1873.   4 vols.   In Collected Works of John
        Stuart Mill.   Volumes 14-17.   Toronto: University
        of Toronto Press; London: Routledge & Kegan Paul,
        1972.

5094.   MUELLER, IRIS WESSEL.   John Stuart Mill and French
        Thought.   Urbana: University of Illinois Press,
        1956.

5095.   MUYSKENS, JAMES LEROY.   "Religion Based on Hope."
        Dissertation Abstracts International 32 (1972),
        6492A (University of Michigan, 1971).   (Chapter 1
        compares Kant and Mill on immortality)

5096.   OUREN, DALLAS VICTOR LIE.   "HaMILLton: Mill on
        Hamilton - a Re-examination of Sir Wm. Hamilton's
        Philosophy."   Dissertation Abstracts 34 (1974),
        7282-83A (University of Minnesota, 1973).

5097.   PACKE, MICHAEL ST. JOHN.   The Life of John Stuart
        Mill.   London: Secker and Warburg, 1954.

5098.   PANKHURST, RICHARD KEIR PETHICK.   The Saint Simonians,
        Mill and Carlyle: A Preface to Modern Thought.
        London: Sidgwick and Jackson, 1957.

5099.   PAUL, WILFORD NOEL.   "The Religious Views of John
        Stuart Mill."   Dissertation Abstracts International
        33 (1973), 3717A (University of New Mexico, 1972).

5100. PRIESTLY, F. E. L. "Introduction." Essays on
       Ethics, Religion and Society by John Stuart Mill.
       In Collected Works of John Stuart Mill. Volume 10.
       Toronto: University of Toronto Press; London:
       Routledge & Kegan Paul, 1969.

5101. RAUCH, LEO. "Mill's Secular Religion." Journal
       of Critical Analysis 3 (Jan. 1972), 178-187.

5102. ROBSON, JOHN MERCEL. The Improvement of Mankind:
       The Social and Political Thought of John Stuart Mill.
       Toronto: University of Toronto Press, 1968.

5103. RYAN, ALAN. J. S. Mill. London and Boston:
       Routledge and Kegan Paul, 1974.

5104. SÄNGER, SAMUEL. "Mills Theodizee." Archiv für
       Geschichte der Philosophie 13, No. 3 (1900), 402-429.

5105. SCHNEEWIND, JEROME B. (Ed.) Mill: A Collection
       of Critical Essays. Garden City, N.Y.: Anchor
       Books, 1968.

5106. SHINE, HILL. "J. S. Mill and an Open Letter to the
       Saint-Simonian Society in 1832." Journal of the
       History of Ideas 6 (1945), 102-108.

5107. SPITZBERG, I. J. "John Stuart Mill and John Henry
       Cardinal Newman: A Comparative Study, with Special
       Reference to Their Views of the Role of Education
       in Society." B.Litt. Dissertation, University of
       Oxford, 1969.

5108. STEINTRAGER, JAMES. "Morality and Belief: The
       Origin and Purpose of Bentham's Writings on Religion."
       Mill News Letter 6, No. 2 (Spring 1971), 3-15.

5109. WENTSCHER, ELSE. "John Stuart Mills Stellung zur
       Religion." Archiv für die gesamte Psychologie 77
       (Aug. 1930), 48-66.

5110. WOODS, THOMAS. Poetry and Philosophy: A Study
       in the Thought of John Stuart Mill. London:
       Hutchinson & Co., 1961.

15.   Mivart, St. George Jackson

See introduction to this category; XVI.   Catholicism.

5111.   ABERCROMBIE, NIGEL J.   "Edmund Bishop and St.
George Mivart."   Month 7 (Mar. 1952), 176-180.
5112.   GRUBER, JACOB W. A.   Conscience in Conflict: The
Life of St. George Jackson Mivart.   New York:
Columbia University Press, 1960.
5113.   HOLMES, J. DEREK.   "Newman and Mivart: Two Attitudes
to a Nineteenth-Century Problem."   Clergy Review 50
(Nov. 1965), 852-867.
5114.   LESLIE, SIR JOHN RANDOLPH SHANE.   "St. George
Mivart: An Angry Victorian."   Wiseman Review 235
(Spring 1961), 48-55.
5115.   MIVART, FREDERICK ST. GEORGE. (Ed.)   "Early Memories
of St. George Mivart."   Dublin Review 174 (Jan.
1924), 1-27.
5116.   MORRISON, JOHN L.   "Orestes Brownson and the Catholic
Reaction to Darwinism."   Duquesne Review 6 (Spring
1961), 75-87.
5117.   WINDLE, SIR BERTRAM COGHILL ALAN.   "Memories of St.
George Mivart."   Dublin Review 173 (July 1923),
70-85.

16.   Newman, Francis William

See introduction to this category.   For his brother,
John Henry, see next section.

5118.   ANON.   "Liberal Newman."   Tablet 211 (June 14,
1958), 559.
5119.   ANON.   "Newman's Brother."   Tablet 200 (Aug. 9,
1952), 113.

5120. ARCHIBALD, RAYMOND C. "Francis William Newman, (1805-1897)." Mathematical Tables and Other Aids to Computation 1, No. 12 (1945), 454-459.

5121. BENNETT, JAMES RICHARD. "F. W. Newman and Religious Freedom." Hibbert Journal 63 (Spring 1965), 106-109.

5122. BENNETT, JAMES RICHARD. "Francis W. Newman and Religious Liberalism in Nineteenth-Century England." Dissertation Abstracts 21 (1961), 3233 (Stanford University, 1961).

5123. BENNETT, JAMES RICHARD. "The Theism of Francis W. Newman: A Reconciliation of Science and Religion." Crane Review 4, No. 1 (Fall 1961), 52-62.

5124. BROWN, W. E. "Francis W. Newman, 1805-1897: A Retrospect." Ph.D. Dissertation, University of Edinburgh, 1954.

5125. GRIBBLE, FRANCIS HENRY. "Francis William Newman." Fortnightly Review 84 (July 1905), 151-161.

5126. HENNIG, J. "Cardinal Newman's Brother in Ireland." Irish Monthly 75 (May 1947), 185-194.

5127. MOZLEY, JOHN RICKARDS. "Francis William Newman." Hibbert Journal 23 (July 1925), 345-360.

5128. RITZ, JEAN-GEORGES. "Trois témoins de la crise religieuse victorienne: les frères Newman." In Religion et politique: les deux guerres mondiales: histoire de Lyon et du Sud-Est: mélanges offerts à M. le Doyen André Latreille. Préf. de M. Pacaut et al. Lyon: Audin, 1972, 177-185.

5129. ROBBINS, WILLIAM M. The Newman Brothers: An Essay in Comparative Intellectual Biography. Cambridge: Harvard University Press, 1966.

5130. ROSS, KENNETH N. "Francis William Newman." Church Quarterly Review 118 (July 1934), 231-244.

5131. SIEVEKING, ISABEL GIBERNE. Memoir and Letters of Francis W. Newman. London: K. Paul, Trench, Trübner & Co., Ltd., 1909.

5132.   SVAGLIC, MARTIN JAMES.   "Charles Newman and His
        Brothers."   PMLA 71 (June 1956), 370-385.   (John
        Henry and Francis William)
5133.   WINSLOW, DONALD F.   "Francis W. Newman's Assessment
        of John Sterling: Two Letters." English Language
        Notes 11 (1974), 278-283.   (To M. Conway, Oct. 6,
        1869 and Oct. 16, 1869).

                    17. Newman, John Henry

        This section contains only a portion of the great
volume of material available on J. H. Newman. See further:
John R. Griffin, Newman: A Bibliography . . .. Front
Royal, Virginia: Christendom Publications, 1980. See
introduction to this category; IV. Education; XIV.
Church of England, 1. General, 3. Oxford Movement;
XVI. Catholicism. For his brother, Francis William,
see previous section.

5134.   ABBOTT, EDWIN ABBOTT.   The Anglican Career of
        Cardinal Newman.   2 vols.   London and New York:
        Macmillan and Co., 1892.
5135.   ACHAVAL, HUGO M. DE.   "An Unpublished Paper by
        Cardinal Newman on the Development of Doctrine."
        Gregorianum 39 (1958), 585-596.
5136.   ALICE, MARY, SISTER.   "From Oxford to Rome."   In K.
        P. K. Menon, M. Manuel and K. Ayyappa Paniker
        (Eds.), Literary Studies: Homage to Dr. A.
        Sivaramasubramonia Aiyer. Trivandrum: A.
        Sivaramasubramonia Aiyer Memorial Committee, 1973,
        173-181.   (Newman's conversion)
5137.   ALLEN, E. L.   "Newman's Apologia Pro Roma."
        Congregational Quarterly 33 (Apr. 1955), 151-159.
5138.   ALLEN, LOUIS.   "Newman and Christopher Wordsworth:
        The Revision of the 'Essay on the Development of

Christian Doctrine'."   In E. T. Dubois et al.
(Eds.), Essays Presented to C. M. Girdlestone.
Newcastle upon Tyne: University of Durham, 1960,
11-26.

5139. ALLEN, LOUIS. "Tract 90 and Durham University."
Notes and Queries 212 (Feb. 1967), 43-47.

5140. ALLISON, W. H. "Was Newman a Modernist?" American
Journal of Theology 14 (Oct. 1910), 552-571.

5141. ALTHOLZ, JOSEF LEWIS. "Newman and History."
Victorian Studies 7 (Mar. 1964), 285-294. (See
Holmes, 5419)

5142. ALTHOLZ, JOSEF LEWIS. "Some Observations on
Victorian Religious Biography: Newman and Manning."
Worship 43 (Aug.-Sept. 1969), 407-415.

5143. ALTHOLZ, JOSEF LEWIS. "Truth and Equivocation:
Liguori's Moral Theology and Newman's 'Apologia'."
Church History 44 (1975), 73-84.

5144. ALVAREZ DE LINERA GRUND, ANTONIO. "La vida y la
mentalidad del Cardinal Newman: su mutuo influjo."
Las Ciencias 13 (1948), 385-407.

5145. ANNAN, NOEL GILROY. "Famous Controversies:
Kingsley and Cardinal Newman." New Statesman and
Nation 27 (Mar. 25, 1944), 209.

5146. ANON. "Cardinal Newman." Edinburgh Review 215
(Apr. 1912), 263-290. (Also in Living Age 273,
Apr. 6, 1912, 48-55)

5147. ANON. "Cardinal Newman's Failure to Blend Rome and
Modernism." Current Literature 52 (June 1912),
678-681.

5148. ANON. "Newman and Manning." Quarterly Review 206
(Apr. 1907), 354-383.

5149. ANTHONY, GERALDINE, SISTER. "Newman's Definition
of Faith from the 'Oxford University Sermons' to
the 'Grammar of Assent'." Cithara 9 (May 1970),
47-63.

5150. ARNOLD, J. W. "Newman: The First Modern Catholic." Ave Maria 89 (Feb. 28, 1959), 22-25.

5151. ARTZ, JOHANNES. "Entstehung und Auswirkung von
       Newmans Theorie der Dogmenentwicklung." Tübinger
       Theologische Quartalschrift 148, Heft 1 (1968),
       63-104; Heft 2 (1968), 167-198.

5152. ARTZ, JOHANNES. "Newman als Brücke zwischen
       Canterbury und Rom." Una Sancta 22 (1967), 173-185.

5153. ARTZ, JOHANNES. "Newman and Intuition." Philosophy
       Today 1 (Mar. 1957), 10-15.

5154. ARTZ, JOHANNES. "Newman as Philosopher." Inter-
       national Philosophical Quarterly 16 (Sept. 1976),
       263-287.

5155. ARTZ, JOHANNES. "Newman's Contribution to Theory
       of Knowledge." Philosophy Today 4 (Sept. 1960),
       12-25.

5156. ATKINS, A. "Newman's 'Apologia' and Lackmann's
       Ecumenism." Journal of Ecumenical Studies 2 (Fall
       1965), 406-425.

5157. ATKINS, GAIUS GLENN. Life of Cardinal Newman. New
       York and London: Harper & Brothers, 1931.

5158. AUBERT, ROGER. Le Problème de l'acte de foi:
       données traditionnelles et résultats des controverses
       récentes. Louvain: E. Warny, 1945. (Newman on
       faith and conscience)

5159. BACCHUS, FRANCIS JOSEPH. "How to Read the 'Grammar
       of Assent'." Month 143 (1924), 106-115.

5160. BACCHUS, FRANCIS JOSEPH. "Newman's Oxford Univer-
       sity Sermons." Month 140 (1922), 1-12.

5161. BAKER, A. E. Prophets for an Age of Doubt: Job,
       Socrates, Pascal, Newman. London: Centenary
       Press, 1934.

5162. BANKERT, M. "Newman in the Shadow of Barchester
       Towers." Renascence 20 (Sept. 1968), 153-161.

5163. BANTOCK, G. H. "Newman and Education." Cambridge
       Journal 4 (1951), 660-678.

5164. BARCUS, JAMES EDGAR. "Structuring the Rage Within:
       The Spiritual Autobiographies of Newman and Orestes
       Brownson." Cithara 15 (Nov. 1975), 45-57.

5165. BARMANN, LAWRENCE F. "Newman and the Theory of Doctrinal Development." American Ecclesiastical Review 143 (Aug. 1960), 121-129.

5166. BARMANN, LAWRENCE F. (Ed.) Newman at St. Mary's. Westminster, Maryland: Newman Press, 1962. (Selections from sermons)

5167. BARMANN, LAWRENCE F. "The Notion of Personal Sin in Newman's Thought." Downside Review No. 268 (1964), 209-221.

5168. BARMANN, LAWRENCE F. "The Spiritual Teaching of Newman's Early Sermons." Downside Review No. 260 (1962), 226-242.

5169. BARRY, WILLIAM FRANCIS. Newman. New York: C. Scribner's Sons, 1904.

5170. BASTABLE, J. D. "Cardinal Newman's Philosophy of Belief." Irish Ecclesiastical Record 83 (Apr.-June 1955), 241-252, 346-351, 436-441.

5171. BASTABLE, J. D. "The Germination of Belief within Probability According to Newman." Philosophical Studies 11 (1961), 8-111.

5172. BAUDIN, E. "La Philosophie de la foi chez Newman." Revue de Philosophie 8 (1906), 571-598; 9 (1906), 20-55, 253-285, 373-390.

5173. BAYART, P. "La Conversion de Newman." Revue de Lille 21 (1903), 747-763, 821-836.

5174. BAYNES, ARTHUR HAMILTON. "From Newman to Gore." Hibbert Journal 32 (1933), 1-8.

5175. BECK, A. "The Teaching of Newman on Church and State." Clergy Review 25 (Oct. 1945), 444-453.

5176. BECK, GEORGE ANDREW. "Newman and Theology in the University." Tablet 218 (Aug. 15, 1964), 909; (Aug. 22, 1964), 937-938.

5177. BECKER, WERNER. "Zu Newmans Stellung zur Welt am Ende seiner evangelikalen Phase." In Helmut Kuhn, Heinrich Kahlefeld and Karl Forster (Eds.), Interpretation der Welt. Würzburg: Echter, 1965, 544-570.

5178.   BENARD, EDMOND DARVIL.   A Preface to Newman's
        Theology.   St. Louis, Mo. and London: B. Herder
        Book Co., 1945.

5179.   BENARD, EDMOND DARVIL.   "The Problem of Belief in
        the Writings of John Henry Newman, William James
        and St. Thomas Aquinas."   Ph.D. Dissertation,
        Catholic University of America, 1950.

5180.   BERGERON, RICHARD.   Les Abus de l'Eglise d'après
        Newman: étude de la préface à la troisième édition
        de la Via Media.   Montréal: Bellarmin, 1971.

5181.   BERNARD, MIRIAM, SISTER.   "John H. Newman: Saint
        of Sincerity."   Catholic World 157 (Apr. 1943),
        66-73.

5182.   BERRANGER, OLIVER DE.   "Dogme et existence dans
        l'oeuvre de Newman."   Revue des Sciences
        Philosophiques et Théologiques 58 (Jan. 1974),
        3-39.

5183.   BIEMER, GÜNTER.   Newman on Tradition.   Trans. and
        Ed. Kevin Smyth.   London: Burns and Oates, 1967.

5184.   BLEHL, VINCENT FERRER.   "The 'Apologia':   Reactions
        1864-5."   Month 31 (1964), 267-277.

5185.   BLEHL, VINCENT FERRER.   "The Council: Newman and
        the Problem of Freedom and Authority."   America 107
        (Oct. 27, 1962), 950-952.

5186.   BLEHL, VINCENT FERRER.   "The Holiness of John Henry
        Newman."   Month 19 (June 1958), 325-334.

5187.   BLEHL, VINCENT FERRER.   "The Letters of John Henry
        Newman: July 1851-June 1852."   Ph.D. Dissertation,
        Harvard University, 1959.

5188.   BLEHL, VINCENT FERRER.   "Newman and the Missing
        Miter."   Thought 35 (Spring 1960), 110-123.

5189.   BLEHL, VINCENT FERRER.   "Newman on Trial."   Month
        27 (1962), 69-80.

5190.   BLEHL, VINCENT FERRER.   "Newman, the Bishops and
        'The Rambler'."   Downside Review 90 (1972), 20-40.

5191.   BLEHL, VINCENT FERRER.   "Newman, the Fathers, and
        Education."   Thought 45 (1970), 196-212.
5192.   BLEHL, VINCENT FERRER.   "Newman's Delation: Some
        Hitherto Unpublished Letters."   Dublin Review 234
        (1960), 296-305.
5193.   BLEHL, VINCENT FERRER.   "The Patristic Humanism of
        John Henry Newman."   Thought 50 (Sept. 1975),
        266-274.
5194.   BLEHL, VINCENT FERRER.   "The Sanctity of Cardinal
        Newman."   America 99 (June 14, 1958), 328-330.
5195.   BLENNERHASSETT, CHARLOTTE JULIA.   John Henry
        Kardinal Newman: Ein Beitrag zur religiösen
        Entwicklungsgeschichte der Gegenwart.   Berlin:
        Gebrüder Paetel, 1904.
5196.   BLENNERHASSETT, SIR ROWLAND.   "Some of My Recol-
        lections of Cardinal Newman."   Cornhill Magazine 11
        (1901), 615-630.
5197.   BLOSS, WILLIAM ESCOTT.   'Twixt the Old and the
        New: A Study in the Life and Times of John Henry,
        Cardinal Newman.   London: Society for Promoting
        Christian Knowledge, 1916.
5198.   BLUNT, HUGH F.   "In Quest of Newman."   American
        Ecclesiastical Review 113 (1945), 253-263.
5199.   BLYTON, W. J.   "Side-lights on Newman."   Month 171,
        No. 886 (Apr. 1938), 301-310.
5200.   BOEKRAAD, ADRIAN J. AND HENRY TRISTRAM.   The
        Argument from Conscience to the Existence of God
        According to J. H. Newman.   Louvain: Editions Nau-
        welaerts, 1961.
5201.   BOEKRAAD, ADRIAN J.   "Newman Studies."   Philosophical
        Studies (Ireland) 20 (1972), 185-202; 22 (1974),
        198-222.
5202.   BOEKRAAD, ADRIAN J.   "Newman's Argument to the
        Existence of God."   Philosophical Studies 6 (Dec.
        1956), 50-71.

5203.   BOEKRAAD, ADRIAN J.   "The Personal Conquest of
        Truth." Louvain Studies 5 (Fall 1974), 136-148.

5204.   BOEKRAAD, ADRIAN J.   The Personal Conquest of
        Truth According to J. H. Newman. Louvain:
        Editions Nauwelaerts, 1955.

5205.   BOUYER, LOUIS.   "Great Preachers: John Henry
        Newman." Theology 55 (Mar. 1952), 87-91.

5206.   BOUYER, LOUIS.   "Newman and English Platonism."
        Monastic Studies 1 (1963), 111-131. (See 5207)

5207.   BOUYER, LOUIS.   "Newman et le platonisme de l'âme
        anglaise." Revue de Philosophie 36, No. 4 (1936),
        285-305. (For translation see 5206)

5208.   BOUYER, LOUIS.   Newman: His Life and Spirituality.
        Trans. James Lewis May. New York: P. J. Kenedy,
        1958.

5209.   BRECHTKEN, JOSEF.   Kierkegaard, Newman: Wahrheit
        und Existenzmitteilung. Meisenheim am Glan: A.
        Hain, 1970.

5210.   BREMOND, HENRI.   The Mystery of Newman. Trans. H.
        C. Corrance. London: Williams and Norgate, 1907.

5211.   BRENDON, PIERS.   "Newman, Keble and Froude's
        'Remains'." English Historical Review 87 (Oct.
        1972), 697-716.

5212.   BRICKEL, A. G.   "Cardinal Newman's Theory of Knowl-
        edge." American Catholic Quarterly Review 43
        (July-Oct. 1918), 507-518, 645-653.

5213.   BRICKEL, A. G.   "Newman's Criteria of Historical
        Evolution." American Catholic Quarterly Review 44
        (Oct. 1919), 588-594.

5214.   BRINKMAN, MARY LEONILLA, SISTER.   "The Analogous
        Image and Its Development in the Thought and Prose
        Style of John Henry Newman, with Special Reference
        to His Anglican Sermons." Dissertation Abstracts
        28 (1968), 5008A (University of Wisconsin, 1968).

5215.   BROICHER, CHARLOTTE.   "Anglikanische Kirche und
        deutsche Philosophie." Preussische Jahrbücher 142
        (1910), 205-233, 457-498. (Includes S. T. Coleridge)

5216.   BROWN, COLIN GREVILLE. "Newman's Minor Critics."
        Downside Review 89 (1971), 13-21.

5217.   BURCH, VACHER. "Newman and the Vision Keble Saw."
        Theology 27 (Sept. 1933), 130-139.

5218.   BURGUM, EDWIN B. "Cardinal Newman and the Complexity
        of Truth." Sewanee Review 38 (July 1930), 310-327.

5219.   BUTLER, B. C. "The Lost Leader." Downside Review
        69 (1951), 62-73.

5220.   BUTLER, FRANCIS JOSEPH. "John Henry Newman's
        'Parochial and Plain Sermons' Viewed as a Critique
        of Religious Evangelicalism." Dissertation Abstracts
        International 33 (1972), 1217-18A (Catholic University
        of America, 1972).

5221.   BUTLER, GIBBON FRANCIS. "John Henry Newman's Use
        of History in His Anglican Career, 1825-1845."
        Microfilm Abstracts 10, No. 4 (1950), 207-208
        (University of Illinois, 1950).

5222.   CADMAN, SAMUEL PARKES. The Three Religious Leaders
        of Oxford and Their Movements: John Wycliffe,
        John Wesley, John Henry Newman. New York: The
        Macmillan Company, 1916.

5223.   CALKINS, ARTHUR BURTON. "John Henry Newman on
        Conscience and the Magisterium." Downside Review
        87 (Oct. 1969), 358-369.

5224.   CAMERON, JAMES MUNRO. "Faith and the Mind."
        Listener 57 (1957), 15-16, 51-52. (Two talks on
        Newman. I. "The Night Battle"; II. "The Logic of
        the Heart")

5225.   CAMERON, JAMES MUNRO. John Henry Newman. London,
        New York: Published for the British Council by
        Longmans, Green, 1956.

5226.   CAMERON, JAMES MUNRO. The Night Battle. London:
        Catholic Book Club, 1962.

5227. CAMERON, JAMES MUNRO. "The Night Battle: Newman
       and Empiricism." Victorian Studies 4 (Dec. 1960),
       99-117.

5228. CAMPBELL, WILLIAM D. "Approaching the Venture of
       Faith: An Apologetical Study of the Rational
       Development and Experience of a Convert Prior to
       the Act of Faith According to the Life and Writings
       of John Henry Cardinal Newman." Dissertation
       Abstracts 25 (1964), 3134-35 (Catholic University
       of America, 1964).

5229. CAPPS, DONALD. "A Biographical Footnote to New-
       man's 'Lead, Kindly Light'." Church History 41
       (1972), 480-486.

5230. CAPPS, DONALD. "John Henry Newman: A Study of
       Religious Leadership." Ph.D. Dissertation, Uni-
       versity of Chicago, 1970.

5231. CAPPS, DONALD. "John Henry Newman: A Study of
       Vocational Identity." Journal for the Scientific
       Study of Religion 9, No. 1 (Spring 1970), 33-51.

5232. CAPPS, DONALD. "Newman Studies: Experiments in
       Crossbreeding." Journal of Religion (University of
       Chicago) 53 (Jan. 1973), 136-140.

5233. CARRY, EUGENE. Les Années anglicanes du Cardinal
       Newman. Genève: A. Garin, 1901.

5234. CASTLE, WILLIAM RICHARDS. "Newman and Coleridge."
       Sewanee Review 17 (Apr. 1909), 139-152.

5235. CAVANAGH, REV. PATRICK EDMUND. "The Doctrine of
       Assent of John Henry Newman." Dissertation Abstracts
       26 (1966), 6772-73 (University of Notre Dame,
       1964).

5236. CHAPMAN, H. J. "Newman and the Fathers." Black-
       friars 14 (July 1933), 578-590.

5237. CHAPMAN, RAYMOND. "The Dean Who Attacked Newman."
       Heythrop Journal 11 (Oct. 1970), 423-426. (Henry
       Cotton, Dean of Lisnore)

5238.   CHESTER, VERA, SISTER.   "The Rhetorician as Theolo-
        gian: A Study of the Sermons of John Henry Newman."
        Dissertation Abstracts International 33 (1972),
        387A (Marquette University, 1971).

5239.   CHESTERTON, CECIL.   "The Art of Controversy:
        Macaulay, Huxley, and Newman." Catholic World
        105 (July 1917), 446-456.

5240.   CHEVALIER, JACQUES. Trois conférences d'Oxford:
        Aristote - Pascal - Newman. Paris: Editions Spes,
        1928. (Revised 1933 with sub-title, St. Thomas -
        Pascal - Newman)

5241.   COATS, R. H. "Birmingham Mystics of the Mid-Victorian
        Era." Hibbert Journal 16 (Apr. 1918), 485-494.

5242.   COGNET, LOUIS. Newman ou la recherche de la vérité.
        Paris: Desclée, 1967.

5243.   COLBY, ROBERT A. "The Structure of Newman's 'Apologia
        Pro Vita Sua' in Relation to His Theory of Assent."
        Dublin Review 227 (1953), 140-156. (Reprinted in
        David J. DeLaura, Ed., Apologia Pro Vita Sua. New
        York: Norton, 1968, 465-480)

5244.   COLLINS, JAMES DANIEL. (Ed.) Philosophical Readings
        in Cardinal Newman. Chicago: H. Regnery Co.,
        1961.

5245.   CONACHER, W. M. "Newman and Liberal Education."
        Queen's Quarterly 54, No. 4 (Winter 1947), 440-450.

5246.   CONNOLLY, FRANCIS X. "Newman and Science." Thought
        38 (Sept. 1963), 107-121.

5247.   CONNOLLY, FRANCIS X. AND VINCENT FERRER BLEHL.
        (Eds.) Newman's Apologia: A Classic Reconsidered.
        New York: Harcourt, Brace and World, 1964.

5248.   CONWAY, M. D. "Newman in Retrospect: A Century of
        Catholic Intellectual Growth in England." America
        74 (Oct. 6, 1945), 6-8.

5249.   CORBISHLEY, P. D. "Outstanding Figure of the
        Nineteenth Century." Heritage 2 (Apr. 1956), 7-8.

5250. CORCORAN, TIMOTHY. "Liberal Studies and Moral
       Aims: A Critical Study of Newman's Position."
       Thought 1 (1926), 54-71.

5251. COULSON, JOHN. "Belief and Imagination." Downside
       Review 90 (Jan. 1972), 1-14.

5252. COULSON, JOHN. "Front-line Theology: A Comment on
       Newman and Lonergan." Clergy Review 58 (Oct.
       1973), 803-811.

5253. COULSON, JOHN, ARTHUR MACDONALD ALLCHIN AND MERIOL
       TREVOR. Newman: A Portrait Restored. An Ecumenical
       Revaluation. London: Sheed and Ward, 1965.

5254. COULSON, JOHN. Newman and the Common Tradition.
       Oxford: Clarendon Press, 1970.

5255. COULSON, JOHN AND ARTHUR MACDONALD ALLCHIN. (Eds.)
       The Rediscovery of Newman: An Oxford Symposium.
       London: Sheed and Ward, 1967.

5256. CRANNY, T. "A Study in Contrasts: Newman and
       Faber." American Ecclesiastical Review 129 (1953),
       300-313.

5257. CRAWFORD, CHARLOTTE ELIZABETH. "Newman's 'Call-
       ista' and the Catholic Popular Library." Modern
       Language Review 45 (1950), 219-221.

5258. CRAWFORD, CHARLOTTE ELIZABETH. "The Novel That
       Occasioned Newman's 'Loss and Gain'." Modern
       Language Notes 45 (1950), 414-418.

5259. CRONIN, JOHN FRANCIS. Cardinal Newman: His Theory
       of Knowledge. Washington, D.C.: The Catholic
       University of America, 1935.

5260. CRONIN, JOHN FRANCIS. "Cardinal Newman's Theory of
       Knowledge." American Catholic Philosophic Associa-
       tion Proceedings 11 (Dec. 1935), 141-149.

5261. CROSS, FRANK LESLIE. John Henry Newman: With
       a Set of Unpublished Letters. London: Allan,
       1928.

5262. CROSS, FRANK LESLIE. "Newman and the Doctrine of
       Development." Church Quarterly Review 115, No. 230
       (Jan. 1933), 245-257.

5263. CULLER, ARTHUR DWIGHT. The Imperial Intellect:
       A Study of Newman's Educational Ideal. New Haven:
       Yale University Press, 1955.

5264. CULLER, ARTHUR DWIGHT. "Newman on the Uses of
       Knowledge." Journal of General Education 4 (1950),
       269-279.

5265. CUNLIFFE-JONES, H. "Newman on Justification."
       Clergy Review 54 (Feb. 1969), 117-124.

5266. CUNNINGHAM, A., SR. "Theology and Humanism:
       Newman Revisited." Chicago Studies 14 (Sept.
       1975), 67-82.

5267. DALE, P. A. "Newman's 'The Idea of a University':
       The Dangers of a University Education." Victorian
       Studies 16 (Sept. 1972), 5-36.

5268. D'ARCY, MARTIN CYRIL. The Nature of Belief.
       London: Sheed and Ward, 1931; new ed., Dublin:
       Clonmore & Reynolds, 1958. (Includes A Grammar of
       Assent; see 5638)

5269. DARK, SIDNEY. Newman. London: Duckworth, 1934.

5270. DAVIS, H. FRANCIS. "Doctrine, Development of." In
       A Catholic Dictionary of Theology. London: Thomas
       Nelson & Sons, 1967, Vol. 2, 177-189.

5271. DAVIS, H. FRANCIS. "English Spiritual Writers:
       IV. Cardinal Newman." Clergy Review 44 (1959),
       132-145.

5272. DAVIS, H. FRANCIS. "Is Newman's 'Idea' Practical?"
       Irish Ecclesiastical Record 78 (1952), 241-254.

5273. DAVIS, H. FRANCIS. "Is Newman's Theory of Develop-
       ment Catholic?" Blackfriars 39 (1958), 310-321.

5274. DAVIS, H. FRANCIS. "Newman and Our Lady." Clergy
       Review 34 (1950), 369-379.

5275. DAVIS, H. FRANCIS. "Newman and the Psychology of
       the Development of Doctrine." Dublin Review 216
       (Apr. 1945), 97-107.

5276. DAVIS, H. FRANCIS. "Newman, Christian or Humanist."
       Blackfriars 37 (1956), 516-526.

5277.   DAVIS, H. FRANCIS.   "Newman on Belief and Unbelief:
        Excerpt from Newman: The Individual and the Church."
        Commonweal 66 (Aug. 9, 1957), 473.

5278.   DAVIS, H. FRANCIS.   "Newman on Educational Method:
        Educating for Real Life." Dublin Review 230, No.
        472 (Winter 1956-57), 101-113.

5279.   DAVIS, H. FRANCIS.   "Newman on Faith and Personal
        Certitude." Journal of Theological Studies 12
        (Oct. 1961), 248-259.

5280.   DAVIS, H. FRANCIS.   "Newman: The Individual and
        the Church." Blackfriars 38 (1957), 214-220.

5281.   DAVIS, H. FRANCIS.   "Newman's Cause." Blackfriars
        33 (Oct. 1952), 396-404.

5282.   DAVIS, H. FRANCIS.   "Parallel between Newman and
        Thomism." Blackfriars (Supplement) 26 (Oct. 1945),
        129-136.

5283.   DAVIS, H. FRANCIS.   "Le Rôle et l'apostolat de la
        hiérarchie et du laïcat dans la théologie de l'Eglise
        chez Newman." Revue des Sciences Religieuses 34
        (1960), 329-349.   (Reprinted in M. Nédoncelle et
        al., L'Ecclésiologie au XIXe siècle. Paris:
        Editions du Cerf, 1960, 329-349)

5284.   DAVIS, H. FRANCIS.   "Was Newman a Disciple of
        Coleridge?" Dublin Review 217 (Oct. 1945), 165-173.

5285.   D'CRUZ, F. A.   Cardinal Newman: His Place in
        Religion and Literature. Madras: The "Good Pastor"
        Press, 1935.

5286.   DEHAVILLAND, JAMES RAINER. Newman, 1801-1890.
        Dijon: Publ. Lumière, 1927.

5287.   DELATTRE, FLORIS. La Pensée de J. H. Newman:
        extraits les plus caractéristiques de son oeuvre.
        Paris: Librairie Payot, 1914.

5288.   DELAURA, DAVID JOSEPH. (Ed.) Apologia Pro Vita
        Sua. New York: W. W. Norton & Co., Inc., 1968.

5289.   DELAURA, DAVID JOSEPH.   "Matthew Arnold and John
        Henry Newman: The 'Oxford Sentiment' and the
        Religion of the Future."   Texas Studies in Liter-
        ature and Language 6 (1965), 573-702.   (Reprinted
        in Hebrew and Hellene in Victorian England: Newman,
        Arnold, and Pater.   Austin, London: University of
        Texas Press, 1969, 5-161)

5290.   DELAURA, DAVID JOSEPH.   "Newman as Prophet."
        Dublin Review 241 (1967), 222-235.

5291.   DELAURA, DAVID JOSEPH.   "Pater and Newman: The
        Road to the Nineties."   Victorian Studies 10 (1966),
        39-69.   (Reprinted in Hebrew and Hellene in Victorian
        England: Newman, Arnold, and Pater.   Austin,
        London: University of Texas Press, 1969, 305-344)

5292.   DE SANTIS, EDWARD.   "Newman's Concept of the Church
        in the World in His 'Parochial and Plain Sermons'."
        American Benedictine Review 21 (June 1970), 268-282.

5293.   DESSAIN, CHARLES STEPHEN.   "Cardinal Newman and
        Ecumenism."   Clergy Review 50 (1965), 119-137,
        189-206, 476, 647.

5294.   DESSAIN, CHARLES STEPHEN.   "Cardinal Newman and the
        Doctrine of Uncreated Grace."   Clergy Review 47
        (1962), 207-225, 269-288.

5295.   DESSAIN, CHARLES STEPHEN.   "Cardinal Newman Considered
        as a Prophet."   Consilium: Internationale Zeitschrift
        für Theologie 7, No. 4 (Sept. 1968), 41-50.

5296.   DESSAIN, CHARLES STEPHEN.   "Cardinal Newman on the
        Laity."   Life Spirit 16 (1961), 51-62.

5297.   DESSAIN, CHARLES STEPHEN.   "Cardinal Newman on the
        Theory and Practice of Knowledge: The Purpose of
        the 'Grammar of Assent'."   Downside Review 75 (Jan.
        1957), 1-23.

5298.   DESSAIN, CHARLES STEPHEN.   "Cardinal Newman's
        Papers: A Complete Edition of His Letters."
        Dublin Review 234 (Winter 1960), 291-296.

5299. DESSAIN, CHARLES STEPHEN. "English Spiritual
       Writers: XVII. Newman's Spirituality and Its
       Value Today." Clergy Review 45 (May 1960), 257-282.
       (Reprinted in Charles Davis, Ed., English Spiritual
       Writers. London: Burns and Oates, 1961, 136-160)

5300. DESSAIN, CHARLES STEPHEN. "'Heart Speaks to Heart':
       Margaret Mary Hollahan and John Henry Newman."
       The Month 34 (1965), 360-367.

5301. DESSAIN, CHARLES STEPHEN. "Infallibility: What
       Newman Taught in Manning's Church." In Infallibility
       in the Church: An Anglican-Catholic Dialogue.
       London: Darton, Longman & Todd, 1968, 59-80.
       (Book listed under title in National Union Catalog)

5302. DESSAIN, CHARLES STEPHEN. John Henry Newman.
       London: Nelson, 1966.

5303. DESSAIN, CHARLES STEPHEN. "Newman and Oxford."
       Wiseman Review 237 (Autumn 1963), 295-302.

5304. DESSAIN, CHARLES STEPHEN. "The Newman Archives and
       the Projected Edition of the Cardinal's Letters."
       Catholic Historical Review 46 (1960), 22-26.

5305. DESSAIN, CHARLES STEPHEN. "Newman's First Con-
       version: A Great Change of Thought, August 1st
       till December 21st 1816." Studies 46 (Spring
       1957), 44-59.

5306. DESSAIN, CHARLES STEPHEN. "Newman's Singleness of
       Purpose." Life of the Spirit 11 (Mar. 1957),
       413-416.

5307. DESSAIN, CHARLES STEPHEN. "An Unpublished Paper by
       Cardinal Newman on the Development of Doctrine."
       Journal of Theological Studies 9 (1958), 324-329.

5308. DEVINE, GEORGE. (Ed.) New Dimensions in Religious
       Experience. Proceedings of the College Theology
       Society for 1970. Staten Island, N.Y.: Alba
       House, 1971. (Includes articles on Newman and
       doctrinal development, Newman and theological
       pluralism, the consensus of the faithful and the
       magisterium)

5309. DIBBLE, ROMUALD A. John Henry Newman: The Concept
        of Infallible Doctrinal Authority. Washington:
        Catholic University of America Press, 1955.

5310. DINSMORE, CHARLES A. "Newman and Bright." Construc-
        tive Quarterly 3 (Mar. 1915), 231-247.

5311. DOLAN, GERALD M., O.F.M. "The Gift of the Spirit
        According to John Henry Newman (1828-1839)."
        Franciscan Studies 30 (1970), 77-130.

5312. DONALD, GERTRUDE. Men Who Left the Movement:
        John Henry Newman, Thomas W. Allies, Henry Edward
        Manning, Basil William Maturin. London: Burns,
        Oates, and Washbourne Ltd., 1933.

5313. DONALDSON, AUGUSTUS BLAIR. Five Great Oxford
        Leaders: Keble, Newman, Pusey, Liddon and Church.
        London: Rivingtons, 1900.

5314. DONOVAN, CHARLES F. "Newman, a Light amid Encircling
        Gloom." American Ecclesiastical Review 113 (1945),
        366-376.

5315. DOWNES, DAVID ANTHONY. The Temper of Victorian
        Belief: Studies in the Religious Novels of Pater,
        Kingsley, and Newman. New York: Twayne Publishers,
        1972.

5316. DRIVER, ARTHUR HARRY. "On Certain Aspects of John
        Henry Newman and Robert William Dale." Congrega-
        tional Quarterly 24, No. 1 (Jan. 1946), 31-40.

5317. DUIVESTEYN, D. "Reflexions on Natural Law: Cardinal
        Newman on Conscience." Clergy Review 52 (1967),
        283-294.

5318. DUPUY, BERNARD-DOMINIQUE. "Bulletin d'histoire des
        doctrines: Newman." Revue des Sciences Philoso-
        phiques et Théologiques 45 (1961), 125-176.

5319. DUPUY, BERNARD-DOMINIQUE. "Bulletin d'histoire des
        doctrines: Newman aujourd'hui." Revue des Sciences
        Philosophiques et Théologiques 56 (1972), 78-126.

5320. EAKER, J. GORDON. "John Henry Newman's Contribution
        to Belief." Forum (Houston) 4, No. 2 (1963),
        22-25.

5321.   ELBERT, JOHN ALOYSIUS. Evolution of Newman's
        Conception of Faith. Philadelphia: Dolphin Press,
        1933.

5322.   ELLIS, JOHN TRACY. "John Henry Newman: A Bridge
        for Men of Good Will." Catholic Historical Review
        56 (Apr. 1970), 1-24.

5323.   EVANS, ARTHUR WILLIAM. (Ed.) Tract Ninety: Or,
        Remarks on Certain Passages in the Thirty-nine
        Articles, by John Henry Newman. Reprinted from
        the Edition of 1841, with an Historical Commentary
        by A. W. Evans. London: Constable and Co., Ltd.,
        1933.

5324.   EVANS, JOHN WHITNEY. "Lonergan and Newman on
        Theology in the University." Religious Education
        64 (May 1969), 184-187.

5325.   FATHERS OF THE BIRMINGHAM ORATORY. (Eds.) Corres-
        pondence of John Henry Newman with John Keble
        and Others, 1839-45. Edited at the Birmingham
        Oratory. London, New York: Longmans, Green, and
        Co., 1917.

5326.   FATHERS OF THE BIRMINGHAM ORATORY. (Eds.) Faith
        and Prejudice, and Other Unpublished Sermons. New
        York: Sheed and Ward, 1956.

5327.   FEMIANO, SAMUEL D. Infallibility of the Laity:
        The Legacy of Newman. New York: Herder and Herder,
        1967.

5328.   FENTON, JOSEPH C. "John Henry Newman and the
        Vatican Definition of Papal Infallibility." American
        Ecclesiastical Review 113 (1945), 300-320.

5329.   FENTON, JOSEPH C. "The Newman Legend and Newman's
        Complaints." American Ecclesiastical Review 139
        (Aug. 1958), 101-121.

5330.   FENTON, JOSEPH C. "Some Newman Autobiographical
        Sketches and the Newman Legend." American
        Ecclesiastical Review 136 (June 1957), 394-410.

5331.   FLANAGAN, PHILIP. Newman: Faith and the Believer.
        London: Sands, 1946.

5332. FLEISSNER, ROBERT F. "'Middlemarch' and Idealism:
       Newman, Naumann, Goethe." Greyfriar 14 (1973),
       21-26.

5333. FLETCHER, JEFFERSON B. "Newman and Carlyle."
       Atlantic Monthly 95 (1905), 669-679.

5334. FLOOD, JOSEPH MARY. Cardinal Newman and Oxford.
       London: I. Nicholson and Watson, 1933.

5335. FOLGHERA, JEAN DOMINIQUE. Newman's Apologetic.
       Trans. P. Hereford. London: Sands & Co., 1928.

5336. FRANDSEN, CARL F. "A Rhetorical Analysis of John
       Henry Cardinal Newman's 'Lectures on the Present
       Position of Catholics in England'." Dissertation
       Abstracts International 36 (1975), 3685A (City
       University of New York, 1975).

5337. FREMANTLE, A. "What, Then, Does Dr. Newman Mean?"
       Commonweal 47 (Dec. 19, 1947), 250-253. (See 5383)

5338. FRIEDEL, FRANCIS J. The Mariology of Cardinal
       Newman. New York, Cincinnati: Benziger Brothers,
       1928.

5339. FRIES, HEINRICH. "Autorität und Freiheit im Leben
       und Denken von John Henry Newman." Catholica 25,
       No. 4 (1971), 249-259.

5340. FRIES, HEINRICH. "J. H. Newmans Beitrag zum
       Verständnis der Tradition." In Michael Schmaus
       (Ed.), Die mündliche Überlieferung. München: M.
       Hueber, 1957, 63-122.

5341. FRIES, HEINRICH. "Neues von und über Newman."
       Theologische Quartalschrift 130 (1950), 54-78.

5342. FRIES, HEINRICH AND WERNER BECKER. (Eds.) Newman-
       Studien. Nürnberg: Glock und Lutz, 1948- .

5343. FRIES, HEINRICH. Die Religionsphilosophie Newmans.
       Stuttgart: Schwaben-Verlag, 1948.

5344. FROST, FRANCIS. "Le Personnalisme dans l'itinér-
       aire théologique et spirituel de J. H. Newman."
       Mélanges de Science Religieuse 32 (Apr. 1975),
       57-70.

5345. FULLER, E. "Arnold, Newman, and Rossetti." Critic
45 (Sept. 1904), 273-276.

5346. GANIM, CAROLE ANN, O.S.U., SISTER. "The Biographical
Writings of John Henry Newman." Dissertation
Abstracts International 31 (1971), 5361A (Fordham
University, 1970).

5347. GARNETT, EMMELINE. Tormented Angel: A Life of
John Henry Newman. New York: Farrar, Straus,
1966.

5348. GERAETS, THEODORE F. "L'Existence de Dieu."
Eglise et Théologie 1 (Jan. 1970), 61-81.

5349. GERARD, T. J. "Bergson, Newman, and Aquinas."
Catholic World 96 (Mar. 1913), 748-762.

5350. GIBERT, J. "Histoire de la publication de
'l'Apologia Pro Vita Sua' dans ses éditions succes-
sives." Etudes Anglaises 27 (1974), 262-274.

5351. GILL, JOHN M. "Newman's Dialectic in 'The Idea of
a University'." Quarterly Journal of Speech 45
(1959), 415-418.

5352. GLADEN, KARL. "Die Erkenntnisphilosophie John Henry
Newmans im Lichte der thomistischen Erkenntnislehre
beurteilt." Ph.D. Dissertation, Universität Bonn,
1933.

5353. GOETZ, J. "Coleridge, Newman and Tyrrell: A
Note." Heythrop Journal 14 (Oct. 1973), 431-436.

5354. GORCE, DENYS. "Etapes de la conversion de Newman."
Revue Générale Belge 89 (1953), 245-258.

5355. GORCE, DENYS. Le Martyre de Newman: à la lumière
par la croix. Paris: Bonne Presse, 1948.

5356. GORCE, DENYS. Newman et les pères (sources de
sa conversion et de sa vie intérieure):   à
l'occasion du centenaire du mouvement d'Oxford
(1833). Juvisy, Seine-et-Oise: Les Editions du
Cerf, 1933.

5357. GORDON, B. "Newman, and the Changing Scene in
University Education." International Review of
Education 9, No. 1 (1963-64), 1-11.

5358.   GORNALL, T.   "Newman on Catholicism: Difficulties
         with the Argument of the Marks of the Church."
         Heythrop Journal 5 (Oct. 1964), 436-439.

5359.   GOUT, RAOUL.   Du protestantisme au catholicisme:
         John-Henry Newman, 1801-1845.   Genève: J.-H.
         Jeheber, 1906.

5360.   GOYAU, MME. LUCIE (FAURE).   Newman: sa vie et
         ses oeuvres.   Paris: Perrin et Cie, 1901.

5361.   GRAEF, HILDA.   God and Myself: The Spirituality
         of John Henry Newman.   London: P. Davies, 1967.

5362.   GRAHAM, THOMAS EMERSON.   "Newman's Idea of Saint-
         liness."   Ph.D. Dissertation, University of
         Pittsburgh, 1933.

5363.   GRAPPE, GEORGES PIERRE FRANCOIS.   J. H. Newman:
         essai de psychologie religieuse.   Préface de P.
         Bourget.   Paris: P.-J. Béduchaud, 1902.

5364.   GREAVES, R. W.   "Golightly and Newman, 1824-1845."
         Journal of Ecclesiastical History 9, No. 2 (1958),
         209-228.

5365.   GREENE, REV. RICHARD ROY, S.T.D.   "John Henry
         Newman's Theology of Doctrinal Development in
         Reference to the Oxford Movement, 1833-1843."
         Dissertation Abstracts International 36 (1976),
         4580A (Catholic University of America, 1975).

5366.   GREENLEAF, W.   "The Book That Changed England's
         Mind: Apologia Pro Vita Sua."   Catholic Digest 29
         (Oct. 1965), 25-29.

5367.   GRIFFIN, JOHN R.   "The Anglican Politics of Cardinal
         Newman."   Anglican Theological Review 55 (Oct.
         1973), 434-443.

5368.   GRIFFIN, JOHN R.   "In Defense of Newman's 'Gentle-
         man'."   Dublin Review 505 (1965), 245-254.

5369.   GUINAN, A.   "Newman and the People."   Commonweal 62
         (July 29, 1955), 418-420.

5370.   GUITTON, JEAN M. P.   La Philosophie de Newman:
         essai sur l'idée du développement.   Paris: Boiven et
         Cie, 1933.

5371. GUNDERSEN, BORGHILD. Cardinal Newman and Apolo-
       getics. Oslo: Dybwad, 1952.

5372. GUNNING, JOHANNES HERMANUS. John Henry, Kardinaal
       Newman: een boek voor protestanten en roomsch-
       katholieken. Amsterdam: H. J. Paris, 1933.

5373. GWYNN, DENIS ROLLESTON. "Dominic Barberi and
       Newman's Conversion." Clergy Review 25 (Feb.
       1945), 49-58.

5374. GWYNN, DENIS ROLLESTON. "Newman, Wiseman and Dr.
       Russell." Irish Ecclesiastical Record 58 (Sept.
       1941), 275-286.

5375. GWYNN, DENIS ROLLESTON. "Was Newman Badly Treated?"
       Clergy Review 25, No. 10 (Oct. 1945), 433-444.

5376. HAMILTON, R. "Darwin and Newman." Pax 42 (1952),
       28-33.

5377. HAMMANS, HERBERT. Die neueren katholischen
       Erklärungen der Dogmenentwicklung. Essen: Ludgerus-
       Verlag, 1965.

5378. HANSEN, N. Kardinal Newman: en historiskliterae
       Skildring. Pauluskredsen: Sønderborg, 1937.

5379. HARDT, GEORG. "John Henry Newman als Prediger."
       Ph.D. Dissertation, Universität Münster, 1928.

5380. HARPER, GORDON HUNTINGTON. Cardinal Newman and
       William Froude, F.R.S. Baltimore: Johns Hopkins
       Press, 1933.

5381. HARROLD, CHARLES FREDERICK. John Henry Newman:
       An Expository and Critical Study of His Mind,
       Thought and Art. London, New York: Longmans,
       Green & Co., Inc., 1945.

5382. HARROLD, CHARLES FREDERICK. "Newman and the
       Alexandrian Platonists." Modern Philology 37 (Feb.
       1940), 279-291.

5383. HARTH, J. S. "What, Then, Does Dr. Newman Mean?
       Reply." Commonweal (Jan. 16, 1948), 348. (See 5337)

5384. HAUGHEY, JOHN C. "The Bishops and Cardinal Newman:
       A Quotation in the Pastoral Letter of the Bishops

of the United States Takes on Added Meaning in Its
Original Context." America 119 (Nov. 30, 1968),
554-555.

5385. HAWKINS, DAVID G. "Cardinal Newman's Social
Philosophy." New Blackfriars 54 (Nov. 1973),
518-524.

5386. HEADLAM, A. C. "John Henry Newman." Church
Quarterly Review 74 (1912), 257-288.

5387. HEGARTY, W. J. "Cardinal Newman's Second Spring:
A Sermon Which Made History." Irish Ecclesiastical
Review 78 (July 1952), 34-43.

5388. HENKEL, WILLI. Die religiöse Situation der Heiden
und ihre Bekehrung nach John Henry Newman. Rome:
Catholic Book Agency, 1967.

5389. HENRY, LAWRENCE JOSEPH. "Newman and Development:
The Genesis of John Henry Newman's Theory of Development and the Reception of His 'Essay on the Development of Christian Doctrine'." Dissertation Abstracts
International 34 (1974), 5971A (University of Texas
at Austin, 1973).

5390. HERMANS, F. "L'Adolescence de Newman." Revue
Générale Belge (1950), 945-959.

5391. HERMANS, F. "Les Familiers du jeune Newman."
Nouvelle Revue Théologique 73 (1951), 43-58.

5392. HERMANS, F. "Newman, curé anglican, 1828-43."
Nouvelle Revue Théologique 71 (1949), 950-966.

5393. HERMANS, F. "Newman est-il un philosophe?"
Nouvelle Revue Théologique 71 (1949), 162-172.

5394. HERRICK, FRANCIS H. "Gladstone, Newman, and
Ireland in 1881." Catholic Historical Review 47
(1961-62), 342-350.

5395. HERRICK, HELEN M. "Cardinal Newman." M.A. Dissertation, University of Liverpool, 1932.

5396. HICK, JOHN. Faith and Knowledge: A Modern
Introduction to the Problem of Religious Knowledge.
Ithaca, N.Y.: Cornell University Press, 1957; 2nd
ed., 1966. (Includes Newman's A Grammar of Assent)

5397.   HICKS, JOHN, ERNEST E. SANDEEN AND ALVAN S. RYAN.
        Critical Studies in Arnold, Emerson, and Newman.
        Iowa City, Ia.: University of Iowa, 1942.
5398.   HILL, ROLAND. "Cardinal Newman's Cause." Ave
        Maria 77 (Feb. 28, 1953), 272-274.
5399.   HILL, ROLAND. "Newman und Manning." Stimmen
        der Zeit 150 (Sept. 1952), 435-439.
5400.   HIRE, R. "Our Catholic Faith: Newman on the
        Development of Doctrine." Our Sunday Visitor
        Magazine 64 (Apr. 4, 1976), 8.
5401.   HODGE, R. "Cardinal Newman: Contemplative."
        Cistercian Studies 11, No. 3 (1976), 192-227.
5402.   HOEFFKEN, THEODORE. John Henry Cardinal Newman
        on Liberal Education. Kirkwood, Missouri:
        Maryhurst Press, 1946.
5403.   HOEFFKEN, THEODORE. "The Role of the Teacher in
        Education according to Newman." The Catholic
        Educational Review 43 (Sept. 1945), 411-416.
5404.   HOGAN, JEREMIAH J. "Newman and Literature."
        Studies 42 (1953), 169-178. (Reprinted in Michael
        Tierney et al., Newman's Doctrine of University
        Education. Dublin: Sealy, Bryers and Walker,
        1954, 49-58)
5405.   HOLLIS, CHRISTOPHER. "Cardinal Newman: The Life
        Completed." Wiseman Review 237 (Spring 1963),
        78-86.
5406.   HOLLIS, CHRISTOPHER. "Cardinal Newman's Letters."
        Wiseman Review 235 (Winter 1961-62), 318-329.
5407.   HOLLIS, CHRISTOPHER. "Newman and Dean Church: A
        Friendship That Endured." Tablet 211 (June 7,
        1958), 528.
5408.   HOLLIS, CHRISTOPHER. Newman and the Modern World.
        London, Sydney: Hollis and Carter, 1967.
5409.   HOLLIS, CHRISTOPHER. "Newman the Man." Wiseman
        Review 236 (Summer 1962), 182-190.

5410. HOLMES, J. DEREK. "Cardinal Newman and the
       Affirmation Bill." Historical Magazine of the
       Protestant Episcopal Church 36 (Mar. 1967), 87-97.

5411. HOLMES, J. DEREK. "Cardinal Newman and the First
       Vatican Council." Annuarium Historiae Conciliorum
       1, Heft 2 (1969), 374-398.

5412. HOLMES, J. DEREK. "Cardinal Newman on the Philosophy
       of History." Tijdschrift voor Filosofie 32 (Sept.
       1970), 521-547.

5413. HOLMES, J. DEREK. "Cardinal Newman's Apologetic
       Use of History." Louvain Studies 4 (Fall 1973),
       338-349.

5414. HOLMES, J. DEREK. "Church and World in Newman."
       New Blackfriars 49 (June 1968), 468-474.

5415. HOLMES, J. DEREK. "Gladstone and Newman." Dublin
       Review 512 (Summer 1967), 141-153.

5416. HOLMES, J. DEREK. "J. H. Newman: History, Liberalism
       and the Dogmatic Principle." Philosophical Studies
       23 (1974), 86-106.

5417. HOLMES, J. DEREK. "John Henry Newman's Attitude
       Towards History and Hagiography." Downside Review
       92 (Oct. 1974), 248-264.

5418. HOLMES, J. DEREK. "Liberal Catholicism and Newman's
       Letter to the Duke of Norfolk." Clergy Review 60
       (Aug. 1975), 498-511.

5419. HOLMES, J. DEREK. "Newman." Victorian Studies 8
       (Mar. 1965), 271-277. (Reply to Altholz, 5141)

5420. HOLMES, J. DEREK. "Newman and Mivart: Two Atti-
       tudes to a Nineteenth-Century Problem." Clergy
       Review 50 (Nov. 1965), 852-867.

5421. HOLMES, J. DEREK. "Newman and Modernism." Baptist
       Quarterly 24 (July 1972), 335-341.

5422. HOLMES, J. DEREK. "Newman and the Kensington
       Scheme." Month 33 (1965), 12-23.

5423. HOLMES, J. DEREK. "Newman and the Use of History."
       Ph.D. Dissertation, University of Cambridge, 1970.

5424. HOLMES, J. DEREK. "Newman, Froude and Pattison: Some Aspects of Their Relations." Journal of Religious History 4 (June 1966), 23-38.

5425. HOLMES, J. DEREK. "Newman, History and Theology." Irish Theological Quarterly 36 (Jan. 1969), 34-45.

5426. HOLMES, J. DEREK. "Newman on Faith and History." Philosophical Studies 21 (1972), 202-216.

5427. HOLMES, J. DEREK. "Newman's Attitude towards Historical Criticism and Biblical Inspiration." Downside Review 89 (1971), 22-37.

5428. HOLMES, J. DEREK. "Newman's Reputation and 'The Lives of the English Saints'." Catholic Historical Review 51 (Jan. 1966), 528-538.

5429. HOLMES, J. DEREK AND ROBERT MURRAY. (Eds.) On the Inspiration of Scripture. Washington: Corpus Books, 1967.

5430. HONORE, JEAN. Itinéraire spirituel de Newman. Paris: Editions du Seuil, 1964.

5431. HONORE, JEAN. John Henry Newman. Paris: Editions Fleurus, 1967.

5432. HONORE, JEAN. Présence au monde et parole de Dieu: la catéchèse de Newman. Paris: Fayard-Mame, 1969.

5433. HOPE, N. V. "Issue between Newman and Kingsley: A Reconciliation and Rejoinder." Theology Today 6 (Apr. 1949), 77-90.

5434. HORGAN, JOHN D. "Newman on Faith and Reason." Studies 42 (June 1953), 132-150. (Reprinted in Michael Tierney et al., Newman's Doctrine of University Education. Dublin: Sealy, Bryers and Walker, 1954, 12-30)

5435. HOSEY, JOSEPH FRANCIS. "Physical Science in Newman's Thought." Dissertation Abstracts 14 (1954), 1411 (University of Pennsylvania, 1954).

5436. HOUGHTON, WALTER EDWARDS. The Art of Newman's Apologia. New Haven: Pub. for Wellesley College

by Yale University Press; London: H. Milford,
Oxford University Press, 1945.

5437.   HOUGHTON, WALTER EDWARDS.   "The Issue between
Kingsley and Newman."   Theology Today 4 (1947),
80-101.   (Reprinted in Robert O. Preyer, Ed.,
Victorian Literature: Selected Essays. New York:
Harper and Row, 1967, 13-36, and in David J. DeLaura,
Ed., Apologia Pro Vita Sua. New York: Norton,
1968, 390-409)

5438.   HOUPPERT, JOSEPH W. (Ed.) John Henry Newman. St.
Louis, Mo.: B. Herder, 1968.

5439.   HUGHES, JOHN JAY. "Two English Cardinals on Anglican
Orders." Journal of Ecumenical Studies 4 (Winter
1967), 1-26. (Newman and Henry Oxenham)

5440.   HUGHES, PHILIP. "Newman and His Age." Dublin
Review 217 (Oct. 1945), 111-136.

5441.   HUMPHRIES, S. "Newman's Analysis of Development."
Blackfriars 14 (July 1933), 620-626.

5442.   HUNT, R. W. "Newman's Notes on Dean Church's
'Oxford Movement'." Bodleian Library Record 8
(1969), 135-137.

5443.   HUNT, WILLIAM C. "Intuition: The Key to John
Henry Newman's Theory of Doctrinal Development."
Dissertation Abstracts 27 (1967), 4331A (Catholic
University of America, 1967).

5444.   JACKSON, J. "John Henry Newman: The Origins and
Applications of His Educational Ideas." Ph.D.
Dissertation, University of Leicester, 1968.

5445.   JAKI, STANISLAS. "J. H. Cardinal Newman." In
Les Tendances nouvelles de l'ecclésiologie. Roma:
Herder, 1957, 35-44.

5446.   JAMES, S. B. "Kingsley vs. Newman." Catholic
Digest 3 (Nov. 1938), 77-80.

5447.   JANSSENS, A. Newman: inleiding tot zijn geest
en zijn werk. Nimeguen: Dekker & Van de Vegt,
1937.

5448. JOHNSON, HUMPHREY JOHN THEWLIS. "The Controversy between Newman and Gladstone over the Question of Civil Allegiance." Dublin Review 217 (1945), 173-182.

5449. JOHNSON, HUMPHREY JOHN THEWLIS. "Leo XIII, Cardinal Newman and the Inerrancy of Scripture." Downside Review 69 (Autumn 1951), 411-427.

5450. JOST, EDWARD F. "Newman and Liberalism: The Later Phase." Victorian Newsletter No. 24 (Fall 1963), 1-5.

5451. JUERGENS, SYLVESTER PETER. Newman on the Psychology of Faith in the Individual. New York: The Macmillan Company, 1928.

5452. KAISER, JAMES. "The Concept of Conscience According to John Henry Newman." Ph.D. Dissertation, Catholic University of America, 1958.

5453. KANDEL, I. L. "Newman's Idea of a University." School and Society 75 (June 21, 1952), 396-397.

5454. KARRER, OTTO. "Newman and the Spiritual Crisis of the Occident." Review of Politics 9 (1947), 230-246. (Translated from the German by Frederick C. Ellert and Alvan S. Ryan)

5455. KELLY, EDWARD EUGENE, S.J. "Cardinal Newman's Unpublished Letters: A Selection from the Year July, 1864 to July, 1865." Dissertation Abstracts 25 (1964), 2493 (Fordham University, 1963).

5456. KELLY, EDWARD EUGENE, S.J. "Newman More Ecumenically Read." Journal of Ecumenical Studies 5 (Spring 1968), 65-70. (See Anthony Stephenson, "Cardinal Newman's Theory of Doctrinal Development: Reply." 5, Spring 1968, 370-377)

5457. KELLY, EDWARD EUGENE, S.J. "Newman, Vatican I and II, and the Church Today." Catholic World 102 (1966), 291-297.

5458. KELLY, EDWARD EUGENE, S.J. "Newman, Wilfrid Ward, and the Modernist Crisis." Thought 48 (Winter 1973), 508-519.

5459.   KELLY, EDWARD EUGENE, S.J.   "Newman's Catholic
        History As Background of the 'Apologia'."   Personalist
        46 (Summer 1965), 382-388.

5460.   KENYON, RUTH.   "Studies in the Social Outlook of
        the Tractarians: 1.   Newman."   Theology 24 (June
        1932), 317-324.   (See Keble and Pusey, 5774)

5461.   KER, I. T.   "Apology for Newman."   Essays in
        Criticism 24 (1974), 319-322.

5462.   KER, I. T.   "Did Newman Believe in the Idea of a
        Catholic University?"   Downside Review 93 (Jan.
        1975), 39-42.

5463.   KERRIGAN, A.   "More about Cardinal Newman: An
        Important Contribution to the History of Scriptural
        Exegesis."   Irish Ecclesiastical Record 81 (June
        1954), 422-435.

5464.   KIENER, MARY ALOYSI, SISTER.   John Henry Newman:
        The Romantic, The Friend, The Leader.   Boston:
        Collegiate Press Corporation, 1933.

5465.   KIM, STAR.   "John Henry Newman as a Church Historian."
        Ph.D. Dissertation, Yale University, 1929.

5466.   KLOCKER, HARRY R.   "Newman and Causality."   Heythrop
        Journal 6 (Apr. 1965), 160-170.

5467.   KLOCKER, HARRY R.   "The Personal God of John Henry
        Newman."   Personalist 57 (Spring 1976), 145-161.

5468.   KNOX, RONALD.   "Many Mansions: The Conversions of
        Newman and Faber."   Tablet 185 (June 30, 1945),
        310.

5469.   LÄPPLE, ALFRED KARL.   Der Einzelne in der Kirche:
        Wesenszüge einer Theologie des Einzelnen nach John
        Henry Kardinal Newman.   München: K. Zink, 1952.

5470.   LAFERRIERE, FRANK VINCENT.   "A Documentary History
        of John Henry Newman's Rectorship of the Catholic
        University of Ireland, 1851-1858."   Dissertation
        Abstracts 26 (1966), 5413-14 (University of California,
        Los Angeles, 1965).

5471. LAFRAMBOISE, JEAN-CHARLES. "La Conversion de
      Newman." Revue de l'Université d'Ottawa 15 (Oct.
      1945), 408-445.

5472. LAFRAMBOISE, JEAN-CHARLES. "La Doctrine spirituelle
      de Newman." Revue de l'Université d'Ottawa 15
      (Jan. 1945), 48-76.

5473. LAHEY, G. F. "Hopkins and Newman." Commonweal 12
      (June 25, 1930), 211-213.

5474. LAMBORN, EDMUND ARNOLD GREENING. "Newman's Church
      at Littlemore." Notes and Queries 190 (Feb. 9,
      1946), 46-49. (Reprinted in English Catholic 15,
      No. 65, Spring 1946, 116-120)

5475. LAMM, WILLIAM ROBERT. The Spiritual Legacy of
      Newman. Milwaukee: The Bruce Publishing Co.,
      1934.

5476. LAMPING, STEPHEN AND SEVERIN. "Cardinal Newman
      versus Liberalism in Religion." Priest 23 (Nov.
      1967), 898-902.

5477. LAPATI, AMERICO D. John Henry Newman. New York:
      Twayne, 1972.

5478. LAROS, MATTHIAS. Kardinal Newman. Mainz: Matthias-
      Grünewald-Verlag, 1921.

5479. LASH, NICHOLAS. "Faith and History: Some Reflections
      on Newman's 'Essay on the Development of Christian
      Doctrine'." Irish Theological Quarterly 38 (July
      1971), 224-241.

5480. LASH, NICHOLAS. "The Notions of Implicit and
      Explicit Reason in Newman's University Sermons: A
      Difficulty." Heythrop Journal 11 (Jan. 1970),
      48-54.

5481. LASH, NICHOLAS. "Was Newman a Theologian?" Heythrop
      Journal 17 (July 1976), 322-325.

5482. LAUER, EUGENE F. "An Historical-Analytical Comparison
      of Newman's Concept of Conscience and Tertullian's
      Testimonium Animae." Dissertation, Gregorian
      University, 1966.

5483. LAWLER, J. G. "No Time To Leave." Commonweal 86
       (Apr. 7, 1967), 87-92.
5484. LEASE, GARY. Witness to the Faith: Cardinal
       Newman on the Teaching Authority of the Church.
       Pittsburgh: Duquesne University Press, 1971.
5485. LEDDY, J. F. "Newman and His Critics: A Chapter
       in the History of Ideas." Canadian Catholic
       Historical Association Report 10 (1942-43), 25-38.
5486. LEIGH, DAVID J. "Newman, Lonergan and Social Sin:
       Centenary Reflections." Month 9 (Feb. 1976),
       41-44.
5487. LESLIE, SIR JOHN RANDOLPH SHANE. "Newman Again."
       Month 9 (1953), 5-13.
5488. LESLIE, SIR JOHN RANDOLPH SHANE. Studies in Sublime
       Failure: Cardinal Newman, Lord Curzon, Charles
       Stewart Parnell, Coventry Patmore, Moreton Frewen.
       London: Benn, 1932. (M. Frewen, 1853-1924)
5489. LIDGETT, JOHN SCOTT. "John Wesley and John Henry
       Newman." London Quarterly Review 146 (July 1926),
       1-10.
5490. LILLY, WILLIAM SAMUEL. "Cardinal Newman and the
       Catholic Laity." Nineteenth Century 71 (Mar.
       1912), 445-465.
5491. LINNAN, JOHN E. The Evangelical Background of
       John Henry Newman, 1816-1826. 2 vols. Louvain:
       Université Catholique de Louvain, Faculté de Théologie,
       1965.
5492. LINNAN, JOHN E. "The Search for Absolute Holiness:
       A Study of Newman's Evangelical Period." The
       Ampleforth Journal 73, No. 2 (1968), 161-174.
5493. LUCAS, HERBERT. "The Life's Work of J. H. Newman."
       Catholic World 112 (Dec. 1920), 303-315.
5494. LUNN, SIR ARNOLD HENRY MOORE. Roman Converts.
       London: Chapman and Hall, 1924. (Newman, H.
       Manning, G. Tyrrell, Ronald Knox, G. K. Chesterton)

5495. LUTZ, JOSEPH A. Kardinal John Henry Newman: Ein
       Zeit- und Lebensbild. Einsiedeln: Benziger, 1948.

5496. LYNCH, T. "The Newman-Perrone Paper on Develop-
       ment." Gregorianum 14 (1935), 402-447.

5497. LYONS, REV. JAMES W. "A Philosophical Critique of
       Certitude According to Newman." Dissertation
       Abstracts International 36 (1975), 349A (Loyola
       University of Chicago, 1975).

5498. MCCORMICK, MARY JAMES, SISTER. "Newman's 'Apologia
       Pro Vita Sua': Its Origin, Composition, and Critical
       Reception." Dissertation Abstracts 26 (1966),
       3926-27 (Fordham University, 1965).

5499. MCDONNELL, T. P. "Newman and the New Criticism."
       America 92 (Mar. 12, 1955), 620-621.

5500. MACDOUGALL, HUGH A. The Acton-Newman Relations:
       The Dilemma of Christian Liberalism. New York:
       Fordham University Press, 1962.

5501. MCELRATH, DAMIAN. "Richard Simpson and John Henry
       Newman: The 'Rambler', Laymen, and Theology."
       Catholic Historical Review 52 (Jan. 1967), 509-533.

5502. MCGRATH, J. S. FERGAL. "The Background of Newman's
       'Idea of a University'." Month 181, No. 946 (July-
       Aug. 1945), 247-258.

5503. MCGRATH, J. S. FERGAL. The Consecration of Learning:
       Lectures on Newman's "Idea of a University." New
       York: Fordham University Press, 1962.

5504. MCGRATH, J. S. FERGAL. Newman's University: Idea
       and Reality. Dublin: Browne and Nolan, 1951.

5505. MCINTOSH, LAWRENCE D. "An Unpublished Letter of
       John Henry Newman." Catholic Historical Review 59
       (Oct. 1973), 429-433.

5506. MCMAHON, F. E. "Brownson and Newman." America 89
       (Apr. 11, 1953), 45-47; (Apr. 18, 1953), 79-80.

5507. MCMANNUS, E. L. "Newman and the Newman Legend."
       American Ecclesiastical Review 139 (Aug. 1958),
       93-100.

5508. MCMANNUS, E. L. "Newman, Parish Priest." Homiletic and Pastoral Review 59 (Jan. 1959), 348-352.

5509. MCMURRY, J. "The Christian Church a Continuation of the Jewish: Newman on the Relationship between Judaism and Christianity." Unitas 18 (Fall 1966), 188-195.

5510. MANN, JOSEF. John Henry Newman als Kerygmatiker: Der Beitrag seiner anglikanischen Zeit zur Glaubensverkündigung und Unterweisung. Leipzig: St. Benno-Verlag, 1965.

5511. MARIELLA, SISTER. "Newman's Anglican Sermons: Influence of Aristotle's Rhetoric." Catholic World 148 (Jan. 1939), 431-437.

5512. MARIZY, LUDWIG. "Die theologische Anthropologie John Henry Newmans." Ph.D. Dissertation, Universität Münster, 1943.

5513. MARTIN, A. "Autobiography in Newman's Novels." Month 23 (1960), 291-302.

5514. MARTINDALE, CYRIL CHARLES. "Sibyl and Sphinx: Newman and Manning in the '80's." Contemporary Review 138 (1930), 470-479.

5515. MAXWELL, J. C. "Newman and Arnold." Notes and Queries 215 (1970), 385.

5516. MAY, JAMES LEWIS. Cardinal Newman. London: G. Bles, 1929.

5517. MAY, JAMES LEWIS. "Newman Once More: The Idea of a University." Catholic World 152 (Mar. 1941), 718-724.

5518. MEEHAN, JAMES T. "The Easter Theme in the Preaching of John Henry Newman." Theol. Diss., Universität Trier, 1969.

5519. MESSENGER, E. C. "Wiseman, the Donatists, and Newman: A Dublin Centenary." Dublin Review 205 (July 1939), 110-119.

5520. MIDDLETON, ROBERT DUDLEY. Newman and Bloxam: An Oxford Friendship. London: Oxford University Press, 1947.

5521. MIDDLETON, ROBERT DUDLEY. Newman at Oxford: His
       Religious Development. London: Oxford University
       Press, 1950.

5522. MISNER, PAUL. "John Henry Newman on the Primacy of
       the Pope." Theol. Diss., Universität München, 1968.
       (See Papacy and Development: John Henry Newman . . . .
       Leiden: E. Brill, 1976)

5523. MISNER, PAUL. "Newman and the Tradition Concerning
       the Papal Antichrist." Church History 42 (1973),
       377-395.

5524. MISNER, PAUL. "Newman's Concept of Revelation and
       the Development of Doctrine." Heythrop Journal 11
       (Jan. 1970), 32-47.

5525. MOIR, J. S. "The Correspondence of Bishop Strachan
       and John Henry Newman." Canadian Journal of Theology
       3 (Oct. 1957), 219-225.

5526. MOODY, JOHN. John Henry Newman. New York: Sheed
       and Ward, 1945.

5527. MOONEY, E. B. "The Formative Evolution of Newman's
       Concept on the Doctrine of Justification." Revue
       de l'Université d'Ottawa (Supplement) 17 (Jan.
       1947), 21-50.

5528. MORSE-BOYCOTT, REV. DESMOND LIONEL. Lead, Kindly
       Light: Studies of the Saints and Heroes of the
       Oxford Movement. London: Centenary Press, 1932.

5529. MOSELEY, D. H. "Lead, Kindly Light." Catholic
       World 137 (June 1933), 298-304.

5530. MOSSNER, ERNEST C. "Cardinal Newman on Bishop
       Butler: An Unpublished Letter." Theology 32
       (1936), 113.

5531. MOZLEY, DOROTHEA. (Ed.) Newman Family Letters.
       London: S.P.C.K., 1962.

5532. MOZLEY, J. F. "Newman in Fetters." Quarterly
       Review 246 (1926), 272-293.

5533. MOZLEY, J. F. "Newman's Opportunity." Quarterly
       Review 246 (1926), 75-92.

5534. MULCAHEY, DONALD CHARLES. "The Prophetic Role of
      the Laity According to John Henry Cardinal Newman."
      Dissertation Abstracts International 30 (1969),
      2609A (Catholic University of America, 1969).

5535. MULCAHY, DANIEL G. "Cardinal Newman's Concept of a
      Liberal Education." Educational Theory 22 (Winter
      1972), 87-98.

5536. MURPHY, T. A. "Newman and Devotion to Our Lady."
      Irish Ecclesiastical Record 72 (1949), 385-396.

5537. MURRAY, PLACID. (Ed.) Newman the Oratorian.
      Dublin: Gill and Macmillan, 1969.

5538. MURRAY, PLACID. "Newman's Views on Anglican Orders."
      Clergy Review 50 (1965), 890-893.

5539. MURRAY, PLACID. "Tower of David: Cardinal New-
      man's Mariology." Furrow 27 (Jan. 1976), 26-34.

5540. MYERS, R. "The Great Cardinal." Homiletic and
      Pastoral Review 75 (Aug.-Sept. 1975), 24-27.

5541. NAULTY, R. A. "Newman's Dispute with Locke."
      Journal of the History of Philosophy 11 (Oct.
      1973), 453-457.

5542. NEDONCELLE, MAURICE GUSTAVE. "L'Apologia de Newman
      dans l'histoire de l'autobiographie et de la théologie."
      In Helmut Kuhn, Heinrich Kahlefeld and Karl Forster
      (Eds.), Interpretation der Welt. Würzburg: Echter,
      1965, 571-585.

5543. NEDONCELLE, MAURICE GUSTAVE. "Chronique Newman-
      ienne." Revue des Sciences Religieuses 45 (Jan.
      1971), 78-89.

5544. NEDONCELLE, MAURICE GUSTAVE. "Une Lettre inédite
      de Newman sur la matière et l'esprit." Revue des
      Sciences Religieuses 41 (Jan. 1967), 39-62.

5545. NEDONCELLE, MAURICE GUSTAVE. "Newman, théologien
      des abus de l'Eglise." Oecumenica, Jahrbuch für
      ökumenische Forschung 2 (1967), 116-134.

5546. NEDONCELLE, MAURICE GUSTAVE. La Philosophie
      religieuse de John Henry Newman. Strasbourg:
      Impr. Société Strasbourgeoise de Librairie, 1946.

5547.   NEILL, T. P.   "Newman's 'Idea' after a Century."
        Catholic World 176 (Nov. 1952), 103-109.
5548.   NEVILLE, WILLIAM PAYNE. (Ed.) Addresses to Cardinal
        Newman with His Replies, 1879-81. London: Longmans,
        Green, 1905. (On his elevation to the cardinalate)
5549.   NEWCOMB, COVELLE. "Newman and Nature." Catholic
        World 149 (June 1939), 338-342.
5550.   NEWCOMB, COVELLE. Red Hat: A Study of John Henry,
        Cardinal Newman. New York: Longmans, Green, 1941.
5551.   NEWMAN, BERTRAM. Cardinal Newman: A Biographical
        and Literary Study. London: G. Bell and Sons
        Ltd., 1925.
5552.   NEWMAN, JAY. "Cardinal Newman's 'Factory-girl
        Argument'." Proceedings of the American Catholic
        Philosophical Association 46 (1972), 71-77.
        (Newman's argument for the existence of God in
        A Grammar of Assent)
5553.   NEWMAN, JAY. "Cardinal Newman's Phenomenology of
        Religious Belief." Religious Studies 10 (June
        1974), 129-140.
5554.   NEWSOME, DAVID H. "Justification and Sanctifica-
        tion: Newman and the Evangelicals." Journal of
        Theological Studies 15 (Apr. 1964), 32-53.
5555.   NEWSOME, DAVID H. "Newman and Manning: Spirit-
        uality and Personal Conflict." In P. Brooks (Ed.),
        Christian Spirituality: Essays in Honour of Gordon
        Rupp. London: SCM Press, 1975, 285-306.
5556.   NEWSOME, DAVID H. "Newmania." Journal of Theological
        Studies 14 (1963), 420-429.
5557.   NICHOLLS, DAVID. "Newman's Anglican Critics."
        Anglican Theological Review 47 (Oct. 1965), 377-395.
        (Refutations by Anglican theologians of Newman's
        Essay on the Development of Christian Doctrine)
5558.   OBERTELLO, LUCA. Conoscenza e persona nel pensiero
        di John Henry Newman. Trieste: 1964.
5559.   O'BRIEN, J. A. "Cardinal Newman." Ecclesiastical
        Review 96 (1937), 45-56, 126-139, 251-263.

5560.   O'BRIEN, ROGER G.   "The Theology of Preaching in
        the Parochial and Plain Sermons of John Henry
        Newman."   Doctoral Dissertation, Katholieke
        Universiteit te Leuven, 1968.

5561.   O'CARROLL, MICHAEL.   "Our Lady in Newman and
        Vatican II."   Downside Review 89 (1971), 38-63.

5562.   O'CONNELL, MARVIN R.   "Newman: The Limits of
        Certitude."   Review of Politics 35 (Apr. 1973),
        147-160.

5563.   O'DONNELL, REV. ROBERT E.   "Newman on Faith and
        Dogma - The Anglican Years: A Critical Analysis of
        Selected Texts 1830-1843."   Dissertation Abstracts
        International 33 (1973), 6441-42A (Catholic University
        of America, 1973).

5564.   O'DONOGHUE, N.   "Newman and the Problem of Privi-
        leged Access to Truth."   Irish Theological Quarterly
        42 (Oct. 1975), 241-258.

5565.   O'DWYER, EDWARD THOMAS, BISHOP OF LIMERICK.   Cardinal
        Newman and the Encyclical "Pascendi Domini Gregis."
        London: Longmans and Co., 1908.

5566.   O'FAOLAIN, SEAN.   Newman's Way: The Odyssey of
        John Henry Newman.   London, New York: Longmans,
        Green, 1952.

5567.   O'FLYNN, J. A.   "Newman and the Scriptures."
        Irish Theological Quarterly 21 (July 1954), 264-269.

5568.   O'HARE, CHARLES M.   "John Henry Newman in Rome,
        1833."   Irish Ecclesiastical Record 36 (Nov. 1930),
        449-458.

5569.   O'MALLEY, FRANK.   "The Thinker in the Church: The
        Spirit of Newman."   Review of Politics 21 (1959),
        5-23.

5570.   O'MEARA, J. J.   "Augustine and Newman: Comparison
        in Conversion."   University Review (Dublin) 1, No.
        1 (Summer, 1954), 27-36.

5571.   ONG, WALTER J.   "Newman and the Religious Life."
        Review for Religious 4 (1945), 230-242.   (Newman's
        reasons for not entering a religious order)

5572.   ONG, WALTER J.   "Newman's Essay on Development in
        Its Intellectual Milieu."   Theological Studies 7
        (1946), 3-45.

5573.   O'ROURKE, JAMES.   "Manning and Newman."   Irish
        Ecclesiastical Record 52 (Nov. 1938), 459-469.

5574.   OVERMANS, JAKOB.   "Harnack und Newman."   Stimmen
        der Zeit 131, Heft 1 (Oct. 1936), 20-31.

5575.   PAGE, FREDERICK.   "Froude, Kingsley, and Arnold, on
        Newman."   Notes and Queries 184 (Apr. 10, 1943),
        220-221.

5576.   PAILIN, DAVID ARTHUR.   The Way to Faith: An Examin-
        ation of Newman's "Grammar of Assent" as a Response
        to the Search for Certainty in Faith.   London:
        Epworth Press, 1969.

5577.   PATTERSON, WEBSTER T.   Newman: Pioneer for the
        Layman.   Cleveland: Corpus Books, 1968.

5578.   PETITPAS, HAROLD M.   "Newman's Idea of Science."
        Personalist 48 (1967), 297-316.

5579.   PETITPAS, HAROLD M.   "Newman's Universe of Know-
        ledge: Science, Literature, and Theology."
        Dalhousie Review 46 (1966), 494-507.

5580.   PHILLIPS, GEORGE E.   "Early Reminiscences of
        Cardinal Newman and of His First Fellow Oratorians."
        Ushaw Magazine 26 (1916), 16-38.

5581.   POL, WILLEM HENDRIK VAN DE.   De kerk in het leven
        en denken van Newman.   Nijkerk: G. F. Callenbach,
        1936.

5582.   POLLEN, JOHN HUNGERFORD. (Ed.)   "Letters of Cardinal
        Newman to Lady Georgiana Fullerton."   Month 129
        (1917), 329-341.

5583.   POTTER, GERALD L.   "The Idea of Preaching According
        to John Henry Newman."   Dissertation, Gregorian
        University, 1963.

5584.   POWELL, JOUETT LYNN.   "Three Uses of Christian
        Discourse in John Henry Newman: An Example of
        Non-Reductive Reflection on the Christian Faith."

Dissertation Abstracts International 33 (1973),
3763A (Yale University, 1972).

5585. PRESCOTT, C. J. "Wesley and Newman: A Problem."
Methodist Quarterly Review 74 (Jan. 1925), 29-42.

5586. PRICE, HENRY HABBERLEY. *Belief: The Gifford
Lectures Delivered at the University of Aberdeen in
1960*. London: Allen and Unwin; New York: Humanities
Press, 1969. (Includes a discussion of the first
part of *A Grammar of Assent*)

5587. PRICKETT, STEPHEN. "Coleridge, Newman and F. D.
Maurice: Development of Doctrine and Growth of
Mind." *Theology* 76 (July 1973), 340-349.

5588. PRZYWARA, ERICH. (Ed.) *J. H. Kardinal Newman:
Christentum, ein Aufbau: Aus seinen Werken zusammen-
gestellt und eingeleitet von Erich Przywara:
Übertragen von Otto Karrer*. 8 vols. Freiburg im
Breisgau: Herder, 1922. (Translated volumes
published as *A Newman Synthesis*; see 5590)

5589. PRZYWARA, ERICH. "Newman." In *Ringen der Gegenwart:
Gesammelte Aufsätze, 1922-1927*. 2 vols. Augsburg:
B. Filser-Verlag, 1929, vol. 2, 802-879.

5590. PRZYWARA, ERICH. (Comp.) *A Newman Synthesis*.
London: Sheed & Ward, 1930. (See 5588)

5591. PRZYWARA, ERICH. *Religionsbegründung, Max Scheler-
J. H. Newman*. Freiburg i. B.: Herder & Co., 1923.

5592. QUINN, J. RICHARD. *The Recognition of the True
Church According to John Henry Newman*. Washington:
Catholic University of America Press, 1954.

5593. RAHNER, KARL AND KARL LEHMANN. "Geschichtlichkeit
der Vermittlung." In Johannes Feiner and Magnus
Löhrer (Eds.), *Mysterium Salutis: Grundriss
heilsgeschichtlicher Dogmatik*. 4 vols. Einsiedeln:
Benziger, 1965, vol. 1, 727-787. (References to
Newman throughout)

5594. READE, FRANCIS VINCENT. "The Spiritual Life of
John Henry Newman." *Dublin Review* 217 (Oct. 1945),
99-111.

5595.   REARDON, BERNARD M. G.   "Newman and the Catholic
        Modernist Movement."   Church Quarterly 4 (July
        1971), 50-60.

5596.   REARDON, BERNARD M. G.   "Newman and the Psychology
        of Belief."   Church Quarterly Review 158 (1957),
        315-332.

5597.   REILLY, JOSEPH JOHN.   "The Present Significance of
        Newman."   Thought 20 (1945), 389-395.

5598.   RENO, STEPHEN J.   "Religious Belief: Continuities
        between Newman and Cirne-Lima."   New Scholasticism
        44 (Fall 1970), 489-514.

5599.   RENZ, WOLFGANG.   Newmans Idee einer Universität:
        Probleme höherer Bildung.   Freiburg, Schweiz:
        Universitätsverlag, 1958.

5600.   REXROTH, KENNETH.   "Newman: Some Preliminary
        Notes."   Continuum 8 (Spring-Summer 1970), 134-140.

5601.   REYNOLDS, ERNEST EDWIN.   Three Cardinals: Newman,
        Wiseman, Manning.   London: Burns & Oates, 1958.

5602.   RIBANDO, WILLIAM, C.S.C.   "The Church as Sacrament
        in the Writings of John Henry Newman."   Disserta-
        tion Abstracts International 31 (1970), 1882A
        (Catholic University of America, 1970).

5603.   RIBANDO, WILLIAM, C.S.C.   "Newman on the Prophetic
        and Pastoral Word."   American Benedictine Review 22
        (1971), 381-386.

5604.   RITZ, JEAN-GEORGES.   "Trois témoins de la crise
        religieuse victorienne: les frères Newman."   In
        Religion et Politique: les deux guerres mondiales:
        histoire de Lyon et du Sud-Est: mélanges offerts à
        M. le Doyen André Latreille.   Préf. de M. Pacaut et
        al.   Lyon: Audin, 1972, 177-185.

5605.   ROBBINS, WILLIAM M.   The Newman Brothers: An
        Essay in Comparative Intellectual Biography.
        Cambridge: Harvard University Press, 1966.

5606.   ROBERTSON, THOMAS LUTHER.   "The Kingsley-Newman
        Controversy and the 'Apologia'."   Modern Language
        Notes 69 (Dec. 1954), 564-569.

5607.   ROBINSON, JONATHAN.   "Did Newman 'Fit In'?"   Dublin
        Review 232 (1958), 245-259.

5608.   ROBINSON, JONATHAN.   "Newman's Use of Butler's
        Arguments."   Downside Review 76 (1958), 161-180.

5609.   ROSS, JOHN ELLIOT.   John Henry Newman: Anglican
        Minister, Catholic Priest, Roman Cardinal.   New
        York: W. W. Norton and Company, Inc., 1933.

5610.   ROWELL, GEOFFREY.   "Some Formative Influences on
        Cardinal Newman's Spirituality."   L'osservatore
        romano No. 39 (Sept. 25, 1975), 5.

5611.   ROWLAND, MARY JOYCE, O.S.F., SISTER.   "The Acts of
        the Mind in Newman's Theory of Assent."   Disserta-
        tion Abstracts 24 (1964), 4239-40 (St. Louis
        University, 1962).

5612.   RUGGLES, ELEANOR.   Journey into Faith: The Anglican
        Life of John Henry Newman.   New York: Norton,
        1948.

5613.   RUGGLES, ELEANOR.   "Newman's Journey into Faith."
        Catholic Digest 12 (Aug. 1948), 33-43.

5614.   RUSSELL, D.   "Cardinal Newman: A Man Ahead of His
        Time."   Liguorian 63 (Nov. 1975), 34-38.

5615.   RUSSELL, H. P.   "The Spirit of John Henry Newman."
        Catholic World 85 (Aug. 1907), 609-620.

5616.   RYAN, ALVAN SHERMAN.   Newman and Gladstone: The
        Vatican Decrees.   Notre Dame: University of Notre
        Dame Press, 1962.

5617.   RYAN, ALVAN SHERMAN.   "Newman on Education."
        Yale Review 45, No. 4 (1956), 597-600.

5618.   RYAN, JOHN K. AND EDMOND D. BENARD. (Eds.)   American
        Essays for the Newman Centennial.   Washington,
        D.C.:   Catholic University of America Press, 1947.

5619.   RYAN, M. J.   "Philosophy of Newman."   American
        Catholic Quarterly Review 33 (Jan. 1908), 77-86.

5620.   RYAN, PATRICK, O.C.S.O.   "Isaac of Stella and
        Newman on Revelation."   Cistercian Studies 5, No. 4
        (1970), 370-387.

5621. RYAN, T. "Newman's Invitation to Orestes A. Brownson
       to Be Lecturer Extraordinary at the Catholic University
       of Ireland." American Catholic Historical Society
       Records 85 (Mar.-June 1974), 29-47.

5622. SAINT-ARNAUD, JEAN GUY. Newman et l'incroyance.
       Paris: Desclée, 1972.

5623. SAROLEA, CHARLES. Cardinal Newman and His Influence
       on Religious Life and Thought. Edinburgh: T. and
       T. Clark, 1908.

5624. SAUNDERS, D. J. "Psychological Reactions before
       Newman's Conversion." Theological Studies 6 (Dec.
       1945), 489-508.

5625. SCHIFFERS, NORBERT. Die Einheit der Kirche nach
       John Henry Newman. Düsseldorf: Patmos-Verlag,
       1956.

5626. SCHIFFERS, NORBERT. "John Henry Kardinal Newman:
       Zum Verhältnis von Lehramt und Theologie: Texthinweise,
       ausgewählt und kommentiert." Internationale katholische
       Zeitschrift No. 2 (Mar.-Apr. 1974), 107-109.

5627. SCHLATTER, R. "Idea of a University." Teachers
       College Record 66 (Apr. 1965), 588-597.

5628. SEEKINGS, H. S. "Permanent Influence of Newman."
       London Quarterly Review 114 (July 1910), 102-114.

5629. SELBY, ROBIN C. The Principle of Reserve in the
       Writings of John Henry Cardinal Newman. London:
       Oxford University Press, 1975.

5630. SENCOURT, ROBERT. The Life of Newman. Westminster,
       London: Dacre Press, 1948.

5631. SENCOURT, ROBERT. "Newman and Pusey." Dublin Review
       217 (Oct. 1945), 156-165.

5632. SEYNAEVE, JAAK. Cardinal Newman's Doctrine of
       Holy Scripture According to His Published Works and
       Previously Unedited Manuscripts. Louvain: Publi-
       cations Universitaires de Louvain, 1953.

5633. SHERAN, W. H. "Newman's Devotion to Our Lady."
       Catholic World 93 (May 1911), 174-183.

5634. SHERIDAN, THOMAS L. Newman on Justification: A
       Theological Biography. Staten Island, N.Y.: Alba
       House, 1967.

5635. SHOOK, LAURENCE. "Newman and the Dialogue."
       Basilian Teacher 5 (1960), 49-52.

5636. SHUTE, GRAHAM J. "Newman's 'Logic of the Heart'."
       Expository Times 78 (May 1967), 232-235.

5637. SILLEM, EDWARD A. "Cardinal Newman: A New
       Discovery." Wiseman Review 237 (Spring 1963),
       66-77.

5638. SILLEM, EDWARD A. "Cardinal Newman as Philosopher."
       Clergy Review 48 (Aug. 1963), 167-85. (A comment
       on 5268; see M. D'Arcy's response, 388-89)

5639. SILLEM, EDWARD A. "Cardinal Newman's Grammar of
       Assent on Conscience as a Way to God." Heythrop
       Journal 5 (Oct. 1964), 377-401.

5640. SILLEM, EDWARD A. (Ed.) The Philosophical Notebook
       of John Henry Newman: Vol. 1. General Introduction
       to the Study of Newman's Philosophy. Louvain:
       Nauwelaerts Pub. House, 1969.

5641. SIMON, P. "Newman and German Catholicism." Dublin
       Review 219 (July 1946), 75-84.

5642. SIMPSON, RICHARD. "A Glimpse of Newman in 1846."
       Downside Review 63 (Oct. 1945), 211-221. (Extracts
       from the diary of Richard Simpson)

5643. SMITH, BASIL A. Dean Church: The Anglican Response
       to Newman. London: Oxford University Press, 1958.

5644. SMITH, FRED. "The Fears and Faith of Newman."
       Homiletic Review 106 (Nov. 1933), 365-367.

5645. SMITH, LEO. "Cardinal Newman and the Benedictines."
       Buckfast Abbey Chronicle 15, No. 2 (Summer 1945),
       67-74.

5646. SMITH, SYDNEY FENN. "The Life of Cardinal Newman."
       Month 119 (1912), 113-126.

5647. SOBRY, PAUL. Newman en zijn Idea of a University.
       Louvain: Bureaux du Recueil, Bibliothèque de
       l'Université, 1934.

5648. SPITZBERG, I. J.   "John Stuart Mill and John Henry
        Cardinal Newman: A Comparative Study, with Special
        Reference to Their Views of the Role of Education
        in Society." B.Litt. Dissertation, University of
        Oxford, 1969.

5649. SPRINGER, HENRI.   "Quelques aspects des rapports
        entre les sciences et la foi dans l'oeuvre du
        cardinal Newman." Nouvelle Revue Théologique 85
        (Mar. 1963), 270-279.

5650. STANFORD, DEREK AND MURIEL SPARK. (Eds.) Letters of
        John Henry Newman: A Selection. London: Owen, 1957.

5651. STARK, WERNER. "John Henry Newman." In Social
        Theory and Christian Thought. London: Routledge,
        Kegan, and Paul, 1958, 106-134.

5652. STEPHENSON, ANTHONY A.   "Cardinal Newman and the
        Development of Doctrine." Journal of Ecumenical
        Studies 3 (Fall 1966), 463-485. (See Kelly, 5456)

5653. STERN, JEAN. Bible et tradition chez Newman,
        aux origines de la théorie du développement.
        Paris: Aubier, 1967.

5654. STERN, JEAN. "Le Culte de la vierge et des saints
        et la conversion de Newman au Catholicisme."
        La Vie Spirituelle 117 (1967), 156-168.

5655. STERN, JEAN. "L'Infaillibilité de l'Eglise dans la
        pensée de J. H. Newman." Recherches de Science
        Religieuse 61 (Apr.-June 1973), 161-185.

5656. STERN, JEAN. "The Institutional Church in Newman's
        Spirituality." L'osservatore romano No. 18 (May 1,
        1975), 8-9.

5657. STERN, JEAN. "Traditions apostoliques et magistère
        selon J. H. Newman." Revue des Sciences Philosophiques
        et Théologiques 47 (Jan. 1963), 35-57.

5658. STOCKLEY, WILLIAM FREDERICK PAUL. "Keble and
        Newman." Irish Ecclesiastical Record 35 (1930),
        40-57, 135-148.

5659. STOCKLEY, WILLIAM FREDERICK PAUL. Newman, Educa-
        tion, and Ireland. London: Sands and Co., 1933.

5660.   STOEL, HARMANUS.   Kardinaal Newman (1801-'90).
        Eerste deel: Zijn strijd om de ware kerk (1801-'45).
        Groningen: Gedr. bij gebroeders Hoitsema, 1914.

5661.   STORK, R.   "The Church and the Individual in the
        Thought of Cardinal Newman." Clergy Review 59 (May
        1974), 353-362.

5662.   STRANGE, R.   "Knowing and Loving Christ: Newman on
        Seeley's Ecce Homo." Clergy Review 61 (Apr. 1976),
        142-147.

5663.   STROEBER, RUDOLF. "Die Idee der Kirche von Coleridge
        bis Newman." Ph.D. Dissertation, Universität
        Erlangen, 1952.

5664.   STRONG, L. A. G.   "Was Newman a Failure?" Nine-
        teenth Century 113 (1933), 620-628.

5665.   STUNT, TIMOTHY C. F.   "John Henry Newman and the
        Evangelicals." Journal of Ecclesiastical History
        21 (1970), 65-74.

5666.   SVAGLIC, MARTIN JAMES.   "Charles Newman and His
        Brothers." PMLA 71 (June 1956), 370-385.   (His
        relationship to John Henry and Francis William)

5667.   SVAGLIC, MARTIN JAMES.   "Light on Newman."
        Commonweal 78 (May 10, 1963), 204-206.

5668.   SVAGLIC, MARTIN JAMES.   "Newman and the Oriel
        Fellowship." PMLA 70 (1955), 1014-32.

5669.   SWISSHELM, G.   "Newman and the Vatican Definition
        of Papal Infallibility." St. Meinrad Essays 12
        (May 1960), 70-88.

5670.   TARDIVEL, FERNANDE. J. H. Newman, éducateur.
        Paris: Imprimeries Les Presses Modernes, 1937.

5671.   TARDIVEL, FERNANDE.   La Personnalité littéraire
        de Newman. Paris: G. Beauchesne et ses Fils,
        1937.

5672.   TAYLOR, DENNIS.   "Some Strategies of Religious
        Autobiography." Renascence 27, No. 1 (1974),
        40-44. (Includes Newman's Apologia)

5673.   THEIS, NICOLAS. John Henry Newman in unserer Zeit.
        Nürnberg: Glock und Lutz, 1972.

5674. THIRLWALL, JOHN CONNOP. "John Henry Newman: His Poetry and Conversion." Dublin Review 242 (1968), 75-88.

5675. THOMAS, J. H. "Is Newman a Mystery?" Scottish Journal of Theology 18 (Dec. 1965), 435-443.

5676. THOMAS, W. H. GRIFFITH. "An Evangelical View of Cardinal Newman." Princeton Theological Review 12 (1914), 23-59.

5677. THUREAU-DANGIN, PAUL. Newman catholique, d'après des documents nouveaux. Paris: Plon-Nourrit et Cie, 1912.

5678. TIERNEY, MICHAEL. "Newman's Doctrine of University Education." Studies 42 (1953), 121-131. (Reprinted in Michael Tierney et al., Newman's Doctrine of University Education. Dublin: Sealy, Bryers and Walker, 1954, 1-11)

5679. TIERNEY, MICHAEL. (Ed.) A Tribute to Newman: Essays on Aspects of His Life and Thought. Dublin: Browne and Nolan, 1945.

5680. TINDAL-ATKINSON, A. "Newman and the National Church." Blackfriars 14 (July 1933), 591-601.

5681. TINSLEY, BEVERLY ANNE BEDSOLE. "John Henry Newman and the 'British Critic'." Dissertation Abstracts International 33 (1973), 5695A (Northwestern University, 1972).

5682. TOON, PETER. "Newman's Essay on Development Revisited." Churchman 89 (Jan.-Mar. 1975), 47-57.

5683. TRACY, CLARENCE R. "Bishop Blougram." Modern Language Review 34 (July 1939), 422-425. (Browning's poem as a comment on Newman and N. Wiseman)

5684. TREDWAY, T. "Newman: Patristics, Liberalism and Ecumenism." Christian Century 82 (Aug. 11, 1965), 987-989. (A. Atkin's reply: 82, Oct. 6, 1965, 1222)

5685. TREVOR, MERIOL. Newman. 2 vols. London: Macmillan, 1962.

5686. TRISTRAM, HENRY. "Cardinal Newman and Baron von
       Hügel." Dublin Review 509 (Autumn 1966), 295-302.
5687. TRISTRAM, HENRY. "Cardinal Newman and the Dublin
       Review." Dublin Review 198 (1936), 221-234.
5688. TRISTRAM, HENRY. "Cardinal Newman at Oxford: An
       Account Using Some Unpublished Sources." Tablet
       174 (1939), 242-243, 274-275, 306-307, 335-336,
       359-360.
5689. TRISTRAM, HENRY. "Cardinal Newman in Birmingham."
       Oscotian 6th Series, 6, No. 3 (Dec. 1945), 84-92;
       7, No. 1 (Apr. 1946), 24-31.
5690. TRISTRAM, HENRY. "Cardinal Newman's Theses de
       Fide and His Proposed Introduction to the French
       Translation of the University Sermons." Gregorianum
       18 (1937), 219-260.
5691. TRISTRAM, HENRY. "The Cause of John Henry, Cardinal
       Newman." Pax 63 (1953), 100-104.
5692. TRISTRAM, HENRY. "The Correspondence between J. H.
       Newman and the Comte de Montalembert." Dublin
       Review 445 (Spring 1949), 118-138.
5693. TRISTRAM, HENRY. "Dr. Russell and Newman's Conversion."
       Irish Ecclesiastical Record 66 (Sept. 1945), 189-200.
       (Charles William Russell)
5694. TRISTRAM, HENRY. "J. A. Möhler et J. H. Newman."
       Revue des Sciences Philosophiques et Théologiques
       27 (1938), 184-204.
5695. TRISTRAM, HENRY. (Ed.) John Henry Newman: Centen-
       ary Essays. Westminster, Maryland: The Newman
       Book Shop, 1945.
5696. TRISTRAM, HENRY. "Lead, Kindly Light - June 16,
       1883." Dublin Review 193 (July 1933), 85-96.
5697. TRISTRAM, HENRY. "Mr. Gibbon, Dr. White, and
       Cardinal Newman: A Tale with Two Morals." Cornhill
       Magazine 70 (Feb. 1931), 209-221. (Edward Gibbon,
       Joseph White)
5698. TRISTRAM, HENRY. "Mr. Newman and Father Clement."
       Dublin Review 196 (Jan. 1935), 100-114.

5699.   TRISTRAM, HENRY. Newman and His Friends. London:
        John Lane, 1933.

5700.   TRISTRAM, HENRY. "Newman and Matthew Arnold."
        Cornhill Magazine 60 (Mar. 1926), 309-319.

5701.   TRISTRAM, HENRY. "Newman and the Novelists."
        Cornhill Magazine 62 (Apr. 1927), 495-509.

5702.   TRISTRAM, HENRY. "Newman's Ancestry: The Origin
        of a Baseless Legend." Month 171, No. 888 (June
        1938), 547-551.

5703.   TRISTRAM, HENRY. "Oxford Background." Dublin
        Review 217 (Oct. 1945), 136-146.

5704.   TRISTRAM, HENRY. "The School-days of Cardinal
        Newman." Cornhill Magazine 58 (1925), 666-677.

5705.   TRISTRAM, HENRY. "Two Leaders: Newman and Carlyle."
        Cornhill Magazine 65 (1928), 367-382.

5706.   TRISTRAM, HENRY. (Ed.) "Two Suppressed Passages
        from Newman's 'Autobiographical Memoir' Relating to
        His Tutorship at Oriel College, Oxford." Revue
        Anglo-Américaine 11, No. 6 (1934), 481-494.

5707.   ULANOV, BARRY. "Newman and Dostoevsky: The
        Politics of Salvation." In Sources and Resources:
        The Literary Traditions of Christian Humanism.
        Westminster, Md.: Newman Press, 1960, 228-269.

5708.   VARGISH, THOMAS. Newman: The Contemplation of
        Mind. Oxford: Clarendon Press, 1970.

5709.   VARGISH, THOMAS. "Studies in Newman's Epistemology."
        Dissertation Abstracts 27 (1966), 1066A (Princeton
        University, 1966).

5710.   VELOCCI, GIOVANNI. Newman mistico. Roma: Libreria
        editrice della Pontificia università lateranense,
        1964.

5711.   VERSFELD, M. "St. Thomas, Newman and the Existence
        of God." New Scholasticism 41 (Winter 1967), 3-30.

5712.   VOGEL, CORNELIA JOHANNA DE. Newmans gedachten
        over de rechtvaardiging: hun zin en recht ten
        opzichte van Luther en het protestantsche Christendom.
        Wageningen: H. Veenman, 1939.

5713.   WALGRAVE, JAN HENRICUS.   "Conscience de soi et
        conscience de Dieu."   Revue Thomiste 71 (Apr.-Sept.
        1971), 367-380.

5714.   WALGRAVE, JAN HENRICUS.   Newman the Theologian.
        Trans. A. V. Littledale.   London: G. Chapman,
        1960.

5715.   WALKER, LUKE, O.P.   "Newman's Approach to the
        Church."   Blackfriars 14 (July 1933), 545-550.

5716.   WALLER, ALFRED RAYNEY AND G. H. S. BARROW.   John
        Henry, Cardinal Newman.   Boston: Small, Maynard
        and Co., 1901.

5717.   WALSH, P.   "Newman in Perspective: The Difficulties
        and Failures He Encountered."   Jubilee 10 (Oct.
        1962), 23-29.

5718.   WAMSLEY, GEOFFREY.   "A New Tract for the Times:
        Newman on Liberalism in Religion."   Clergy Review
        56 (Sept. 1971), 665-671.

5719.   WARD, MAISIE. (Ed.)   The English Way: Studies
        in English Sanctity from St. Bede to Newman.
        London and New York: Sheed and Ward, 1933.

5720.   WARD, MAISIE.   Young Mr. Newman.   London: Sheed
        and Ward, 1948.

5721.   WARD, WILFRID PHILIP.   "Cardinal Newman's Sensi-
        tiveness."   Dublin Review 150 (Apr. 1912), 217-229.
        (Reprinted in Men and Matters.   New York, London:
        Longmans, Green, and Co., 1914, 273-289)

5722.   WARD, WILFRID PHILIP.   "The Genius of Cardinal
        Newman: A Criticism of Popular Misconceptions."
        In Last Lectures: Being the Lowell Lectures, 1914,
        and Three Lectures Delivered at the Royal Institution,
        1915.   London, New York: Longmans, Green, and Co.,
        1918, 1-149.

5723.   WARD, WILFRID PHILIP.   The Life of John Henry
        Cardinal Newman: Based on Private Journals and
        Correspondence.   2 vols.   London, New York:
        Longmans, Green and Co., 1912.

5724.   WARD, WILFRID PHILIP.   "Newman and Sabatier."
        Fortnightly Review 75 (May 1901), 808-822.

5725.   WARD, WILFRID PHILIP.   "Two Views of Cardinal
        Newman." Dublin Review 141 (July 1907), 1-15.

5726.   WATKIN-JONES, HOWARD.   "Two Oxford Movements:
        Wesley and Newman." Hibbert Journal 31 (Oct.
        1932), 83-96.

5727.   WEATHERBY, HAROLD L.   Cardinal Newman in His Age:
        His Place in English Theory and Literature. Nashville,
        Tennessee: Vanderbilt University Press, 1973.

5728.   WEATHERBY, HAROLD L.   "The Encircling Gloom:
        Newman's Departure from the Caroline Tradition."
        Victorian Studies 12 (Sept. 1968), 57-82.

5729.   WEATHERBY, HAROLD L.   "Newman and the Origins of a
        'High Church' Left." Modern Age 12 (1968), 58-64.

5730.   WEATHERBY, HAROLD L.   "Newman and Victorian Liberalism:
        The Failure of Influence." Critical Quarterly 13
        (Autumn 1971), 205-213.

5731.   WEBB, CLEMENT CHARLES JULIAN.   "Two Philosophers of
        the Oxford Movement." Philosophy 8 (1933), 273-284.
        (Newman and W. G. Ward)

5732.   WHALEN, JOHN PHILIP.   "The Notion of Tradition in
        the Writings of John Henry Newman." Dissertation
        Abstracts 26 (1965), 2899 (Catholic University of
        America, 1965).

5733.   WHEELER, T. S.   "Newman and Science." Studies 42
        (1953), 179-196. (Reprinted in Michael Tierney et
        al., Newman's Doctrine of University Education.
        Dublin: Sealy, Bryers and Walker, 1954, 59-76)

5734.   WHITE, GAVIN DONALD.   "Newman's Missionary Dream."
        Modern Churchman 14 (July 1971), 267-272.

5735.   WHITE, NEWPORT JOHN DAVIS.   John Henry Newman.
        London: Society for Promoting Christian Knowledge,
        1925.

5736.   WHITE, W. D.   "John Henry Newman, Anglican Preacher:
        A Study in Theory and Style." Dissertation Abstracts
        29 (1969), 3213A (Princeton University, 1968).

5737. WHITE, W. D. "John Henry Newman's Critique of Popular Preaching." South Atlantic Quarterly 69 (1970), 108-117.

5738. WHITE, W. D. (Ed.) The Preaching of John Henry Newman. Philadelphia: Fortress Press, 1969.

5739. WHYTE, ALEXANDER. Newman: An Appreciation. New York: Longmans, Green, and Co., 1901.

5740. WHYTE, JOHN HENRY. "Newman in Dublin: Fresh Light from the Archives of Propaganda." Dublin Review 234 (Spring 1960), 31-39.

5741. WILBERFORCE, ROBERT. "J. H. Newman: His Prophetic Sense." Catholic World 162 (Jan. 1946), 336-342.

5742. WILDMAN, J. H. "Newman's First Apologia." Ecclesiastical Review 102 (June 1940), 525-530. (Loss and Gain)

5743. WILHELMSEN, FREDERICK D. "New Dimension for Newman." Commonweal 66 (May 3, 1957), 123-125.

5744. WILLAM, FRANZ MICHEL. Aristotelische Erkenntnis-lehre bei Whately und Newman, und ihre Bezüge zur Gegenwart. Freiburg: Herder, 1960.

5745. WILLEBRANDS, JAN G. M. "Het Christelijk Platonisme van Kardinaal Newman." Studia Catholica 17 (1941), 373-388.

5746. WILLIAMS, J. "John Henry Newman's Consideration of Physical Education as a Liberal Pursuit." Peabody Journal of Education 42 (May 1965), 343-348.

5747. WILLIAMS, PHILIP. (Ed.) "Some Unpublished Letters of Cardinal Newman, in Oscott Museum." Oscotian Series 3, 2 (1902), 117-122.

5748. WILLIAMSON, CLAUDE. (Ed.) Great Catholics. London: Nicholson and Watson, Limited, 1938.

5749. WILSON, ROBERT F. Newman's Church in Dublin. Dublin: Irish Industrial Printing and Publishing Co., 1916.

5750. WINSLOW, DONALD F. "Some Unpublished Newman Autographs." Anglican Theological Review 55 (Jan. 1973), 43-52.

5751. WISE, JOHN E. "Newman and the Liberal Arts."
       Thought 20 (1945), 253-270.
5752. WRIGHT, CUTHBERT. "Newman and Kingsley." Harvard
       Graduate Magazine 40 (1931), 127-134.
5753. WRIGHT, JOHN, CARDINAL. "Cardinal Newman and
       Theological Clarity." Homiletic and Pastoral
       Review 75 (Apr. 1975), 6-10.
5754. WRIGHT, JOHN, CARDINAL. "Reflection of Cardinal
       Newman on the Evangelization by the First Apostles."
       L'osservatore romano No. 51 (Dec. 19, 1974), 9, 12.
5755. YANITELLI, VICTOR R. (Ed.) A Newman Symposium:
       Report on the Tenth Annual Meeting of the Catholic
       Renascence Society. New York: Fordham University,
       1953.
5756. YAO-SHAN, C. "Newman the Dreamer." Clergy Review
       60 (June 1975), 368-373.
5757. YEARLEY, LEE HOWARD. "Natural and Supernatural
       Activity in the Tradition Represented by St. Thomas
       and Cardinal Newman." Ph.D. Dissertation, University
       of Chicago, 1969.
5758. YEARLEY, LEE HOWARD. "Newman's Concrete Specifica-
       tion of the Distinction between Christianity and
       Liberalism." Downside Review 93 (Jan. 1975),
       43-57.
5759. YOUNG, GEORGE MALCOLM. "Sophist and Swashbuckler."
       In Daylight and Champaign. London: Jonathan Cape,
       1937, 102-111. (C. Kingsley and Newman)
5760. ZALE, ERIC MICHAEL. "The Defenses of John Henry
       Newman." Dissertation Abstracts 23 (1962), 637
       (University of Michigan, 1962).
5761. ZENO, P., O.F.M. "Een blik in Newmans zieleleven
       voor de Oxford Movement." Dietsche Warande en
       Belfort 53 (1953), 477-488, 517-532.
5762. ZENO, P., O.F.M. "An Introduction to Newman's
       'Grammar of Assent'." Irish Ecclesiastical Record
       103 (June 1965), 389-406.

5763.   ZENO, P., O.F.M.   John Henry Newman, Our Way to
        Certitude: An Introduction to Newman's Psychological
        Discovery, the Illative Sense and His Grammar of
        Assent. Leiden: E. J. Brill, 1957. (Translation
        of Newmans leer over het menselijk denken: inleiding
        op Newmans "Grammar of Assent" en zijn psychologische
        ontdekking: de illatieve zin. Utrecht: Dekker &
        Van de Vegt, 1943)
5764.   ZENO, P., O.F.M. (Ed.) "The Newman-Meynell Corres-
        pondence." Franciscan Studies 12 (1952), 301-348.
        (1869-71)
5765.   ZENO, P., O.F.M.   "Newman's Inner Life As Shown by
        Some of His Poems." Irish Ecclesiastical Record 81
        (1954), 245-259.
5766.   ZENO, P., O.F.M.   "Newman's Inner Life up to His
        Election as Fellow of Oriel College." Irish
        Ecclesiastical Record 78 (Oct. 1952), 255-275.
5767.   ZENO, P., O.F.M.   "Reliability of Newman's Autobio-
        graphical Writings." Irish Ecclesiastical Record
        86 (Nov. 1956), 297-305; 87 (Jan. 1957), 25-37.

18.   Pusey, Edward Bouverie

See introduction to this category; XIV.   Church of
England, 1.   General, 3.   Oxford Movement.

5768.   BOOTH, STEPHEN PAUL. "'Essays and Reviews': The
        Controversy As Seen in the Correspondence and
        Papers of Dr. E. B. Pusey and Archbishop Archibald
        Tait." Historical Magazine of the Protestant
        Episcopal Church 38 (1969), 259-279.
5769.   DONALDSON, AUGUSTUS BLAIR. Five Great Oxford
        Leaders: Keble, Newman, Pusey, Liddon and Church.
        London: Rivingtons, 1900.

5770. FORRESTER, D. W. F. "The Intellectual Development
of E. B. Pusey, 1800-50." Ph.D. Dissertation,
University of Oxford, 1967.

5771. GORCE, DENYS. "La Spiritualité d'Edward Bouverie
Pusey, Regius Professor d'hébreu à Oxford, d'après
ses lettres inédites à sa fiancée." Revue des
Sciences Religieuses 26 (1952), 30-58. (1827-28)

5772. GRIFFIN, JOHN R. "Dr. Pusey and the Oxford Movement."
Historical Magazine of the Protestant Episcopal
Church 42 (1973), 137-153.

5773. HUNTER-BLAIR, SIR DAVID OSWALD. "Great Men of
Bygone Oxford." In Victorian Days and Other Papers.
London, New York: Longmans, Green and Co., 1939,
94-106. (Includes J. Ruskin, Max Müller)

5774. KENYON, RUTH. "Two Studies in the Social Outlook
of the Tractarians: 2. Keble and Pusey."
Theology 25 (July 1932), 24-34. (See Newman, 5460)

5775. LEWIS, FRANK R. "New Letters of E. B. Pusey."
Times Literary Supplement No. 3573 (Aug. 21, 1970),
928. (Ten hitherto unknown letters, written in
1873-74)

5776. MEHLIS, RITA. "Die religiöse Entwicklung des
jungen Edward B. Pusey mit besonderer Berücksichtigung
seiner Beziehungen zu Deutschland." Ph.D. Dissertation, Universität Bonn, 1954.

5777. O'ROURKE, JAMES. "Edward Bouverie Pusey." Irish
Ecclesiastical Record 47 (May 1936), 474-482.

5778. PRESTIGE, GEORGE LEONARD. Pusey: The Tractarian
Series. London: P. Allan, 1933.

5779. RICHARDS, GEORGE CHATTERTON. "Pusey and the Oxford
Movement." Durham University Journal 28 (1932-34),
161-178, 245-257.

5780. RUSSELL, GEORGE WILLIAM ERSKINE. Dr. Pusey.
London: A. R. Mowbray & Co., Limited, 1907.

5781. RUSSELL, GEORGE WILLIAM ERSKINE. Dr. Pusey.
       London: A. R. Mowbray & Co., Limited, 1912.
5782. SENCOURT, ROBERT. "Newman and Pusey." Dublin
       Review 217 (Oct. 1945), 156-165.

                    19. Stephen, Leslie

       See introduction to this category.

5783. ANNAN, NOEL GILROY. Leslie Stephen: His Thought
       and Character in Relation to His Time. London:
       MacGibbon and Kee, 1951.
5784. APPLEMAN, PHILIP. "Darwin and the Literary Critics."
       Dissertation Abstracts 15 (1955), 1618 (Northwestern
       University, 1955).
5785. APPLEMAN, PHILIP. "Evolution and Two Critics of
       Art and Literature." In Proceedings of the Third
       International Congress on Aesthetics. Atti del III
       congresso internazionale di estetica, Venezia, 3-5
       settembre 1956. Torino: Edizioni della Rivista di
       estetica, Istituto di estetica dell'Università di
       Torino, 1957, 237-240. (J. Symonds and Stephen)
5786. BATESON, F. W. "God-Killer." New Statesman 65
       (June 21, 1963), 945-946.
5787. BICKNELL, JOHN W. "Leslie Stephen's 'English
       Thought in the Eighteenth Century': A Tract for
       the Times." Victorian Studies 6 (1962-63), 103-120.
5788. HIMMELFARB, GERTRUDE. "Mr. Stephen and Mr. Ramsay:
       The Victorian as Intellectual." Twentieth Century
       152 (Dec. 1952), 513-525. (Mr. Ramsay - the father
       in To the Lighthouse, 1927, by Virginia Woolf,
       Stephen's daughter)
5789. HOOPER, HENRY T. "Leslie Stephen's Agnosticism."
       London Quarterly Review 108 (July 1907), 108-111.

5790.   IRVING, JOHN A.   "Evolution and Ethics."   Queen's
        Quarterly 55 (1948), 450-463.
5791.   MACCARTHY, DESMOND.   Leslie Stephen.   Cambridge,
        England: The University Press, 1937.
5792.   SANDERS, CHARLES RICHARD.   "Sir Leslie Stephen,
        Coleridge and Two Coleridgeans."   PMLA 55 (1940),
        795-801.   (F. D. Maurice, J. Dykes Campbell)
5793.   TANGL, MARIA REGINA.   "Leslie Stephens Weltanschauung."
        Ph.D. Dissertation, Universität Hamburg, 1961.
5794.   TOLLESON, FLOYD CLYDE, JR.   "The Relation between
        Leslie Stephen's Agnosticism and Voltaire's Deism."
        Dissertation Abstracts 15 (1955), 2218-19 (University
        of Washington, 1955).
5795.   WOOLF, VIRGINIA.   "My Father: Leslie Stephen."
        Atlantic Monthly 185 (Mar. 1950), 39-41.
5796.   ZINK, DAVID D.   Leslie Stephen.   New York: Twayne
        Publishers Inc., 1972.

20.   Tennyson, Alfred

See introduction to this category; VI.   Geology . . .;
VII.   Evolution, 3.   Chambers, Robert.

5797.   AIMEE, SISTER.   "The Religious Beliefs of Three
        Victorian Poets: Tennyson, Browning, and Arnold,
        and Their Influence on English Literature."   Ph.D.
        Dissertation, University of Ottawa, 1942.
5798.   ANON.   "'In Memoriam' and the Christian Faith."
        Contemporary Review 93, Literary Supplement No. 8
        (May 1908), 18-20.
5799.   ANON.   "Tennyson and the Science of the Nineteenth
        Century."   Popular Science 75 (Sept. 1909), 306.
5800.   APPLEMAN, PHILIP.   "The Dread Factor: Eliot,
        Tennyson, and the Shaping of Science."   Columbia
        Forum 3 (1974), 32-38.

5801. AUGUST, EUGENE R. "Tennyson and Teilhard: The
        Faith of 'In Memoriam'." PMLA 84 (1969), 217-226.
5802. BARNES, S. D. "The Faith of Tennyson." Methodist
        Review 82 (1900), 582-591.
5803. BATTENHOUSE, HENRY M. Poets of Christian Thought.
        New York: Ronald Press, 1947. (Includes R. Browning)
5804. BOEGNER, ANDRE. La Pensée religieuse de Tennyson
        dans "In Memoriam." Cahors: Coueslant, 1905.
5805. BOYD, JOHN DOUGLAS. "Tennyson's Poetry of Reli-
        gious Debate: Rhetoric and the Problem of Belief."
        Dissertation Abstracts 29 (1968), 1531-32A (Cornell
        University, 1968).
5806. COFFIN, LAWRENCE I. "Empiricism and Mysticism in
        Tennyson's Poetry." Dissertation Abstracts Inter-
        national 31 (1970), 2870-71A (State University of
        New York at Albany, 1970).
5807. COLLINS, JOSEPH J. "Tennyson and Kierkegaard."
        Victorian Poetry 11 (1973), 345-350.
5808. DEATRICK, W. WILBURFORCE. "The Religious Signi-
        ficance of Tennyson's 'In Memoriam'." Reformed
        Quarterly Review 58 (Oct. 1909), 481-498.
5809. EMERY, CLARK. "The Background of Tennyson's 'Airy
        Navies'." Isis 35 (1944), 139-147. ("Locksley
        Hall")
5810. EVANS, JESSIE RUTH. "The Religious Philosophy of
        Alfred Lord Tennyson." Masters Essay, George
        Washington University, 1925.
5811. EVERETT, C. C. "Tennyson and Browning as Spiritual
        Forces." New World 2 (1893), 240-256.
5812. GLISERMAN, SUSAN M. "Early Victorian Science
        Writers and Tennyson's 'In Memoriam': A Study in
        Cultural Exchange." Victorian Studies 18 (1975),
        277-308, 437-459.
5813. GLISERMAN, SUSAN M. "Literature as Historical
        Document: Tennyson and the Nineteenth-Century
        Science Writers, 1830-1854." Dissertation Abstracts

International 34 (1974), 4201A (Indiana University, 1973).

5814. GRANT, STEPHEN A. "The Mystical Implications of 'In Memoriam'." Review of English Studies 23 (1947), 244-256.

5815. GREENBERG, ROBERT A. "A Possible Source of Tennyson's 'Tooth and Claw'." Modern Language Notes 71 (1956), 491-492.

5816. HARDIN, M. C. "Theology in Tennyson's Poetry." Methodist Review (South) 44 (1896-97), 315-319.

5817. HARDWICK, JOHN CHARLTON. "Tennyson's Religion." Modern Churchman 39 (Dec. 1949), 317-329.

5818. HARRISON, JAMES ERNEST. "Tennyson and Embryology." Humanities Association Bulletin 23, No. 2 (1972), 28-32.

5819. HARRISON, JAMES ERNEST. "Tennyson and Evolution." Durham University Journal 64 (1971), 26-31.

5820. HASKELL, H. B. "Tennyson as a Religious Teacher." Masters Essay, University of Maine, 1906.

5821. HAYES, J. W. Tennyson and Scientific Theology. London: Stock, 1909.

5822. HEGNER, ANNA. "Die Evolutionsidee bei Tennyson und Browning." Ph.D. Dissertation, Universität Freiburg, 1931.

5823. HOBSON, WILLIAM A. "The Religion of Tennyson." Calcutta Review 5 (Apr. 1917), 189-199.

5824. HOUGH, GRAHAM. "The Natural Theology of 'In Memoriam'." Review of English Studies 23 (July 1947), 244-256.

5825. HUMBERT, GERALD VERNON. "Scientific Thought in Tennyson." Masters Essay, University of Nebraska, 1930.

5826. JONES, D. M. "The Religious Teaching of Tennyson." Wesleyan Methodist Magazine 115 (1892), 876-880.

5827. JUMP, JOHN D. "Tennyson's Religious Faith and Doubt." In D. J. Palmer (Ed.), Writers and Their

Background: Tennyson. Athens: Ohio University
Press, 1973, 89-114.

5828. LANDOW, GEORGE P. "Closing the Frame: Having
Faith and Keeping Faith in Tennyson's 'The Passing
of Arthur'." Bulletin of the John Rylands Library
56 (1974), 423-442.

5829. LASKI, MARGHANITA. Ecstasy: A Study of Some
Secular and Religious Experiences. London: Cresset
Press, 1961. (Includes the Brontës)

5830. LAWRY, J. S. "Tennyson's 'The Epic': A Gesture of
Recovered Faith." Modern Language Notes 74 (1959),
400-403.

5831. LEE, GEORGE. "Tennyson's Religion." American
Catholic Quarterly Review 25 (Jan. 1900), 119-132.

5832. LESTER, G. "Concerning Lord Tennyson's Knowledge
and Use of the Bible." Methodist Review (South) 45
(1897), 163-170.

5833. LIBERA, SHARON MAYER. "John Tyndall and Tennyson's
'Lucretius'." Victorian Newsletter No. 45 (Spring
1974), 19-22.

5834. LOCKWOOD, FRANK C. "Tennyson's Religious Faith."
Methodist Review 91 (Sept. 1909), 783-787.

5835. LOCKYER, SIR JOSEPH NORMAN AND WINIFRED L. LOCKYER.
Tennyson as a Student and Poet of Nature. London:
Macmillan, 1910.

5836. LODGE, SIR OLIVER JOSEPH. "The Attitude of Tennyson
towards Science." In Modern Problems. London:
Methuen, 1912, 301-307.

5837. LOWBER, JAMES WILLIAM. "Tennyson's Science of
Religion." In World Wide Problems: Or, Macrocosmus.
Cincinnati: Standard Publishing Co., 1923, 300-311.

5838. LOWE, RALPH FERNALD. "Relation of Tennyson to
Movements of Religious Thought in England, 1830-
1890." Masters Essay, Wesleyan University, 1908.

5839. MCCORKINDALE, T. B. "Some Elements in the Religious
Teaching of Tennyson." Queen's Quarterly 21 (Apr.-
June 1914), 449-455.

5840. MACEWEN, V. Knights of the Holy Eucharist. London:
Gardner, 1912. (On Tennyson's "The Holy Grail" and
its spiritual teaching)

5841. MCKEAN, G. R. "Faith in 'Locksley Hall'." Dalhousie
Review 19 (1940), 472-478.

5842. MACKIE, ALEXANDER. Nature Knowledge in Modern
Poetry, Being Chapters on Tennyson, Wordsworth,
Matthew Arnold, and Lowell as Exponents of Nature-
Study. London: Longmans, Green, and Co., 1906.
(Includes chapter on Tennyson as botanist, entomol-
ogist, ornithologist, and geologist)

5843. MASTERMAN, CHARLES FREDERICK GURNEY. Tennyson
as a Religious Teacher. London: Methuen, 1899.

5844. MATTES, ELEANOR B. "The Religious Influences upon
Tennyson's 'In Memoriam'." Ph.D. Dissertation,
Yale University, 1945.

5845. MERCER, ARTHUR. "'In Memoriam' as a Revelation of
the Religious Philosophy of Tennyson." New Church
Review 16 (Oct. 1909), 540-559.

5846. MILLHAUSER, MILTON. Fire and Ice: The Influence
of Science on Tennyson's Poetry. Lincoln: The
Tennyson Society, 1971.

5847. MILLHAUSER, MILTON. "A Plurality of After-Worlds:
Isaac Taylor and Alfred Tennyson." Hartford Studies
in Literature 1 (1969), 37-49.

5848. MILLHAUSER, MILTON. "Tennyson, 'Vestiges', and the
Dark Side of Science." Victorian Newsletter No. 35
(1969), 22-25.

5849. MILLHAUSER, MILTON. "Tennyson's 'Princess' and
'Vestiges'." PMLA 69 (1954), 337-343.

5850. MITCHELL, WILLIAM RICHARD. "Theological Origins of
the Christ-Image in Victorian Literature with
Special Reference to 'In Memoriam'." Dissertation
Abstracts International 30 (1969), 1990A (University
of Oklahoma, 1969).

5851.   MOONEY, EMORY A., JR.   "A Note on Astronomy in
        Tennyson's 'The Princess'."   Modern Language Notes
        64 (1949), 98-102.

5852.   MOONEY, EMORY A., JR.   "Tennyson and Modern Science."
        Ph.D. Dissertation, Cornell University, 1938.

5853.   MOORE, CARLISLE.   "Faith, Doubt, and Mystical
        Experience in 'In Memoriam'."   Victorian Studies 7
        (Dec. 1963), 155-169.

5854.   MORITZ, ALBERT FRANK.   "Tennyson and the Defense of
        Romantic Faith."   Dissertation Abstracts Interna-
        tional 36 (1976), 6709A (Marquette University,
        1975).

5855.   MOUNGER, SAMUEL GWIN.   "Will within Will: Religious
        Thought in the Poetry of Tennyson."   Dissertation
        Abstracts International 34 (1974), 5113A-14A
        (University of Virginia, 1973).

5856.   PARSONS, E.   "Tennyson's Attitude towards Skept-
        icism."   Homiletic Review 32 (1896), 205-213.

5857.   PARSONS, E.   "Tennyson's Theology."   Methodist
        Review 76 (1894), 917-927.

5858.   PEAKE, LESLIE SILLMAN.   "Tennyson and Faith."
        London Quarterly and Holborn Review 157 (Apr.
        1932), 182-189.

5859.   POTTER, GEORGE REUBEN.   "Tennyson and the Biological
        Theory of Mutability of Species."   Philological
        Quarterly 16 (1937), 321-343.

5860.   PRIESTLEY, F. E. L.   Language and Structure in
        Tennyson's Poetry.   London: Deutsch, 1973.

5861.   RADER, WILLIAM.   The Elegy of Faith: A Study of
        Tennyson's "In Memoriam."   New York: Crowell,
        1902.

5862.   RICE, JESSIE FOLSOM.   "The Influence of Frederick
        Dennison [sic] Maurice on Tennyson."   Dissertation,
        University of Chicago, 1913.

5863.   ROSENBERG, JOHN D.   "The Two Kingdoms of 'In
        Memoriam'."   Journal of English and German Philology
        58 (1959), 228-240.

5864. RUDMAN, HARRY W. "Keats and Tennyson on 'Nature, Red in Tooth and Claw'." Notes and Queries 199 (1954), 293-294.

5865. RUTLAND, WILLIAM R. The Becoming of God. Oxford: Blackwell, 1972.

5866. RUTLAND, WILLIAM R. "Tennyson and the Theory of Evolution." Essays and Studies by Members of the English Association 26 (1941), 7-29.

5867. RYALS, CLYDE DE L. "The 'Heavenly Friend': The 'New Mythus' of 'In Memoriam'." Personalist 43 (1962), 383-402. (Tennyson's spiritual crisis in "In Memoriam" parallels Carlyle's in Sartor Resartus)

5868. SANDERS, CHARLES RICHARD. "Carlyle and Tennyson." PMLA 76 (1961), 82-97.

5869. SCOTT, PATRICK GREIG. "Tennyson and Charles Kingsley." Tennyson Research Bulletin 2, No. 3 (1974), 135-136.

5870. SINCLAIR, WILLIAM MACDONALD. "The Religious Poetry of Tennyson." Churchman 12 (1897-98), 435-443, 475-483.

5871. SNEATH, ELIAS HERSHEY. The Mind of Tennyson: His Thoughts on God, Freedom, and Immortality. Westminster: Constable, 1900.

5872. SOLIMINE, JOSEPH, JR. "The Dialectics of Church and State: Tennyson's Historical Plays." Personalist 47 (1966), 218-233.

5873. SONN, CARL ROBINSON. "Poetic Vision and Religious Certainty in Tennyson's Earlier Poetry." Modern Philology 57 (1959), 83-93.

5874. SPANGENBERG, ALICE. "Tennyson's Attitude toward Science." Masters Essay, Boston University, 1925.

5875. STEVENSON, MORLEY. The Spiritual Teaching of Tennyson's "In Memoriam": Six Lenten Addresses. London: Gardner, Darton, 1904.

5876. STOCKLEY, WILLIAM FREDERICK PAUL. "The 'Faith' of 'In Memoriam'." Catholic World 120 (Mar. 1925), 801-809.

5877. TALLCOTT, ROLLO ANSON. "Tennyson's Fluxuations of Doubt and Faith from 1820-1850." Masters Essay, Syracuse University, 1920.

5878. TENNYSON, SIR CHARLES. Alfred Tennyson. London: Macmillan & Co.; New York: Macmillan Co., 1949.

5879. TIETZE, FREDERICK INGLEBRIT. "Tennyson, Science, and the Poetic Sensibility." Summaries of Doctoral Dissertations, University of Wisconsin 14 (1954), 452-453 (University of Wisconsin, 1953).

5880. TOLLEMACHE, LIONEL ARTHUR. "Jowett and Tennyson." Spectator 119 (Oct. 20, 1917), 411.

5881. TURNER, PAUL. "The Stupidest English Poet." English Studies 30 (1949), 1-12. (Tennyson's theology)

5882. VAN DEN NOORT, JUDOKUS. "Theology in Tennyson." Masters Essay, Boston University, 1923.

5883. WARD, MAISIE. "Wilfrid Ward and Tennyson." Commonweal 21 (Nov. 16, 1934), 87-88. (See also M. Earls, Commonweal 21, Nov. 23, 1934, 124)

5884. WARD, WILFRID PHILIP. "Tennyson Centenary: Tennyson's Religious Poetry." Dublin Review 145 (Oct. 1909), 306-322. (Also in Living Age 263, Nov. 27, 1909, 523-533)

5885. WEATHERHEAD, LESLIE DIXON. "A New Projection of Christian Thought Born of the Fear of Death - Tennyson." In The After-World of the Poets. London: Epworth Press, 1929, 79-122.

5886. WHEELER, P. M. "Tennyson, a Victorian Astronomer." Popular Astronomy 56 (Dec. 1948), 527-540.

5887. WHITE, F. E. "Unorthodox Tendencies in Tennyson." Review of Religion 15 (Nov. 1950), 19-28.

5888. WILEY, MARGARET LEE. "The Religious Poems of Tennyson, with Special Reference to Present Day Unrest." Masters Essay, University of Texas, 1924.

5889. WILLIAMS, MELVIN G. "'In Memoriam': A Broad Church Poem." Costerus 4 (1972), 223-233.

5890. WOOD, HOMA. "The Religion of Tennyson as Shown by
His Works." Masters Essay, University of Oklahoma,
1915.

5891. WYMER, THOMAS LEE. "Romantic to Modern: Tennyson's
Aesthetic and Religious Development." Dissertation
Abstracts 29 (1968), 242-243A (University of Oklahoma,
1968).

21. Tyndall, John

See introduction to this category; I. Science,
1. General.

5892. ANON. AND E. H. STEVENS. "Scientist and Pioneer:
The Discoveries of John Tyndall." Times Literary
Supplement (June 23, 1945), 294; (July 7, 1945),
319.

5893. ANON. "Tyndall and Evolution." Scientific American
Supplement 54 (Dec. 20, 1902), 22547.

5894. BLINDERMAN, CHARLES S. "John Tyndall and the
Victorian New Philosophy." Bucknell Review 9 (Mar.
1961), 281-290.

5895. BRAGG, SIR WILLIAM LAWRENCE. "Tyndall's Experiments
on Magne-Crystallic Action." Royal Institution
Library of Science 9 (1927), 131-154. (Also in
Scientific Monthly 25, July 1927, 65-79)

5896. CONANT, JAMES BRYANT. "Pasteur's and Tyndall's
Study of Spontaneous Generation." In James Bryant
Conant (Ed.), Harvard Case Histories in Experi-
mental Science. 2 vols. Cambridge, Mass.: Harvard
University Press, 1953, Vol. 2, 487-539.

5897. CRELLIN, J. K. "Airborne Particles and the Germ
Theory, 1860-1880." Annals of Science 22 (1966),
49-60.

5898.   CRELLIN, J. K.   "The Problem of Heat Resistance of
        Micro-organisms in the British Spontaneous Genera-
        tion Controversies of 1860-1880."   Medical History
        10 (1966), 50-59.

5899.   EVE, ARTHUR STEWART AND CLARENCE HAMILTON CREASY.
        Life and Work of John Tyndall.   London: Macmillan,
        1945.

5900.   FRIDAY, JAMES R., ROY M. MACLEOD AND PHILIPPA
        SHEPHERD.   John Tyndall, Natural Philosopher, 1820-
        1893:   Catalogue of Correspondence, Journals and
        Collected Papers.   London: Mansell, 1974.

5901.   HAUGRUD, RAYCHEL A.   "Tyndall's Interest in Emerson."
        American Literature 41 (1970), 507-517.

5902.   HUXLEY, LEONARD.   "John Tyndall: A Centenary
        Sketch, 1820-1893."   Cornhill Magazine 49 (1920),
        627-640.

5903.   JAMES, T. E.   "Rumford and the Royal Institution:
        Tyndall."   Nature 128 (Sept. 19, 1931), 481.

5904.   LIBERA, SHARON MAYER.   "John Tyndall and Tennyson's
        'Lucretius'."   Victorian Newsletter No. 45 (Spring
        1974), 19-22.

5905.   MURAS, T. H.   "John Tyndall's Radiation Experiment."
        Nature 158 (Aug. 10, 1946), 203.

5906.   SARTON, GEORGE.   "Faraday to Tyndall."   Isis 31
        (1940), 303-304.

5907.   SATTERLY, J.   "Tyndall and Stefan's Radiation Law."
        Nature 157 (June 1, 1946), 737.   (See also June 29,
        1946, 879)

5908.   SHIPLEY, M.   "Forty Years of a Scientific Friendship:
        Herbert Spencer and John Tyndall."   Open Court 34
        (Apr. 1920), 252-255.

5909.   SMITH, ARTHUR WHITMORE.   "John Tyndall (1820-1893)."
        Scientific Monthly 11 (1920), 331-340.

5910.   SOPKA, KATHERINE.   "An Apostle of Science Visits
        America: John Tyndall's Journey of 1872-1873."
        Physics Teacher 10 (1972), 369-375.

5911. THOMPSON, D. "John Tyndall and the Royal Insti-
        tution." Annals of Science 13 (1957), 9-21.
5912. THOMPSON, D. "John Tyndall (1820-1893): A Study
        in Vocational Enterprise." Vocational Aspects of
        Secondary and Further Education 9 (1957), 38-48.
5913. WATSON, E. C. "Vanity Fair Caricature of John
        Tyndall." American Journal of Physics 17 (Feb.
        1949), 86-88.
5914. WEED, LYLE A. "John Tyndall and His Contribution
        to the Theory of Spontaneous Generation." Annals of
        Medical History 4 (1942), 55-62.
5915. WILLIAMSON, D. E. "The Double-Beam Infrared Gas
        Analyzer of John Tyndall." American Scientist 39
        (1951), 672-681, 687.
5916. WISEMAN, E. J. "John Tyndall: His Contributions
        to the Defeat of the Theory of Spontaneous Gener-
        ation of Life." School Science Review 159 (1965),
        362-367.
5917. YOUNG, HENRY. A Record of the Scientific Work
        of John Tyndall. London: Chiswick Press, 1935.

                22. Ward, Wilfrid Philip

        See introduction to this category; XIV. Church of
England, 1. General, 3. Oxford Movement; XVI. Catholicism.

5918. BRAYBROOKE, NEVILLE. "Two Editors: Wilfrid Ward
        and Wilfrid Meynell." Dublin Review 228 (1954),
        46-52.
5919. COCKSHUT, ANTHONY O. J. "Ward's Life of Newman: A
        Great Biography." In William Wallace Robson (Ed.),
        Essays and Poems Presented to Lord David Cecil.
        London: Constable, 1970, 126-139.
5920. HANBURY, M. "The Lesson of Wilfrid Ward." Pax 28
        (May 1938), 35-40.

5921. HUBY, J. "Wilfrid Ward et ses souvenirs: le
       catholicisme en Angleterre après le mouvement
       d'Oxford." Etudes: Revue Catholique d'Intérêt
       Général 223 (1935), 721-742.
5922. KELLY, EDWARD EUGENE, S.J. "Newman, Wilfrid Ward,
       and the Modernist Crisis." Thought 48 (Winter
       1973), 508-519.
5923. KING, REV. WILLIAM FRANCIS, S.J., S.T.D. "Authority
       in the Church: A Study of the Interrelationship of
       the Authority of Revelation, the Authority of
       Theologians, and Ruling Authority in the Thought of
       Wilfrid Ward." Dissertation Abstracts International
       34 (1974), 7877A (Catholic University of America,
       1974).
5924. POLE, F. "A Great Liaison Officer of the Catholic
       Church." Downside Review 53 (Apr. 1935), 153-163.
5925. TRISTRAM, HENRY. "'The Wilfrid Wards and the Trans-
       ition,' by M. Ward." Dublin Review 196 (Apr. 1935),
       292-303. (See Ward, Maisie, 5929)
5926. WARD, MAISIE. Insurrection "versus" Resurrection.
       London: Sheed & Ward, 1937.
5927. WARD, MAISIE. "W. G. Ward and Wilfrid Ward."
       Dublin Review 198 (1936), 235-252.
5928. WARD, MAISIE. "Wilfrid Ward and Tennyson."
       Commonweal 21 (Nov. 16, 1934), 87-88. (See also M.
       Earls, Commonweal 21, Nov. 23, 1934, 124)
5929. WARD, MAISIE. The Wilfrid Wards and the Transition.
       2 vols. London: Sheed & Ward, 1934-37. (See
       Tristram, Henry, 5925)
5930. WARD, WILFRID PHILIP. "Some Recollections, 1882-
       1887." Life and Letters 10 (1934), 677-690.

23.   Wilberforce, Samuel

See introduction to this category; XIV. Church of
England, 1. General, 5. Biblical Criticism and Essays and
Reviews (1860).

5931.   ARMSTRONG, A. MACC.   "Samuel Wilberforce vs. T. H.
        Huxley: A Retrospect." Quarterly Review 296
        (1958), 426-437.

5932.   BIBBY, HAROLD CYRIL.   "The Huxley-Wilberforce
        Debate: A Postscript." Nature 176 (1955), 363.

5933.   BLINDERMAN, CHARLES S.   "The Oxford Debate and
        After." Notes and Queries 202 (Mar. 1957), 126-
        128.

5934.   FOSKETT, D. J.   "Wilberforce and Huxley on Evolution."
        Nature 172 (1953), 920.

5935.   HARDWICK, JOHN CHARLTON.   "Father of 'Good' Churchmen."
        Modern Churchman 20 (Apr. 1930), 13-25.

5936.   HARDWICK, JOHN CHARLTON.   Lawn Sleeves: A Short
        Life of Samuel Wilberforce. Oxford: B. Blackwell,
        1933.

5937.   JENKINS, CLAUDE.   "Bishop Wilberforce Centenary."
        Theology 49, No. 308 (Feb. 1946), 43-46.

5938.   MEACHAM, STANDISH.   Lord Bishop: The Life of
        Samuel Wilberforce 1805-1873. Cambridge, Mass.:
        Harvard University Press, 1970.

5939.   NEWSOME, DAVID H.   "The Churchmanship of Samuel
        Wilberforce." Studies in Church History 3 (1966),
        23-47.

5940.   NEWSOME, DAVID H.   "How Soapy Was Sam? A Study of
        Samuel Wilberforce." History Today 13 (Sept.
        1963), 624-632.

5941.   NEWSOME, DAVID H.   The Parting of Friends: A
        Study of the Wilberforces and Henry Manning.
        London: Murray, 1966.

5942.   PHELPS, LYNN A. AND EDWIN COHEN.   "The Wilberforce-
        Huxley Debate."   Western Speech 37 (Winter 1973),
        56-64.

5943.   PUGH, RONALD K.   "The Episcopate of Samuel Wilber-
        force, Bishop of Oxford, 1845-69, and of Winchester,
        1869-73, with Special Reference to the Administration
        of the Diocese of Oxford."   Ph.D. Dissertation,
        University of Oxford, 1957.

5944.   PUGH, RONALD K. AND J. F. A. MASON. (Eds.)   The
        Letter-Books of Samuel Wilberforce 1843-68.   London:
        Oxfordshire Record Society, 1970.

5945.   PUGH, RONALD K.   "Samuel Wilberforce: Unfortunate
        Orthodoxy."   Listener 63 (Apr. 28, 1960), 760-761,
        774.

5946.   SMYTH, CHARLES HUGH EGERTON.   "Samuel Wilberforce,
        D.D."   Times Educational Supplement 2087 (May 20,
        1955), 516.

5947.   WAND, JOHN WILLIAM CHARLES.   "Samuel Wilberforce."
        Historical Magazine of the Protestant Episcopal
        Church 16 (1947), 302-308.

5948.   WILBERFORCE, REGINALD GARTON.   Bishop Wilberforce.
        Oxford, London: A. R. Mowbray and Co., 1905.

5949.   WILSON, DAVID.   "Huxley and Wilberforce at Oxford
        and Elsewhere."   Westminster Review 167 (Mar.
        1907), 311-316.

5950.   WINDLE, SIR BERTRAM COGHILL ALAN.   "Not Peace, But
        a Sword."   Catholic World 122 (Feb. 1926), 630-633.
        (The Wilberforce family)

24. Others

This section deals with the religious and (occasion-
ally) scientific outlook of various writers, including
Robert Browning, William Kingdon Clifford, Charles Dickens,
Thomas Hardy, Harriet Martineau, John Morley, Walter
Horatio Pater, W. Winwood Reade, John Ruskin, Henry Sidgwick,
Anthony Trollope, James Ward, and Joseph Blanco White.
See introduction to this category; III. Ideas . . .;
VIII. Evolution and . . . Thought, 6. Spencer, Herbert.

5951. ADRIAN, ARTHUR A. "Dickens and the Brick-and-Mortar
        Sects." Nineteenth-Century Fiction 10 (1955-56),
        188-201. (Dislike of Nonconformity)

5952. ALEXANDER, B. J. "Anti-Christian Elements in
        Thomas Hardy's Novels." Dissertation Abstracts
        International 36 (1975), 2803A (North Texas State
        University, 1975).

5953. ALEXANDER, B. J. "Thomas Hardy's 'Jude the Obscure':
        A Rejection of Traditional Christianity's 'Good'
        God Theory." Southern Quarterly 3 (1964-65),
        74-82.

5954. ARMSTRONG, MARY. "The Writings of Thomas Hardy
        Considered as an Illustration of the Influence of
        the Darwinian Ideas." Masters Essay, University of
        California, 1924.

5955. BADGER, KINGSBURY. "'See the Christ Stand!':
        Browning's Religion." Boston University Studies in
        English 1 (1955), 53-73. (From "Saul")

5956. BAKER, JOSEPH ELLIS. "Religious Implications in
        Browning's Poetry." Philological Quarterly 36
        (Oct. 1957), 436-452. (Browning's religion as
        "Puritan Romanticism")

5957. BIRRELL, OLIVE. "The Early Days of Joseph Blanco
        White." Contemporary Review 94 (1908), 477-489.

5958. BLOUNT, TREVOR. "The Chadbands and Dickens' View
       of Dissenters." Modern Language Quarterly 25
       (1964), 295-307.
5959. BOYLE, SIR EDWARD. "Harriet Martineau." In
       Biographical Essays: 1790-1890. London: Oxford
       University Press, 1936, 160-192.
5960. BUCKINGHAM, MINNIE SUSAN. "The Use of Religious
       Elements in the Fiction of Margaret Wilson Oliphant."
       Cornell University Abstracts of Theses (1939), 22-25
       (Ph.D. Dissertation, Cornell University, 1938).
5961. BURTIS, MARY ELIZABETH. Moncure Conway, 1832-1907.
       New Brunswick, N.J.: Rutgers University Press,
       1952.
5962. CHAPMAN, WALTER R. C. "Walter Horatio Pater's
       Contacts with the Religious Thought of the Nine-
       teenth Century." Ph.D. Dissertation, Universität
       Innsbruck, 1934.
5963. CHARLTON, H. B. "Browning as Poet of Religion."
       Bulletin of the John Rylands Library 27 (1943),
       271-307.
5964. CONNELL, JAMES MACLUCKIE. "Dickens' Unitarian
       Minister, Edward Tagart." Transactions of the
       Unitarian Historical Society 8, No. 2 (Oct. 1944),
       68-83.
5965. CONNELL, JAMES MACLUCKIE. "The Religion of Charles
       Dickens." Hibbert Journal 36 (1938), 225-234.
5966. COOLIDGE, ARCHIBALD C., JR. "Dickens and Latit-
       udinarian Christianity." Dickensian 59 (1963),
       57-60.
5967. DALHOFF, RUDOLF. "Studien über die Religiosität
       John Ruskins, insbesondere ihre Entstehung,
       Entwicklung und Bedeutung für sein Leben und
       Schaffen." Ph.D. Dissertation, Universität Marburg,
       1935.
5968. DRUMMOND, ANDREW LANDALE. "Blanco White: Spanish
       Priest, Refugee, Celebrity, English Clergyman,

Unitarian and Sceptic, 1775-1841." Hibbert Journal
42 (1944), 263-272.

5969. DUFFEY, BERNARD. "The Religion of Pater's 'Marius'."
Texas Studies in Literature and Language 2 (1960),
103-114.

5970. EDWARDS, RALPH. "Trollope on Church Affairs."
Times Literary Supplement No. 2229 (Oct. 21, 1944),
516.

5971. ELIOT, THOMAS STEARNS. "The Place of Pater." In
Walter De La Mare (Ed.), The Eighteen-Eighties.
Cambridge: Cambridge University Press, 1930,
93-106. (Reprinted as "Arnold and Pater." In Eliot,
Selected Essays, 1917-1932. New York: Harcourt,
Brace and Company, 1932, 346-357)

5972. ELLER, V. "Robert Browning: A Promising Theologian." Religion in Life 38 (Summer 1969), 256-263.

5973. FÖRSTER, META. "Robert Brownings Religiosität."
Ph.D. Dissertation, Universität Berlin, 1940.

5974. GOLDSMITH, RICHARD WEINBERG. "The Relation of
Browning's Poetry to Religious Controversy 1833-
1868." Dissertation Abstracts 19 (1959), 2612
(University of North Carolina, 1958).

5975. GOSE, ELLIOTT B., JR. "Psychic Evolution: Darwinism and Initiation in 'Tess of the D'Urbervilles'."
Nineteenth Century Fiction 18 (1963), 261-272.

5976. GOYNE, GROVER CLEVELAND, JR. "Browning and the
Higher Criticism." Dissertation Abstracts 28
(1968), 4128A (Vanderbilt University, 1967).

5977. HAMER, DAVID ALAN. John Morley: Liberal Intellectual in Politics. Oxford: Clarendon Press,
1968.

5978. HARPER, JAMES SEBASTIAN. "Trollope's Ecclesiastical
Concern." Dissertation Abstracts International 33
(1973), 6870-71A (Boston University Graduate School,
1973).

5979.   HARPER, JAMES WINTHROP.   "Browning and the Evangel-
        ical Tradition."   Dissertation Abstracts 21 (1961),
        3089-90 (Princeton University, 1960).
5980.   HICKS, GEORGE DAWES.   "James Ward's Philosophical
        Approach to Theism."   Hibbert Journal 24 (1925-26),
        49-63.
5981.   HICKS, GEORGE DAWES.   "The Philosophy of James
        Ward."   Mind 34 (1925), 281-289.
5982.   HIRT-REGER, HELLA.   "Die Entwicklung von Ruskins
        Ästhetik, Religiosität und Moralauffassung und ihre
        Bedeutung für seine Kritik an der Kunst des Quattro-
        und Cinquescento."   Ph.D. Dissertation, Universität
        Hamburg, 1952.
5983.   HOLLAND, NORMAN, JR.   "'Jude the Obscure': Hardy's
        Symbolic Indictment of Christianity."   Nineteenth-
        Century Fiction 9 (1954), 50-60.
5984.   HOUGH, GRAHAM.   "Books in General."   New Statesman
        and Nation 35 (Mar. 20, 1948), 237.   (W. Reade and
        Victorian agnosticism)
5985.   JACKSON, GEORGE.   "Lord Morley and the Christian
        Faith."   London Quarterly and Holborn Review 122
        (July, 1914), 36-51.
5986.   JAMES, DAVID GWILYM.   Henry Sidgwick: Science and
        Faith in Victorian England.   London, New York:
        Oxford University Press, 1970.   (Includes A. H.
        Clough)
5987.   JOHANSON, ARNOLD E.   "'The Will to Believe' and the
        Ethics of Belief."   Transactions of the Charles
        S. Peirce Society 11 (1975), 110-127.   (William
        James and W. K. Clifford)
5988.   KEGEL, CHARLES H.   "William Morris and the Religion
        of Fellowship."   Western Humanities Review 12
        (1958), 233-240.
5989.   KENNEY, MRS. DAVID J.   "Anthony Trollope's Theology."
        American Notes and Queries 9 (Dec. 1970), 51-54.

5990. KING, DANIEL P. "The Most Remarkable Book."
       Baker Street Journal 25, No. 2 (1975), 77-79. (W.
       Reade, The Martyrdom of Man)

5991. KIRCHHOFF, FREDERICK THOMAS. "Ruskin and Natural
       Science." Ph.D. Dissertation, Harvard University,
       1969.

5992. KLEINPETER, HANS. "Kant und die naturwissenschaft-
       liche Erkenntniskritik der Gegenwart (Mach, Hertz,
       Stallo, Clifford)." Kant-Studien 8 (1903),
       258-320.

5993. KNICKERBOCKER, FRANCES W. Free Minds: John Morley
       and His Friends. Cambridge: Harvard University
       Press, 1943.

5994. KNOEPFLMACHER, U. C. "Pater's Religion of Sanity:
       'Plato and Platonism' as a Document of Victorian
       Unbelief." Victorian Studies 6 (Dec. 1962),
       151-168.

5995. LAMPRECHT, STERLING P. "James Ward's Critique of
       Naturalism." Monist 36 (1926), 136-152.

5996. LANGFORD, THOMAS A. "The Ethical and Religious
       Thought of Walter Pater." Ph.D. Dissertation,
       Texas Christian University, 1968.

5997. LARSON, GEORGE STANLEY. "Religion in the Novels of
       Charles Dickens." Dissertation Abstracts Inter-
       national 30 (1969), 328-329A (University of
       Massachusetts, 1969).

5998. LASKEY, DALLAS. "Practical Reason in the Moral
       Philosophy of Henry Sidgwick." Ph.D. Dissertation,
       Harvard University, 1961.

5999. LAWSON, E. LEROY. Very Sure of God: Religious
       Language in the Poetry of Robert Browning.
       Nashville: Vanderbilt University Press, 1974.

6000. LEE, JAMES M. "Trollope's Clerical Concerns: The
       Low Church Clergyman." Hartford Studies in Litera-
       ture 1 (1969), 198-208.

6001. LIVINGSTON, JAMES C.   "The Religious Creed and
      Criticism of Sir James Fitzjames Stephen."   Victorian
      Studies 17 (Mar. 1974), 279-300.
6002. LUNDEEN, THOMAS BAILEY.   "Trollope and the Mid-
      Victorian Episcopate."   Historical Magazine of the
      Protestant Episcopal Church 30 (1961), 55-67.
6003. MCCARTHY, BERNARDIN D., O.P.   "Browning and the
      Roman Catholic Church."   Ph.D. Dissertation, Yale
      University, 1940.
6004. MCCRAW, HARRY WELLS.   "Walter Pater's 'Religious
      Phase':   The Riddle of 'Marius the Epicurean'."
      Southern Quarterly 10 (1972), 245-273.
6005. MACFARLANE, ALEXANDER.   "William Kingdon Clifford
      (1845-1879)."   In Lectures on Ten British Mathe-
      maticians of the Nineteenth Century.   New York:
      John Wiley & Sons, Inc., 1916, 78-91.
6006. MANTRIPP, J. C.   "Florence Nightingale and Religion."
      London Quarterly and Holborn Review 157 (1932),
      318-325.
6007. MARTIN, HUGH.   The Faith of Robert Browning.
      London: S.C.M. Press, 1963.
6008. MAY, CHARLES EDWARD.   "The Loss of God and the
      Search for Order: A Study of Thomas Hardy's
      Structure and Meaning in Three Genres."   Dissert-
      ation Abstracts 27 (1967), 2535A (Ohio University,
      1966).
6009. MAYNARD, JOHN.   "Robert Browning's Evangelical
      Heritage."   Browning Institute Studies 3 (1975),
      1-16.
6010. MICKLEWRIGHT, FREDERICK HENRY AMPHLETT.   "The
      Humanism of Edward Carpenter."   Friends' Quarterly
      Examiner No. 320 (1946), 224-228.
6011. MILNER, GAMALIEL.   "The Religion of Thomas Hardy."
      Modern Churchman 30 (1940), 157-164.
6012. MORLANG, WILHELM.   "Die Beziehungen zwischen Kunst
      und Religion in den Werken John Ruskins."   Ph.D.
      Dissertation, Universität Marburg, 1935.

6013. MORTON, PETER R. "'Tess of the D'Urbervilles': A Neo-Darwinian Reading." Southern Review (University of Adelaide) 7 (1974), 38-50.

6014. MURRAY, ANDREW HOWSON. The Philosophy of James Ward. Cambridge, England: Cambridge University Press, 1937.

6015. NELSON, HARLAND STANLEY. "Evangelicalism in the Novels of Charles Dickens." Dissertation Abstracts 20 (1959), 2295-96 (University of Minnesota, 1959).

6016. NEVILL, JOHN CRANSTOUN. Harriet Martineau. London: F. Muller, Ltd., 1943.

6017. NEWMAN, JAMES R. "William Kingdon Clifford." Scientific American 188 (Feb. 1953), 78-84.

6018. OLIVERO, FEDERICO. Il pensiero religioso ed estetico di Walter Pater. Torino: S.E.I., 1939.

6019. OSBOURN, R. V. "'Marius the Epicurean'." Essays in Criticism 1 (1951), 387-403. (W. Pater)

6020. PALMER, C. H. "The Religion of Thomas Hardy's Essex." American Church Monthly 21 (1927), 425-430.

6021. PALMER, C. H. "Thomas Hardy and the Church." American Church Monthly 22 (1928), 159-165.

6022. PEAKE, LESLIE SILLMAN. "Browning and God in Nature." Saturday Review 153 (1932), 293, 316.

6023. PETZOLD, GERTRUD VON. Harriet Martineau und ihre sittlich-religiöse Weltschau. Bochum-Langendreer: H. Pöppinghaus, 1941.

6024. PICKERING, SAMUEL F., JR. "'Dombey and Son' and Dickens' Unitarian Period." Georgia Review 26 (1972), 438-454.

6025. PILKINGTON, F. "Religion in Hardy's Novels." Contemporary Review 188 (1955), 31-35.

6026. PIÑEYRO, ENRIQUE. "Blanco White." Bulletin Hispanique 12 (1910), 71-100, 163-200.

6027. PLYBON, IRA FERNANDO. "John Morley: The Victorian Rationalist as Literary Critic." Dissertation

Abstracts International 35 (1975), 6107A (University
of Maryland, 1974).

6028.   POWER, E. A.   "Exeter's Mathematician - W. K.
Clifford, F.R.S., 1845-79." Advancement of Science
(London) 26 (1970), 318-328.

6029.   PRIESTLEY, F. E. L.   "Blougram's Apologetics."
University of Toronto Quarterly 15 (1946), 139-147.
(Browning)

6030.   PRIESTLEY, F. E. L.   "A Reading of 'La Saisiaz'."
University of Toronto Quarterly 25 (1955), 47-59.
(Browning)

6031.   PRIESTLEY, F. E. L.   "Some Aspects of Browning's
Irony." In Clarence R. Tracy (Ed.), Browning's
Mind and Art. Edinburgh and London: Oliver Boyd,
1968, 123-142.

6032.   RAYMOND, WILLIAM ODBER. The Infinite Moment and
Other Essays in Robert Browning. Toronto:
University of Toronto Press, 1950. (Includes
Browning and higher criticism)

6033.   RENNER, STANLEY WILLIAM. "Joseph Conrad and
Victorian Religion." Dissertation Abstracts
International 31 (1970), 1289-90A (University of
Iowa, 1970).

6034.   RICE, THERESA ANGILEE. "The Religious and Moral
Ideas in the Novels of Charles Dickens." Summaries
of Doctoral Dissertations, University of Wisconsin
14 (1954), 445-446 (University of Wisconsin, 1953).

6035.   RILEY, KATHERINE MORAN. "Anthony Trollope's Barset
Clergyman." Dissertation Abstracts International
36 (1975), 284A (University of Pennsylvania, 1974).

6036.   RIVENBURG, NAROLA E. Harriet Martineau: An Example
of Victorian Conflict. Philadelphia: The Author,
1932. (Ph.D. Dissertation, Columbia University,
1932)

6037.   ROBB, JOHN W. "An Examination of James Ward's
Philosophy as a Basis for a Philosophy of

Religion."   Ph.D. Dissertation, University of
Southern California, 1953.

6038.   ROUSSEAU, FRANCOIS.   "Blanco White: souvenirs d'un
proscrit espagnol réfugié en Angleterre, 1775-1815."
Revue Hispanique 12 (1910), 615-647.

6039.   SCHNEEWIND, JEROME B.   "Sidgwick and the Cambridge
Moralists."   Monist 58 (July 1974), 371-404.

6040.   SCOTT, NATHAN A., JR.   "Literary Imagination and
the Victorian Crisis of Faith: The Example of
Thomas Hardy." Journal of Religion 40 (Oct. 1960),
267-281.

6041.   SHAW, VALERIE A.   "Ruskin and Science." Dissertation Abstracts International 32 (1972), 6943A
(Yale University, 1971).

6042.   SHETTLE, G. T.   Dickens and the Church.   London:
Churchman Pub. Co., 1946.

6043.   SMALLWOOD, OSBORN T.   "In Quest of a Faith: John
Ruskin's Theological Searchings." Cresset 13
(1950), 7-13.

6044.   SMALLWOOD, OSBORN T.   "Theological Influence in the
Prose of Ruskin." Ph.D. Dissertation, New York
University, 1948.

6045.   SORLEY, WILLIAM RITCHIE.   "James Ward." Mind 34
(1925), 273-279.

6046.   SORLEY, WILLIAM RITCHIE.   "Ward's Philosophy of
Religion." Monist 36 (1926), 56-69.

6047.   SOUTHWELL, SAM BEALL.   "Religion in the Life and
Thought of John Ruskin." Ph.D. Dissertation,
University of Texas at Austin, 1956.

6048.   STAEBLER, WARREN.   The Liberal Mind of John Morley.
Princeton: Princeton University Press (for University
of Cincinnati), 1943.

6049.   STASNY, JOHN F.   "W. Winwood Reade's 'The Martyrdom
of Man': A Darwinian History." West Virginia
University Bulletin: Philological Papers 13 (1962),
37-49.

6050. STONE, WILFRED HEALEY.   "Browning and 'Mark Ruther-
       ford'."   Review of English Studies 4 (July 1953),
       249-259.

6051. SULLIVAN, THOMAS RICHARD.   "A 'Way to the Better':
       Hardy's Two Views of Evolution."   Dissertation
       Abstracts 29 (1968), 276A (University of Iowa,
       1968).

6052. TIMKO, MICHAEL.   "Browning upon Butler: Or Natural
       Theology in the English Isle."   Criticism 7 (1965),
       141-150.

6053. TITCHENER, E. B. AND W. S. FOSTER.   "A List of the
       Writings of James Ward."   Monist 36 (1926), 170-176.

6054. TRACY, CLARENCE R.   "Bishop Blougram."   Modern
       Language Review 34 (July 1939), 422-425.   (Browning's
       poem as a comment on J. H. Newman and N. Wiseman)

6055. TRACY, CLARENCE R.   "Browning and the Religious
       Rationalism of His Time."   Ph.D. Dissertation, Yale
       University, 1935.

6056. TRACY, CLARENCE R.   "Browning's Heresies."   Studies
       in Philology 33 (1936), 610-625.

6057. TRACY, CLARENCE R.   "'Caliban upon Setebos'."
       Studies in Philology 35 (July 1938), 487-499.
       (Browning and natural theology)

6058. TUTTLETON, JUNE MARTIN.   "Thomas Hardy and the
       Christian Religion."   Dissertation Abstracts 26
       (1965), 1637 (University of North Carolina at
       Chapel Hill, 1964).

6059. VANSON, FREDERIC.   "Robert Browning - Christian
       Optimist."   London Quarterly and Holborn Review 28
       (1959), 331-335.

6060. VAN TASSEL, DANIEL ELLSWORTH.   "The Christian
       Church as a Motif in the Novels of Thomas Hardy."
       Dissertation Abstracts International 31 (1970),
       2892A (University of Iowa, 1970).

6061. VOGELER, MARTHA SALMON.   "The Religious Meaning of
       'Marius the Epicurean'."   Nineteenth-Century Fiction
       19 (1964), 287-299.   (W. Pater)

6062.   WARD, HAYDEN WIGHTMAN, JR.   "The Religious Aesthetic
        of Walter Pater."   Dissertation Abstracts Inter-
        national 30 (1969), 2503A   (Columbia University,
        1969).

6063.   WARD, W. A.   "The Idea in Nature: A Study of the
        Thought of Walter Pater, with Reference to Hegel
        and the Theory of Evolution."   Ph.D. Dissertation,
        St. John's College, University of Cambridge, 1964.

6064.   WEATHERFORD, WILLIS DUKE.   "Fundamental Religious
        Principles in Browning's Poetry."   Ph.D. Dissertation,
        Vanderbilt University, 1907.

6065.   WEBB, ROBERT KIEFER.   Harriet Martineau: A Radical
        Victorian.   New York: Columbia University Press,
        1960.

6066.   WHEATLEY, VERA.   The Life and Work of Harriet
        Martineau.   London: Secker and Warburg, 1957.

6067.   WHITLA, WILLIAM.   The Central Truth: The Incarna-
        tion in Robert Browning's Poetry.   Toronto:
        University of Toronto Press; London: Oxford
        University Press, 1964.

6068.   WHITTAKER, EDMUND.   "Renewal of a Classic."   Nature
        159 (1947), 248.   (Review of W. K. Clifford, The
        Common Sense of the Exact Sciences)

6069.   WILSON, JOHN ROBERT.   "Dickens and Christian
        Mystery."   South Atlantic Quarterly 73 (1974),
        528-540.

XX. The Agnostic Novel

Studies of novels dealing with doubt and unbelief
are assembled here. Among the authors included are James
Anthony Froude, George MacDonald, William Hurrell Mallock,
Olive Schreiner, Mrs. Humphry Ward, and William Hale
White ("Mark Rutherford"). See III. Ideas . . .;
XIX. Varieties of Belief . . ., 7. Eliot, George,
8. Froude, James Anthony, 24. Others.

6070. ADAMS, AMY BELLE. The Novels of William Hurrell
        Mallock. Orono, Maine: The University Press,
        1934.

6071. ADDISON, WILLIAM GEORGE. "Weg on Bobbie." Theology
        52 (1949), 1-8. (W. E. Gladstone's reaction to
        Mrs. Humphry Ward's Robert Elsmere)

6072. ANON. "Mrs. Humphry Ward." Times Literary Sup-
        plement (June 15, 1951), 372.

6073. BECCARD, MARIA. "Religiöse Fragen in den Romanen
        von Mrs. Humphry Ward." Ph.D. Dissertation,
        Universität Münster, 1935.

6074. BERESFORD, ROSEMARY. "Mark Rutherford and Hero-
        Worship." Review of English Studies New Series 6
        (1955), 264-272.

6075. BULLOCH, JOHN MALCOLM. "A Bibliography of George
        MacDonald." Aberdeen University Library Bulletin 5
        (Feb. 1925), 679-747.

6076. CECIL, LADY ROBERT. "Mark Rutherford." Nation
        and Athenaeum 34 (Oct. 27, 1923), 151-152.

6077. CHUMLEY, CHARLES EDWARD. "The Religious Problem in
        the Works of Mark Rutherford." M.A. Thesis, Uni-
        versity of Illinois, 1947.

6078. COGHLAN, KATHRYN ALBERTA. "Mrs. Humphry Ward,
        Novelist and Thinker." Dissertation Abstracts 17
        (1957), 2606-07 (Boston University Graduate School,
        1957).

6079. COUCH, WARREN DAVID. "William Hurrell Mallock's
'The New Republic' and the Crisis of Faith."
Dissertation Abstracts International 35 (1974),
3731-32A (University of Maryland, 1974).
6080. CUNNINGHAM, VALENTINE D. "William Hale White and
Samuel Bamford." Notes and Queries 219 (Dec.
1974), 454-461.
6081. DAVIDSON, BASIL. "In Memory of Olive Schreiner."
New Statesman and Nation 49, No. 1255 (Mar. 26,
1955), 426-428.
6082. DAVIS, W. EUGENE. "William Hale White: An
Annotated Bibliography." English Literature in
Transition 10 (1967), 97-117, 150-160.
6083. DUNBAR, GEORGIA DOLFIELD SHERWOOD. "The Faithful
Recorder: Mrs. Humphry Ward and the Foundation of
Her Novels." Dissertation Abstracts 14 (1954),
122-123 (Columbia University, 1953).
6084. GARNETT, DAVID. "Books in General." New Statesman and Nation 12, No. 285 (Aug. 8, 1936), 193; 13,
No. 306 (Jan. 2, 1937), 17. (Mark Rutherford)
6085. GREGG, LYNDALL. Memories of Olive Schreiner.
London: Chambers, 1957.
6086. HANNA, WILLARD ANDERSON. "'Robert Elsmere': A
Study in the Controversy between Science and
Religion in the Nineteenth Century." Ph.D. Dissertation, University of Michigan, 1940.
6087. HARRISON, ARCHIBALD HAROLD WALTER. "Mark Rutherford and J. A. Froude." London Quarterly and Holborn
Review 164 (1939), 40-44.
6088. HARWOOD, H. C. "Mark Rutherford." London Mercury
3 (1921), 388-397.
6089. HEIN, ROLLAND NEAL. "Faith and Fiction: A Study
of the Effects of Religious Convictions in the
Adult Fantasies and Novels of George MacDonald."
Dissertation Abstracts International 32 (1971),
919-920A (Purdue University, 1970).

6090.   HINES, JOYCE ROSE.   "Getting Home: A Study of
        Fantasy and the Spiritual Journey in the Christian
        Supernatural Novels of Charles Williams and George
        MacDonald."   Dissertation Abstracts International
        33 (1972), 755-756A (CUNY, 1972).

6091.   HOBMAN, D. L.   Olive Schreiner: Her Friends and
        Times.   London: Watts, 1955.

6092.   HORDER, REV. W. GARRETT.   "George MacDonald: A
        Nineteenth Century Seer."   Review of Reviews 32
        (Oct. 1905), 357-362.

6093.   HUGHES, LINDA K.   "Madge and Clara Hopgood:
        William Hale White's Spinozan Sisters."   Victorian
        Studies 18 (1974), 57-75.

6094.   JEFFS, ERNEST H.   "Hale White."   In William Ebor
        (Ed.), Great Christians.   London: Ivor Nicholson
        and Watson, 1933, 607-615.

6095.   JONES, ENID HUWS.   Mrs. Humphry Ward.   London:
        William Heinemann, 1973.

6096.   KLINKE, HANS.   William Hale White (Mark Rutherford):
        Versuch einer Biographie, mit besonderer Berücksichtigung
        der Einflüsse von Dichtern, Denkern und Ereignissen
        mit vielem unveröffentlichten Material dargestellt.
        Frankfurt: W. Bohn, 1930.

6097.   LASKI, MARGHANITA.   "Words from 'Robert Elsmere'."
        Notes and Queries 206 (June 1961), 229-230.

6098.   LEWIS, NAOMI.   "Books in General."   New Statesman
        and Nation 34, No. 859 (Aug. 23, 1947), 152.   (Mrs.
        Humphry Ward)

6099.   LOW, FLORENCE B.   "Walks and Talks with Mark
        Rutherford."   Contemporary Review 187 (1955),
        405-409.

6100.   LUCAS, JOHN.   "Tilting at the Moderns: W. H.
        Mallock's Criticisms of the Positivist Spirit."
        Renaissance and Modern Studies 10 (1966), 88-143.

6101.   MACCRACKEN, H. N.   "Book Reviews: Pages from a
        Journal, with Other Papers."   Yale Review New
        Series 3 (1913), 189-194.   (Mark Rutherford)

6102. MCCRAW, HARRY WELLS. "Two Novelists of Despair: James Anthony Froude and William Hale White." Southern Quarterly 13 (1974), 21-51.

6103. MCCRAW, HARRY WELLS. "The Victorian Novel of Religious Controversy: Five Studies." Dissertation Abstracts International 31 (1971), 4725-26A (Tulane, 1970). (T. Carlyle, Sartor Resartus, Mrs. Humphry Ward, Robert Elsmere, J. H. Newman, Loss and Gain, J. A. Froude, The Nemesis of Faith, and W. H. White, The Autobiography of Mark Rutherford)

6104. MACDONALD, GREVILLE. George Macdonald and His Wife. London: George Allen and Unwin, 1924.

6105. MACLEAN, CATHERINE MACDONALD. Mark Rutherford: A Biography of William Hale White. London: Macdonald, 1955.

6106. MAN OF KENT, A. "Mark Rutherford among the Spiritualists." British Weekly 54 (Apr. 3, 1913), 13.

6107. MANTRIPP, J. C. "Mark Rutherford." Holborn Review 13 (Apr. 1922), 182-194.

6108. MARVIN, FRANCIS SYDNEY. "'Robert Elsmere': Fifty Years After." Contemporary Review 156 (1939), 196-202.

6109. MASSINGHAM, H. W. "Memorial Introduction." In [W. H. White] The Autobiography of Mark Rutherford. London: Humphrey Milford, 1936, vii-xxxii. (See 6110)

6110. MEISSNER, P. "The Novels of 'Mark Rutherford'." Deutsche Literaturzeitung 59 (July 31, 1938), 1092-96. (Review in German of 6109)

6111. MERTON, EGON STEPHEN. "The Autobiographical Novels of Mark Rutherford." Nineteenth-Century Fiction 5 (1950), 189-207.

6112. MERTON, EGON STEPHEN. "George Eliot and William Hale White." Victorian Newsletter No. 25 (1964), 13-15.

6113. MERTON, EGON STEPHEN. "Mark Rutherford: The World
       of His Novels." Bulletin of the New York Public
       Libraries 67 (1963), 470-478.

6114. MERTON, EGON STEPHEN. Mark Rutherford (William
       Hale White). New York: Twayne Publishers, 1967.

6115. MERTON, EGON STEPHEN. "The Personality of Mark
       Rutherford." Nineteenth-Century Fiction 6 (1951),
       1-20.

6116. MICHIE, JAMES A. "The Wisdom of Mark Rutherford."
       London Quarterly and Holborn Review Series 6, 28
       (1959), 124-128.

6117. MIDDEL, WALTER. "William Hale Whites religiös-
       weltanschauliche Entwicklung: Ein Beitrag zur
       Geschichte des Puritanismus im 19. Jahrhundert."
       Ph.D. Dissertation, Universität Kiel, 1940.

6118. MORTON, ARTHUR LESLIE. "The Last Puritan." In
       Language of Men. London: Cobbett Press, 1945,
       49-57. (W. H. White)

6119. MURRY, JOHN MIDDLETON. "The Religion of Mark
       Rutherford." The Adelphi 2 (July 1924), 93-104.

6120. NEWTON, JUDY LOWDER. "Experiments in Dissent: A
       Study of Form in Five 'Novels of Disbelief'."
       Dissertation Abstracts International 31 (1971),
       6562-63A (University of California, Berkeley, 1970).
       (J. A. Froude, The Nemesis of Faith, William
       Arnold, Oakfield or Fellowship in the East,
       Samuel Butler, Ernest Pontifex or The Way of All
       Flesh, William Hale White, Autobiography of Mark
       Rutherford and Mark Rutherford's Deliverance)

6121. NICKERSON, CHARLES C. "W. H. Mallock's Contributions
       to 'The Miscellany'." Victorian Studies 6 (Dec.
       1962), 169-177.

6122. NICOLL, SIR WILLIAM ROBERTSON. A Bookman's Letters.
       London: Hodder and Stoughton, 1913. (Mark Rutherford)

6123. NICOLL, SIR WILLIAM ROBERTSON. Introduction to
       the Novels of Mark Rutherford. London: T. Fisher
       Unwin, 1924.

6124.   O'FLAHERTY, GERALD V.   "In Search of the Self: The
        Quest for Spiritual Identity in Five Nineteenth-
        Century Religious Novels."   Dissertation Abstracts
        International 34 (1974), 7717A (University of
        Pennsylvania, 1973).   (J. A. Froude, The Nemesis
        of Faith, Charlotte Yonge, The Heir of Redclyffe,
        Joseph Henry Shorthouse, John Inglesent, W. Pater,
        Marius the Epicurean, Mrs. Humphry Ward, Robert
        Elsmere)

6125.   PATRICK, J. MAX.   "The Portrait of Huxley in
        Mallock's 'New Republic'."   Nineteenth-Century
        Fiction 11 (1956), 61-69.

6126.   PETERSON, WILLIAM S.   "Gladstone's Review of
        'Robert Elsmere':   Some Unpublished Correspondence."
        Review of English Studies 21 (Nov. 1970), 442-461.

6127.   PETERSON, WILLIAM S.   "Mrs. Humphry Ward on 'Robert
        Elsmere':   Six New Letters."   Bulletin of the
        New York Public Library 74 (1970), 587-597.

6128.   PLOMER, WILLIAM CHARLES FRANKLYN.   "Olive Schreiner:
        Her Life and Ideals."   Listener 53 (Mar. 24, 1955),
        521-522.

6129.   PRAZ, MARIO.   "L'autobiografia di Mark Rutherford."
        Anglica 1 (Apr.-June 1946), 49-65.

6130.   PRITCHETT, V. S.   "Books in General."   New States-
        man and Nation 25, No. 638 (May 15, 1943), 323.
        (W. W. Reade, Martyrdom of Man)

6131.   RENIER, OLIVE.   "A South African Rebel."   Listener
        53 (Apr. 7, 1955), 613-614.   (O. Schreiner)

6132.   REYNOLDS, LOU AGNES.   "Mrs. Humphry Ward and the
        Arnold Heritage."   Summaries of Doctoral Dissert-
        ations, Northwestern University 20 (1953), 29-34
        (Northwestern University, 1952).

6133.   ROWSE, A. L.   "Books in General."   New Statesman
        and Nation 25 (Mar. 20, 1943), 191.   (J. A. Froude)

6134.   SCHIEDER, R. M.   "Loss and Gain?   The Theme of
        Conversion in Late Victorian Fiction."   Victorian

Studies 9 (Sept. 1965), 29-44. (Includes Mrs.
Humphry Ward, W. H. White, George MacDonald)

6135. SCHMELLER, HILDEGARD. "Die religiösen und sozialen
Probleme in den Romanen von Mrs. Humphry Ward."
Ph.D. Dissertation, Universität Wien, 1936.

6136. SCHREINER, S. C. The Life of Olive Schreiner.
London: T. Fisher Unwin, 1924.

6137. SECCOMBE, THOMAS. "Mark Rutherford." New Witness
1 (Mar. 27, 1913), 658-659. (Reprinted as "The
Literary Work of Mark Rutherford." Living Age 277,
May 24, 1913, 498-501)

6138. SELBY, THOMAS G. "George MacDonald and the Scottish
School." In The Theology of Modern Fiction.
London: C. H. Kelly, 1896, 131-172.

6139. SPEEGLE, KATHERINE SLOAN. "God's Newer Will: Four
Examples of Victorian 'Angst' Resolved by Humani-
tarianism." Dissertation Abstracts International
36 (1975), 2857A (North Texas State University,
1975). (Mrs. Lynn Linton, Olive Schreiner, W. H.
White, Mrs. Humphry Ward)

6140. SPERRY, WILLARD LEAROYD. "Mark Rutherford."
Harvard Theological Review 7 (Apr. 1914), 166-192.

6141. STOCK, IRVIN. "André Gide, William Hale White, and
the Protestant Tradition." Accent 12 (1952),
205-215.

6142. STOCK, IRVIN. W. Hale White: A Critical Study.
New York: Columbia University Press, 1956.

6143. STONE, WILFRED HEALEY. "Browning and 'Mark
Rutherford'." Review of English Studies 4 (July
1953), 249-259.

6144. STONE, WILFRED HEALEY. Religion and the Art of
William Hale White (Mark Rutherford). Stanford:
Stanford University Press, 1954.

6145. SUCKOW, RUTH. "Robert Elsmere." Georgia Review 9
(1955), 344-348.

6146. TAYLOR, ALFRED EDWARD. "The Novels of Mark
       Rutherford." Essays and Studies by Members of the
       English Association 5 (1914), 51-74.

6147. TEMPEST, E. VINCENT. "Optimism in Mark Rutherford."
       Westminster Review 180 (1913), 174-184.

6148. THOMSON, PATRICIA. "The Novels of Mark Rutherford."
       Essays in Criticism 14 (1964), 256-267.

6149. TREVELYAN, JANET. "Mrs. Humphry Ward and Robert
       Elsmere." Spectator (June 8, 1951), 745. (See
       comment by J. A. T., June 15, 1951, 786)

6150. UNIKEL, GRAHAM. "The Religious Arnoldism of
       Mrs. Humphry Ward." Ph.D. Dissertation, University
       of California, Berkeley, 1951.

6151. WALTERS, J. STUART. Mrs. Humphry Ward: Her Work
       and Influence. London: Kegan, Paul, 1912.

6152. WARNER, ALAN. "Mark Rutherford: The Puritan as
       Novelist." Theoria 8 (1956), 37-45.

6153. WEBSTER, N. W. "Mrs. Humphry Ward: A Retrospect."
       Cornhill Magazine 177, No. 1059 (1969), 223-233.

6154. WEE, DAVID LUTHER. "The Forms of Apostasy: The
       Rejection of Orthodox Christianity in the British
       Novel, 1880-1900." Dissertation Abstracts 28
       (1967), 205-206A (Stanford University, 1967).
       (Includes W. H. White, S. Butler, O. Schreiner,
       Mrs. Humphry Ward)

6155. WHITE, DOROTHY VERNON. The Groombridge Diary.
       London: H. Milford, Oxford University Press, 1924.
       (W. H. White)

6156. WILLCOX, LOUISE COLLIER. "A Neglected Novelist."
       North American Review 183 (Sept. 1906), 394-403.
       (George MacDonald)

6157. WILLEY, BASIL. "How 'Robert Elsmere' Struck Some
       Contemporaries." Essays and Studies by Members
       of the English Association 10 (1957), 53-68.

6158. WILLIAMS, KENNETH EARL. "Faith, Intention, and
       Fulfillment: The Religious Novels of Mrs. Humphry

Ward."   Dissertation Abstracts International 30
(1969), 2553-54A (Temple University, 1969).

6159.   WILSON, SAMUEL LAW.   The Theology of Modern Literature.
Edinburgh: T. and T. Clark, 1899.   (R. Emerson,
T. Carlyle, R. Browning, George Eliot, George
MacDonald, Mrs. Humphry Ward, T. Hardy, George
Meredith)

6160.   WOLFF, RENATE CHRISTINE.   "Currents in Naturalistic
English Fiction, 1880-1920:   With Special Emphasis
on 'Mark Rutherford'."   Dissertation Abstracts 13
(1952), 398-399 (Bryn Mawr College, 1951).

6161.   WOLFF, ROBERT LEE.   The Golden Key: A Study of
the Fiction of George MacDonald.   New Haven: Yale
University Press, 1961.

6162.   WOODRING, CARL R.   "Notes on Mallock's 'The New
Republic'."   Nineteenth-Century Fiction 6 (1951),
71-74.

6163.   WOODRING, CARL R.   "W. H. Mallock: A Neglected
Wit."   More Books, Bulletin of the Boston Public
Library 12 (1947), 243-256.

6164.   YARKER, P. M.   "Voltaire among the Positivists: A
Study of W. H. Mallock's 'The New Paul and Virginia'."
Essays and Studies New Series 8 (1955), 21-39.

6165.   YARKER, P. M.   "W. H. Mallock's Other Novels."
Nineteenth-Century Fiction 14 (1959), 189-205.

XXI. Atheism, Secularism, Rationalism, Materialism

The majority of works in this category deal with
organized secularism and those dedicated to its cause,
including George Jacob Holyoake, Charles Bradlaugh, and
Annie Besant. See III. Ideas . . .; appropriate sections
of XIX. Varieties of Belief . . .; XX. The Agnostic
Novel; XXII. Positivism.

6166. ARNSTEIN, WALTER L. "The Bradlaugh Case: A
      Reappraisal." Journal of the History of Ideas 18
      (1957), 254-269.
6167. ARNSTEIN, WALTER L. The Bradlaugh Case: A Study
      in Late Victorian Opinion and Politics. Oxford:
      Clarendon Press, 1965.
6168. ARNSTEIN, WALTER L. "Gladstone and the Bradlaugh
      Case." Victorian Studies 5 (June 1962), 303-330.
6169. BANKS, JOSEPH A. AND OLIVE BANKS. "The Bradlaugh-
      Besant Trial and the English Newspapers." Population
      Studies 8 (July 1954), 22-34.
6170. BENN, ALFRED WILLIAM. The History of English
      Rationalism in the Nineteenth Century. 2 vols.
      London: Longmans & Co., 1906.
6171. BESTERMAN, THEODORE. Mrs. Annie Besant: A Modern
      Prophet. London: Kegan Paul, 1934.
6172. BLYTH, JACK A. "The Origins of Secularism in Early
      Nineteenth-Century Britain." Dalhousie Review 52
      (1972), 237-250.
6173. BONNER, HYPATIA BRADLAUGH AND JOHN MACKINNON ROBERTSON.
      Charles Bradlaugh. London: T. F. Unwin, 1908.
6174. BONNER, HYPATIA BRADLAUGH. "Charles Bradlaugh as a
      Freemason." Notes and Queries 166 (May 26, 1934),
      370; 166 (June 9, 1934), 411-412.
6175. BUCKLEY, JEROME HAMILTON. "The Revolt from Ration-
      alism in the Seventies." In Hill Shine (Ed.),

Booker Memorial Studies: Eight Essays on Victorian
Literature in Memory of John Manning Booker.
Chapel Hill: University of North Carolina Press,
1950, 122-132.

6176.   BUDD, SUSAN. "The British Humanist Movement,
1860-1966." Ph.D. Dissertation, University of
Oxford, 1969.

6177.   BUDD, SUSAN. "Reasons for Unbelief among Members
of the Secular Movement in England, 1850-1950."
Past and Present 36 (Apr. 1967), 106-125.

6178.   BURY, JOHN BAGNELL. A History of Freedom of Thought.
London: H. Holt & Co., 1913.

6179.   FOOTE, GEORGE A. "Mechanism, Materialism and
Science in England, 1800-1850." Annals of Science
8 (1952), 152-161.

6180.   GILMOUR, JAMES PINKERTON. (Ed.) Champion of Liberty:
Charles Bradlaugh. London: C. A. Watts & Co., and
the Pioneer Press, 1933.

6181.   GOULD, FREDERICK JAMES. The History of the Leicester
Secular Society. Leicester: Leicester Secular
Society, 1900.

6182.   GRANT, A. CAMERON. "Combe on Phrenology and Free
Will: A Note on XIXth Century Secularism." Journal
of the History of Ideas 26 (1965), 141-147.

6183.   HARTZELL, KARL DREW. "The Origins of the English
Secularist Movement, 1817-1846." Harvard University
Summaries of Theses (1935), 158-162. (Ph.D. Dis-
sertation, Harvard University, 1934)

6184.   ILARDO, J. A. "Charles Bradlaugh: Victorian
Atheist Reformer." Today's Speech: Journal of the
Speech Association of the Eastern States 17 (Nov.
1969), 25-34.

6185.   KACZKOWSKI, CONRAD JOSEPH. "John Mackinnon Robertson:
Freethinker and Radical." Dissertation Abstracts
26 (1965), 2163-64 (St. Louis University, 1964).

6186. KNIGHT, DAVID M. "Chemistry, Physiology and Materialism in the Romantic Period." Durham University Journal 64 (1972), 139-145.

6187. KRANTZ, CHARLES KRZENTOWSKI. "The British Secularist Movement: A Study in Militant Dissent." Dissertation Abstracts 25 (1964), 2946-47 (University of Rochester, 1964).

6188. LANGE, FRIEDRICH ALBERT. History of Materialism and Criticism of Its Present Importance. Trans. E. Thomas. 3 vols. London: Trübner & Co., 1877-81.

6189. LECKY, WILLIAM EDWARD HARTPOLE. History of the Rise and Influence of the Spirit of Rationalism in Europe. 2 vols. London: Longman, Green, Longman, Roberts, & Green, 1865.

6190. MCCABE, JOSEPH. Life and Letters of George Jacob Holyoake. 2 vols. London: Watts and Co., 1908.

6191. MCGEE, JOHN EDWIN. History of the British Secular Movement. Girard, Kansas: Haldeman-Julius Publications, 1948.

6192. MCLAREN, ANGUS. "George Jacob Holyoake and the Secular Society: British Popular Freethought, 1851-1858." Canadian Journal of History 7 (Dec. 1972), 235-251.

6193. MATHEWSON, GEORGE. "Shelley's Atheism: An Early Victorian Explanation." Keats-Shelley Journal 15 (1966), 7-9. (Attitudes towards P. B. Shelley's atheism, including views of Thomas De Quincey and the Rev. George Gilfillan)

6194. MICKLEWRIGHT, FREDERICK HENRY AMPHLETT. "The Local History of Victorian Secularism." Local Historian 8 (1969), 221-227.

6195. NELSON, WALTER D. "British Rational Secularism: Unbelief from Bradlaugh to the Mid-Twentieth Century." Dissertation Abstracts 24 (1964), 4659-60 (University of Washington, 1963).

6196.   NETHERCOT, ARTHUR HOBART.   The Life of Annie Besant.
        2 vols.   Chicago: University of Chicago Press,
        1960-63.   (Volume 1:   The First Five Lives of Annie
        Besant; Volume 2:   The Last Four Lives of Annie
        Besant)

6197.   NOEL-BENTLEY, PETER CHARLES.   "The Religious Poetry
        of James Thomson (B.V.)."   Dissertation Abstracts
        International 34 (1973), 1250A   (University of
        Toronto, 1972).

6198.   PLUMPTRE, CONSTANCE E.   On the Progress of Liberty
        of Thought during Queen Victoria's Reign.   London:
        Watts and Co., 1902.

6199.   ROBERTSON, JOHN MACKINNON.   Charles Bradlaugh.
        London: Watts and Co., 1920.

6200.   ROBERTSON, JOHN MACKINNON.   A History of Freethought
        in the Nineteenth Century.   London: Watts and Co.,
        1929.

6201.   ROYLE, EDWARD.   "George Jacob Holyoake and the
        Secularist Movement in Britain, 1841-61."   Ph.D.
        Dissertation, University of Cambridge, 1968.

6202.   ROYLE, EDWARD.   Radical Politics 1790-1900:
        Religion and Unbelief.   Harlow: Longman, 1971.

6203.   ROYLE, EDWARD.   Victorian Infidels: The Origins
        of the British Secularist Movement 1791-1866.
        Manchester: Manchester University Press, 1974.

6204.   SMITH, FRANCIS BARRYMORE.   "The Atheist Mission,
        1840-1900."   In Robert Robson (Ed.), Ideas and
        Institutions of Victorian Britain.   London: G.
        Bell & Sons Ltd., 1967, 205-235.

6205.   SMITH, WARREN SYLVESTER.   "Charles Bradlaugh and
        the Secularists."   Christian Century 80 (Mar. 13,
        1963), 331-334.

6206.   STEELE, MICHAEL RHOADS.   "Secularist Literature of
        Victorian England: 1870-1880."   Dissertation
        Abstracts International 36 (1976), 6121A (Michigan
        State University, 1975).

6207.   TRIBE, DAVID H.   A Hundred Years of Free Thought.
        London: Elek, 1967.
6208.   TRIBE, DAVID H.   President Charles Bradlaugh, MP.
        Hamden, Conn.:   Archon Books, 1971.
6209.   TRIBE, DAVID H.   "Secular Centenary."   Contemporary
        Review 209 (1966), 200-205.   (On the Secular Society
        founded in 1866)
6210.   TURNER, FRANK MILLER.   "Lucretius among the Victor-
        ians."   Victorian Studies 16 (1973), 329-348.
6211.   WATSON, JOHN GILLARD.   "From Secularism to Human-
        ism: An Aspect of Victorian Thought."   Hibbert
        Journal 60 (Jan. 1962), 133-140.
6212.   WHYTE, A. GOWANS.   "Victorian Rationalism and
        Religion."   Rationalist Annual (1949), 81-88.
6213.   WILLIAMS, GERTRUDE LEAVENWORTH (MARVIN).   The
        Passionate Pilgrim: A Life of Annie Besant.   New
        York: Coward-McCann, 1931.

XXII.   Positivism

1.   General

This section centers on the influence of Auguste
Comte in England and on the activities of the English
Positivists (excluding Frederic Harrison, who is given a
separate section below).   See III.   Ideas . . .; XIX.
Varieties of Belief . . ., 7. Eliot, George, 14.   Mill,
John Stuart; XX.   The Agnostic Novel.

6214.   ACTON, H. B.   "Comte's Positivism and the Science
        of Society."   Philosophy 26 (1951), 291-310.
        (Includes J. S. Mill)
6215.   ANNAN, NOEL GILROY.   The Curious Strength of
        Positivism in English Political Thought.   London:
        Oxford University Press, 1959.
6216.   BAKER, WILLIAM J.   "The Kabbalah, Mordecai, and
        George Eliot's Religion of Humanity."   Yearbook of
        English Studies 3 (1973), 216-221.
6217.   BRYSON, GLADYS.   "Early English Positivists and the
        Religion of Humanity."   American Sociological Review
        1 (June 1936), 343-362.
6218.   COHEN, VICTOR.   "Auguste Comte."   Contemporary
        Review 195 (1959), 630-663.   (Includes his English
        disciples)
6219.   DUCASSE, PIERRE.   Essai sur les origines intuitives
        du positivisme.   Paris: F. Alcan, 1939.
6220.   DUCASSE, PIERRE.   "Le Positivisme comme philosophie
        intuitive."   Thalès 4 (1939), 79-90.
6221.   DUCASSE, PIERRE.   "La Synthèse positiviste: Comte
        et Spencer."   Revue de Synthèse 67 (1950), 155-187.
6222.   EISEN, SYDNEY.   "Herbert Spencer and the Spectre of
        Comte."   Journal of British Studies 7 (Nov. 1967),
        48-67.

6223. EISEN, SYDNEY. "Huxley and the Positivists."
       Victorian Studies 7 (June 1964), 337-358. (Includes
       R. Congreve, F. Harrison)

6224. EVERETT, EDWIN MALLARD. The Party of Humanity:
       The "Fortnightly Review" and Its Contributors,
       1865-1874. Chapel Hill: University of North
       Carolina Press, 1939. (Includes J. Morley, J. S.
       Mill, G. H. Lewes)

6225. FARMER, MARY E. "The Positivist Movement and the
       Development of English Sociology." Sociological
       Review 15 (1967), 5-20.

6226. GOUHIER, HENRI GASTON. La Jeunesse d'Auguste
       Comte et la formation du positivisme. 3 vols.
       Paris: J. Vrin, 1933-41.

6227. GREENE, JOHN C. "Biology and Social Theory in the
       Nineteenth Century: Auguste Comte and Herbert
       Spencer." In Marshall Clagett (Ed.), Critical
       Problems in the History of Science. Madison,
       Wisconsin: University of Wisconsin Press, 1959,
       419-446.

6228. HARRISON, ROYDEN J. Before the Socialists: Studies
       in Labour and Politics, 1861-1881. Toronto:
       University of Toronto Press; London: Routledge and
       Kegan Paul, 1965. (English Positivists and Labour)

6229. HARRISON, ROYDEN J. "E. S. Beesly and Karl Marx."
       International Review of Social History 4 (1959),
       22-58, 208-238.

6230. HAYEK, FRIEDRICH AUGUST VON. The Counter-Revolution
       of Science: Studies on the Abuse of Reason.
       Glencoe: Free Press, 1952. (Includes English
       Positivism and Saint-Simonism)

6231. JUNG, WALTER. "Der Einfluss des Positivismus auf
       George Eliot." Ph.D. Dissertation, Universität
       Leipzig, 1923.

6232. LEVIN, RUDOLF. Der Geschichtsbegriff des Positiv-
       ismus unter besonderer Berücksichtigung Mills und der

rechtsphilosophischen Anschauungen John Austins.
Leipzig: Buchdruckerei Joh. Moltzen, 1935.

6233. LINS, IVAN. "L'Oeuvre d'Auguste Comte et sa signi-
fication scientifique et philosophique au XIXe
siècle." Cahiers d'Histoire Mondiale 11 (1969),
675-711.

6234. LIVEING, SUSAN. A Nineteenth-Century Teacher:
John Henry Bridges, M.B., F.R.C.P. London:
K. Paul, Trench, Trübner and Co., 1926.

6235. LUCAS, JOHN. "Tilting at the Moderns: W. H.
Mallock's Criticisms of the Positivist Spirit."
Renaissance and Modern Studies 10 (1966), 88-143.

6236. MCGEE, JOHN EDWIN. A Crusade for Humanity: The
History of Organized Positivism in England. London:
Watts & Co., 1931.

6237. MARVIN, FRANCIS SYDNEY. Comte: The Founder of
Sociology. London: Chapman and Hall, 1936.

6238. MILNE, H. J. M. "A Positivist Archive." British
Museum Quarterly 13 (1938-39), 54-55.

6239. MURPHY, JAMES M. "Positivism in England: The
Reception of Comte's Doctrines, 1840-1870."
Dissertation Abstracts International 30 (1969),
1601-02A (Columbia University, 1968).

6240. NYLAND, THOMAS. "The English Positivists." M.A.
Thesis, University of London, 1937.

6241. PARIS, BERNARD J. "George Eliot's Religion of
Humanity." English Literary History 29 (Dec.
1962), 418-443.

6242. RESTAINO, FRANCO. "La fortuna di Comte in Gran
Bretagna." Rivista critica di storia della
filosofia 23 (1968), 171-201, 391-409; 24 (1969),
148-178, 374-381.

6243. SANTILLANA, GIORGIO DE. "Positivism and the
Technocratic Ideal in the Nineteenth Century." In
M. F. Ashley Montagu (Ed.), Studies and Essays
Offered to George Sarton. New York: Schuman,
1947, 247-259.

6244.   SARTON, GEORGE. "Auguste Comte, Historian of
        Science: With a Short Digression on Clotilde de
        Vaux and Harriet Taylor." Osiris 10 (1952),
        328-357.
6245.   SCOTT, JAMES F. "George Eliot, Positivism, and the
        Social Vision of 'Middlemarch'." Victorian Studies
        16 (Sept. 1972), 59-76.
6246.   SIMON, WALTER M. "Auguste Comte's English Disciples."
        Victorian Studies 8 (Dec. 1964), 161-172.
6247.   SIMON, WALTER M. European Positivism in the Nine-
        teenth Century: An Essay in Intellectual History.
        Ithaca: Cornell University Press, 1963.
6248.   THOMAS, PHILIP. Auguste Comte and Richard Congreve.
        London: Watts and Co., 1910.
6249.   TORLESSE, FRANCES HARRIET. Some Account of John
        Henry Bridges and His Family. London: Privately
        Printed, 1912.
6250.   WHITTAKER, THOMAS. "Comte and Mill." In Reason:
        A Philosophical Essay with Historical Illustrations.
        Cambridge: Cambridge University Press, 1934,
        31-80.
6251.   YARKER, P. M. "Voltaire among the Positivists: A
        Study of W. H. Mallock's 'The New Paul and Virginia'."
        Essays and Studies New Series 8 (1955), 21-39.

2. Harrison, Frederic

See introduction to this category.

6252.   ADELMAN, PAUL. "Frederic Harrison on Mill."
        Mill News Letter 5, No. 2 (Spring 1970), 2-4.
6253.   ADELMAN, PAUL. "The Social and Political Ideas of
        Frederic Harrison in Relation to English Thought
        and Politics, 1855-86." Ph.D. Dissertation,
        University of London, 1968.

6254.   BUSEY, GARRETTA HELEN.   "The Reflection of Positi-
        vism in English Literature to 1880:   The Positivism
        of Frederic Harrison."   Ph.D. Dissertation,
        University of Illinois at Urbana-Champaign, 1924.

6255.   EISEN, SYDNEY.   "Frederic Harrison and Herbert
        Spencer: Embattled Unbelievers."   Victorian Studies
        12 (1968), 33-56.

6256.   EISEN, SYDNEY.   "Frederic Harrison and the Religion
        of Humanity."   South Atlantic Quarterly 66 (Autumn,
        1967), 574-590.

6257.   EISEN, SYDNEY.   "Frederic Harrison: The Life and
        Thought of an English Positivist."   Ph.D. Dissert-
        ation, Johns Hopkins University, 1957.

6258.   HARRISON, AUSTIN.   Frederic Harrison: Thoughts and
        Memories.   London: William Heinemann Ltd., 1926.

6259.   MCCREADY, HERBERT W.   "Frederic Harrison and the
        British Working-Class Movement, 1860-1875."   Ph.D.
        Dissertation, Harvard University, 1953.

6260.   MANEIKIS, WALTER.   "Frederic Harrison: Positivist
        Critic of Society and Literature."   Ph.D. Dissert-
        ation, Northwestern University, 1943.

6261.   MARANDON, S. O.   "Frederic Harrison (1831-1923)."
        Etudes Anglaises 13 (1960), 415-426.

6262.   MARVIN, FRANCIS SYDNEY.   "Frederic Harrison."
        Humanity 33 (1925), 4-8.

6263.   RICKS, C. B.   "Frederic Harrison and Bergson."
        Notes and Queries 204 (1959), 175-178.

6264.   SALMON, MARTHA.   "Frederic Harrison: The Evolution
        of an English Positivist, 1831-1881."   Dissertation
        Abstracts 20 (1959), 1356 (Columbia University,
        1959).   (See Vogeler, below)

6265.   VOGELER, MARTHA SALMON.   "Frederic Harrison and
        John Ruskin: The Limits of Positivist Biography."
        Texas Quarterly 18, No. 2 (1975), 91-98.   (See
        Salmon, above)

6266. VOGELER, MARTHA SALMON. "Matthew Arnold and
       Frederic Harrison: The Prophet of Culture and
       the Prophet of Positivism." Studies in English
       Literature 2 (1962), 441-462. (See Salmon, above)
6267. ZUCKER, LOUIS C. "Frederic Harrison: Positivist
       Victorian." Ph.D. Dissertation, University of
       Wisconsin, 1929.

AUTHOR INDEX

BASTABLE, J. D. 5170, 5171
BASTIN, JOHN. 19, 20
BATE, WALTER JACKSON. 4623
BATEMAN, CHARLES THOMAS.
3444, 3445, 3446
BATESON, F. W. 5786
BATESON, W. 4790
BATHER, FRANCIS ARTHUR.
1119
BATTARBEE, KEITH J. 4428
BATTENHOUSE, HENRY M. 5803
BATTISCOMBE, GEORGINA.
2462, 2993, 2994, 2995
BATZEL, VICTOR MERLYN.
2082
BAUDIN, E. 5172
BAUM, PAULL F. 4595
BAUMEL, H. B. 1964
BAUMER, FRANKLIN L. 672
BAYART, P. 5173
BAYFORD, E. G. 21, 22, 23
BAYLEN, JOSEPH O. 4383
BAYNES, ARTHUR HAMILTON.
3385, 5174
BEACH, ARTHUR G. 2996
BEACH, JOSEPH WARREN. 2206
BEAN, WILLIAM JACKSON. 24
BEARDSLEE, CLAUDE GILLETTE.
4515
BEATTIE, DAVID JOHNSTONE.
3447
BEBB, EVELYN DOUGLAS. 3448
BEBBINGTON, D. W. 3449,
3450
BECCARD, MARIA. 6073
BECHER, HARVEY W. 285
BECHER, SIEGFRIED. 286
BECK, A. 5175
BECK, GEORGE ANDREW. 3994,
3995, 5176
BECKER, WERNER. 5177, 5342
BECKERLEGGE, OLIVER
AVEYARD. 3451
BECKETT, J. C. 2463
BECKINSALE, ROBERT P. 54
BECQUET, T. 3996
BEDDALL, BARBARA G. 1421,
1965, 1966
BEEK, WILLEM JOSEPH ANTOINE
MARIE. 2997
BEER, GILLIAN. 4888
BEHNKEN, ELOISE MARJORIE.
4550
BEIDELMAN, THOMAS O. 2276,
3452

BELL, ENID HESTER CHATAWAY
MOBERLY. 871
BELL, GEORGE KENNEDY ALLEN.
2464, 2465
BELL, P. M. H. 2466
BELL, PETER ROBERT. 1422
BELL, WALTER LYLE. 287
BELLESHEIM, A. 3997
BELLOC, ELIZABETH. 2998
BELPAIRE, T. 3386
BELSEY, ANDREW. 288
BELTRAN, ENRIQUE. 1249,
1250
BENARD, EDMOND DARVIL.
5178, 5179, 5618
BENAS, BERTRAM B. 2999
BENFEY, OTTO THEODOR. 1035
BENN, ALFRED WILLIAM. 673,
6170
BENNE, K. D. 2132
BENNETT, FREDERICK. 3000
BENNETT, GARETH VAUGHAN.
416
BENNETT, JAMES RICHARD.
5121, 5122, 5123
BENNETT, JOAN. 4718
BENNETT, RAYMOND M. 4765
BENNETT, SCOTT. 3998
BENSON, ARTHUR CHRISTOPHER.
674
BENSON, EDWARD FREDERIC.
2467, 2468
BENTLEY, ANNE. 2859
BENTLEY, JAMES. 2469
BENTLEY, M. 2338
BENZIGER, JAMES. 417
BERESFORD, ROSEMARY. 6074
BERGERON, RICHARD. 5180
BERGONZI, BERNARD. 2207, 3999
BERKELEY, HUMPHRY. 4000
BERLIN, ISAIAH. 5066
BERMAN, MORRIS. 25, 26, 27
BERNAL, JOHN DESMOND. 28
BERNARD, L. L. 2133, 2134
BERNARD, MIRIAM, SISTER.
5181
BERNSTEIN, J. A. 2860
BERRANGER, OLIVER DE. 5182
BERRILL, N. J. 1423
BERRY, MARGARET ANN, D.C.,
SISTER. 4001
BERTHELOT, RENE. 1251
BERTOCCI, PETER ANTHONY.
621

BURTCHAELL, JAMES TUNSTEAD.
4017, 4018
BURTIS, MARY ELIZABETH.
5961
BURTON, JEAN. 4387
BURY, JOHN BAGNELL. 2261,
4019, 6178
BURY, SHIRLEY. 3020
BUSEY, GARRETTA HELEN.
6254
BUSH, DOUGLAS. 1361, 2208
BUSH, R. C. 40
BUSS, FREDERICK HAROLD.
3483
BUSSEY, O. 427
BUTCHER, R. W. 41
BUTLER, B. C. 5219
BUTLER, BARBARA J. 4353
BUTLER, EDWARD CUTHBERT.
4020, 4021
BUTLER, FRANCIS JOSEPH.
5220
BUTLER, GIBBON FRANCIS.
2262, 5221
BUTLER, HENRY MONTAGU. 689
BUTLER, P. 3021, 3484
BUTLER, WILLIAM FRANCIS
THOMAS. 4022
BUTTERFIELD, HERBERT. 4023
BUTTERWORTH, H. 887
BUTTMAN, GUNTHER. 295
BUTTS, DENIS. 4434
BUTTS, ROBERT E. 296, 297,
298, 299, 300
BYL, SIMON. 1446
BYNUM, WILLIAM FREDERICK.
2280
BYRNE, M. B. 888
BYRT, GEORGE WILLIAM. 3485

CACOULLOS, ANN R. 690
CADMAN, SAMUEL PARKES.
691, 5222
CAHILL, GILBERT ALOYSIUS.
4024, 4025, 4026
CAHN, THEOPHILE. 1262
CAIRNS, D. 1045
CALDECOTT, ALFRED. 428
CALDER, GRACE J. 1263, 4552
CALDER-MARSHALL, ARTHUR.
2502
CALKINS, ARTHUR BURTON.
5223
CAMERON, ALLAN THOMAS.
2503

CAMERON, HECTOR CHARLES.
42
CAMERON, JAMES MUNRO.
5224, 5225, 5226, 5227
CAMERON, THOMAS W. M. 1194
CAMP, BURTON H. 2103
CAMPBELL, ALLAN WALTER.
2846
CAMPBELL, BERNARD. 1546
CAMPBELL, H. M. 4435
CAMPBELL, JOHN ANGUS.
1447, 1448, 1449, 1450
CAMPBELL, JOHN LORNE. 4388
CAMPBELL, MARY FRANCES T.,
SISTER. 3022
CAMPBELL, REGINALD JOHN.
3270
CAMPBELL, ROBERT ALLAN.
4553, 4893, 4991
CAMPBELL, WILLIAM D. 5228
CANADY, CHARLES ELI. 3023
CANGUILHEM, GEORGES. 1195,
1264, 1451, 1452
CANN, ELSIE. 3944
CANNAN, GILBERT. 4525
CANNON, HERBERT GRAHAM.
1265
CANNON, WALTER F. 43, 44,
301, 302, 692, 693, 694,
695, 1046, 1047, 1048,
1049, 1086, 1121, 1122,
1453, 1454, 3271
CANNON, WILLIAM RAGSDALE.
3486
CANTERBURY, DEAN OF. 2858
CANTOR, G. N. 303, 2345
CAPONIGRI, A. R. 3024
CAPPS, DONALD. 5229, 5230,
5231, 5232
CARBAUGH, DANIEL CARTER.
2140
CARDNO, J. A. 2346, 2347
CARDWELL, DONALD STEPHEN
LOWELL. 45
CAREY, GLENN O. 4526
CARLILE, JOHN CHARLES.
3487, 3488, 3489, 3952
CARMICHAEL, LEONARD. 2348
CARNEIRO, ROBERT L. 2141,
2142
CAROZZI, ALBERT V. 1123,
1266, 1267
CARPENTER, E. S. 2263
CARPENTER, EDWARD FREDERICK.
2504, 2592

ROUSSEAU, GEORGES. 1332,
1834
ROUSSEAU, RICHARD W. 1835
ROUTH, HAROLD VICTOR. 810
ROUTLEY, ERIK. 3825, 3826
ROWAN, EDGAR. 2747
ROWDON, HAROLD HAMLYN.
3827
ROWELL, GEOFFREY. 650,
3828, 5610
ROWLAND, MARY JOYCE, O.S.F.,
SISTER. 5611
ROWLANDS, M. J. 1509, 1510
ROWLEY, NORBERT, F.S.C.,
BROTHER. 4284
ROWSE, A. L. 4929, 6133
ROYCE, J. 2186
ROYLE, EDWARD. 6201, 6202,
6203
RUBAILOVA, N. G. 1836,
1837
RUDMAN, HARRY W. 5864
RUDWICK, MARTIN J. S. 221,
222, 1098, 1099, 1100,
1101, 1178, 1179, 1180,
1181, 1182, 1838
RUGGLES, ELEANOR. 5612,
5613
RULE, PHILIP CHARLES.
4679, 4930
RUMNEY, JUDAH. 2187
RUNCORN, STANLEY KEITH.
223
RUNKLE, G. 1839
RUPP, ERNEST GORDON. 3310,
3527, 3829, 3830, 3831,
3832
RUSE, MICHAEL. 811, 1840,
1841, 1842, 1843, 1844,
1845
RUSLING, G. W. 3833
RUSSELL, SIR A. 224
RUSSELL, A. J. 573
RUSSELL, ARTHUR JAMES.
574, 1846
RUSSELL, C. A. 812, 1333
RUSSELL, D. 5614
RUSSELL, ELBERT. 3834
RUSSELL, GEORGE WILLIAM
ERSKINE. 2748, 2749, 2750,
2938, 3206, 5780, 5781
RUSSELL, H. P. 5615
RUTHERFORD, HENRY W. 1847
RUTLAND, WILLIAM R. 5865,
5866

RYALS, CLYDE DE L. 4486,
4574, 5867
RYAN, ALAN. 371, 372, 5103
RYAN, ALVAN SHERMAN. 4285,
4460, 5616, 5617
RYAN, CHARLES JAMES. 4413
RYAN, GUY. 4286
RYAN, JOHN K. 5618
RYAN, M. J. 5619
RYAN, PATRICK, O.C.S.O.
5620
RYAN, T. 5621
RYANS, D. G. 2076
RYDER, CYRIL. 4287

SÄNGER, SAMUEL. 5104
SAFFIN, N. W. 996
SAILLENS, EMILE. 3835
SAINT-ARNAUD, JEAN GUY.
5622
ST. JOHN, H. 3206a
ST. JOHN-STEVAS, NORMAN.
575, 813, 2033, 2034
SALMON, MARTHA. 6264. See
also Vogeler, Martha
Salmon
SALOMON, R. G. 2751
SALTER, FRANK REYNER.
3836, 3837
SALTER, WILLIAM HENRY.
4414
SAMBROOK, A. J. 3311
SANDEEN, ERNEST E. 4460
SANDEEN, ERNEST ROBERT.
3838, 3839
SANDEMAN, C. 225
SANDERS, CHARLES RICHARD.
3312, 4575, 4576, 4577,
4680, 4681, 4682, 4683,
4684, 4685, 4686, 5045,
5046, 5047, 5048, 5049,
5050, 5792, 5868
SANDERS, J. N. 2752
SANDERSON, DAVID R. 4687
SANDFORD, ERNEST GREY.
3313, 3314
SANDOW, ALEXANDER. 1848
SAN JUAN, EPIFANIO, JR. 4487
SANKEY, BENJAMIN. 4688
SANTILLANA, GIORGIO DE.
6243
SARJEANT, W. A. S. 1137
SAROLEA, CHARLES. 5623
SARTON, GEORGE. 1849, 1850,
1851, 2188, 5906, 6244

SMITH, BASIL A. 3213, 3214, 5643
SMITH, BRIAN C. 2099
SMITH, C. F. 586
SMITH, DAVID S. 1870
SMITH, EDWARD. 242
SMITH, FRANCIS BARRYMORE. 6204
SMITH, FRED. 5644
SMITH, H. K. 2771, 2943
SMITH, HENRY. 3869
SMITH, JONATHAN ZITTELL. 2321, 2322
SMITH, LEO. 5645
SMITH, R. E. 4960
SMITH, ROGER. 1985
SMITH, SUSY. 1986
SMITH, SYDNEY. 1871, 1872
SMITH, SYDNEY FENN. 4299, 5646
SMITH, WARREN SYLVESTER. 818, 3870, 6205
SMOKLER, HOWARD EDWARD. 380
SMYTH, CHARLES HUGH EGERTON. 2772, 2773, 2774, 2944, 2945, 5946
SMYTH, KEVIN. 5183
SMYTHE, B. H. 1001
SNEAD-COX, JOHN GEORGE. 3215, 4300
SNEATH, ELIAS HERSHEY. 5871
SNELDERS, H. A. M. 1340
SNELL, FREDERICK JOHN. 3317
SNELL, WILLIAM E. 243
SNIEGOWSKI, DONALD CHESTER. 3318
SNODDY, ELMER ELLSWORTH. 1873
SNOW, DOROTHY M. B. 3216
SNYDER, A. D. 4692
SNYDER, EMILY EVELETH. 1102
SOBOL', S. L. 1341, 1874, 1875
SOBRY, PAUL. 5647
SOCKMAN, RALPH WASHINGTON. 2775
SOKAL, MICHAEL M. 2415
SOKOLOFF, NANCY BOYD. 5052
SOLIMINE, JOSEPH, JR. 5872
SOLOWAY, RICHARD ALLEN. 587, 2776, 2777
SOMERVELL, D. C. 819
SOMERVILLE, H. 4301
SOMKIN, FRED. 1876, 2323
SOMMER, J. J. 3751
SONN, CARL ROBINSON. 5873
SOPER, DON LAURENCE. 820
SOPKA, KATHERINE. 5910
SORLEY, WILLIAM RITCHIE. 821, 2254, 6045, 6046
SOUTHWELL, SAM BEALL. 6047
SOWERBY, ALICE MURIEL. 244
SOWERBY, ARTHUR DE CARLE. 244
SPANGENBERG, ALICE. 5874
SPARK, MURIEL. 5650
SPARKES, DOUGLAS C. 3871
SPARROW, JOHN HANBURY ANGUS. 3319
SPECTOR, BENJAMIN. 1072
SPEEGLE, KATHERINE SLOAN. 6139
SPENCE, M. 4934
SPENCER, ROBERT F. 2273
SPENCER, T. J. B. 1877
SPERRY, WILLARD LEAROYD. 6140
SPILLER, GUSTAV. 822, 1878, 2077
SPILSBURY, RICHARD. 1879
SPIRO, MELFORD E. 1153
SPITZBERG, I. J. 5107, 5648
SPOERL, HOWARD D. 2416
SPOKES, SIDNEY. 245
SPRING, DAVID. 588, 2946
SPRINGER, HENRI. 5649
SPURGEON, THOMAS, PASTOR. 3803
STACKHOUSE, REGINALD FRANCIS. 1880, 3320
STAEBLER, WARREN. 6048
STAFF, RUDOLF. 4544
STAFLEU, FRANS A. 1342
STAGEMAN, PETER. 246
STAHL, GARY. 336
STALKER, JAMES. 3872
STAM, JOHANNES JACOBUS. 4302
STANFIELD, J. F. 235
STANFORD, DEREK. 5650
STANLEY, OMA. 4867
STAPLETON, MARY WINIFRED FRIDESWIDE. 4303
STARK, WERNER. 2194, 5651
STARZYK, LAWRENCE J. 589, 4492, 4583

SUBJECT INDEX

This bibliography (as the Table of Contents and the headnotes indicate) is arranged by topics, some broad, others narrow. While the Subject Index adds a good deal of detail, it is not intended to be exhaustive. To a considerable extent, we have been guided by "key words" in titles, and general works, for the most part, have not been broken down. Some cross references have been added for entries that overlap (e.g., "Anglicanism" and "Church of England"); nevertheless, it is advisable to pursue subjects under more than one heading. In citing individuals, first names are generally given in full, except for those persons who appear in the table of contents, for whom only initials are provided; "Darwin" without initials is always Charles. Where entries consist of two components (e.g., "Science and J. Tyndall"), it may be useful to look further under the major listing of the second (in this case, "Tyndall, J."). In determining the breadth or specificity of entries, we have tended to make finer distinctions in those areas that would be of special interest to students of Victorian science and religion.

GOSSE, PHILIP H.   7, 31, 34, 38,
   66, 73, 82, 84, 110, 126, 142,
   214, 246, 279, 280, 3576;
   Omphalos 46, 1196
GOTHIC REVIVAL.   435, 2839
GOTT, JOHN.   2551
GOVERNMENT AND SCIENCE.   17,
   166, 170, 172, 173, 174, 179
GOVERNMENT GRANT.   172; see
   Government and Science
GRACE.   See Doctrine of
   Uncreated Grace
GRAHAM, BILLY.   520
GRAMMAR SCHOOLS.   946, 992
GRANT, DR. ROBERT E.   1682
GRAY, ASA.   1425
GRAY, JOHN E.   117, 118
GREAT FAMINE.   3754
GREECE AND CHRISTIANITY.   615
GREEN, JOHN R.   686, 2438, 2496
GREEN, THOMAS H.   664, 686, 690,
   738, 739, 764, 784, 797, 800,
   801, 820
GREENOUGH, GEORGE.   1179
GREG, WILLIAM R.   670, 3378
GREY, EARL.   3453
GRIFFITH, WILLIAM.   202
GRIMSHAW, WILLIAM.   2914
GROVE, WILLIAM R.   61
GUILDFORD.   2792
GULICK, JOHN T.   1220
GUNNING LECTURES.   628
GUNTHER, ALBERT.   117
GURNEY, EDMUND.   4394
GURNEY, JOSEPH J.   2950, 3878,
   3879
GUTTERY, ARTHUR T.   3466

HACK, MARIA.   3524
HADLEIGH, SUFFOLK.   3015, 3216
HAECKEL, ERNST.   1377, 1883,
   4865
HAGIOGRAPHY.   5417
HALDANE, JOHN B. S.   765, 4880
HALDANE, RICHARD B.   61
HALEVY THESIS.   3851
HALIFAX, LORD.   2539, 2651
HALL, SIR JAMES.   88
HALLAM, ARTHUR H.   683
HALLIWELL, JAMES O.   189
HAMELIN, OCTAVE.   776
HAMILTON, BISHOP.   2796
HAMILTON, SIR WILLIAM.   681,
   755, 776, 798, 5096

HAMPDEN, RENN D.   654, 2528,
   2804, 3072
HAMPTON, RICHARD.   3855
HANCOCK, THOMAS.   617, 2555,
   4372
HARCOURT, ARCHBISHOP VERNON.
   2782
HARDY, THOMAS.   547, 743, 770,
   814, 854, 2241, 4719, 5952,
   5954, 6008, 6011, 6020, 6021,
   6025, 6040, 6051, 6058, 6060,
   6159; Jude the Obscure 5953,
   5983; Tess of the D'Urbervilles
   5975, 6013
HARE, JULIUS C.   912, 3264,
   3294, 3295, 3296, 3297, 3298,
   3307, 4680
HARGROVE, CHARLES.   3639
HARNACK, ADOLF VON.   5574
HARRISON, F.   XXII, sec. 2;
   3378, 6223
HARRISON, JAMES.   156
HARROW.   2582
HARROW WEALD.   3101
HARTLEY, DAVID.   2404, 2425,
   4651, 4700
HARTLEY, SIR WILLIAM.   3769
HARVEY, WILLIAM H.   202, 270
HASELBURY BRYAN.   2476
HATCH, EDWIN.   2629, 3276
HATCHAM.   2524
HAUGHTON, SAMUEL.   148
HAWEIS, THOMAS.   2962
HAWKER, ROBERT S.   2609, 3019,
   3069
HAWKESWORTH, JOHN.   4654
HAWKINS, EDWARD.   2804
HAWKINS, JOHN.   224
HEADLAM, ARTHUR C.   2619
HEADLAM, STEWART.   2555, 4350,
   4355, 4372
HEAT RESISTANCE.   5898
HEATHEN.   5388
HEAVEN.   536, 2268, 3077, 5030
HEBREW AND HELLENE.   4459
HEBREW - REGIUS PROFESSOR.   5771
HEDLEY, BISHOP.   4338
HEGEL, GEORG W. F.   698, 776,
   2229, 4743, 6063
HEGELIANS.   782
HEGEMONY.   26
HEINE, HEINRICH.   4459
HELL.   650, 2268, 5030
HELLENE AND HEBREW.   See Hebrew
   and Hellene